Forecasting CO$_2$ Sequestration with Enhanced Oil Recovery

Forecasting CO$_2$ Sequestration with Enhanced Oil Recovery

Editors

William Ampomah
Brian McPherson
Robert Balch
Reid Grigg
Martha Cather

MDPI • Basel • Beijing • Wuhan • Barcelona • Belgrade • Manchester • Tokyo • Cluj • Tianjin

Editors
William Ampomah
New Mexico Institute of
Mining and Technology
USA

Brian McPherson
University of Utah
USA

Robert Balch
New Mexico Institute of
Mining and Technology
USA

Reid Grigg
New Mexico Institute of
Mining and Technology
USA

Martha Cather
New Mexico Insitute of
Mining and Technology
USA

Editorial Office
MDPI
St. Alban-Anlage 66
4052 Basel, Switzerland

This is a reprint of articles from the Special Issue published online in the open access journal *Energies* (ISSN 1996-1073) (available at: https://www.mdpi.com/journal/energies/special_issues/forecasting_co2_sequestration).

For citation purposes, cite each article independently as indicated on the article page online and as indicated below:

LastName, A.A.; LastName, B.B.; LastName, C.C. Article Title. *Journal Name* **Year**, *Volume Number*, Page Range.

ISBN 978-3-0365-5915-5 (Hbk)
ISBN 978-3-0365-5916-2 (PDF)

Contents

About the Editors

William Ampomah

William Ampomah is an Assistant Professor of Petroleum Engineering at New Mexico Tech. Dr. Ampomah is also the Section Head of the REACT group at the Petroleum Recovery Research Center. Dr. Ampomah was appointed by the NM Energy Secretary as the Designee to the NM Oil Conservation Commission in 2022. He has been a Principal Investigator and Co-Principal Investigator on several United States Department of Energy grants in relation to improved oil recovery, CO_2 storage, Wellbore integrity, and subsurface stress projects. Dr. Ampomah obtained his MS and PhD degrees at New Mexico Tech in 2012 and 2016, respectively, in petroleum engineering. He completed his BS in petroleum engineering in Kwame Nkrumah University of Science and Technology (KNUST), Ghana in 2009.

Brian McPherson

Brian McPherson is a USTAR Professor of Civil and Environmental Engineering at the University of Utah. For the past 20 years, Dr. McPherson has conducted carbon management and engineering research, especially geological sequestration studies. Technical focus areas include groundwater and reservoir simulation, multiphase flow analysis and simulation, rock deformation, and subsurface chemically reactive transport analysis and simulation. Other research that Dr. McPherson continues to pursue includes the study of the coupled heat and fluid processes in sedimentary basins and geothermal systems, and petroleum generation and migration processes.

Robert Balch

Robert Balch is the Director of the Petroleum Recovery Research Center, located on the campus of New Mexico Tech. At the university he also holds the position of Adjunct Professor positions in petroleum engineering and geophysics and has been research advisor to more than 40 graduate students. During his 20 years at the PRRC, he has been the principal investigator on a range of enhanced oil recovery projects focused on developing and applying solutions to problems at many scales using geological, geophysical, and engineering data. Dr. Balch is the Principal Investigator of the Southwest Partnerships Phase III demonstration project, where 1,000,000 metric tonnes of anthropogenic CO_2 is being injected for combined storage and EOR into a mature waterflood in North Texas.

Reid Grigg

Reid Grigg heads the Gas Flooding Processes and Flow Heterogeneities Group at the PRRC and is an Adjunct Professor in the Petroleum and Chemical Engineering Department at New Mexico Tech. His research interests include enhanced oil recovery (EOR), carbon sequestration, gas hydrates and phase behavior. Before joining PRRC in 1992, Grigg worked for Conoco for over 10 years as a research scientist and chemist, spending a year with Core Laboratories before joining the PRRC. He obtained his PhD in physical chemistry from Brigham Young University and has authored over 100 publications. He is a member of the Society of Petroleum Engineers, the American Chemical Society, and Sigma Xi.

Martha Cather

Martha Cather works with the Industry Service and Outreach group. She earned her B.S. and

M.S. degrees in Geology from the University of Texas at Austin. Her research interests range from examinations of the effects of brine composition on the wettability of reservoir rocks to petrographic analysis of rocks from a number of different oil and gas fields, and to reservoir characterization of various formations.

Preface to "Forecasting CO$_2$ Sequestration with Enhanced Oil Recovery"

Funding for this project is provided by the U.S. Department of Energy's (DOE) National Energy Technology Laboratory (NETL) through the Southwest Partnership on Carbon Sequestration (SWP) under Award No. DE-FC26-05NT42591.

William Ampomah, Brian McPherson, Robert Balch, Reid Grigg, and Martha Cather

Editors

 *energies*

Editorial

Forecasting CO₂ Sequestration with Enhanced Oil Recovery

William Ampomah [1,*], Brian McPherson [2], Robert Balch [1], Reid Grigg [1] and Martha Cather [1]

[1] Petroleum Recovery Research Center, New Mexico Tech, Socorro, NM 87801, USA
[2] Energy & Geoscience Institute (EGI), University of Utah, Salt Lake City, UT 84108, USA
* Correspondence: william.ampomah@nmt.edu

Citation: Ampomah, W.; McPherson, B.; Balch, R.; Grigg, R.; Cather, M. Forecasting CO₂ Sequestration with Enhanced Oil Recovery. *Energies* **2022**, *15*, 5930. https://doi.org/10.3390/en15165930

Received: 9 August 2022
Accepted: 14 August 2022
Published: 16 August 2022

Publisher's Note: MDPI stays neutral with regard to jurisdictional claims in published maps and institutional affiliations.

1. Introduction

Over the years, naturally occurring CO_2 has been used in many enhanced oil recovery (EOR) projects in the United States. There is opportunity to supplement and gradually replace scarce and regionally limited natural CO_2 sources with anthropogenic sources, giving incentive for operators to become involved in the storage of anthropogenic CO_2 within partially depleted reservoirs. Aside from the incremental produced oil revenues, incentives include a wider availability of anthropogenic sources in regions distant from natural CO_2 sources, and a reduction in emissions to meet regulatory requirements, tax incentives, and favorable public relations. The US Department of Energy through its Carbon Storage Program has sponsored several Regional Carbon Sequestration Partnerships (RCSPs) that have conducted field demonstrations for both EOR and saline aquifer storage. This Special Issue highlights some of the observations and lessons learned through one of these RCSP programs, that of the Southwest Regional Partnership on Carbon Sequestration (SWP). This Special Issue includes scientific output from the RCSP program on key topics related to CCUS including reservoir characterization, simulation, monitoring, verification and accounting (MVA), and risk assessment.

This Special Issue reports some of the work performed by the Southwest Regional Partnership on Carbon Sequestration (SWP) as part of the United States Department of Energy (DOE) National Energy Technology Laboratory (NETL) Regional Carbon Sequestration Partnerships (RCSPs) Phase III demonstration project. The ultimate goal of the RCSPs was to support the development of regional infrastructure for carbon capture and storage (CCS). The program had three phases: characterization (Phase I), validation (Phase II), and development (Phase III). The primary focus of Phase III was on large-scale field laboratories in saline formations and oil and gas fields, with a target of injecting at least 1 million metric tons (MMT) of CO_2 per project. For the SWP, the Phase III project's objective has been to characterize and evaluate an active commercial-scale carbon capture, utilization and storage (CCUS) operation, and demonstrate the associated effective site characterization, MVA, and risk assessment techniques. In sum, this project contributes to the development of future commercial CCUS projects in the United States by demonstrating all aspects of an actual commercial CCUS field operation, including effective reservoir engineering, characterization, monitoring, and simulation technologies.

In our introduction, we briefly describe Phase I and II findings, set the stage for our Phase III project, and summarize the papers included in this Special Issue.

2. Phase Summary

2.1. Phase I: Summary

The SWP commenced work on Phase I in 2003 [1]. The main objective of the SWP Phase I project was to evaluate and demonstrate the means for achieving an 18% reduction in carbon intensity by 2012. Many other goals were accomplished on the way to this objective, including (1) analysis of CO_2 storage options in the region, including characterization of storage capacities and transportation options, (2) analysis and summary of CO_2 sources,

(3) analysis and summary of CO_2 separation and capture technologies employed in the region, (4) evaluation and ranking of the most appropriate sequestration technologies for capture and storage of CO_2 in the Southwest region, (5) dissemination of existing regulatory/permitting requirements, and (6) assessing and initiating public knowledge and acceptance of possible sequestration approaches.

Results of the Southwest Partnership's Phase I evaluation suggested that the most convenient and practical "first opportunities" for sequestration would lie along existing CO_2 transportation networks in the region. From this study, six Phase II validation tests in the region were developed, with a portfolio that included four geologic pilot tests distributed among Utah, New Mexico, and Texas, along with a regional terrestrial sequestration pilot program focused on improved terrestrial monitoring, verification, and accounting (MVA) methods and reporting approaches specific for the Southwest region. Phase II also included a local-scale terrestrial sequestration pilot study using desalinated water from one of the pilot tests to restore.

2.2. Phase II Validation: Summary

The SWP carried out five field pilot tests in its Phase II Carbon Sequestration Demonstration effort [2]. Field-testing demonstrated the efficacy of proposed sequestration technologies to reduce or offset greenhouse gas emissions in the region. Risk MVA protocols and effective outreach and communication were additional critical goals of these field validation tests. The program included geologic pilot tests located in Utah, New Mexico, and Texas, and a region-wide terrestrial analysis. Each geologic sequestration test site was planned to be injected with a minimum of 75,000 tons/year CO_2, with a minimum injection duration of one year. These medium-scale validation tests were sited in sinks that have the capacity for possible larger-scale sequestration operations in the future. Tests demonstrated a broad variety of carbon sink targets and multiple value-added benefits, including the testing of enhanced oil recovery and sequestration, enhanced coalbed methane production, and a geologic sequestration test combined with a local terrestrial sequestration pilot.

2.3. Phase III Demonstration: Summary

In Phase III, the SWP's work was more closely focused on a single field laboratory sited in the Farnsworth Unit (FWU) Field, a mature active oilfield in Ochiltree County in the far northeastern Texas panhandle. The site operator began injection of anthropogenic CO_2 in the field in late 2010, starting with several five-spot well patterns with the intent to add several more each year, up to a total of 25. Planned net CO_2 injection at Farnsworth was 10 MMscf/D (million standard cubic feet per day), ~190,000 tons/year, not including recycled CO_2. The actual delivered volumes averaged slightly less, ~9.3 to 9.4 MMscf/D. The SWP began working at this site in 2013, establishing baselines for surface and subsurface metrics, drilling, logging, and coring three science wells, collecting a variety of 2D and 3D seismic data, and devising long-term monitoring protocols. The field operator allowed access to a wealth of legacy data and the SWP was able to evaluate surface and subsurface areas of the field with varying degrees of CO_2 exposure, from none to 22 months at project inception. The access to data and continued monitoring efforts have been maintained for almost nine years to date of this publication, providing an unprecedented look at an active commercial CCUS project. The ability to compare regions with and without CO_2 exposure provides an invaluable opportunity for the calibration of tools and techniques.

The injection target at FWU is the informally named Morrow B sandstone, a regionally important rock unit that has produced more than 100 million barrels of oil and 500 billion cubic feet of gas [3]. Several regional and local studies [3–6] provided excellent baseline information; however, few have been specifically concerned with using the sandstone as a target of CO_2-EOR or storage and none have had the rich and deep dataset afforded by this project.

3. Summary of Publications

This Special Issue presents work accomplished as part of the Phase III of the SWP project at the FWU field site. The work presented can, in some cases, be used as a guideline for what would need to be carried out for any successful CCUS project. Selected publications include those on site characterization, simulation, monitoring, verification, and accounting (MVA), and risk assessment. The project utilized the FWU as the study site and, unless noted, all papers relate to this area. The Special Issue also received an additional publication presenting aspects of CO_2 storage in Poland which was included as relevant.

3.1. Characterization

Cather et al. [7] present a geological description of the rocks comprising the reservoir that is a target for both oil production and CO_2 storage, as well as the overlying units that make up the primary and secondary seals. Core descriptions and petrographic analyses were used to determine depositional setting, general lithofacies, and a diagenetic sequence for reservoir and caprock at FWU. This paper synthesizes multiple studies conducted to determine the FWU capacity and suitability for long-term carbon storage. A rich dataset including core data and core descriptions, petrographic analyses, petrophysical and geomechanical data from the core, legacy logs from 149 wells, and a very complete suite of modern logs for three characterization wells, as well 2D and 3D seismic survey data, were all used in this effort.

Van Wijk et al. [8] report on the analyses of natural, geologic CO_2 migration paths in and near the FWU on the western flank of the Anadarko Basin. The paper interprets 2D and 3D seismic reflection datasets from the study site and compares seismic interpretations with results from a tracer study. The authors conclude that CO_2 escape in Farnsworth Field via geologic pathways such as tectonic faults is unlikely. Analysis of 2D legacy and 3D seismic datasets do reveal depth and thickness variations of the Morrow B reservoir rock; the interpretation is that they are related to erosional events and the creation of paleotopographic features that underlie the Morrow sandstone and are unlikely to be faults or fractures within the reservoir.

To assess the multiscale sealing integrity of the caprock system that overlies the Morrow B sandstone reservoir, Farnsworth Unit (FWU), Texas, USA, Trujillo et al. [9] combine pore-to-core observations, laboratory testing, well logging results, and noble gas analysis. A cluster analysis using many parameters defined lithologic classes within the upper Morrow shale and Thirteen Finger limestone caprock units, and geomechanical properties were calculated for each class. Several lines of evidence indicate that the overlying shale and limestone seal rocks have excellent sealing capacity with both strength and elasticity. The Morrow B sands are weaker than the overlying lithologies and any fracture initiation around the injection well would not be expected to propagate into the overlying sealing units. Noble gas analysis from fresh core shows that the caprock lithologies show no degree of leakage from historical water and CO_2 flooding in the FWU, whereas the Morrow B sandstone shows an impact from historical EOR activities.

Asante et al. [10] present probabilistic methods to estimate the quantity of CO_2 that can be stored in a mature oil reservoir and analyze the uncertainties associated with the estimation. The results of the estimation of the CO_2 storage capacity of the reservoir are presented with both an expectation curve and log probability plot. From the probabilistic output generated by both techniques, at least 7.68 MMtons can be stored, 17.79 MMtons of CO_2 can probably be stored, and it may be possible to store as much as 40.58 MMtons of CO_2 in the Morrow B reservoir. From the relative impact plot, the net thickness, storage efficiency factor, and area contributed about 95% to the total uncertainty for both techniques. Any further estimation of the storage capacity of the Morrow B reservoir should focus on reducing the uncertainty of these parameters.

3.2. Simulation

Relative permeability curves assumed for simulations can introduce a large source of uncertainty, significantly impacting forecasts of all aspects of the reservoir simulation, from CO_2 trapping efficiency and phase behavior to volumes of oil, water, and gas produced. Moodie et al. [11] evaluate the impacts on CO_2-EOR model forecasts of a wide range of relevant relative permeability curves, from the near linear to highly curved. Small variations in the shape of the relative permeability curve have a significant impact on the model forecasts; thus, selecting an appropriate relative permeability curve for the reservoir of interest is critical for CO_2-EOR model design. If measured laboratory relative permeability data are not available or limited for the study domain, the relative permeability curve should be considered a significant source of model uncertainty and accounted for as part of the simulation effort.

Sun et al. [12] present a hybrid numerical machine-learning workflow to solve various optimization problems. By coupling the expert machine-learning proxies with a global optimizer, the workflow successfully solves the history-matching and CO_2-water-alternative-gas (WAG) design problem with low computational overheads. The history-matching work considers the heterogeneities of multiphase relative characteristics, and the CO_2-WAG injection design takes multiple techno-economic objective functions into account. This work trained an expert response surface, a support vector machine, and a multilayer neural network as proxy models to effectively learn the high dimensional nonlinear data structure. The selection of the machine-learning algorithm may comprehensively consider the dimension of the problem and the demand of error margin. The RSM, SVM, and MLNN are suitable for different types of datasets and a wise choice of method could essentially enhance the prediction performance of the proxy model. The Pareto front optimum protocol provides an alternative way to address multiobjective optimization problems.

Kutsienyo et al. [13] assess the fate and impact of CO_2 injected into the Morrow B sandstone in the Farnsworth Unit (FWU) through numerical non-isothermal reactive transport modeling, and compare the performance of three major reactive solute transport simulators, TOUGHREACT, STOMP-EOR, and GEM, under the same input conditions. Model results show several broad similarities, such as the pattern of reservoir cooling caused by the injected fluids, a large initial pH drop followed by gradual pH neutralization, the long-term persistence of an immiscible CO_2 gas phase, the continuous dissolution of calcite, very small decreases in porosity, and the increasing importance over time of carbonate mineral CO_2 sequestration. The results of the study show the usefulness of numerical simulations in identifying broad patterns of behavior associated with CO_2 injection, but also point to significant uncertainties in the numerical values of many model output parameters.

3.3. MVA

Will et al. [14] present the current status of time-lapse seismic integration at the FWU. The efficacy of seismic time-lapse monitoring depends on a number of key factors which vary widely from one application to another. Most important among these are the thermophysical properties of the original fluid in place and the displacing fluid, followed by the petrophysical properties of the rock matrix, which together determine the effective elastic properties of the rock fluid system. They present a systematic analysis of fluid thermodynamics and the resulting thermophysical properties, petrophysics and rock frame elastic properties, and elastic property modeling through fluid substitution using data collected at FWU. The resulting fluid/rock physics models are applied to the output from the calibrated FWU compositional reservoir simulation model to forward model the time-lapse seismic response. Modeled results are compared with field time-lapse seismic measurements and strategies for numerical model feedback/updates are discussed.

Morgan et al. [15] analyze greenhouse gas (GHG) emissions related to FWU's EOR operations through a gate-to-gate life cycle assessment (LCA). The analysis yielded a net negative (positive storage) of 1.31×10^6 tonnes of CO_2 equivalent, representing 79% of purchased CO_2. An optimized 18-year forecasted analysis estimated 86% storage of the forecasted 3.21×10^6 tonnes of purchased CO_2 with an equivalent 2.90×10^6 tonnes of crude oil produced by 2038. The work presented provides a potential roadmap to others for performing these assessments and, in this case, indicates that the integration of CO_2-EOR and carbon storage is a valid approach to minimizing net GHG emissions.

3.4. Risk

Lee et al. [16] summarize the risk assessment and management workflow developed and used at the FWU. The SWP employed quantitative methods of risk analysis including the Response Surface Method (RSM), Polynomial Chaos Expansion (PCE), and National Risk Assessment Partnership (NRAP) toolset. Tools and workflows used provided useful methods of risk quantification. However, simulation processes (especially geological ones) inherently contain aleatory uncertainty. Thus, it would be most helpful to correctly define the ranges and distribution of uncertain parameters to significantly reduce the uncertainty.

Wei at al. [17] present a simplified model used to screen representative cases from many mineral reactive surface area (RSA) combinations to reduce computational cost. Three selected cases with low, mid, and high RSA values were used for the FWU model. Results suggest that the impact of RSA values on CO_2 mineral trapping is more complex than it is on individual reactions. The impact of mineral RSA values on CO_2 mineral trapping, on the whole, is more complex than it is on individual geochemical reactions. Additionally, the presence of hydrocarbons affects geochemical reactions and can lead to net CO_2 mineral trapping, whereas mineral dissolution is forecasted when hydrocarbons are removed from the system.

Xiao et al. [18] present a quantified risk assessment case study of the FWU that identifies water chemistry indicators for early leak detection and includes the use of response surface methodology (RSM) to quantify potential risks of CO_2 and brine leakage to the overlying USDW quality. Salient findings include: (1) with a leakage flux up to 0.4% of injected CO_2 and brine from a conceptual leaky well with failure, it is likely that the impacted area is limited to within 50 m from the well after 200 years; (2) toxic trace metals may be considered an insignificant long-term concern because of clay adsorption; (3) site-specific, no-impact thresholds could be a preferable reference for groundwater quality evaluations; and (4) pH is suggested as a likely geochemical indicator for early detection of a leakage, due to its easy testing and sensitivity aspects.

3.5. Other

Slota-Valim et al. [19] provide the first study of a Polish oil reservoir as a potential candidate for CCUS. Capacity and integrity were examined using numerical methods that combined geomechanical and reservoir fluid flow modelling with a standard two-way coupling procedure. The long-term simulations resulted in a comprehensive assessment of the total amount of CO_2 leakage as a function of time and the leaked CO_2 distribution within the caprock.

4. Conclusions

The storage of CO_2 as an incidental byproduct of EOR projects has been happening in the U.S. for almost fifty years with related research going back about a century. The first CO_2-EOR projects used exclusively anthropogenic CO_2, but as demand far outpaced anthropogenic sources in the Permian Basin, natural CO_2 became the dominant source of CO_2. With increasingly urgent demands to reduce GHG emissions, as well as new incentives offered by tax credits and incremental oil recovery, there is renewed interest in using depleted reservoirs for carbon storage. Carbon storage can be a bridge between a carbon-based energy economy and a renewable low-carbon energy economy. The experience and data collected from EOR projects are vital to the further development of a viable carbon storage industry. The body of work presented in this Special Issue provides a real-world example of the techniques and methodologies used to develop and execute a successful CCUS project.

Author Contributions: Conceptualization, R.G., B.M. and R.B.; methodology, R.G., B.M. and R.B.; resources, R.G., B.M. and R.B.; writing—original draft preparation, W.A., R.G. and M.C.; writing—review and editing, R.G., B.M. and R.B.; project administration, B.M. and R.B.; funding acquisition, R.G., B.M. and R.B. All authors have read and agreed to the published version of the manuscript.

Funding: Funding for this project is provided by the U.S. Department of Energy's (DOE) National Energy Technology Laboratory (NETL) through the Southwest Partnership on Carbon Sequestration (SWP) under Award No. DE-FC26-05NT42591.

Acknowledgments: Additional support has been provided by Schlumberger Ltd. We would also like to acknowledge the support of current and former operators of FWU in providing data, field access, and logistical support.

Conflicts of Interest: The authors declare no conflict of interest.

References

1. McPherson, B. *Southwest Regional Partnership on Carbon Sequestration*; New Mexico Institute of Mining and Technology: Socorro, NM, USA, 2006.
2. Grigg, R.; McPherson, B.; Lee, R. *Phase II Final Scientific/Technical Report*; New Mexico Institute of Mining and Technology: Socorro, NM, USA, 2011.
3. Bowen, D.W.; Weimer, P. Reservoir geology of Nicolas Liverpool Cemetery fields (Lower Pennsylvanian), Stanton County Kansas, their significance to regional interpretation of Morrow Formation incised-valley-fill systems in eastern Colorado western Kansas. *AAPG Bull.* **2004**, *88*, 47–70. [CrossRef]
4. Sonnenberg, S.A. Tectonic sedimentation model for Morrow sandstone deposition, Sorrento Field area, Denver Basin, Colorado. *Mt. Geol.* **1985**, *22*, 180–191. [CrossRef]
5. Swanson, D. Deltaic deposits in the Pennsylvanian Upper Morrow Formation in the Anadarko Basin. In *Pennsylvanian Sandstones of the Mid-Continent*; Special Publication; Tulsa Geological Society: Tulsa, OK, USA, 1979; pp. 115–168.
6. Munson, T. Depositional, diagenetic, production history of the upper Morrowan Buckhaults Sandstone, Farnsworth Field, Ochiltree County Texas. Oklahoma City Geological Society. In *The Shale Shaker Digest*; Times-Journal Pub. Co: Oklahoma City, OK, USA, 1989; pp. 2–20.
7. Cather, M.; Rose-Coss, D.; Gallagher, S.; Trujillo, N.; Cather, S.; Hollingworth, R.S.; Mozley, P.; Leary, R.J. Deposition, Diagenesis, and Sequence Stratigraphy of the Pennsylvanian Morrowan and Atokan Intervals at Farnsworth Unit. *Energies* **2021**, *14*, 1057. [CrossRef]
8. van Wijk, J.; Hobbs, N.; Rose, P.; Mella, M.; Axen, G.; Gragg, E. Analysis of Geologic CO_2 Migration Pathways in Farnsworth Field, NW Anadarko Basin. *Energies* **2021**, *14*, 7818. [CrossRef]
9. Trujillo, N.; Rose-Coss, D.; Heath, J.E.; Dewers, T.A.; Ampomah, W.; Mozley, P.S.; Cather, M. Multiscale Assessment of Caprock Integrity for Geologic Carbon Storage in the Pennsylvanian Farnsworth Unit, Texas, USA. *Energies* **2021**, *14*, 5824. [CrossRef]
10. Asante, J.; Ampomah, W.; Rose-Coss, D.; Cather, M.; Balch, R. Probabilistic Assessment and Uncertainty Analysis of CO_2 Storage Capacity of the Morrow B Sandstone—Farnsworth Field Unit. *Energies* **2021**, *14*, 7765. [CrossRef]
11. Moodie, N.; Ampomah, W.; Jia, W.; McPherson, B. Relative Permeability: A Critical Parameter in Numerical Simulations of Multiphase Flow in Porous Media. *Energies* **2021**, *14*, 2370. [CrossRef]
12. Sun, Q.; Ampomah, W.; You, J.; Cather, M.; Balch, R. Practical CO_2—WAG Field Operational Designs Using Hybrid Numerical-Machine-Learning Approaches. *Energies* **2021**, *14*, 1055. [CrossRef]
13. Kutsienyo, E.J.; Appold, M.S.; White, M.D.; Ampomah, W. Numerical Modeling of CO_2 Sequestration within a Five-Spot Well Pattern in the Morrow B Sandstone of the Farnsworth Hydrocarbon Field: Comparison of the TOUGHREACT, STOMP-EOR, and GEM Simulators. *Energies* **2021**, *14*, 5337. [CrossRef]
14. Will, R.; Bratton, T.; Ampomah, W.; Acheampong, S.; Cather, M.; Balch, R. Time-Lapse Integration at FWU: Fluids, Rock Physics, Numerical Model Integration, and Field Data Comparison. *Energies* **2021**, *14*, 5476. [CrossRef]
15. Morgan, A.; Grigg, R.; Ampomah, W. A Gate-to-Gate Life Cycle Assessment for the CO_2-EOR Operations at Farnsworth Unit (FWU). *Energies* **2021**, *14*, 2499. [CrossRef]
16. Lee, S.-Y.; Hnottavange-Telleen, K.; Jia, W.; Xiao, T.; Viswanathan, H.; Chu, S.; Dai, Z.; Pan, F.; McPherson, B.; Balch, R. Risk Assessment and Management Workflow—An Example of the Southwest Regional Partnership. *Energies* **2021**, *14*, 1908. [CrossRef]
17. Jia, W.; Xiao, T.; Wu, Z.; Dai, Z.; McPherson, B. Impact of Mineral Reactive Surface Area on Forecasting Geological Carbon Sequestration in a CO_2-EOR Field. *Energies* **2021**, *14*, 1608. [CrossRef]

18. Xiao, T.; McPherson, B.; Esser, R.; Jia, W.; Dai, Z.; Chu, S.; Pan, F.; Viswanathan, H. Chemical Impacts of Potential CO_2 and Brine Leakage on Groundwater Quality with Quantitative Risk Assessment: A Case Study of the Farnsworth Unit. *Energies* **2020**, *13*, 6574. [CrossRef]
19. Słota-Valim, M.; Gołąbek, A.; Szott, W.; Sowiżdżał, K. Analysis of Caprock Tightness for CO_2 Enhanced Oil Recovery and Sequestration: Case Study of a Depleted Oil and Gas Reservoir in Dolomite, Poland. *Energies* **2021**, *14*, 3065. [CrossRef]

 *energies*

Project Report

Deposition, Diagenesis, and Sequence Stratigraphy of the Pennsylvanian Morrowan and Atokan Intervals at Farnsworth Unit

Martha Cather [1,*], Dylan Rose-Coss [2], Sara Gallagher [3], Natasha Trujillo [4], Steven Cather [5], Robert Spencer Hollingworth [6], Peter Mozley [6] and Ryan J. Leary [6]

1   Petroleum Recovery Research Center, New Mexico Institute of Mining and Technology, Socorro, NM 87801, USA
2   New Mexico Oil Conservation Division, Santa Fe, NM 87505, USA; DylanH.Rose-Coss@state.nm.us
3   California Geological Survey, Eureka, CA 95503, USA; Sara.gallagher@conservation.ca.gov
4   Pioneer Natural Resources, Irving, TX 75039, USA; natasha.trujillo@pxd.com
5   New Mexico Bureau of Geology and Mineral Resources, New Mexico Institute of Mining and Technology, Socorro, NM 87801, USA; Steven.Cather@nmt.edu
6   Department of Earth and Environmental Science, New Mexico Institute of Mining and Technology, Socorro, NM 87801, USA; shollingworth1722@gmail.com (R.S.H.); peter.mozley@nmt.edu (P.M.); ryan.leary@nmt.edu (R.J.L.)
*   Correspondence: martha.cather@nmt.edu

Citation: Cather, M.; Rose-Coss, D.; Gallagher, S.; Trujillo, N.; Cather, S.; Hollingworth, R.S.; Mozley, P.; Leary, R.J. Deposition, Diagenesis, and Sequence Stratigraphy of the Pennsylvanian Morrowan and Atokan Intervals at Farnsworth Unit. *Energies* **2021**, *14*, 1057. https://doi.org/10.3390/en14041057

Academic Editor: Nikolaos Koukouzas

Received: 4 January 2021
Accepted: 9 February 2021
Published: 17 February 2021

Publisher's Note: MDPI stays neutral with regard to jurisdictional claims in published maps and institutional affiliations.

Abstract: Farnsworth Field Unit (FWU), a mature oilfield currently undergoing CO_2-enhanced oil recovery (EOR) in the northeastern Texas panhandle, is the study area for an extensive project undertaken by the Southwest Regional Partnership on Carbon Sequestration (SWP). SWP is characterizing the field and monitoring and modeling injection and fluid flow processes with the intent of verifying storage of CO_2 in a timeframe of 100–1000 years. Collection of a large set of data including logs, core, and 3D geophysical data has allowed us to build a detailed reservoir model that is well-grounded in observations from the field. This paper presents a geological description of the rocks comprising the reservoir that is a target for both oil production and CO_2 storage, as well as the overlying units that make up the primary and secondary seals. Core descriptions and petrographic analyses were used to determine depositional setting, general lithofacies, and a diagenetic sequence for reservoir and caprock at FWU. The reservoir is in the Pennsylvanian-aged Morrow B sandstone, an incised valley fluvial deposit that is encased within marine shales. The Morrow B exhibits several lithofacies with distinct appearance as well as petrophysical characteristics. The lithofacies are typical of incised valley fluvial sequences and vary from a relatively coarse conglomerate base to an upper fine sandstone that grades into the overlying marine-dominated shales and mudstone/limestone cyclical sequences of the Thirteen Finger limestone. Observations ranging from field scale (seismic surveys, well logs) to microscopic (mercury porosimetry, petrographic microscopy, microprobe and isotope data) provide a rich set of data on which we have built our geological and reservoir models.

Keywords: morrow; Farnsworth; Anadarko; incised valley

1. Introduction

A detailed understanding of reservoir and caprock lithologies is important for any CO_2-EOR project and is crucial for carbon storage projects. Characterization entails developing knowledge of the reservoir from the pore to the field scale. Such knowledge must include understanding of the various lithofacies (how they were formed and what diagenetic processes have they undergone), understanding of depositional systems that dictate reservoir architecture and heterogeneity, and an understanding of the relationships between rock composition and resultant geomechanical and flow properties that feed into the reservoir models used for simulation and prediction of reservoir behavior.

Most reservoir studies of Morrow reservoirs in the Anadarko basin have focused on sandstones because of their importance in conventional oil production. Although this study examines the sandstones within the field, an important feature of our work is detailed characterization of both the overlying and underlying confining layers, as these are critical to containment of injected CO_2.

The significance of this paper is twofold: we present a synthesis of extensive work by several researchers that has not previously been published, and we provide a brief view of some of the rich dataset that we have collected for the project, which will be archived with public repositories at the conclusion of the project.

2. Materials and Methods

A variety of datasets were used for the work presented here. Cores from several wells, including characterization wells drilled specifically for this project, were described and analyzed. Cored intervals from five wells on the western side of the field (8-5, 9-8, 13-10, 13-14, and 13-10A) and three from the eastern side of the field (32-2, 32-6, and 32-8), were described (Figure 1). Wells 13-10A, 13-14, and 32-8 were drilled as characterization wells for this study, and extensive datasets were collected from them. No destructive testing was allowed on the legacy cores, so observations were limited to visual inspection only. Limited core was available for each of the legacy wells and almost none from non-reservoir intervals. Approximately 250 ft of core was obtained from each of the new characterization wells. Cored intervals include the entire Morrow B reservoir interval, as well as Morrow shale that underlies and overlies the Morrow B, the B1 sandstone interval, and the Thirteen Finger limestone that makes up the remainder of the primary seal.

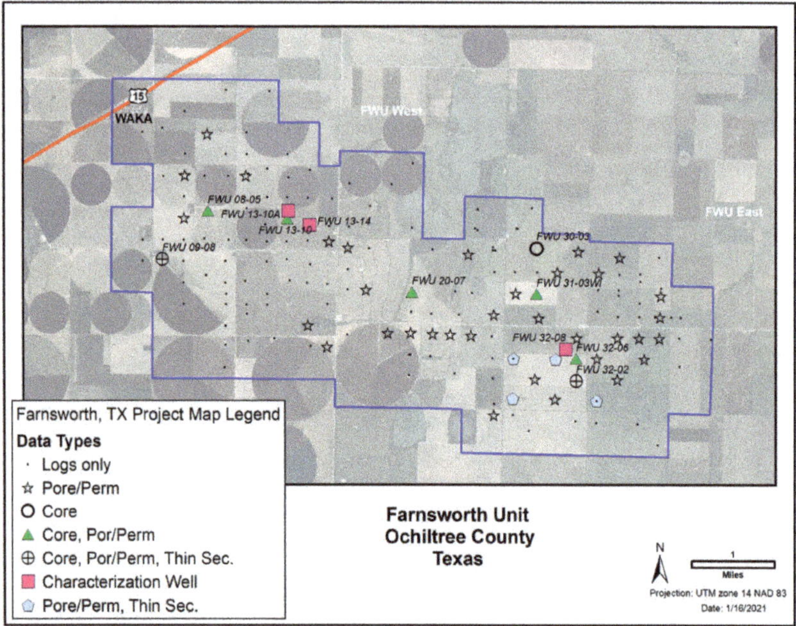

Figure 1. Map of Farnsworth Field Unit (FWU), Ochiltree County, Texas, showing locations of wells and various data types collected from wells. FWU 13-10A, 13-14, and 32-8 were drilled for this project.

The core descriptions were the basis for formation identification, development of detailed stratigraphic columns, facies classification, and interpretation of depositional environment and sequence stratigraphy. Samples from the three characterization wells were used for measurement of porosity, permeability, geomechanical properties, petrographic analysis, X-ray diffraction (XRD), stable isotope and electron microprobe studies for characterization of rock composition, textural relationships, diagenetic alteration, and studies of pore structure and networks. Well logs including borehole imaging logs were run in characterization wells to aid in interpretation of features such as fractures, lithology changes, and orientation for interpretation of bedding and sediment flow direction. Samples were also taken for U–Pb and ^{40}Ar/^{39}Ar dating of zircon and muscovite to understand regional sediment routing and provide new constraints for the age of the Morrow B. Additional data included a small set of thin sections and core plug porosity and permeability data from older wells in the field.

Formation-top data and wire-line logs wells in a 95 mile^2 area around FWU were made available from the field operator. An advanced suite of logs for the three characterization wells was also available. Formation top-picks in legacy wells were checked and standardized with respect to the newly-acquired advanced logs, and all well data were compiled into Schlumberger's Petrel software program. Correlation of formation tops allowed construction of formation surfaces and thickness maps. Combining well and log data with 3D seismic data for FWU allowed creation of a detailed and more accurate model of this reservoir.

3. Results and Discussion

The processes used during this project are now used as a blueprint for other carbon sequestration characterization efforts that have been initiated elsewhere within the western USA. Results of the work include detailed descriptions of cores and thin sections, field maps, and cross-sections that have been used to establish a robust and increasingly sophisticated geological model for the field [1]

3.1. Geologic Setting

3.1.1. Regional Stratigraphic Framework

The Farnsworth Field Unit (FWU) is currently the site of a large-scale carbon dioxide (CO_2) storage and enhanced oil recovery (EOR) project. The field is located in the northwestern part of the Texas panhandle in Ochiltree County, near the town of Perryton. The FWU is situated within the northwestern shelf of the Anadarko basin and is one of many reservoirs that produce from a Pennsylvanian sequence of alternating mudstone and sandstone intervals [2]. Production at FWU is from the operationally-named Morrow B sandstone—the uppermost sandstone encountered below the Thirteen Finger limestone (Figure 2). The Morrow B sandstone has been previously interpreted to be Morrowan in age-based lithostratigraphic correlation and biostratigraphy of overlying units [3]. Analysis of fusulinids and references therein] and conodonts [4] recovered from the Thirteen Finger limestone in southeast Colorado and Kansas suggests an Atokan age that provides a minimum biostratigraphic constraint for the Morrow B. However, Hollingworth et al. [5] report a new U–Pb detrital zircon maximum depositional age of 310.9 ± 4.9 Ma, suggesting that the age of the Morrow B is closer to the Atokan–Desmoinesian boundary. This also has implications for the age of the primary caprock intervals at FWU, the upper Morrow shale, previously interpreted as Morrowan in age, and the Thirteen Finger limestone, previously interpreted as Atokan in age [3]. Additional geochronologic and biostratigraphic work is necessary to resolve the apparent conflict between depositional ages, but in this study, we rely on the existing biostratigraphic framework [3,6]. The Thirteen Finger limestone is an informal name for a series of approximately thirteen predominantly limestone intervals that are intercalated with mudstone and coal layers. Similar Morrowan deposits in the Anadarko basin throughout the Texas and Oklahoma panhandles, southeastern Colorado, and western Kansas, have been studied extensively due to their importance as oil-producing reservoirs [7–12]. Overlying rocks have received less attention but are a potential target for unconventional production elsewhere in the basin [4]. Our interest in the Thirteen Finger interval is primarily as a seal and caprock for CO_2 storage.

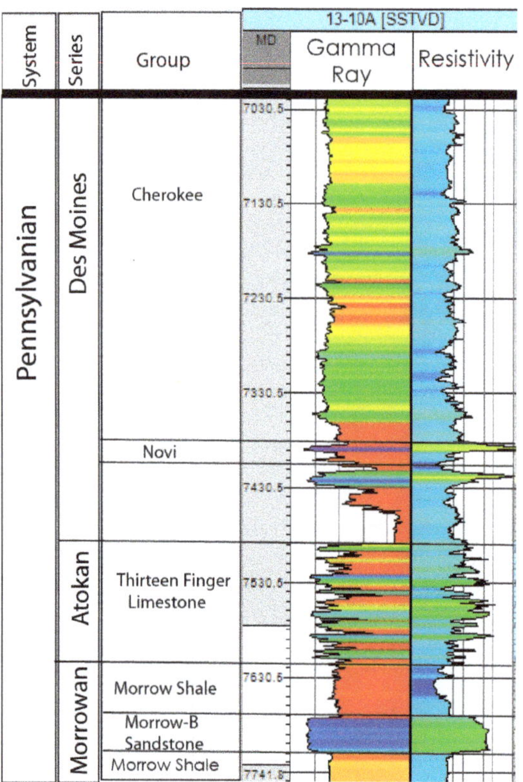

Figure 2. Stratigraphic column for Upper Morrow and Atokan intervals cored at FWU.

The Morrowan and Atokan intervals were deposited during the early to middle Pennsylvanian [13]. During this time, the African, South American, and Antarctic cratons were coalescing in the southern latitudes to form the Gondwanan land mass [14,15]. This coalescence created a setting in which large ice sheets grew and receded on time scales corresponding to orbital Milankovitch cycles [16,17]. The fluctuations in ice volume in turn caused global sea levels to rise and fall, creating the rapid facies changes typical of Pennsylvanian and Permian stratigraphy [8,9,18]. During this period, North America was at equatorial latitudes and drifting northward, causing the area to transition from humid to subhumid climates during the Pennsylvanian [19]. The collision of the North American and Gondwanan, and/or activity on the western and southwestern Laurentian, margins drove deformation across southwestern and central Laurentia [20–24]. Uplift associated with this deformation shed substantial volumes of clastic sediment to adjacent basins (e.g., [12,25–28]). The subsequent filling of these basins is responsible for most of the overlying rock column at FWU. Overlying strata includes upper Pennsylvanian through the middle Permian shales and limestones, with lesser amounts of dolomite, sandstone, and evaporites [12,18].

3.1.2. Tectonic Setting

The FWU is located on the northwest shelf of the Anadarko basin (Figure 3). From FWU, the basin plunges to the southeast where it reaches depths of over 40,000 ft (12,192 m) adjacent to the Amarillo–Wichita Uplift [29]. The Anadarko basin formed as the result of flexural loading of the lithosphere by the adjacent Amarillo–Wichita Uplift [23,29,30]. The Amarillo–Wichita Uplift formed as the result of reactivation of basement faults in a region known as the Southern Oklahoma Aulacogen that are associated with the Neoproterozoic breakup of the Rodinian supercontinent [29,31] However, the tectonic drivers for the uplift on the Amarillo–Wichita are the subject of debate. Most interpretations attribute uplift to either collision between Gondwana and Laurentia [20,23,32] or stress generated along the southwest margin of Laurentia [21,24]. Positive features that might have influenced deposition within the region include the Ancestral Front Range to the northwest [33], the Central Kansas uplift to the northeast [34], and the Amarillo–Wichita uplift to the south [35–37].

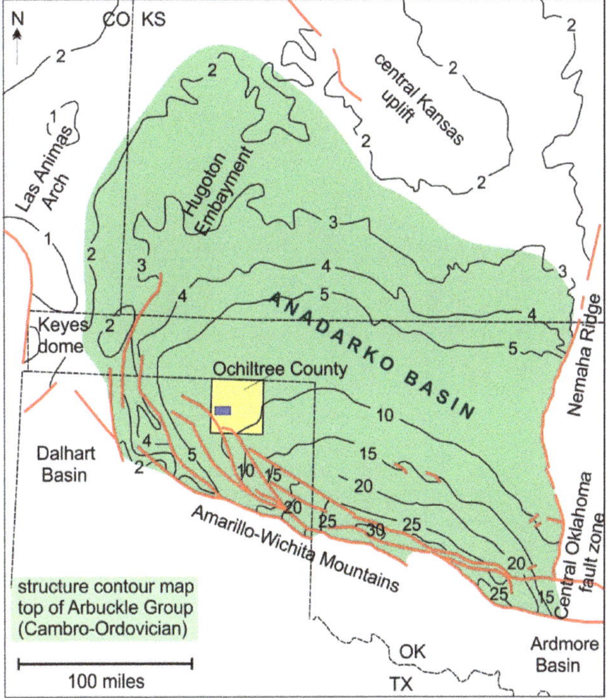

Figure 3. Anadarko basement structure map with major basin bounding faults and tectonic provinces (from Gragg [38], modified from Davis and Northcutt [39]). Contour intervals are in thousands of feet. Blue square is approximate location of FWU northwest of the deepest portions of the basin.

Anadarko basin subsidence and Amarillo–Wichita basement uplift were approximately synchronous, beginning in the Chesterian–Morrowan and continuing through the Pennsylvanian and ending in the Wolfcampian [21]. Maximum rates of subsidence occurred during Morrowan to Atokan times [12,29,35]. Tectonic activity slowed after the Atokan and the region was quiescent by the end of the Pennsylvanian. The uplifts and associated basins combined with the climate at time of deposition set the stage for the stratigraphic sequence seen in FWU cores.

3.1.3. Depositional Environment

Interpretations of the depositional setting of the Morrow B have evolved over the decades. Many previous workers have recognized upper Morrowan sandstones in the northwestern Anadarko basin as fluvial deposits [2,7–10,36,40–46]. Early regional depositional and stratigraphic models were developed using core and electric log data from wells across the Anadarko basin [3,40,41,47]. Swanson [40] proposed relatively synchronous deposition of sands and muds as different parts of the same fluvial and deltaic systems.

In contrast, Sonnenberg [41,48] proposed an incised valley-fill (IVF) model to describe Morrowan sandstones. In the Sonnenberg IVF model, sand distribution is confined within the walls of previously incised valleys. In this model, lateral changes from sandstone to mudstone are interpreted to mark the edges of those paleovalleys. A major difference between this and the earlier-proposed Swanson model is that in the Sonnenberg model, the laterally equivalent mudstones that encase the reservoir sands were deposited under marine conditions during the previous high-stand system tract (HST) and are older than the IVF deposits. Our work at FWU refines the IVF depositional model, and provides a sequence stratigraphic framework for the reservoir and seal (see below).

In contrast to the predominantly fluvial to estuarine environment proposed for the Morrow sandstones and shales, the overlying Atokan Thirteen Finger limestone was deposited in an estuarine to marginal marine environment representing several cycles of transgression and regression during its deposition [48]. The main lithologies include mudstone, interlayered with limestone (cementstone), and some coal [49].

3.2. Lithofacies

3.2.1. Log Identification

In FWU, the Morrow B reservoir is identified in logs by a characteristic blocky shape created by low gamma ray (GR) values (Figure 2), or negative deviations in spontaneous potential (SP) measurements, both associated with clean reservoir sand [2,3,45]. The Morrow B reservoir is the first blocky, low-GR signature below the Thirteen Finger limestone [36]. If a sand signature is present below the Morrow B it is termed the Morrow B1, and Morrow-D if there is a third. The Morrow shale is those parts of the GR curve having consistently high GR values immediately above and below the Morrow B reservoir (Figure 2). These intercalated sand and shale packages are all within the upper Morrow, an informal subdivision used through many parts of the Anadarko basin [48].

The Thirteen Finger limestone has a distinctive wireline log signature showing 12–17 sharp fluctuations in the GR curve (Figure 2). Spikes of low gamma ray readings correspond to limestone beds that give the unit its name, and the intervening GR highs are associated with mudstone or shale beds. The base of the Thirteen Finger limestone is picked at the top of the second low GR spike above the Morrow B reservoir. This corresponds to a coal bed at the top that is a reliable regional stratigraphic marker for the top of the uppermost Morrow shale. The top of the Thirteen Finger limestone is less uniformly designated, but for this project was chosen using the same criteria as geologists from the industry partner [50]. The top of the Thirteen Finger is picked from the GR curve at a prominent low GR spike, above which are relatively high GR values, consistent for 5–15 ft (1.5–4.5 m), above which the next decrease in GR is less dramatic, and finally above which are two prominent GR lows associated with limestone beds of the lower Cherokee Group (Figure 2).

3.2.2. Core Description

Core descriptions are based on works of Gallagher [46] and Rose-Coss [49]. Gallagher's descriptions used observations of legacy cores (Figure 1) that were made available for viewing, and core from well 13-10A, the first of three characterization wells drilled during the project (Figure 1). Gallagher's work focused strictly on the Morrow B and described the four principal lithofacies encountered in core: fine-grained sandstone, coarse-grained sandstone, and conglomerate. Rose-Coss [49] was only able to view photographs of the legacy core, but had access to all three cores for wells drilled for this project, which included not only the Morrow reservoir rock but portions of underlying Morrow shale, as well as overlying Morrow sandstone, Morrow shale, Thirteen Finger, and Marmaton units. Rose-Coss [49] described the entire cored interval of the Morrow sandstone, as well as the overlying Morrow shale and Atokan Thirteen Finger caprock intervals, subdividing the section into 10 different lithofacies based on composition, sedimentary structures, grain size, and color (Tables 1–3). Miall's 1985 classification [51] was used for the Morrow B sandstone, and the classification of Lazar et al. [52] for mudstones in the overlying Morrow shale caprock. Detailed core descriptions and photographs are found in Gallagher [46] for the reservoir intervals in legacy wells and well 13-10A, and Rose-Coss [49] for full core descriptions of the three characterization wells: 13-10A, 13-14, and 32-8.

Figure 4 presents a generalized paleoenvironment reconstruction based on lithofacies noted in cores for the Morrow B and B1 sandstone at FWU, along with typical examples of each. Similar features and sequences of lithofacies were noted in all the cores examined, with small variations in thicknesses, clast types, and sedimentary structures. The general sequence of lithofacies for the Morrow noted in all FWU cores, from deeper to shallower, is marine mudstone, channel lag conglomerate, fluvial coarse-grained sandstone, estuarine fine-grained sandstone, and marine mudstone.

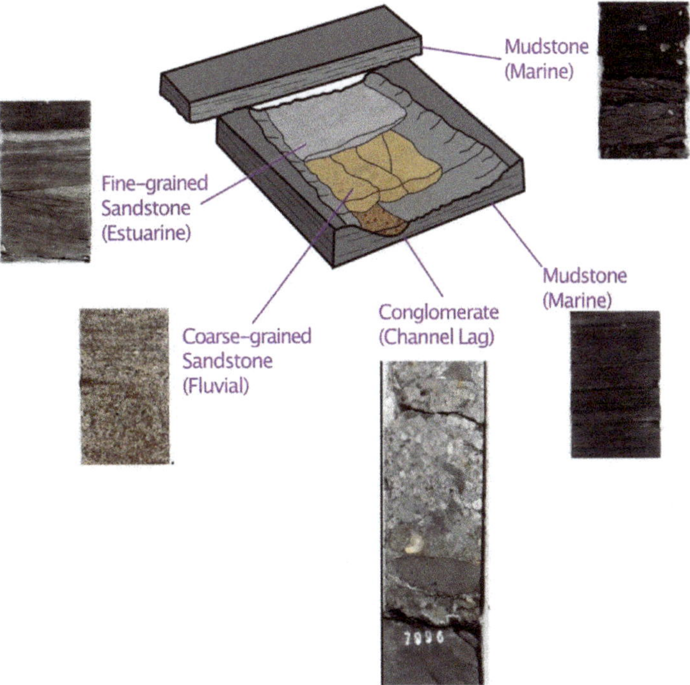

Figure 4. Generalized sequence of depositional environments and lithofacies seen in all Morrow B and B1 cores. All core slabs in the image are approximately 10 cm wide. Conglomerate slab image is approximately 22 cm in length.

Morrow Sandstone

The base of the Morrow B sandstone interval is an abrupt, irregular contact above underlying Morrow shale. The lowest sandstone interval noted in many of the cores is a highly-indurated calcite-cemented basal lag conglomerate that ranges in thickness from 0.3–0.9 m. In most cores, the conglomerate is clast-supported with subrounded clasts, primarily composed of quartzite and granitic rock fragments, as well as some mudstone and siderite concretion clasts [52]. Maximum clast size is 5 cm; however, average clast size is 1–2 cm. Well 13-10A contains ~15 cm of matrix-supported conglomerate (paraconglomerate) approximately 20 cm above the contact between the Morrow shale and the basal conglomerate (Figure 5). The conglomerate is made up of a lower coarse-grained light gray sandstone matrix surrounding larger clasts, many elongate, that are typically subrounded, 2-mm to 3-cm long, and consist of mudstone, sandstone, and pyritized rip-up clasts. The matrix-supported conglomerate was not noted in the other cored wells. The basal conglomerate section in well 13-14 does not contain mudstone clasts and is finer-grained than in the other two wells. Conglomerates tend to be highly-cemented with a variety of carbonate cements, including calcite and ankerite.

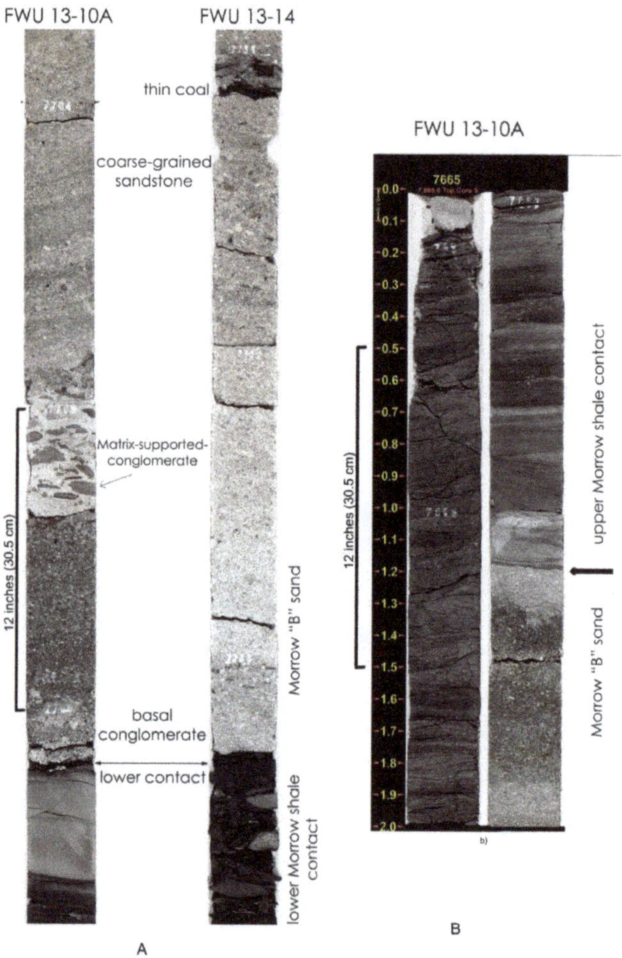

Figure 5. (**A**) Lower part of Morrow B core from wells 13-10A, left, and 13-14, right, showing contact with the underlying Morrow shale and the basal conglomerate grading up into coarse sandstone. Core slabs are ~10 cm (4 inches) wide. One-foot intervals marked on cores; (**B**) sharp contact between Morrow B sand and overlying upper Morrow shale in well 13-10A. There is approximately 11.5 m (38 ft) of Morrow B reservoir sandstone between the upper and lower shale contacts in well 13-10A.

Basal conglomerates grade upwards into an overlying coarse-grained sandstone facies. In all characterization cores there is a very thin (<2.5-cm thick) coal layer 15 to 20 cm above the conglomerate section. The remainder of the Morrow B interval is composed of brown to dark brown, moderately to poorly sorted, subrounded to subangular, very coarse sands and fine gravels. The uppermost portion of the Morrow B sandstone is generally finer-grained, ranging from fine to upper-medium sand.

Primary depositional textures and structures are similar in most of the cores. Using Miall's classification for fluvial sediments [51], lithofacies are first described by grain size, and then bedding. Within the Morrow B, grain size is described as either gravel (G) or sand (S), and bedding either massive (m), laminated (l), or irregular (i). Thus, an interval of predominantly massive gravels is denoted Gm, and an interval of laminated gravels is Gl. Lithofacies codes and descriptions for Morrow B and B1 sandstones are noted in Table 1.

Table 1. Morrow B and B1 sandstone lithofacies.

Facies	Sedimentary Features	Description
(Gm) Gravel massive	Massive	Light grey to light brown, matrix supported, granules to pebbles, moderately to poorly sorted, angular to subrounded, massive bedding, with mudstone rip-up clasts, and calcite cementation in lowermost section.
(Gl) Gravel laminated	Planar to low-angle cross-bedding	Light brown, matrix supported granules, 2–4-cm beds/lamina create fine-scale fining and coarsening upward sections that alternate between very coarse upper and coarse lower with minor fines or clay seams. Facies Gl is modified from Miall [51] facies Gp.
(Gi) Gravel irregularly bedded	Massive to crudely bedded.	Finer-grained than Gm with less-developed and finer (<1 cm) bedding than Gi. Continuously poorly sorted and possible low-angle streaks of clay or fine sand.
(Sm) Sand massive	Massive, to faint laminations	Light brown, massively bedded, moderately sorted, lower medium to upper coarse grains.
(Sl) Sand laminated	Low-angle (<10°) cross-bedding	Light brown, lower medium to fine granules, moderate to poorly sorted Low-angle bedding, approximately 10 degrees; clay streaks often present.

In FWU cores, the thickness of sandstone intervals with massive, laminated, or irregular bedding ranges from 15–61 cm [49]. Laminated intervals are planar to low-angle with 2–10-cm thick fining or coarsening upward sequences. Clay seams, thin mudstone interbeds, and stylolites were noted in most wells (e.g., Figure 5, depth 7733 ft.) and examples in [46,49]. Sandstones are mostly fine- to coarse-grained sandstone and exhibit fining upward sequences. Rounded mudstone intraclasts occur locally, some with desiccation cracks. A fining upward sequence 15–46-cm thick caps the Morrow B sandstone section, however internal grain size sequence intervals do not appear to repeat in any recognizable pattern [49]. Some overall variation in grain size was noted through the field [46]. The sandstone facies in wells 9-8 (far west side) and 32-8 (east side) is finer-grained than observed in other wells. Well 32-6 contains cross-bedding that exhibits considerable variation in grain size between laminae, alternating in size between coarse sand and conglomerate; such well-defined bedding was not noted in any of the other cores examined [46].

Morrow Shale

The Morrow B (and B1, where present) sandstones are encased above and below by shales. Contacts with shale both below and above the sandstone are sharp and irregular (Figure 5). Morrow shale facies are described in Table 2. Only a few feet of underlying shale were cored, so this work describes the shale sequence encountered in the shale overlying the Morrow B sandstone, except as noted. Similar facies are seen in the underlying shale section. The overlying shale section starts above a fining-upward sequence at the top of the Morrow B sandstone (Figure 6 and [46]). The Morrow shale generally fines upwards in a series of thin beds 2.5–5-cm thick that alternate between upper fine sands and fine to medium muds. Sand content decreases upwards through the section. The lowest lithofacies of the Morrow shale (facies gbM) is olive to grey, weakly to moderately bioturbated, with either massive bedding, or discontinuous parallel and non-parallel laminations. The rock is friable, argillaceous fine- to medium-grained mud, with minor organic and detrital content. Facies gbM ranges from 4.3–6.7-m thick in the cored wells. This facies terminates abruptly in the black, fissile-laminated, mudstone facies blM. Facies blM is argillaceous and siliceous, fine to coarse muds with minor organic content. Interspersed with the mudstone are intervals containing fossil hash beds and scattered pyritized shells, sparse calcite concretions and rare pyrite nodules. Continuing up-section, facies blM transitions to a greenish grey, friable calcareous mudstone, facies cM, topped by a thin coal layer. Although facies cM closely resembles facies gbM seen lower in the section, it has a greater amount of organic and calcareous content, as compared to facies gbM. Facies cM is the highest stratigraphic interval of the upper Morrow shale and was not noted in the interval of Morrow shale below the Morrow B sandstone, which is predominately facies blM with minor gbM.

Table 2. Morrow shale lithofacies. Total organic carbon (TOC) data from TerraTek Schlumberger core services.

Facies	Sedimentary Features	Description	TOC%	Fossils
(Mfms) Friable, bioturbated fine sands to fine mudstone	Low-angle to planar laminations, low to moderate bioturbation.	Olive to grey, moderately bioturbated laminated to massive, friable.	0.3–1%	Absent to rare.
(Mfml) Black, finely laminated fine to medium mudstone	Low-angle to planar laminations, carbonaceous concretions, fossil hash, pyrite.	black, laminated, fissile.	0.53–2.67	Scattered disarticulated pyritized shell fragments. thick to very thin shell fragments disarticulated in a convex up position, fossil hash beds.
(Mc) Calcareous mudstone	fossil hash, laminations, bioturbation coal.	Brown to grey, green, laminated to massive, broken and bioturbated sections, friable.	0.44–10.7%	13-14 thin section at 7641 ft, abundant microfossils including foraminifera, gastropods, ostracods, bryozoans, mollusks, and fish scales.

The uppermost part of the Morrow shale has a distinctive sequence of facies that is noted in all of the characterization wells (Figure 6). Approximately 45 cm below the contact with the Thirteen Finger limestone, there is a 2.5–7.5-cm thick section of rock that is smooth and well-indurated with irregular to rounded bounding surfaces. This distinct interval contains a diverse microfaunal assemblage including foraminifera, gastropods, ostracods, bryozoans, mollusks, and fish scales (Figure 6, and [44] Figure 30). Other noteworthy aspects within the short interval include phosphate nodules and possible dewatering structures. Above the well-indurated section is another 30–45 cm of facies cM. The facies terminates at a sharp contact with an 18-cm thick coal bed. The coal is followed by a 15–20-cm grey, calcite-cemented layer (appearing as a concretion in well 13-10A), and then a 5-cm, highly burrowed interval. The burrowed interval is topped by another 5–7.7-cm fossil hash bed (Figure 7), then a black carbonaceous mudstone (bcM) that is one of the signature facies of the Thirteen Finger limestone [49].

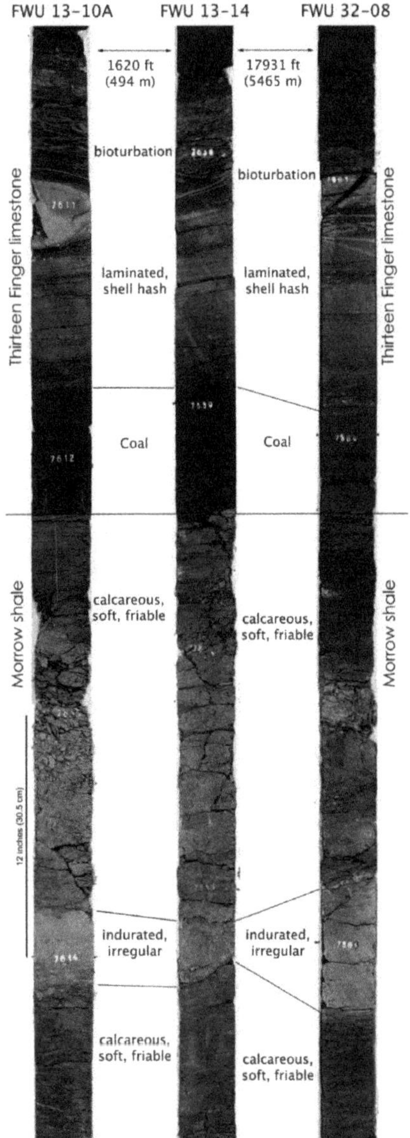

Figure 6. Uppermost sequence of Morrow shale that overlies the Morrow B, showing consistent sequence of lithofacies seen in all three characterization well cores. All core slabs are approximately 10-cm wide and 122-cm height. One-foot intervals marked on core.

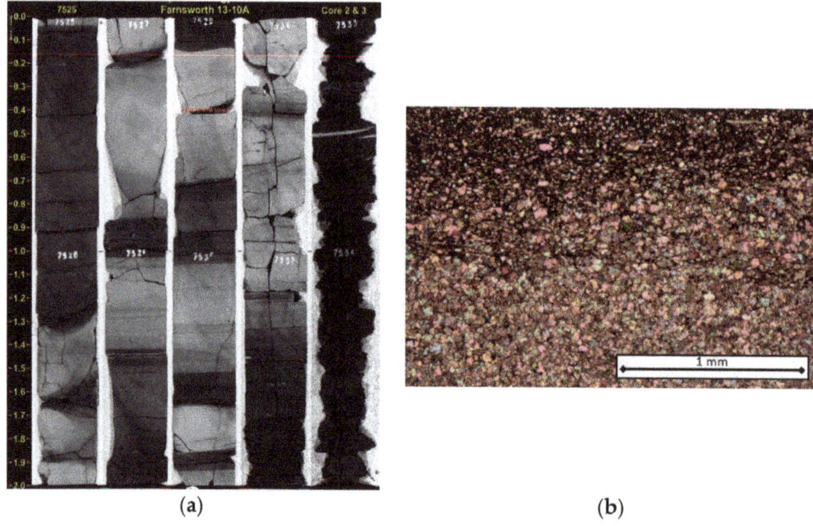

(a)   (b)

Figure 7. Core from well 13-10A (**a**, left) showing finely interbedded cementstone (light gray) and mudstone (dark or black) lithologies of the Thirteen Finger limestone; (**b**, right) thin section photomicrograph in cross-polarized light, showing the finely interbedded nature of the mudstones (upper portion of slide) and cementstones [53] (lower portion).

Thirteen Finger Limestone

The Thirteen Finger limestone is not a single limestone bed, but a series of intercalated black, carbonaceous mudstones (bcM), coals, and limestone intervals that are primarily diagenetic in origin (cC) (Figure 7 and Table 3). Individual limestone beds are 10–60 cm in thickness and are separated by 2–10-cm mudstone intervals. The limestones are clustered in 0.2–2.7-m thick intervals, separated by 0.3–1.2-m mudstone beds. The entire Thirteen Finger interval is 39.6-m thick, with approximately 41% of the thickness composed of mudstone, 4% coal, and 46% is limestone. The number of limestone and mudstone beds varies from well to well; in well 13-10A, 60-70 individual limestone beds were counted [49].

Table 3. Thirteen Finger limestone facies. Total organic carbon (TOC) data from TerraTek Schlumberger core services.

Facies	Sedimentary Features	Description	TOC (%)	Fossils
(Cc) Carbonate cementstone	Massive to faint stratification.	Grey to white, well indurated, smooth, sparse cemented fractures, abrupt to gradational bounding surfaces.	2.29%	No macroscopic fossils but microscopic sponge spicules are present.
(Mfmc) Pyrite- and fossil-bearing fine to medium mudstone and coal	pyrite nodules, fossil hash, bioturbation, bedding parallel fibrous calcite veins.	Black to grey, smooth, well indurated fine to medium mudstone.	0.44–10.70%	Disarticulated and comminuted shells. Shells in lower part of section are pyritized.

Limestone beds are grey to dull white and are dominated by diagenetic calcite and dolomite cements with sparse sponge spicules and pyrite framboids [54]. This facies is more accurately described as a cementstone; consequently, the limestone beds in the Thirteen Finger limestone are classified herein as facies cC. In core, facies cC may have gradational or sharp contacts with mudstone facies and is predominantly massively bedded with only faint signs of stratification. Some cC sections have irregular rounded upper and lower contacts, and are laterally discontinuous, so are concretion-like in appearance. Mineralized fractures are present locally. The mudstone, facies bcM, is black to grey, massive and smooth in some sections, laminated and fissile in others. Some layers are poorly consolidated and broken, in part due to mechanical fracturing during coring, whereas others are well-indurated. Facies bcM contains abundant clays and authigenic calcite, dolomite, and phosphate. Sedimentary features include pyrite nodules, fossil hash, bioturbation, and bedding parallel fibrous veins (Figure 8) referred to as "calcite beef" [55,56]. Coal beds within the Thirteen Finger limestone are between 0.15–0.6-m thick and separated by 3–6-m intervals. Total organic carbon (TOC) within the Thirteen Finger limestone facies is high, with one measurement at 49.0% within the coal at the base of the formation and another at 10.7% [49].

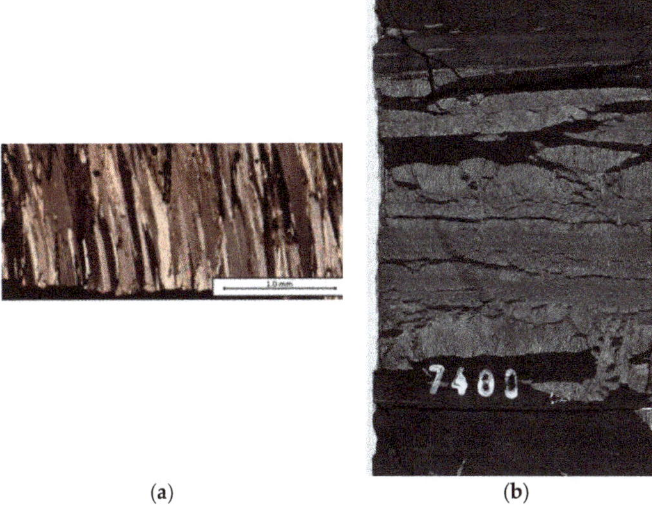

(a) (b)

Figure 8. Thin section image (**a**) and core slab image (**b**) from well 13-10A showing fibrous calcite-filled fracture or "beef". Core slab is approximately 10 cm in width.

3.2.3. Petrographic Analysis and Interpretation

Details of petrographic analysis for FWU cored samples examined in this project, including data and extensive photomicrographs, can be found in Gallagher [46], Cather [57,58], Cather and Cather [54], Rose-Coss [19], and Trujillo [59]. A summary is included here to support the interpretations of facies model and sequence stratigraphy. Upper Morrow sandstones are mostly subarkosic, with an average framework grain composition of 78% quartz, 7% feldspar, and 15% rock fragments. Feldspars are predominately alkali feldspar. Grain and clast types are interpreted to indicate derivation primarily from granitic sources [54]. The relatively coarse grain size of these sandstones suggests proximal sources, probably in the Sierra Grande uplift to the west or the Amarillo–Wichita uplift to the south. Silicic volcanic sources were a nearly ubiquitous, but minor, contributor to most of the studied sandstones. Cambrian and Mesoproterozoic rhyolite is common in the Amarillo–Wichita uplift region, and these rocks are interpreted to have been a source for silicic volcanic grains in the Morrow B [60,61]. Recycled sedimentary detritus is a minor component of about half of the samples; much of this may be intraformational. No vertical trends in the abundance of detrital components are apparent [54]. Composition of the mudstones both within the Morrow shale and the mudstone beds of the Thirteen Finger limestone is predominately illite/smectite clays with trace to minor amounts of quartz. Mudstones may contain a significant amount of carbonate, mainly dolomite, and the amount of carbonate observed in core and in log analyses increases up section from the base of the Morrow shale that overlies the Morrow B sandstone [62]. The limestones within the Thirteen Finger limestone are almost pure calcium carbonate, primarily diagenetic calcite, and can be described as cementstones [49].

Documenting the diagenetic processes observed within FWU rocks is important to the characterization of the reservoir and seal for a variety of reasons. Discussions of diagenesis and paragenetic sequence can be found in Gallagher [46] (Figure 41) and [54]. Within the Morrow B, diagenetic processes such as precipitation of cements and clays, and dissolution of various mineral phases appear to exert the greatest controls on petrophysical properties, and they overprint primary depositional processes [46,54]. Authigenic cements within Morrow B reservoir sandstone include quartz and feldspar overgrowths, calcite in the form of poikilotopic or sparry cement that fills pores and often replaces feldspar grains, siderite (sphaerosiderite), dolomite and ankerite, kaolinite, illite and other clay minerals, and residual oil or bitumen [54]. Pyrite occurs both as crystals and nodules within the mudstones, and as replacement of fossils.

Diagenetic processes within the Thirteen Finger limestone and upper Morrow shale influence mechanical properties and seal behavior. Mechanical testing has shown the cementstones of the Thirteen Finger are the most brittle rocks in the primary seal and the most likely to fracture under stress; therefore, they were a focus of more detailed study [59]. Although most of the cementstones encountered within the cores appear to be continuous through the core diameter, several cemented zones were also noted with rounded or cigar-shaped morphology indicative of concretions, thus raising questions concerning the lateral extent of the cementstones. Isotopic analysis to determine the geometry of the cementstones (following the approach of Klein et al. [63]) was inconclusive, so for purposes of modeling and caprock simulation studies, the cementstones are currently treated as continuous layers [59].

3.2.4. Paragenesis

Morrow B—The paragenetic sequence for the Upper Morrow sandstones is presented in Figure 9. Rare early cements, including pyrite and phosphate, were noted in a few thin sections, but siderite and calcite are the most prevalent early cements. Siderite occurs as sphaerosiderite, commonly associated with pedogenesis, and individual or clusters of small rhombic crystals. All appear to have precipitated before significant compaction, and a high Fe/Mg ratio in some microprobe analyses indicate that formation likely occurred in a freshwater environment. In some samples, multiple stages of siderite with differing Fe/Mg ratios may document changing pore water chemistry resulting from marine transgression [46,64,65]. Poikilotopic calcite was precipitated early in the diagenetic sequence, filling primary porosity and replacing feldspars [54]. Extensive calcite cementation prevented significant compaction in some intervals but occludes almost all porosity, creating some of the lowest permeability intervals within the Morrow B [49].

Feldspar alteration and precipitation of authigenic clay, predominantly kaolinite, are important within the Morrow B, as these processes probably had the greatest effect on the evolution of the porosity and permeability trends seen in the reservoir. Microprobe and X-ray diffraction analyses suggest that most detrital feldspar has been replaced by diagenetic albite [53]. Grains are commonly vacuolized, kaolinized, and/or sericitized. In many samples, wholesale dissolution of feldspar can create large pore spaces, or pore space that has subsequently been partially to completely filled with kaolinite. It is clear from the presence of many delicate skeletal feldspars and large kaolinite-filled pores that most feldspar dissolution and replacement occurred after compaction. Other authigenic clays such as illite/smectite and chlorite were noted but were not significant components of any of the samples examined. Gypsum cement is also a rare constituent. Its position in the paragenetic sequence is unclear, but seems to be associated with the oxidation of pyrite. Petrographic evidence [54] shows it follows at least one stage of calcite precipitation in fracture fills. The studied samples are from below the water table in a reduced, petroleum-bearing stratigraphic interval, so oxidation of pyrite to form gypsum late in the paragenetic sequence is unlikely. The degree of compaction within Morrow B sands varies from very little in samples that have extensive early cementation to significant in a few samples where stylolites and long, concavo-convex, or sutured grain contacts are noted. Hydrocarbon migration appears to be the last major diagenetic event affecting the Morrow B sandstone, although an additional, relatively minor episode of feldspar dissolution after hydrocarbon migration was noted in a few thin sections [46,54,57].

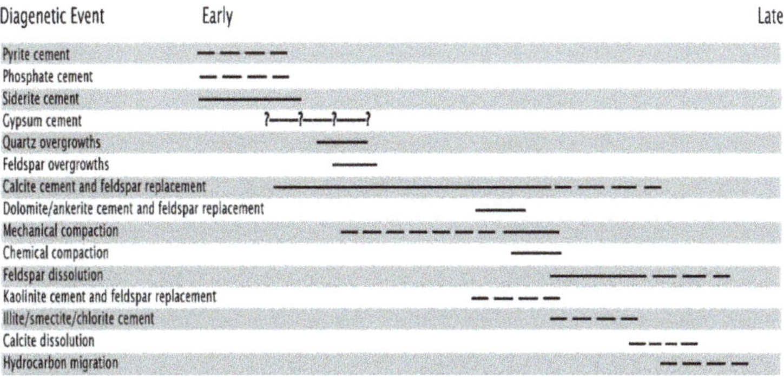

Figure 9. Paragenetic sequence for Morrow B sandstones.

Less emphasis was placed on study of paragenetic sequence in the Morrow shale and Thirteen Finger limestone. For the purposes of this project, the greatest emphasis was placed on the effect of diagenetic events on mechanical properties, and the overall effectiveness of the caprock as a seal for the injected CO_2. Several lines of investigation were pursued, including fracture analysis, geomechanical studies, and mercury porosimetry studies; more information can be found in Trujillo [59] and in Trujillo et al. [64]. Isotopic studies demonstrate that the cementstones, the calcite fracture fillings, and fibrous calcite "beef" fracture filling each represent different diagenetic events (Figure 10). Cementstone carbonates were precipitated at temperatures ranging from 15 to 27 °C (59 to 81 °F), which corresponds to a range of depths from 582 to 1077 m (1909 to 3533 ft), assuming a geothermal gradient of 25 °C/km (1 °F/70 feet). Using the same assumptions, the fibrous calcite beef would have precipitated at higher temperatures, ranging from 30 to 32 °C (86 to 90 °F) and at depths of 1188 to 1289 m (3896 to 4228 ft) [59]. The depths for precipitation of the fibrous calcite beef match the depth at the onset of rapid subsidence within the basin [63] and coincide with those for the generation and migration of hydrocarbons from the Thirteen Finger limestone [38]. It is theorized that migration of the hydrocarbons from a tightly-compacted shale could cause overpressurization within the Thirteen Finger limestone and might have created the horizontal fibrous calcite-filled fractures [64,66].

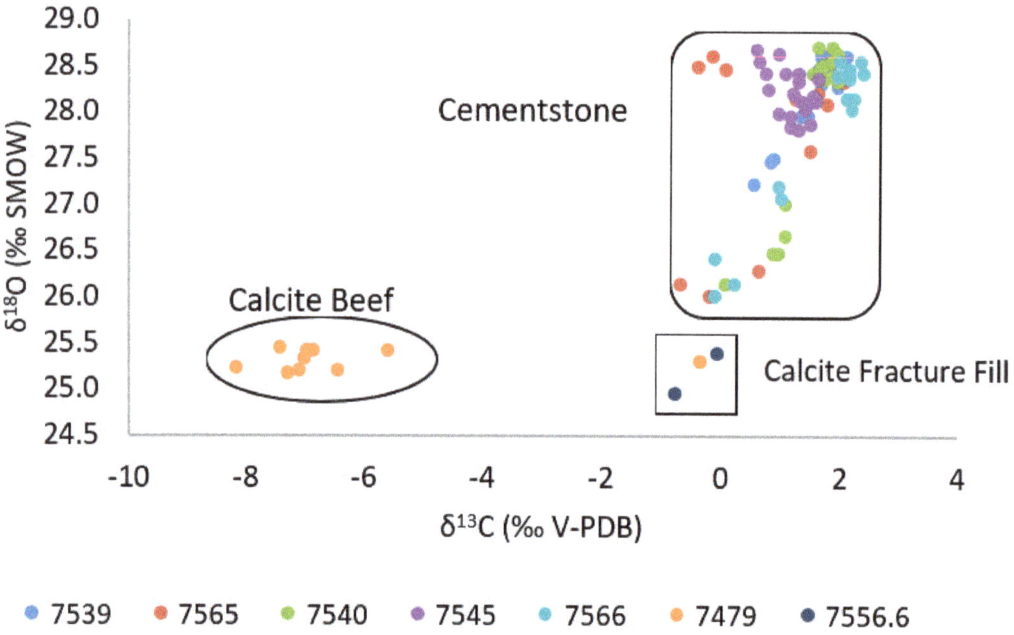

Figure 10. $\delta^{13}C$ versus $\delta^{18}O$ values for all cemented grid and calcite fracture samples [59].

4. Discussion

4.1. Facies Model

The upper Morrowan facies in FWU, with their sequences of basal conglomerate, coarse-grained sandstone, and fine-grained sandstone, appear to be typical of IVF deposits (Figure 4), as described by many previous workers [2,7,42,46,67,68]. The FWU lies southeast of the area where many of these studies were conducted, but still exhibits many of the same characteristics, and is interpreted to have been deposited in a similar setting. However, detrital geochronology [5] suggests that the deposits in FWU were sourced from the Amarillo–Wichita Uplift to the south of the study area, rather than from the northwest (Figure 11) as has previously been interpreted for similar deposits [2,8].

Characteristics of Morrowan sands in southeast Colorado and western Kansas (a region known as the State Line Trend) suggest deposition in fluvial that varied from sluggish and meandering systems to fast and braided ones. The distinction is not trivial, as reservoir models that represent the system would need to be constructed differently depending on the fluvial style. The State Line Trend Morrow fluvial systems contain braided components; however, bedding dip measurements and facies trends that suggest that meandering fluvial and estuarine deposits are more volumetrically important [42]. At FWU, the relatively large grain size, poor sorting, and lack of fines indicate a generally high-energy fluvial environment of deposition for the Morrow B sandstone. A lack of any indication of marine deposition (e.g., fossils or glauconite), along with analysis of reservoir architecture and dip measurements from borehole image logs provides strong evidence for braided river deposition of the Morrow B [45,49]. The patterns observed in cores and dip measurements indicate deposition produced by a braided river flowing in an easterly direction with three to four aggradational events [69].

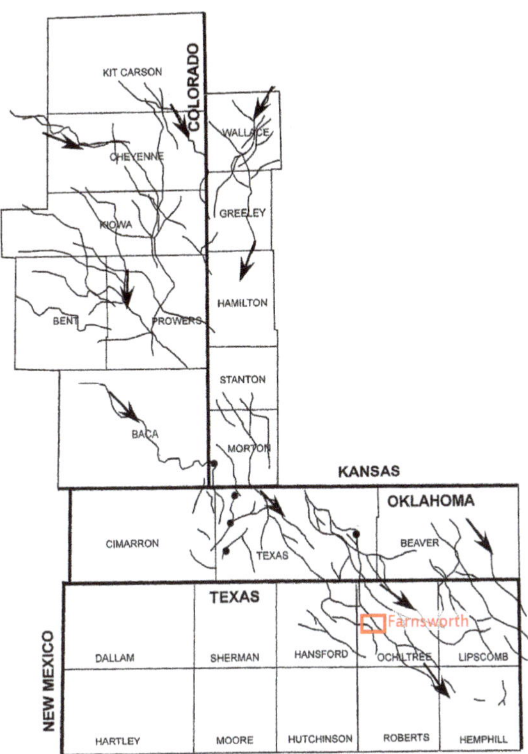

Figure 11. Sediment dispersal systems for late Morrowan deposition (modified from Puckette et al., 2008).

The reservoir architecture and geometry at FWU are well-described by IVF models [49]; however, depositional patterns do not fit as neatly into the paleogeographic interpretations of Bowen and Weimer [8,9] and Puckette et al. [2]. In these studies, connecting productive Morrowan IVF trends reflect paleo-drainage patterns within the greater Hugoton Embayment. Morrow fields of southeast Colorado and western Kansas derived source material from the Ancestral Rocky Mountains to the northwest [2]. In this model, FWU would be considered downstream from these fields, but still receiving sediment from the northwest. Similarly, Bowen and Weimer [8,9] divide Morrowan reservoirs into upstream and downstream facies tracts, where upstream is to the northwest and downstream is to the southeast. Here, coastal inundation influenced more southeasterly fields in the downstream tract earlier during transgression, and reservoir sands would show a stronger estuarine influence. Again, Farnsworth field would be in the downstream or distal facies tract in these models, and thus would be expected to have finer-grained, more mature sands showing evidence of marine or estuarine influence. The Morrow B sands at FWU are rather coarse-grained and poorly sorted, with little evidence of marine influence. Our observations suggest the Amarillo–Wichita uplift to the south likely provided source material, and not the Ancestral Rocky Mountains to the north, which is consistent with detrital zircon studies [5,46,53]. While the Morrow B at FWU fits the general facies model of the incised valley river system, the coarse grain size, general composition, and lack of maturity reflect a proximal position to a more local source of sediment. The transition to the fine mudstones and carbonate facies noted in the Thirteen Finger limestone demonstrates a gradual submergence (transgression) of the fluvial facies and a general transition to a marine environment.

4.2. Sequence Stratigraphy

4.2.1. Depositional Environments

Morrow B

In addition to the core facies descriptions already discussed, Rose-Coss [49] used bedding dip measurements from characterization well logs, isopach and contour maps, and cross-sections generated from the many legacy wells to aid in interpretation of depositional environments and reservoir architecture. Most evidence suggests that the Morrow B sandstone formed as a series of stacked mid-channel bar forms within a northwest-to-southeast-trending braided river system. Isopach contour mapping and cross-sections confirm an incised valley geometry. Coastal processes probably influenced the top of the reservoir interval, and a ravinement surface separates coarse clastics of the reservoir from fine sands and muds of the overlying (upper) Morrow shale.

Morrow Shale

In general, the Morrow shale facies are interpreted as having been deposited in an increasingly marine setting. The finest-grained sand facies at the top of the Morrow B sand is interpreted to be estuarine, deposited in a mixed energy setting [46,49]. Estuaries receive sediment from fluvial and marine sources and are influenced by tide, wave, and fluvial processes [70]. A continued fining upwards into the Morrow shale facies gbM represents the progression from a mixed energy environment to low-energy deposition within central basin estuarine conditions. Continuing upwards, facies gbM transitions abruptly to the black laminated fissile mudstone (facies blM). Thin beds of fossil hash, pyritized shells, and with occasional calcite concretion indicate deposition under primarily anoxic marine conditions with generally low sediment input.

Facies blM gradually transitions to a more friable and calcareous mudstone, facies cM, an olive-colored friable mudstone that superficially resembles facies gbM lower in the shale section. The increase in calcareous and carbonaceous content may result from deposition in deeper water that was more favorable for carbonate-secreting organisms. Within facies cM there is a well-indurated section with irregular bounding surfaces, noted in all three characterization wells, which, following several lines of evidence, is interpreted as a hardground—a sediment cemented on the sea floor [49]. Hardgrounds often form hiatal surfaces that can be traced over vast areas and record periods of very little to no sediment accumulation [71]. The increase in carbonate content and the presence of the hardground suggest that of a shallow marine depositional environment for the uppermost part of the Morrow shale. At the top of the interval is a coal bed, indicating a swamp setting [72]. The hardground and the coal layers are good marker beds because they are present in all the characterization wells, despite a 3.5-mile (5.6-km) separation. The coal bed is used as a formation top marker for the Morrow, above which depositional cycles of the Atokan differ markedly.

Thirteen Finger Limestone

The base of the coal at the top of facies cM marks the transition to the Thirteen Finger limestone. Above the coal is a distinctive section of bioturbated mudstone indicative of a shift to deeper water and better-oxygenated conditions. The Thirteen Finger limestone alternates between cementstones, mudstones, and coal beds. Sedimentary features within the mudstone intervals include pyrite framboids, sparse fossil hash, and bedding parallel fibrous veins, also known as "beef" [56]. These features, excluding the beef, can be used to help interpret depositional setting and sequence stratigraphic context. The repeating coal intervals suggest that the setting was near a coast with low detrital sediment input. Diagenetic pyrite typically forms in reducing conditions with limited oxygen diffusion due to water column restriction, stratification, or high demand from organic matter. Other conditions that favor the formation of pyrite are an abundance of organic matter, and sufficient availability of iron, generally derived from iron oxyhydroxide coatings on detrital grains (especially clays) and in particulate and colloidal form [52,73]. The presence of coal and the pyrite indicate low clastic input in paralic to shallow marine conditions.

Carbonate cementstones comprise what have typically been referred to as the limestones of the Thirteen Finger limestone. Cementstones are of diagenetic origin; some occurrences have been linked to intervals of low sediment accumulation [52,74,75]. They are thought to form below the sediment–water interface during depositional hiatuses [74]. The presence of sponge spicules noted within the cementstones indicates a shallow-marine shelf environment for the host sediment [76].

In general, facies of the Thirteen Finger limestone lack the high clastic input of the Morrowan. Fluvial environments of Morrowan time are increasingly submerged, becoming first estuarine and subsequently marine. During the time of Thirteen Finger deposition, the area was largely cut off from clastic input and was a mix of organic-rich coastal swamps and shallow marine shelves, depending on water depth. The numerous facies changes indicate that water depths were oscillating significantly; one of many observations that tie the Farnsworth Field interpretations into the sequence stratigraphic context of late Pennsylvanian deposition elsewhere in the Anadarko basin and greater Midcontinent region described below.

4.2.2. Stratigraphic Sequences

Incorporating the core analysis into a sequence stratigraphic context facilitates a broader geologic understanding of the formations and helps predict lateral and vertical facies trends. Figure 12 presents a sequence stratigraphic representation of the cored and logged intervals of the Morrow and Thirteen Finger Limestone at FWU. Regional sequence stratigraphy has previously been described by many [2,3,8,9,19,67,77–80]. Previously published works on FWU have not addressed this aspect of reservoir or seal rocks at FWU. In this study, we note some small differences from regional models that may reflect the specific paleogeographic location of the FWU.

Fluctuations of sea level during Pennsylvanian time resulted in development of unconformity-bounded depositional sequences including IVF deposits distributed widely over the Midcontinent region [78]. Depending on paleogeographic location with respect to the encroaching late Paleozoic marine seaway, sequences can show alternation between subaerial and subaqueous depositional settings, resulting in sequences of sandstone intercalated with mudstone, or shallower/deeper marine settings that could produce intercalated shale/limestone sequences [78,80]. The lowest cored interval at FWU (below the Morrow B sandstone) contains mudstones interpreted to have been deposited during a high-stand system tract (HST) through the falling stage system tract (FST). The contact between the underlying mudstones and the basal Morrow B is sharp and erosive, where fluvial sediments sit unconformably on marine sediments. The contact is interpreted to represent a low-stand surface of erosion (LSE) and a sequence boundary. The lowest interval of the Morrow B is presumed to represent a period of transition from fluvial incision into underlying marine mudstones during the low-stand systems tract (LST) into a period of fluvial aggradation. This aggradational stage is interpreted to represent the transgressive system tract (TST) paleovalleys that were previously erosive and were backfilled with clastic sediments. In some areas of FWU, up to five intervals of coarse-grained sandstone are evident from the wireline logs [49], suggesting that this cycle of sea level rise and fall occurred numerous times during upper Morrowan time at FWU.

The contact between the Morrow B reservoir and the overlying Morrow shale caprock is sharp and erosive (Figure 5b). The erosive contact is interpreted as a flooding surface and possible wave ravinement surface (WRS) and separates the Morrow B reservoir from the overlying Morrow shale caprock. The continued fining-upward succession in the lowest portion of the Morrow shale in facies gbM documents continued sea level rise, as the depositional environment transitions from fluvial to shoreface and estuarine settings. The contact between facies gbM and blM within the Morrow shale is also interpreted as a flooding surface and a transition to a deeper water environment. As compared with the underlying facies gbM, facies blM (black mudstone) is finer-grained and inferred to be deposited under deeper, quieter, and primarily anoxic paralic to marine conditions. The maximum flooding zone is inferred to be near the middle of the Morrow shale in facies blM, marking the end of the TST and start of an FST as water depth decreased.

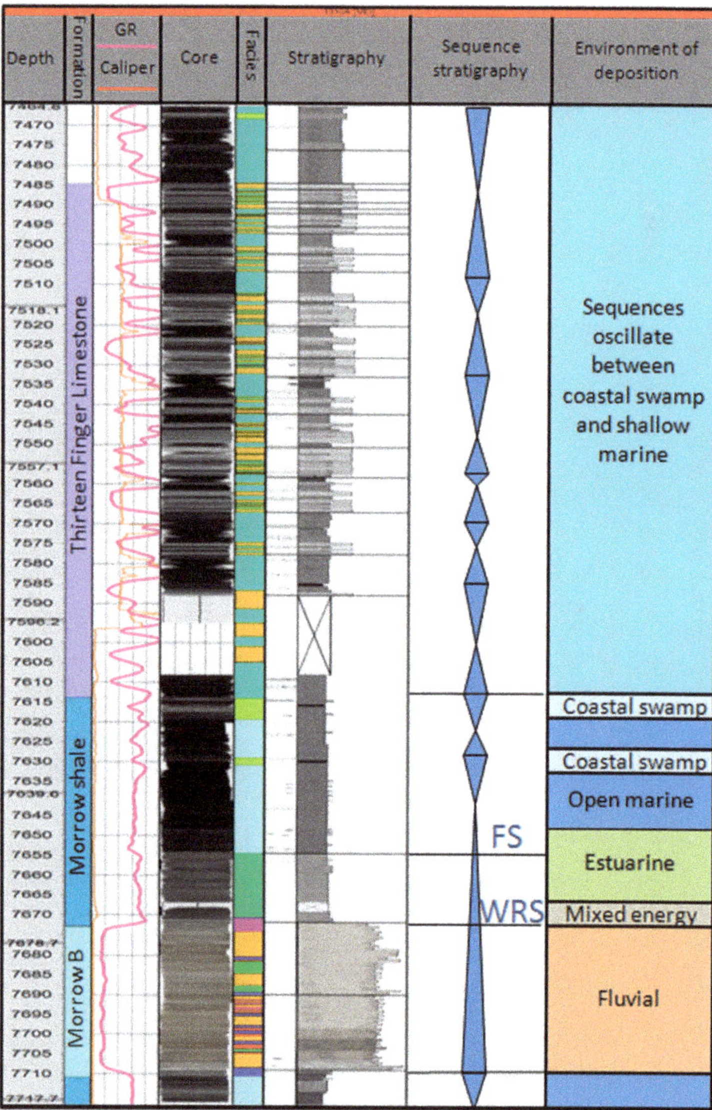

Figure 12. Sequence stratigraphy of the FWU cored reservoir and caprock interval in well 13-10A. From left to right columns illustrate depth, formation, gamma ray (GR) signature, facies, stratigraphic column, sequence stratigraphic surfaces, and depositional environments. Narrow points at the apex of blue triangles represent marine flooding surfaces, triangles narrowing upwards are high-stand systems tracts, while those broadening upwards are low-stand systems tracts. WRS = possible wave ravinement surface; SF = flooding surface.

The decrease in water depth is indicated by the facies transition from blM to a more friable calcareous mudstone having higher calcareous and carbonaceous content, indicating shallowing water (facies cM) overlain by a thin coal bed. Deposition of facies cM is interpreted to have occurred in a restricted shallow marine setting during the FST and the coal within a swamp during the LST.

The coal bed is topped by facies cM, followed by another interval of facies blM, suggesting an additional parasequence where the coal bed represents the LST followed by facies cM and blM deposited during the TST and HST. Continuing up section, facies blM is again followed by facies cM, indicating another parasequence FST. Midway through a second interval of facies cM, there is a well-indurated section with irregular bounding surfaces, which, following several lines of evidence (see above), is interpreted as a hardground. This previously undocumented hardground was identified in all characterization wells and is thought to represent a period of very little to no deposition (paraconformity) during the FST to LST. However, with no absolute chronological data, it is impossible to confirm a period of missing time, only a suggestion of very slow sedimentation rates. Regardless, the hardground is followed by another interval of facies cM, and then a coal bed.

The second coal bed is often used as the marker for the top of the Morrow and base of the Thirteen Finger limestone. It marks a transition from a system of relatively high clastic input that oscillated between fluvial and marine depositional environments, in the Morrowan, to a system of limited siliciclastic input that varied from a shallow marine environment during high-stands to a coastal marsh setting during low-stands in the Atokan formation. These coals provide a robust stratigraphic marker that is seen elsewhere in the basin and is often used as an anchor in correlating Pennsylvanian cyclothems [81].

The Thirteen Finger limestone was deposited during the early-to-mid Pennsylvanian at a time of global transgression and regional basin subsidence [12,18], ultimately culminating with the Late Pennsylvanian Midcontinent Sea stretching from present day Colorado to Pennsylvania [19,82,83]. Eustatic sea level changes, possibly interwoven with climatic variability [80], caused deposition in the FWU area to alternate between limestone and mudstone intervals resulting from variations in water depth and oxygenation levels [18,79].

5. Conclusions

This paper synthesizes work done over the course of six years by multiple researchers working on characterization of the reservoir and seal rocks at the Farnsworth Unit to determine capacity and suitability for long-term carbon storage. A rich dataset including core and core descriptions, petrographic analyses, petrophysical and geomechanical data from core, legacy logs from 149 wells, and a very complete suite of modern logs for three characterization wells, as well 2D and 3D seismic survey data were all used in this effort.

From characterization work presented in part in this paper we have:

- Confirmed that the Morrow B reservoir at Farnsworth resembles incised valley deposits described elsewhere in the western Anadarko basin and was probably deposited as a series of stacked mid-channel barforms deposited in a braided stream environment in an incised valley fluvial system. Many noted features are common to IVF deposits worldwide and are useful in recognition of these deposits, which can provide important carbon storage reservoirs [68].
- Described the sequence of facies from the basal conglomerate to the coal layers at the top of the Morrow that appear to represent the end of incised valley deposition and a change to the more marine environment seen in Thirteen Finger limestone and other units that overly the Morrow shale. The layers are a robust stratigraphic marker that can serve as a useful correlation across the region.
- Presented one of the most thorough descriptions of the Morrow shale and Thirteen Finger limestone caprock facies available for this part of the Anadarko basin. The core and facies descriptions laid a framework for subsequent geochemical and geomechanical investigations of previously undescribed seal rocks.

- Determined that limestones of the Thirteen Finger limestone are most accurately described as cementstones; an important distinction, because cementstones are of diagenetic, rather than primary depositional origin, and not composed of skeletal material. These cementstones were not previously recognized or described in the region.
- Provided evidence of a potentially younger age of deposition than previously thought.

With this geologic framework, researchers have constructed increasingly detailed reservoir models that are the basis for much of the ongoing work conducted by reservoir modeling, simulation, and risk analysis groups for the Southwest Partnership's Phase III research project [84–86].

Beyond providing insights into Morrowan incised valley-fill reservoirs, the study has provided rarely available data on depositional environments and sequence stratigraphy of the upper Morrow shale through the Thirteen Finger limestone caprock. Findings from the work may be applicable to geologically-similar fields regionally and worldwide; the processes used can be duplicated in other characterization projects regardless of field geology.

Author Contributions: Conceptualization: M.C., P.M., and R.J.L.; methodology: D.R.-C., S.G., N.T., S.C., and R.S.H.; formal analysis: M.C., R.S.H., and N.T.; investigation: D.R.-C., S.G., N.T., S.C., and R.S.H.; data curation: S.G. and D.R.-C.; writing—original draft preparation: M.C.; writing—review and editing: S.C., R.J.L., and R.S.H.; supervision: M.C., R.J.L., and P.M.; project administration: M.C. All authors have read and agreed to the published version of the manuscript.

Funding: Funding for this project is provided by the U.S. Department of Energy's (DOE) National Energy Technology Laboratory (NETL) through the Southwest Partnership on Carbon Sequestration (SWP) under Award No. DE-FC26-05NT42591.

Institutional Review Board Statement: Not applicable.

Informed Consent Statement: Not applicable.

Data Availability Statement: No new data were created or analyzed in this study. Data sharing is not applicable to this article.

Acknowledgments: Additional support has been provided by Schlumberger Ltd. We would also like to acknowledge support of current and former operators of FWU in providing data, field access, and logistical support.

Conflicts of Interest: The authors declare no conflict of interest.

References

1. Ampomah, W.; Rose-Coss, D.; Cather, M.; Balch, R.S. Reservoir Characterization Probabilistic Reserve Assessment of Farnsworth Field CO2-EOR Site. In Review.
2. Puckette, J.; Al-Shaieb, Z.; Van Evera, E. Sequence stratigraphy, lithofacies, and reservoir quality, Upper Morrow sandstones, northwestern shelf, Anadarko Basin. In *Morrow Springer in the Southern Midcontinent, 2005 Symposium*; Circular 111; Andrews, R.D., Ed.; Oklahoma Geological Survey: Norman, OK, USA, 2008; pp. 81–97.
3. Rascoe, B., Jr.; Adler, F.J. Permo-Carboniferous Hydrocarbon Accumulations, Mid-Continent, U.S.A. *Am. Assoc. Pet. Geol.* **1983**, *67*, 979–1001.
4. Bagley, M.E. Lithofacies, Geochemical Trends, and Reservoir Potential, Thirteen Finger Limestone, Hugoton Embayment. Master's Thesis, Oklahoma State University, Stillwater, OK, USA, 2012.
5. Hollingworth, R.S.; Leary, R.J.; Heizler, M.T. Detrital U-Pb zircon 40Ar/39Ar muscovite geochronology from Early Pennsylvanian strata in the Anadarko basin, Texas Panhandle, USA. *Palaeogeogr. Palaeoclimatol. Palaeoecol.* In Revision.
6. Rascoe, B., Jr. Regional Stratigraphic Analysis of Pennsylvanian Permian Rocks in Western Mid-Continent, Colorado, Kansas, Oklahoma, Texas. *Am. Assoc. Pet. Geol.* **1962**, *46*, 1345–1370.
7. Bowen, D.W.; Krystinik, L.F.; Grantz, R.E. Geology reservoir characteristics of the Sorrento-Mt. Pearl field complex, Cheyenne County, Colorado. In *Morrow Sandstones of Southeast Colorado Adjacent Areas*; Sonnenberg, S.A., Shannon, L.T., Rader, K., von Drehle, W.F., Martin, G.W., Eds.; Rocky Mountain Association of Geologists: Denver, CO, USA, 1990; pp. 67–77.
8. Bowen, D.W. Weimer, Regional sequence stratigraphic setting reservoir geology of Morrow incised-valley sandstones (lower Pennsylvanian), eastern Colorado western Kansas. *Am. Assoc. Pet. Geol.* **2003**, *87*, 781–815.

9.	Bowen, D.W. Weimer, Reservoir geology of Nicolas Liverpool Cemetery fields (Lower Pennsylvanian), Stanton County Kansas, their significance to regional interpretation of Morrow Formation incised-valley-fill systems in eastern Colorado western Kansas. *Am. Assoc. Pet. Geol.* **2004**, *88*, 47–70.

10.	Devries, A.A. Sequence Stratigraphy Micro-Image Analysis of the Upper Morrow Sandstone in the Mustang East Field, Morton County, Kansas. Master's Thesis, Oklahoma State University, Stillwater, OK, USA, 2005.

11.	Munson, T. Depositional, Diagenetic, Production History of the Upper Morrowan Buckhaults Sandstone, Farnsworth Field, Ochiltree County Texas. Master's Thesis, West Texas State University, Canyon, TX, USA, 1988; 117p.

12.	Higley, D.K. Petroleum systems assessment of undiscovered oil gas in the Anadarko Basin Province, Colorado, Kansas, Oklahoma, Texas-Mississippian through Permian Assessment Units. In *Petroleum Systems Assessment of Undiscovered Oil Gas in the Anadarko Basin Province, Colorado, Kansas, Oklahoma, Texas-USGS Province 58*; Higley, D.K., Ed.; USGS Digital Data Series DDS-69-EE; U.S. Geological Survey: Reston, VA, USA, 2014; pp. 1–60.

13.	Richards, B.C. Current status of the international Carboniferous time scale. In *The Carboniferous-Permian Transition*; Lucas, S.G., DiMichele, W.A., Berrick, J.E., Schneider, J.W., Spielmann, J.A., Eds.; New Mexico Museum of Natural History Science: Albuquerque, NM, USA, 2013; Volume 60, pp. 348–352.

14.	Scotese, C.R.; Boucot, A.J.; McKerrow, W.S. Gondwanana Palaeogeography Palaeoclimatology. *J. Afr. Earth Sci.* **1999**, *28*, 99–114. [CrossRef]

15.	Domeier, M.; Torsvik, T.H. Plate tectonics in the late Paleozoic. *Geosci. Front.* **2014**, *5*, 303–350. [CrossRef]

16.	Heckel, P.H. Evaluation of evidence for glacio-eustatic control over marine Pennsylvanian cyclothems in North America consideration of possible tectonic effects. In *SEPM Concepts in Sedimentology Paleontology*; Dennison, J.M., Ettensohn, F.R., Eds.; SEPM: Tulsa, OK, USA, 1994; Volume 4, pp. 65–87.

17.	Heckel, P.H. Significance of midcontinent Pennsylvanian cyclothems to deciphering global Pennsylvanian stratigraphy. In *Predictive Stratigraphic Analysis: Concept Application*; Cecil, C.B., Edgar, N.T., Eds.; USGS Bulletin 2110; U.S. Geological Survey: Reston, VA, USA, 1994; pp. 37–42.

18.	Ross, C.A.; Ross, J.P. *Late Paleozoic Transgressive-Regressive Deposition*; Special Publication; Society of Economic Paleontologists Mineralogists (SEPM): Tulsa, OK, USA, 1988; Volume 42, pp. 227–247.

19.	Algeo, T.J.; Heckel, P.H. The Late Pennsylvanian Midcontinent Sea of North America: A Review. *Palaeogeogr. Palaeoclimatol. Palaeoecol.* **2008**, *268*, 205–221. [CrossRef]

20.	Kluth, C.F.; Coney, P.J. Plate tectonics of the Ancestral Rocky Mountains. *Geology* **1981**, *9*, 10–15. [CrossRef]

21.	Ye, H.; Royden, L.; Burchfiel, C.; Schuepbach, M. Late Paleozoic deformation of interior North America: The greater ancestral Rocky Mountains. *Am. Assoc. Pet. Geol.* **1996**, *80*, 1397–1432.

22.	Dickinson, W.R.; Lawton, T.F. Carboniferous to Cretaceous assembly fragmentation of Mexico. *Geol. Soc. Am. Bull.* **2001**, *113*, 1142–1160. [CrossRef]

23.	Soreghan, G.; Keller, G.R.; Gilbert, M.C.; Chase, C.G.; Sweet, D.E. Load-induced subsidence of the Ancestral Rocky Mountains recorded by preservation of Permian landscapes. *Geosphere* **2012**, *8*, 654–658. [CrossRef]

24.	Leary, R.J.; Umhoefer, P.; Smith, M.E.; Riggs, N. A three-sided orogen: A new tectonic model for Ancestral Rocky Mountain uplift basin development. *Geology* **2017**, *45*, 735–738. [CrossRef]

25.	Wengerd, S.A.; Strickland, J.W. Pennsylvanian stratigraphy of Paradox salt basin, Four Corners region, Colorado Utah. *Am. Assoc. Pet. Geol.* **1954**, *38*, 2157–2199.

26.	Johnson, S.Y. *Stratigraphic Sedimentologic Studies of Late Paleozoic Strata in the Eagle Basin Northern Aspen Sub-Basin, Northwest Colorado*; U.S. Geological Survey Open File Report 87–286; U.S. Geological Survey: Reston, VA, USA, 1987; pp. 1–82.

27.	Hoy, R.G.; Ridgway, K.D. Syndepositional thrust-related deformation sedimentation in an Ancestral Rocky Mountains basin, Central Colorado trough, Colorado, USA. *Geol. Soc. Am. Bull.* **2002**, *114*, 804–828. [CrossRef]

28.	Cather, S.M.; Heizler, M.T.; Williamson, T.E. Laramide fluvial evolution of the San Juan Basin, New Mexico Colorado: Paleocurrent detrital-sanidine age constraints from the Paleocene Nacimiento Animas formations. *Geosphere* **2019**, *15*, 1641–1664. [CrossRef]

29.	Perry, W.J. *Tectonic Evolution of the Anadarko Basin Region, Oklahoma*; U.S. Geological Survey Bulletin 1866-A; U.S. Geological Survey: Reston, VA, USA, 1989; p. 19.

30.	Johnson, K.S. Geologic Evolution of the Anadharko Basin. *Okla. Geol. Surv. Circ.* **1989**, *90*, 3–12.

31.	Keller, G.R.; Stephenson, R.A. The Southern Oklahoma Dniepr-Donets aulacogens. *Geol. Soc. Am. Mem.* **2007**, *200*, 127–143.

32.	Budnik, R.T. Left-lateral intraplate deformation along the Ancestral Rocky Mountains: Implications for late Paleozoic plate motions. *Tectonophysics* **1986**, *132*, 195–214. [CrossRef]

33.	Sweet, D.E.; Soreghan, G.S. Late Paleozoic tectonics paleogeography of the ancestral Front Range: Structural, stratigraphic, sedimentologic evidence from the Fountain Formation (Manitou Springs, Colorado). *Geol. Soc. Am. Bull.* **2010**, *122*, 575–594. [CrossRef]

34.	Merriam, D.F. *The Geologic History of Kansas*; Kansas Geological Survey Bulletin 162; The University of Kansas: Lawrence, KS, USA, 1963.

35.	Evans, J.L. Major structural stratigraphic features of the Anadarko Basin. In *Pennsylvanian sandstones of the Mid-Continent*; Hyne, N.J., Ed.; Special Publication; Tulsa Geological Society: Tulsa, OK, USA, 1979; pp. 97–113.

36. Munson, T. Depositional, diagenetic, production history of the upper Morrowan Buckhaults Sandstone, Farnsworth Field, Ochiltree County Texas. Oklahoma City Geological Society. In *The Shale Shaker Digest*; Times-Journal Pub. Co: Oklahoma City, OK, USA, 1989; pp. 2–20.
37. Chapin, C.E.; Kelley, S.A.; Cather, S.M. The Rocky Mountain Front, southwestern USA. *Geosphere* **2014**, *10*, 1043–1060. [CrossRef]
38. Gragg, E. Petroleum System Modeling in the Northwest Anadarko Basin: Implications for Carbon Storage. Master's Thesis, New Mexico Institute of Mining Technology, Socorro, NM, USA, 2016.
39. Davis, H.G.; Northcutt, R.A. The greater Anadarko Basin—An overview of petroleum exploration development. In *Anadarko Basin symposium. Norman, Oklahoma*; Circular 90; Johnson, K.S., Ed.; Oklahoma Geological Survey: Norman, OK, USA, 1989; pp. 13–24.
40. Swanson, D. *Deltaic deposits in the Pennsylvanian Upper Morrow Formation in the Anadarko Basin, In Pennsylvanian Sandstones of the Mid-Continent*; Special Publication; Tulsa Geological Society: Tulsa, OK, USA, 1979; pp. 115–168.
41. Sonnenberg, S.A. Tectonic sedimentation model for Morrow sandstone deposition, Sorrento Field area, Denver Basin, Colorado. *Mt. Geol.* **1985**, *22*, 180–191. [CrossRef]
42. Krystinik, L.; Blakeney, B. Sedimentology of the Upper Morrow Formation in Eastern Colorado Western Kansas. In *Morrow Sandstones of Southeast Colorado Adjacent Areas: Rocky Mountain Association of Geologists Guidebook*; Sonnenberg, S.A., Shannon, L.T., Rader, K., von Drehle, W.F., Martin, G.W., Eds.; Rocky Mountain Association of Geologists: Denver, CO, USA, 1990; pp. 37–50.
43. Al-Shaieb, Z.; Puckette, J.; Abdalla, A. Influence of sea-level fluctuation on reservoir quality of the upper Morrowan sandstones, Northwestern Shelf of the Anadarko Basin. In *Sequence Stratigraphy of the Mid-Continent*; Hyne, N.J., Ed.; Tulsa Geological Society Special Publication: Tulsa, OK, USA, 1995; pp. 249–268.
44. McKay, R.H.; Noah, J.T. Integrated perspective of the depositional environment reservoir geometry, characterization, performance of the Upper Morrow Buckhaults sandstone in the Farnsworth Unit, Ochiltree County, Texas. In *Deltaic Reservoirs in the Southern Midcontinent, 1993 Symposium*; Circular 98; Johnson, K.S., Ed.; Oklahoma Geological Survey: Norman, OK, USA, 1996; pp. 101–114.
45. Puckette, J.; Azhari, A.; Rice, A.; Al-Shaieb, Z. The Upper Morrow reservoirs: Complex fluvio-deltaic depositional systems. In *Deltaic Reservoirs in the Southern Midcontinent, 1993 Symposium*; Circular 98; Johnson, K.S., Ed.; Oklahoma Geological Survey: Norman, OK, USA, 1996; pp. 47–84.
46. Gallagher, S.R. Depositional Diagenetic Controls on Reservoir Heterogeneity: Upper Morrow Sandstone, Farnsworth Unit, Ochiltree County, Texas. Master's Thesis, New Mexico Institute of Mining Technology, Socorro, NM, USA, 2014.
47. Forgotson, J.M.; Statler, A.T.; David, M. Influence of regional tectonics local structure on deposition of Morrow formation in western Anadarko Basin. *Am. Assoc. Pet. Geol.* **1966**, *50*, 518–532.
48. Sonnenberg, S.A.; Shannon, L.T.; Rader, K.; Von Drehle, W.F.; Martin, G.W. Regional structure stratigraphy of the Morrowan series, southeast Colorado adjacent areas. In *Morrow Sandstones of Southeast Colorado Adjacent Areas: Rocky Mountain Association of Geologists Guidebook*; Sonnenberg, S.A., Shannon, L.T., Rader, K., von Drehle, W.F., Martin, G.W., Eds.; Rocky Mountain Association of Geologists: Denver, CO, USA, 1990; pp. 37–50.
49. Rose-Coss, D. A Refined Depositional Sequence Stratigraphic Structural Model for the Reservoir Caprock Intervals at the Farnsworth Unit, Ochiltree County, TX. Master's Thesis, New Mexico Institute of Mining Technology, Socorro, NM, USA, 2017.
50. Boyd, R.; ret. Chaparral Energy, Oklahoma City, OK, USA. Personal Communication, 2014.
51. Miall, A.D. *Architectural-Element Analysis: A New Method of Facies Analysis Applied to Fluvial Deposits, In Recognition of Fluvial Depositional Systems Their Resource Potential. SEPM Short Course No. 19*; Flores, R.M., Ethridge, F.G., Miall, A.D., Galloway, W.E., Fouch, T.D., Eds.; SEPM: Tulsa, OK, USA, 1985; pp. 33–81.
52. Lazar, O.R.; Bohacs, K.M.; Schieber, J.; Macquaker, J.H.; Demko, T.M. *Mudstone Primer: Lithofacies Variations, Diagnostic Criteria, Sedimentologic-Stratigraphic Implications at Lamina to Bedset Scales*; SEPM Concepts in Sedimentology Paleontology No. 12; SEPM: Tulsa, OK, USA, 2015; 194p.
53. Wright, V.P. A revised classification of limestones. *Sediment. Geol.* **1992**, *76*, 177–185. [CrossRef]
54. Cather, S.M.; Cather, M.E. *Comparative Petrography Paragenesis of Pennsylvanian (Upper Morrow) Sandstones From the Farnsworth Unit 13-10A, 13-14, 32-8 wells, Ochiltree County, Texas*; PRRC Report 16-01; Petroleum Recovery Research Center: Socorro, NM, USA, 2016.
55. Heath, J.; Dewers, T.; Mozley, P. *Petrologic, Petrophysical, Geomechanical Characteristics of the Farnsworth Unit, Ochiltree County, Texas, Southwest Partnership CO2 Storage-EOR Project*. 2015; Unpublished report.
56. Cobbold, P.R.; Zanella, A.; Rodriguez, N.; Løseth, H. Bedding-Parallel fibrous veins (beef cone-in-cone): Worldwide occurrence possible significance in terms of fluid overpressure, hydrocarbon generation mineralization. *Mar. Pet. Geo.* **2013**, *43*, 1–20. [CrossRef]
57. Cather, S.M. *Preliminary Petrographic Analysis of Pennsylvanian (Upper Morrow) Strata from the Farnsworth Unit 13-10A Well, Ochiltree, County, Texas*, Quarterly Progress Report of the Southwest Regional Partnership on Carbon Sequestration. 2014; Unpublished report.
58. Cather, S.M. *Preliminary Petrographic Analysis of Pennsylvanian (Upper Morrow) Strata from the Farnsworth Unit 13-10A Well, Ochiltree, County, Texas-Mudstones Limestones*, Quarterly Progress Report of the Southwest Regional Partnership on Carbon Sequestration. 2015; Unpublished report.

59. Trujillo, N.A. Influence of lithology diagenesis on mechanical sealing properties of the Thirteen Finger Limestone Upper Morrow Shale, Farnsworth Unit Ochiltree County, Texas. Master's Thesis, New Mexico Institute of Mining Technology, Socorro, NM, USA, 2017.

60. Hanson, R.E.; Puckett, R.E., Jr.; Keller, G.R.; Brueseke, M.E.; Bulen, C.L.; Mertzman, S.A.; Finegan, S.A.; McCleery, D.A. Intraplate magmatism related to opening of the southern Iapetus Ocean: Cambrian Wichita igneous province in the Southern Oklahoma rift zone. *Lithos* **2013**, *174*, 57–70. [CrossRef]

61. Bickford, M.; Van Schmus, W.; Karlstrom, K.; Mueller, P.; Kamenov, G. Mesoproterozoic trans-Laurentian magmatism; a synthesis of continent-wide age distributions, new SIMS U/Pb ages, zircon saturation temperatures, Hf Nd isotopic compositions. *Precambrian Res.* **2015**, *265*, 286–312. [CrossRef]

62. Unpublished Well Reports for Wells 13-10A, 13-14, 32-08, Prepared by TerraTek for Southwest Partnership. 2014.

63. Klein, J.S.; Mozley, P.S.; Campbell, A.; Cole, R. Spatial distribution of carbon oxygen isotopes in laterally extensive carbonate cemented layers: Implications for mode of growth subsurface identification. *J. Sediment. Res.* **1999**, *69*, 184–201. [CrossRef]

64. Trujillo, N.; Heath, J.E.; Mozley, P.S.; Dewers, T.A.; Ampomah, W.; Cather, M. Lithofacies Diagenetic Controls on Formation-scale Mechanical Transport Sealing Behavior of Caprocks: A Case Study of the Morrow Shale Thirteen Finger Limestone Farnsworth Unit Texas. In *Proceedings of the 2016 American Geophysical Union (AGU) Fall Meeting*; San Francisco, CA, USA, 12–16 December 2016. Available online: https://www.osti.gov/servlets/purl/1414640 (accessed on 16 February 2021).

65. Mozley, P.S. Relation between depositional environment the elemental composition of early diagenetic siderite. *Geology* **1989**, *17*, 704–706. [CrossRef]

66. Stoneley, R.S. Fibrous calcite veins, overpressures, primary oil migration. *Am. Assoc. Pet. Geol.* **1983**, *67*, 1427–1428.

67. Wheeler, D.; Scott, A.; Coringrato, V.; Devine, P. Stratigraphy depositional history of the Morrow Formation, southeast Colorado southwest Kansas. In *Morrow Sandstones of Southeast Colorado Adjacent Areas*; Sonnenberg, S.A., Ed.; Rocky Mountain Association of Geologists: Denver, CO, USA, 1990; pp. 9–35.

68. Blakeney, B.A.; Krystinik, L.F.; Downey, A.A. Reservoir heterogeneity in Morrow valley fills, Stateline trend: Implications for reservoir management field expansion. In *Morrow Sandstones of Southeast Colorado Adjacent Areas*; Sonnenberg, S.A., Shannon, L.T., Rader, K., von Krehle, W.F., Martin, G.W., Eds.; Rocky Mountain Association of Geologists: Denver, CO, USA, 1990; 236p.

69. Brown, J. Borehole Image Whole Core Interpretation Overview. Unpublished report prepared for TerraTek Southwest Partnership. 2014.

70. Dalrymple, R.W.; Zaitlin, B.A.; Boyd, R. Estuarine facies models; conceptual basis stratigraphic implications. *J. Sediment. Res.* **1992**, *62*, 1130–1146. [CrossRef]

71. Flugel, E. *Microfacies of Carbonate Rocks: Analysis, Interpretation Application*; Springer: Berlin/Heidelberg, Germany, 2004; Volume 206, 976p.

72. Diessel, C.F. *Coal-Bearing Depositional Systems*; Springer: Berlin/Heidelberg, Germany, 1992; 721p.

73. Schieber, J. Iron Sulfide Formation. In *Encyclopedia of Geobiology*; Springer: Berlin/Heidelberg, Germany, 2011; pp. 486–502.

74. Taylor, K.G.; Gawthorpe, R.L.; Wagoner, J.C. Stratigraphic control on laterally persistent cementation, Book Cliffs, Utah. *J. Geol. Soc. Lond.* **1995**, *152*, 225–228. [CrossRef]

75. Taylor, K.G.; Gawthorpe, R.L.; Curtis, C.D.; Marshall, J.D.; Awwiller, D.N. Carbonate cementation in a sequence-stratigraphic framework: Upper Cretaceous sandstones, Book Cliffs, Utah-Colorado. *J. Sediment. Res.* **2000**, *70*, 360–372. [CrossRef]

76. Scholle, P.A.; Ulmer-Scholle, D.S. *A Color Guide to the Petrography of Carbonate Rocks: Grains, Textures, Porosity, Diagenesis*; American Association of Petroleum Geologists: Tulsa, OK, USA, 2003; 473p.

77. Bagley, M.; Puckette, J.; Boardman, D.; Watney, W. Lithofacies reservoir assessment for the Thirteen Finger limestone, Hugoton Embayment, Kansas. In Proceedings of the AAPG Mid-Continent Section Meeting, Oklahoma City, OK, USA, 1–4 October 2011. AAPG Search Discovery Article #50513.

78. Watney, W.L.; Bhattacharya, S.; Byrnes, A.; Dovton, J.; Victorine, J. Geo-engineering modeling of Morrow/Atoka incised-valley-fill deposits using web-based freeware. In *Morrow Springer in the Southern Midcontinent, 2005 Symposium*; Circular 111; Andrews, R.D., Ed.; Oklahoma Geological Survey: Norman, OK, USA, 2008; pp. 121–134.

79. Drake, W.; Pollock, C.; Kenk, B.; Anderson, J.; Clemons, K. Evolution of the southwestern Midcontinent Basin during the Middle Pennsylvanian: Evidence from sequence stratigraphy, core XRF in southeastern Colorado. In Proceedings of the AAPG Annual Convention & Exhibition, Denver, CO, USA, 31 May 31–3 June 2015. Search Discovery Article #10753.

80. Youle, J.C.; Watney, W.L.; Lambert, L.L. Stratal hierarch sequence stratigraphy; Middle Pennsylvanian, southwestern Kansas, U.S.A. In *Pangea: Paleoclimate, Tectonics, Sedimentation during Accretion, Zenith, Breakup of a Supercontinent*; Klein, G.D., Ed.; GSA Special Paper 288; Geological Society of America: Boulder, CO, USA, 1994; pp. 267–285.

81. Rasmussen, D.L.; Sandia National Laboratory, Albuquerque, NM, USA. Personal Communication, 2020.

82. Blakey, R. Paleogeography Map Website. Available online: https://www2.nau.edu/rcb7/ (accessed on 4 April 2016).

83. Cecil, C.B.; Di Michele, W.A.; Elrick, S.D. Middle Late Pennsylvanian cyclothems, American Midcontinent: Ice-age environmental changes terrestrial biotic dynamics. *Geoscience* **2014**, *346*, 159–168. [CrossRef]

84. Rose-Coss, D.; Ampomah, W.; Cather, M.; Balch, R.S.; Mozley, P. An improved approach for Sandstone reservoir characterization, paper SPE-180375-MS. In Proceedings of the SPE Western Regional Meeting, Anchorage, AK, USA, 23–26 May 2016. [CrossRef]

85. Ampomah, W.; Balch, R.; Cather, M.; Rose-Coss, D.; Gragg, E. *Numerical Simulation of CO$_2$-EOR Storage Potential in the Morrow Formation, Ochiltree County, Texas*; SPE Oklahoma City Oil Gas Symposium; Society of Petroleum Engineers: Tulsa, OK, USA, 2017; SPE-185086-MS.
86. Ampomah, W.; Balch, R.; Grigg, R.B.; Cather, M.; Gragg, E.; Will, R.A.; White, M.; Moodie, N.; Dai, Z. Performance assessment of CO$_2$-enhanced oil recovery storage in the Morrow reservoir. *Geomech. Geophys. Geo-Energy Geo-Resour.* **2017**, *3*, 245–263. [CrossRef]

Article

Analysis of Geologic CO$_2$ Migration Pathways in Farnsworth Field, NW Anadarko Basin

Jolante van Wijk [1,2,*], Noah Hobbs [1], Peter Rose [3], Michael Mella [3], Gary Axen [1] and Evan Gragg [1,4]

[1] Department of Earth and Environmental Science, New Mexico Institute of Mining and Technology, 801 Leroy Place, Socorro, NM 87801, USA; noah.hobbs@student.nmt.edu (N.H.); gary.axen@nmt.edu (G.A.); evangragg@gmail.com (E.G.)

[2] Los Alamos National Laboratory, Computational Earth Science Group, Los Alamos, NM 87544, USA

[3] Energy & Geoscience Institute at the University of Utah, Salt Lake City, UT 84108, USA; prose@egi.utah.edu (P.R.); mmella@egi.utah.edu (M.M.)

[4] SM Energy Company, 1775 Sherman St., Suite 1200, Denver, CO 80203, USA

[*] Correspondence: jolante.vanwijk@nmt.edu

Abstract: This study reports on analyses of natural, geologic CO$_2$ migration paths in Farnsworth Oil Field, northern Texas, where CO$_2$ was injected into the Pennsylvanian Morrow B reservoir as part of enhanced oil recovery and carbon sequestration efforts. We interpret 2D and 3D seismic reflection datasets of the study site, which is located on the western flank of the Anadarko basin, and compare our seismic interpretations with results from a tracer study. Petroleum system models are developed to understand the petroleum system and petroleum- and CO$_2$-migration pathways. We find no evidence of seismically resolvable faults in Farnsworth Field, but interpret a karst structure, erosional structures, and incised valleys. These interpretations are compared with results of a Morrow B well-to-well tracer study that suggests that inter-well flow is up-dip or lateral. Southeastward fluid flow is inhibited by dip direction, thinning, and draping of the Morrow B reservoir over a deeper, eroded formation. Petroleum system models predict a deep basin-ward increase in temperature and maturation of the source rocks. In the northwestern Anadarko Basin, petroleum migration was generally up-dip with local exceptions; the Morrow B sandstone was likely charged by formations both below and overlying the reservoir rock. Based on this analysis, we conclude that CO$_2$ escape in Farnsworth Field via geologic pathways such as tectonic faults is unlikely. Abandoned or aged wellbores remain a risk for CO$_2$ escape from the reservoir formation and deserve further monitoring and research.

Keywords: carbon sequestration; Farnsworth Field; petroleum system modeling; CO$_2$ migration

Citation: van Wijk, J.; Hobbs, N.; Rose, P.; Mella, M.; Axen, G.; Gragg, E. Analysis of Geologic CO$_2$ Migration Pathways in Farnsworth Field, NW Anadarko Basin. *Energies* **2021**, *14*, 7818. https://doi.org/10.3390/en14227818

Academic Editor: João Fernando Pereira Gomes

Received: 20 August 2021
Accepted: 11 November 2021
Published: 22 November 2021

Publisher's Note: MDPI stays neutral with regard to jurisdictional claims in published maps and institutional affiliations.

1. Introduction

Underground injection of CO$_2$ is a proven technology for reducing CO$_2$ emissions into the atmosphere [1–3]. In Farnsworth Field, northern TX (Figures 1 and 2), CO$_2$ was injected into an existing porous petroleum reservoir (the Morrow B sandstone, at a depth of about 8000 feet) as part of an enhanced oil recovery operation in which the CO$_2$ displaces and mobilizes oil. A portion of the CO$_2$ is extracted; the remainder is stored in the subsurface [4]. To protect future generations from environmental impacts, the storage time of the sequestered CO$_2$ needs to be of a timescale of 100–1000s of years or longer [2]. Dependent on the trapping mechanism, and in the absence of leakage, the residence time can be millions of years [5].

Leakage of CO$_2$ to shallow aquifers and the atmosphere may occur through abandoned or aged wellbores e.g., [6–10], along natural and induced faults and fractures e.g., [11,12], along igneous or sedimentary injections (chimneys [13]) and by diffuse leakage through the overburden rock e.g., [13,14]. In Farnsworth Field, northern TX (Figures 1 and 2) the risk of leakage through abandoned or aged wellbores is present because about 150–213 wells have been drilled into the Morrow B reservoir, of which an unknown percentage has been abandoned. Petroleum production in this field started in 1952 with the drilling of a gas well that was completed at a depth of 8096 feet. The first oil well was drilled in 1955 and completed in the Morrow B sandstone [15], at a depth of ~7965 feet. Secondary oil recovery began in 1964 by waterflooding, and tertiary recovery is currently underway through CO_2 injection [4]. In Farnsworth Field, leakage through abandoned or aged wellbores is monitored by the Southwest Regional Partnership (SWP) through CO_2 atmosphere monitoring and soil gas measurements. Farnsworth Oil Field is a study site selected by U.S. Department of Energy to study carbon management strategies. The Southwest Regional Partnership monitors and researches CO_2 movement through the Morrow B reservoir in Farnsworth Oil Field [4].

This study focuses on geologic leakage pathways of CO_2 in the study site through igneous and sedimentary intrusions (chimneys) and natural fractures and faults, and aims to understand natural lateral and vertical migration pathways of CO_2. We interpret and review 2D legacy- and 3D seismic reflection datasets of Farnsworth Oil Field and surrounding areas to locate chimneys, faults and fractures in the Morrow B and its seal, the Atokan Thirteen Finger Limestone. Seismic interpretations are combined with a tracer study to understand well-to-well flow. Petroleum system modeling is used to identify natural (lateral and vertical) migration pathways for CO_2. Because CO_2 is injected into a petroleum reservoir, an understanding of the petroleum system in which the CO_2 is being stored is part of our evaluation. A large-scale petroleum system model of the entire western Anadarko Basin was published previously [16]; here we develop 1D and a 2D smaller-scale models for the study site, using geochemical, geological and geophysical calibrations collected at Farnsworth Field. This petroleum system model provides insight into the burial, thermal, and petroleum and CO_2 migration history of the CO_2 reservoir.

The next section summarizes the tectonic history and stratigraphy of the northwestern Anadarko Basin. During the tectonic history of the study site, phases of subsidence alternated with periods of sometimes significant tectonic uplift, which have had an important influence on the petroleum system.

2. Anadarko Basin Overview

The Anadarko Basin is a mature, deep (as deep as ~12 km or ~40,000 ft, Figure 1) sedimentary basin in the North American craton that has been a prolific source of oil and gas since the early- to mid-1900s [17,18]. Its tectonic history starts in the Pre-Cambrian, and includes a series of orogenic and basin-forming events [17–25]. Pre-Cambrian basement rocks (Figure 3) consist of igneous and metasedimentary rocks emplaced in a basin of unknown origin [18]. In the Early- to Mid- Cambrian, a system of faults formed that has been interpreted either as indicating extensional deformation, or as a system of strike-slip faults [26], resulting in the Southern Oklahoma Aulacogen (a failed rift [17,25]. During this time, sedimentary and igneous rocks were emplaced in this basin [18]. Rifting ceased by Middle Cambrian time.

Subsequently, a phase of thermal subsidence occurred that led to the deposition of alternating carbonates, shales and sandstones in the Southern Oklahoma Trough [27]. These are the Arbuckle and Ellenburger Groups, deposited during the Middle Cambrian-Early Ordovician, mainly consisting of carbonates (Figure 3). The Simpson Group with sandy carbonates and clastic rocks was deposited in the Middle Ordovician, followed by the Late Ordovician Viola limestone and the Sylvan shale. The Silurian-Devonian Hunton Group mainly consists of shale and limestone; the Late Devonian- Early Mississippian Woodford Shale Formation overlays this group (Figure 3) [17,27–35]. Finally, the Woodford Shale was deposited; this is one of the source rocks for the Morrow B reservoir.

Flexural subsidence in the Anadarko Basin began in the Mississippian and continued through Late Pennsylvanian- Early Permian (Figure 3). From Middle Mississippian to Early Pennsylvanian (Morrowan), more than 2 km of sediments were deposited in the Anadarko Basin [36], including the Morrow B reservoir rock which is the focus of this study, and the

Middle Pennsylvanian Kansas City Group located above the Atokan Thirteen Finger (cap rock for the Farnsworth Field petroleum system).

Cambrian through Mississippian sediments were deposited in large epicontinental sea environments and consist of a mixed clastic-carbonate system, siltstone, and shallow marine carbonate facies with minor sandstones and shales [18,36–39]. The Woodford Shale, one of the Anadarko's major source rocks, was deposited during the Devonian; it thickens to over 375 ft in the basin center while pinching out along basin margins [16]. A pre-Pennsylvanian unconformity marks the contact between Mississippian and Pennsylvanian units and is present across most of the basin [21]. Paleo highs and lows during this time helped to control the deposition and drainage systems of early Pennsylvanian Morrowan incised valley fluvial systems [21].

Intrusive igneous Cambrian rocks were exposed south and southwest of the Anadarko basin as a result of Early to Middle Pennsylvanian tectonic activity, and are a source for Morrow B sandstones [40] as well as forming 'granite wash' deposits in proximity of the southern margin of the Anadarko Basin [41,42]. Subsidence of the Anadarko Basin slowed during Middle-Late Pennsylvanian. Pennsylvanian-Permian organic rich sediments deposited in the Anadarko Basin formed, together with the Woodford Shale, source rocks for the Farnsworth Field petroleum system according to our analysis (Section 5).

Strata of the Early Pennsylvanian Morrow B, which is the CO_2 injection reservoir, are up to 4000 ft thick in the deepest southern part of the Anadarko Basin, and thin northward to less than 100 ft in the study site and on the shelf (Figure 3). Deeper Morrow strata consist of shallow marine shales, sandstones and limestones; Upper Morrow strata are shales, discontinuous sandstones, and deltaic deposits, see Figure 3 [18]. As discussed above, Morrowan deposits are the primary target interval for CO_2 injection. An incised valley model is in agreement with the depositional character of many Morrowan fields, including Farnsworth Field [43–53]. The formation dips southeastward, and the Morrowan sandstones exhibit field-scale reflector offsets in seismic data that could indicate facies changes (discussed below).

Overlying the Morrowan rocks is the Thirteen Finger formation belonging to the Atokan deposits (Early Pennsylvanian) that forms the seal of the CO_2 reservoir. They consist of marine shales, sandstones and limestones [18] (Figure 3). The seal directly overlying the Morrowan sandstone reservoir is the Morrowan black shale. It formed as sea levels rose along the basin margins [40]. Both the Morrowan and Thirteen Finger black shales contain appreciable TOC values [24]. The lower and Upper Morrow shales and Thirteen Finger Limestone form, together with the Woodford Shale, the source rocks for the Morrow B reservoir (Section 5). Timing of maturation of these source rocks is discussed in Section 5 of this manuscript.

Deposition of shales, carbonates and sandstones continued into the Triassic and Permian periods. The early Cenozoic Laramide Orogeny affected the Anadarko Basin by fault reactivation along the Wichita-Amarillo uplift, and tilted the basin eastward [19]. Additionally, associated epeirogenic uplift caused 1–3 km of erosion during the Cenozoic and brought the Anadarko into its present state. Cretaceous deposits in the basin have largely been eroded.

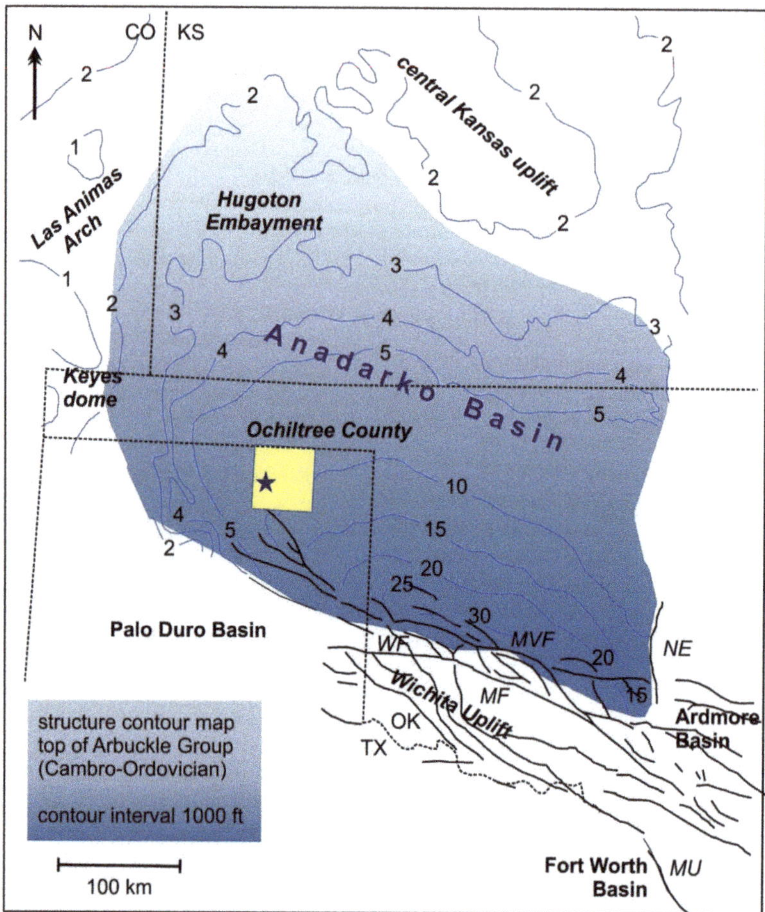

Figure 1. Location of the study site (dark blue star) in Ochiltree County (yellow) in the western Anadarko Basin (blue). Modified from [6]. Contours: depth to top Arbuckle Group. The Arbuckle Group is below the Morrow B reservoir, and pre-dates the basin's flexural subsidence phase. Contour interval 1000 ft below surface. Black solid lines: major faults. MF = Meers fault zone, MU = Muenster Arch, MVF = Mountain View fault zone, NE = Nemaha uplift, WF = Willow fault zone. Dashed black lines are state boundaries: CO = Colorado, KS = Kansas, OK = Oklahoma, TX = Texas. See Figure 2 for details and the location of Farnsworth Field.

3. Subsurface Structure of the Western Anadarko Basin

Faults and fractures provide natural pathways for CO_2 migration towards Earth's surface. Regional and field-scale faults have been documented across the Oklahoma and Texas portions of the Anadarko Basin [54–58]. In the western Anadarko Basin, north of our study area in western Oklahoma, a NW-SE trending fault is reported in Beaver County [59]. In the Texas portion of the basin, a similar trending fault is inferred southeast of our study area [58]. Marsh and Holland [58] do not list faults in Farnsworth Field.

The potential presence of faults in proximity to the CO_2 sequestration study area motivates further research on these possible CO_2 migration pathways. Below, we discuss our interpretation of 2D legacy seismic data (Figures 2 and 4), that allow us to search for faults directly east and north of Farnsworth Field, as well as an interpretation of 3D seismic data of the Farnsworth Field study site (Figures 5–7).

3.1. Interpretation of 2D Seismic Lines

The SWP purchased over 100 miles of 2D legacy seismic data from Seismic Exchange Inc. covering an area from near Farnsworth Field to the east and southeast (locations shown in Figure 2). The 2D lines were acquired by Seisdata Services, Inc. from 1984 to 1986 and all sources were generated by Vibroseis except for line DC-NEP-10 which was generated using Primacord. The lines were reprocessed in 2014 by Seismic Exchange, Inc. Interpretations from Gragg [60] (interpreted lines DC-NEP-10 and DC-NEP-33) are shown in Figure 4. Mis-tie corrections were applied to crossing 2D lines to account for the inconsistencies in quality, static solutions, and vintages [60]. A seismic well tie was made from the Killingsworth well because it had sonic and density logs, a checkshot from surface to approximately 8600 ft., and was close to the seismic line [60]. Gragg [60] constructed a velocity model from the Killingsworth well tie, and the seismic lines were converted to the depth domain. The interpreted 2D seismic data also provided the geometry input for the petroleum system model (Section 5).

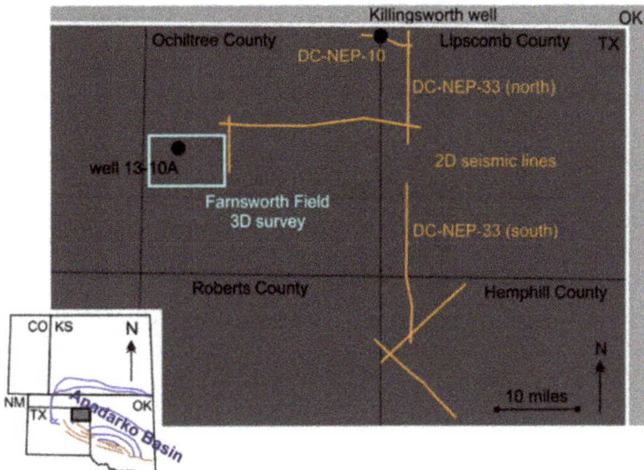

Figure 2. Overview map of Farnsworth Field study site within Ochiltree County, seismic lines and well data used in this study, and locations of 2D seismic lines and 3D seismic survey. The inset shows the location (grey) of the map. The Killingsworth well has been used for the well-tie with DC-NEP-10. Well 13-10A is a CO_2 injection well in Farnsworth Field.

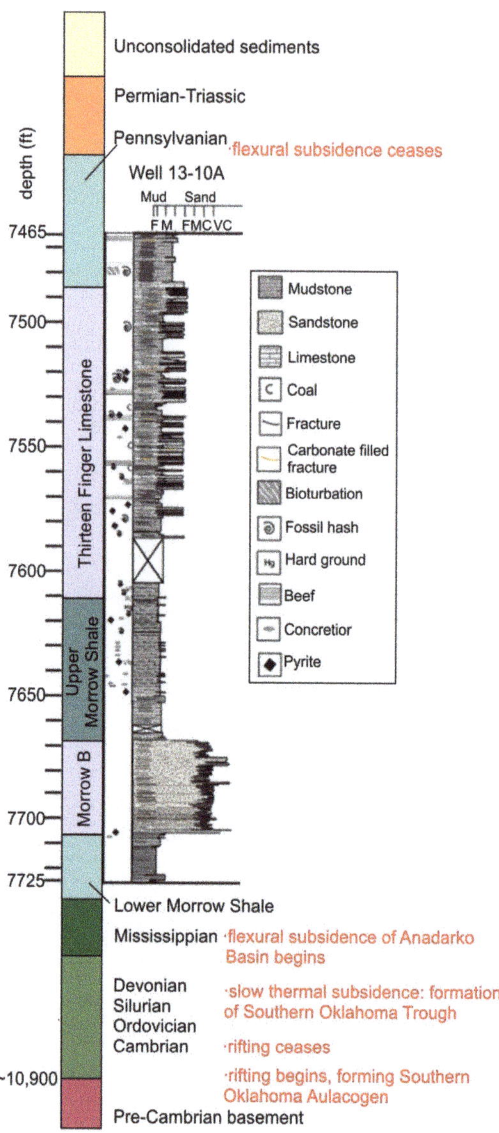

Figure 3. Simplified stratigraphic column of Farnsworth Field area, with details of CO_2 reservoir (Morrow B) and cap rock at well 13–10A (location of this well is shown in Figure 2), and major tectonic events (in red). Grain size: F is fine, M is medium, C is coarse, VC is very coarse.

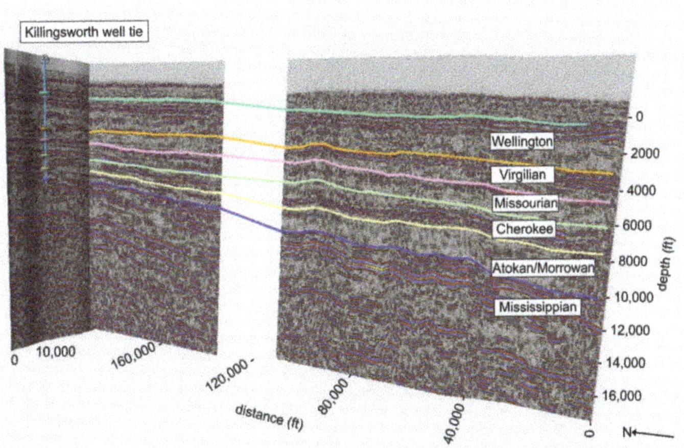

Figure 4. Interpretation of seismic lines DC-NEP-10 (short east–west line) and DC-NEP-33 (north and south), and well-tie with the Killingsworth well (light blue). Locations of lines and Killingsworth well shown in Figure 2. The Atokan/Morrowan formations are mostly transparent, and their boundaries cannot be resolved in these data. The Missourian Kansas City Group and Late Devonian-Early Mississippian Woodford shale are interpreted in the 3D seismic data (Figure 5). No seismically resolvable faults are observed on the 2D seismic lines in the Atokan or Morrowan formations.

Formations generally dip southeastward in our study area, which is in agreement with regional trends [22,61,62]. The dip increases with stratigraphic age and units tend to thicken toward the deep basin. The Atokan and Morrowan formations are continuous, relatively transparent seismically, and cannot be resolved individually on these low-resolution lines. The top Mississippian is interpreted as an angular erosional unconformity, formed as a result of the Wichita orogeny [24]. There are no indications for faulting or seismic chimneys in both the east–west and north–south lines in the Atokan and Morrowan formations.

3.2. Interpretation of 3D Seismic Survey

3D Reflection seismic data (Figures 2, 5 and 6) were acquired in 2013 by SWP through WesternGeco over an approximately 42 mi^2 surface area, with full fold covering 27 mi^2. The geophones had 33 ft. spacing and dense vibroseis source points with a sweep frequency of 2–100 Hz. Processing steps and survey characteristics are described by [63]. Preliminary interpretations of this dataset suggested that the Morrow B reservoir could be faulted, with faults striking E, S, and SE [64]. Our interpretations show that the seismic discontinuities that were interpreted as faults in White et al. [64] are erosional features, incised channels, and karst structures.

Wells 13-10A and 32-8 (locations shown in Figure 5) were used for well ties. The Morrow B reservoir rock reaches a maximum thickness of ~70 ft in Farnsworth Field, and its boundaries cannot be resolved in the seismic data. Since the exact location of the Morrow B horizon in the seismic data is thus uncertain, a reflector in the proximity of the Morrow B (Figure 5a) was traced confidently in the seismic data. In further analyses, we used an isopach map of the Morrow B that we created from interpolation of 346 well logs. Figure 5 shows the position of the Morrow B reservoir layer with respect to the overlaying Thirteen Finger Limestone and Kansas City Group, and the underlying Woodford and Hunton Formations. In Farnsworth Field, paleozoic formations dip slightly toward the southeast (Figure 6).

We applied several seismic attributes [65–67] to the 3D seismic dataset to detect faults and fractures that may act as migration pathways for sequestered CO_2. Seismic attributes were used to help identify features such as fractures, faults, and stratigraphic changes that may not be easily discerned in the original data. We generated edge-detection attributes that measure waveform similarity (Variance and Amplitude Contrast) and Ant Tracking volumes that track continuous features in an effort to illuminate possible fault structures. Parameters were varied within Petrel® software to best highlight any discontinuities while keeping the parameters in a reasonable range.

Variance is an edge enhancement attribute used to estimate localized variance in the seismic signal [65]. The Amplitude Contrast attribute analyzes derivatives in all three components [66]. Petrel® software 2017 allows the user to apply dip corrections, vertical smoothing filters, and the ability to steer the volume along an azimuth. The default values resulted in the best results, as discussed below. The edge detection attribute Ant Tracking uses either the Amplitude Contrast or Variance volumes as input. The attribute attempts to improve the signal-to-noise ratio of discontinuities. To generate the best Ant Tracking volume for highlighting discontinuities, multiple variations of Variance and Amplitude Contrast volumes were generated.

The edge-detection attributes did not illuminate any features that could be interpreted as faults (Figure 6). We did identify channels, a karst-collapse structure, and erosional features; these are discussed next.

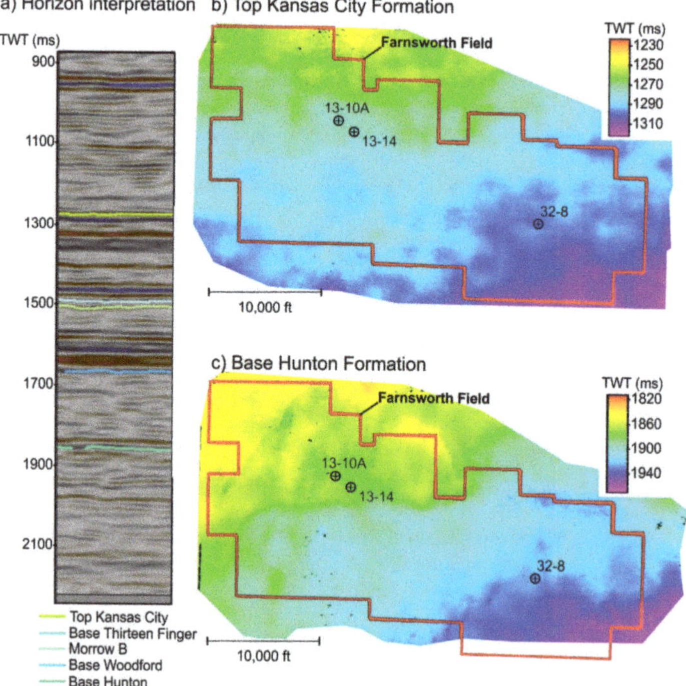

Figure 5. Interpreted horizons at location of well 13-10A (**a**); Top Kansas City Formation (**b**); and Base Hunton Formation (**c**). Farnsworth Field is outlined in red; also shown are locations of wells 13-10A, 13-14, and 32-8. Based on well ties, the "Morrow B reflector" is located within the Morrow B.

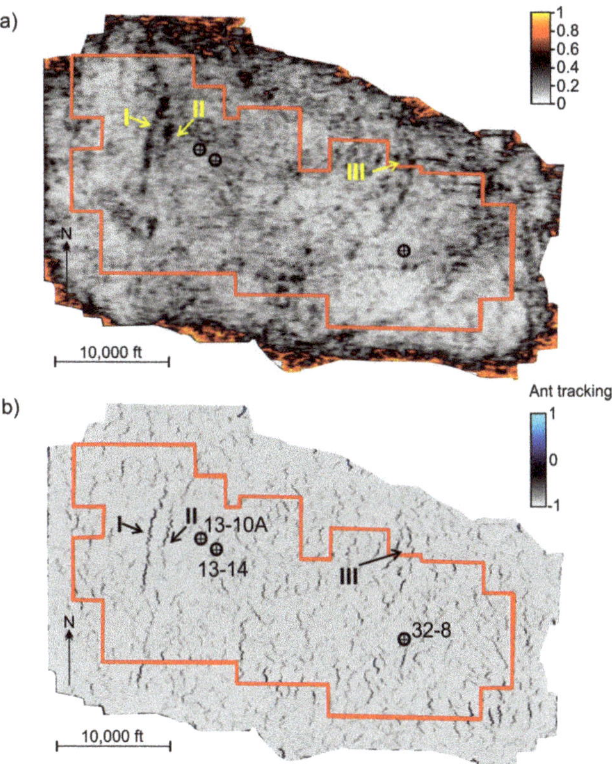

Figure 6. The Morrow B surface displaying the three characterization wells 13-10A, 13-14, and 32-8; (**a**) the Variance attribute overlain on the Morrow B surface; and (**b**) the Ant Tracking attribute overlain on the Ant Tracking volume. Features discussed in text are labeled as I, II, and III. In (**a**), the concentric rings in the eastern side of the survey have not been identified in other attributes and there is no evidence that they are geologic features.

We identified an elongated area associated with north–south trending linear features visible in the Variance and Ant Tracking volumes (discontinuities I and II in Figure 6), which were previously interpreted as faults [64]. In the Variance volume, the N-S features bounding this area appear wide (approximately 100 to 1000 ft. wide) while they look sharp in the Ant Tracking volume (Figure 6). Two vertical cross sections through this feature (Figure 7) illustrate that this is probably caused by differential compaction of shales above an erosional feature in the Hunton limestone that developed prior to the deposition of the overlying horizon (which has tentatively been marked "Woodford"). This resulted in draping over the deeper structure. The Ant Tracking discontinuities and the wide banding on the Variance volume likely identified changes in the seismic signature. Surfaces interpolated from well-logs supported this interpretation; a small gradient that is compatible with draping, and vertical offsets are only about 9 ft to 48 ft at the top of the Morrow B [56].

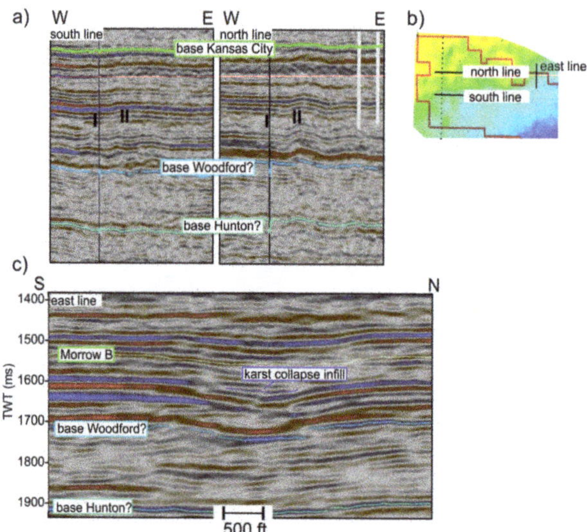

Figure 7. (**a**) Two seismic lines through the feature (labeled I and II in Figure 6) in the western part of Farnsworth Field; locations of lines indicated in (**b**); (**b**) portion of Farnsworth Field (outlined in red) and locations of seismic lines; (**c**) karst collapse infill structure in the eastern part of Farnsworth Field; location of east line indicated in (**b**). Vertical lines in (**a**) mark the dashed line in (**b**).

The second structure that we discuss here is the feature on the edge of the NE side of the field (Figure 7c). This feature appears very faintly on the Variance cube (Figure 6a, indicated with III), and does not appear on the Ant Tracking attribute (Figure 6b). This feature was interpreted as an E-striking fault in White et al. [64]. Offset of the top of the Morrow B directly above this feature, based on well log tops, is approximately 30 to 60 ft. The feature could be identified easily as a pronounced down-warping of the Morrowan, Atokan, and basal Woodford reflectors. We interpreted this as a karst collapse structure. The collapse structure possibly formed in the Hunton limestone during a period of low sea level.

We have identified several channel features (Figure 8) in the western Farnsworth Field area. Our interpretation is consistent with the incised valley model proposed by Krystinik and Blakeney [44] in which the Morrow B sandstones were deposited in incised valleys on the basin margin as the sea level rose. Although not resolvable with the 3D seismic data, it is possible that such facies changes or channels could form preferential flow paths for CO_2. The channel in Figure 8b is particularly well defined, but most of the reflectors that could be identified as channel fills seem laterally discontinuous (Figure 8c,d). This is due to the fact that the thickness of many of these channel infills is below the 3D seismic resolution.

The Kansas City Formation, Morrow B, and Woodford Formation drop down a few tens of ft southward in the center of Farnsworth Field (Figure 5). They drape over an erosional edge in the Hunton limestone, which was previously interpreted as an east-striking fault (fault #3 in White et al., 2017, their Figure 3).

We were not able to identify faults in the Morrow B or overlying formations. We note that this does not necessarily mean that faults and fractures are not present, but they can simply not be resolved with our seismic datasets. To understand possible flow paths for CO_2 within the Morrow B, and between the Morrow B and surrounding formations, we will compare our interpretations of the Morrow B to results of a tracer study and with a petroleum modeling study.

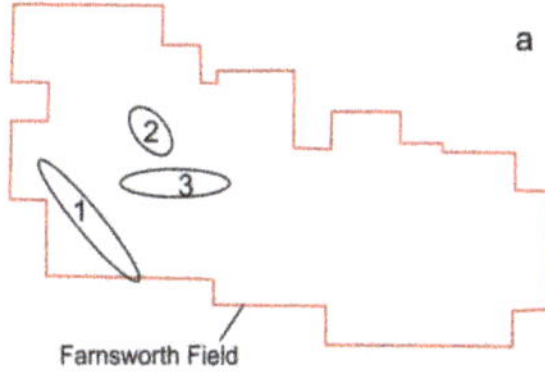
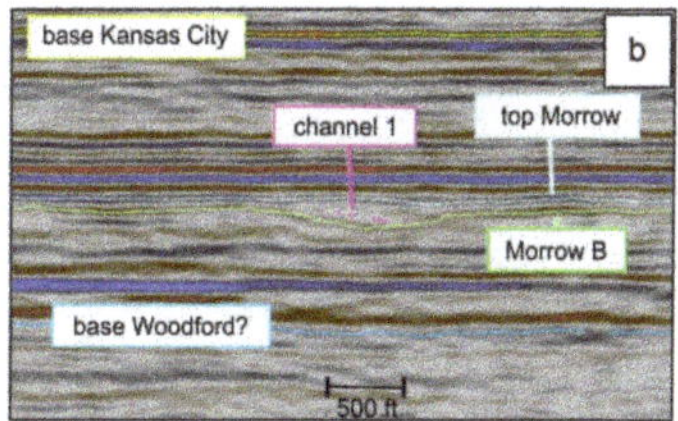
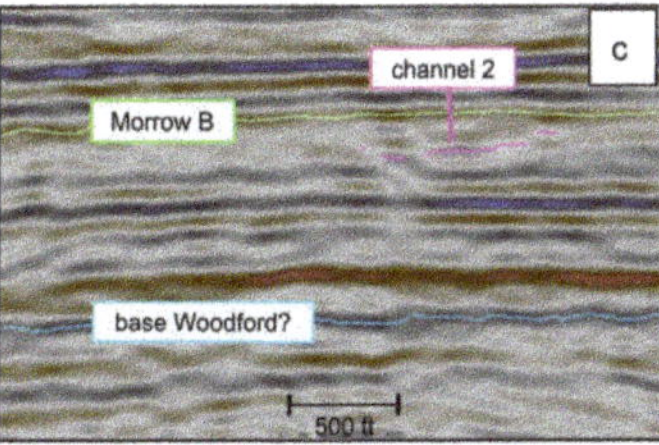
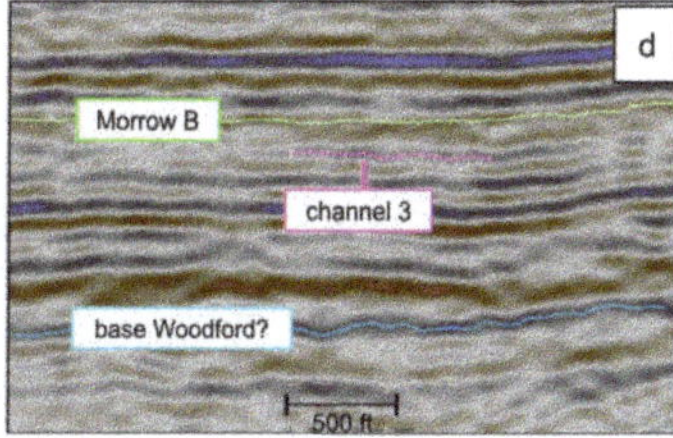

Figure 8. (**a**) Location of three channel-like features in the western Farnsworth Field; (**b**) details of seismic line intersecting channel 1; (**c**) details of seismic line intersecting channel 2; (**d**) details of seismic line intersecting feature 3.

4. Flow Paths in the CO_2 Reservoir

Seal analyses of the Morrow B and Thirteen-Finger Limestone Formations [68] suggest that upward CO_2 migration through the seal, by flow through permeable pathways or stress-induced breakage of the seal, is unlikely in Farnsworth Field. The absence of detectable faults in the Morrow B and overlying formations suggests that no major fault zones are present along which CO_2 could migrate. Here, we discuss lateral migration through the Morrow B reservoir.

Aqueous-phase tracer studies were conducted in the western half of Farnsworth Field in 2014, 2015, and 2017. Naphthalene sulfonates were injected in wells 13-3 (2017), 13-5 (2014), 13-10A (2014), 13-13 (2014), and 14-1 (2015), Figure 9, Figure 10, Figure 11. These are a family of organic compounds that have successfully been used as tracers in geothermal and petroleum reservoirs [69–73], and ground water studies [74].

4.1. Tracer Study Setup

The naphthalene sulfonate compounds that were used as tracers in this study were obtained from YickVic Chemicals, Hong Kong, China. The three tracer tests described below were initiated between May 2014 and June 2017, with sampling and analysis continuing into 2018. In each case, a concentrated aqueous solution (~5%) of a naphthalene sulfonate was mixed with fresh water at the wellhead and injected as a slug over a duration of approximately 0.5 h. Table 1 summarizes the tracer injection scheme.

Table 1. Tracer injection parameters. Well locations shown in Figure 12.

Tracer	Mass Injected (kg)	Well	Injection Date
1,6-naphthalene disulfonate (1,6-nds)	27.5	13-13	2 May 2014
1,3,6-naphthalene trisulfonate (1,3,6-nts)	50	13-10A	2 May 2014
1,5-naphthalene disulfonate (1,5-nds)	25	13-5	2 May 2014
2,7-naphthalene disulfonate (2,7-nds)	100	14-1	13 October 2015
2,6-naphthalene disulfonate (2,6-nds)	100	13-3	15 June 2017
2-naphthalene sulfonate (2-ns)	80	13-3	15 June 2017

The surrounding production wells were sampled over the subsequent four years for tracer analysis by High-Performance Liquid Chromatography (HPLC) and fluorescence detection with detection limits of approximately 100 parts per trillion [68,69]. In order to separate matrix interferences from the tracer analytes, the samples were subjected to solid phase extraction prior to analysis. Decay kinetics studies have shown that all of the naphthalene sulfonate compounds used in this study are suitable for use in geothermal reservoirs having temperatures up to 250 °C, and a subset are suitable for use in reservoirs as hot as 300 °C [68–72].

4.2. Tracer Study Results

4.2.1. Tracer Tests of WAG Injection Wells 13-13, 13-10A, and 13-5

On 2 May 2014, 27.5 kg of 1,6-naphthalene disulfonate (1,6-nds); 50 kg of 1,3,6-naphthalene trisulfonate (1,3,6-nts); and 25 kg of 1,5-naphthalene disulfonate (1,5-nds) were each mixed with water and injected into the three injection wells 13-13, 13-10A, and 13-5, respectively. All three wells were receiving alternating injections of water and CO_2. The aqueous phase of all surrounding wells was sampled regularly over the subsequent four years and analyzed for the presence of the naphthalene sulfonate tracers.

The well showing the most significant returns of the three tracers was 11-2 (Figure 9). Due to a long sampling hiatus between about day 500 and day 1100, it is not known when breakthrough first occurred. However, given the shape of the curve, it was probably not before about day 1000. Thus, it was almost three years before the tracers injected into wells 13-13, 13-10A, and 13-5 broke through to the northeast well 11-2. A second well showing returns of tracers injected during the 2014 test was 13-17. The other two tracers in which tracer was injected in 2014 showed negligible concentrations in 13-17.

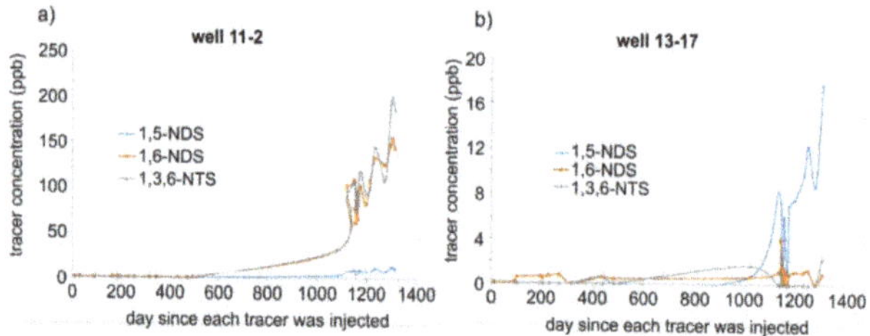

Figure 9. (**a**) Returns to well 11-2 of tracers 1,6-nds, 1,3,6-nts, and 1,5-nds that were injected into wells 13-13, 13-10A, and 13-5, respectively, on 2 May 2014; (**b**) returns to well 13-17 of the tracer 1,5-nds that was injected into well 13-5 during the 2014 tracing campaign.

4.2.2. Tracer Test of Water-Injection Well 14-1

On 13 October 2015, 100 kg of 2,7-nds was mixed with water and injected as a slug into water-injection well 14-1. Surrounding production wells were sampled and analyzed over the subsequent 2.5 years. Figure 10 shows the return curves for the wells that showed tracer returns.

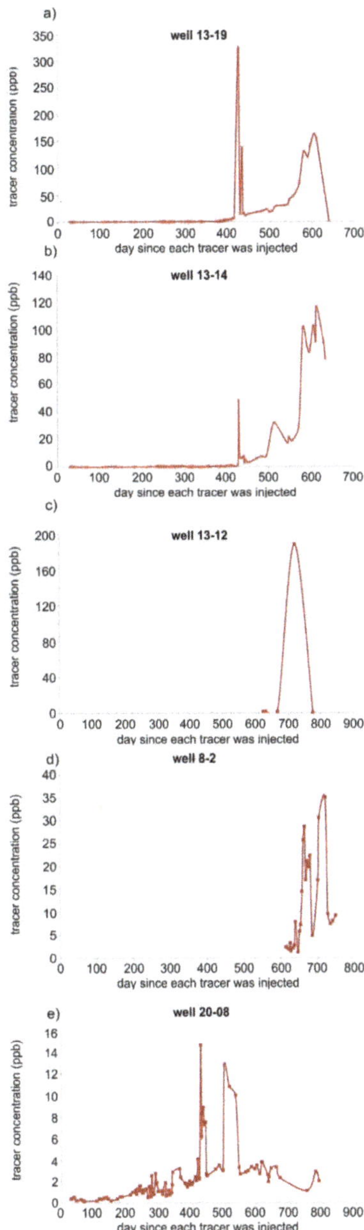

Figure 10. (**a**) Tracer 2,7-nds concentrations measured in well 13–19. 100 kg of this tracer was injected into water-injection well 14-1 on 13 October 2015; (**b**) 2,7-nds concentrations measured in well 13-14; (**c**) 2,7-nds concentrations measured in well 13-12. A paucity of data points on this plot reflects the fact that this well was sampled infrequently; (**d**) 2,7-nds concentrations measured in well 8-2. The paucity of data points on this plot reflects the fact that this well was sampled relatively infrequently; (**e**) 2,7-nds concentrations measured in well 20-8.

4.2.3. Tracer Test of WAG Injection Well 13-3

On 15 June 2017, 100 kg of 2,6-nds and 80 kg of 2-ns were mixed with water and injected as a pulse into well 13-3. The surrounding wells were sampled over the subsequent six months, but the only well showing returns was 8-2, which is located directly west of 13-3. Shown in Figure 11 are the very strong returns of these tracers to 8-2.

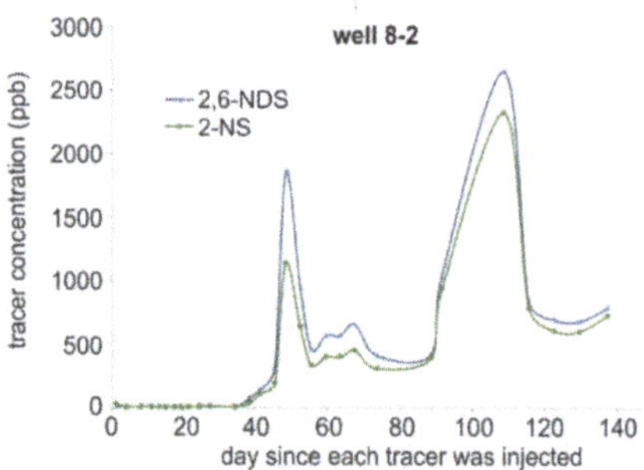

Figure 11. Returns of the tracers 2,6-nds and 2-ns, which were injected into well 13-3.

4.3. Discussion: Interpretation of Tracer Study Results

The line of wells running north to south near the middle of the study site (13-13, 13-10A, and 13-5) were all tagged in May 2014. These three wells were subjected to water-alternating-gas (WAG), whereby brine would be injected for a few weeks followed by the injection of CO_2 for a few weeks. The first arrivals of tracer were all approximately 1000 days after injection in spite of the fact that the distances between wellheads varied significantly. In the case of well 13-5, the first arrival of tracer was approximately the same for tracer arriving at adjacent well 13-17 as it was for the much more distal well 11-2. In contrast, the first arrival of tracer injected into well 14-1 was never more than 625 days in spite of traveling a greater distance—at least in the case of well 8-2. This difference in aqueous-phase flow velocity might be explained by the fact that 14-1 was never subjected to WAG but was only on water injection. In the case of well 13-3, the first arrival time to adjacent well 8-2 was much shorter (~40 days) with concentrations that were more than 10 times greater than those of any of the other tracers. The two peaks in this return curve (Figure 11), with maxima at 50 and 110 days reveal the very heterogeneous flow patterns between this pair of wells. Flow heterogeneity was likewise observed in many of the other return–curve plots (Figures 9 and 10).

Although we have only a limited number of datapoints available, a well-to-well flow pattern can be recognized. Flow of the tracers is generally west to northwest, except for tracers injected into well 14-1, which were also detected toward the east of the injection well (Figure 12). Close inspection of the depth of the Morrow B reservoir in and between these well locations suggests that inter-well flow occurs between wells where the Morrow B is at the same depth, or in an up-dip direction; all wells where tracers were detected are up-dip from the injection wells. The Morrow B isochore map (Figure 12) shows that the Morrow B is of irregular thickness, and locally thins significantly just south of wells 13-17 and 13-19. The Morrow B deepens southward along an approximately EW-striking "step" (previously interpreted as fault #3 [64]), and drapes over the deeper Woodford and Hunton formations; the Hunton formation is locally eroded. It is possible that tracers injected north of the step (indicated with the dashed contour in Figure 12) in the Morrow B will not flow south of this step. If inter-well flow is horizontal or up-dip, the impermeable cap rock of the Morrow B (Thirteen Finger) would prohibit down-dip, or (generally) southeastward flow from the tracer injection wells, since the step in Morrow B depth is of the same magnitude as the thickness of the Morrow B.

This interpretation differs from earlier preliminary work that ascribed inter-well flow directions to presumed faults in the Morrow B [64]; our updated interpretations discussed above suggest that erosional features play a role in inter well flow. The seismic data do not provide sufficient resolution to link flow directions with incised valleys.

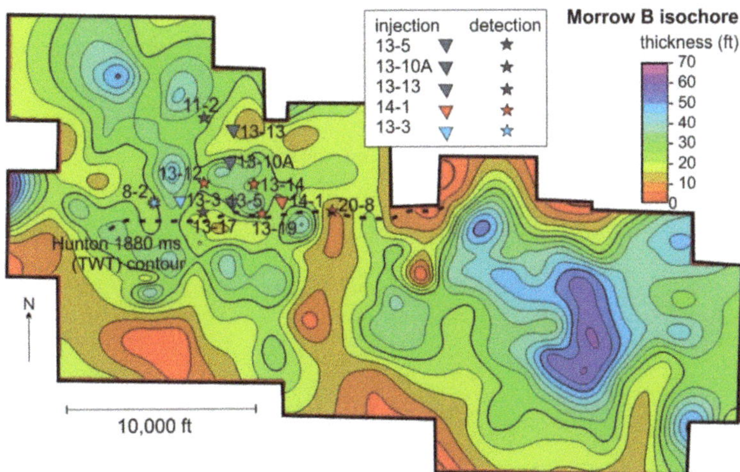

Figure 12. Farnsworth Field map of Morrow B isochore from well logs (well locations shown in White et al., 2017, Figure 3); tracer injection wells (triangles) and detection wells (stars) from tracer studies conducted in 2014, 2015, and 2017 indicated in grey, red, and blue, respectively. Stars mark wells where tracers were detected: injections in 13-13, 13-10A, and 13-5 were detected in 11-2; injections in 13-5 were detected in 13-17; injections in 13-13 were detected in 11-2; injections in 14-1 were detected in 13-12, 13-14, 13-19, 8-2, and 20-8; injections in 13-3 were detected in 8-2.

5. Petroleum System Model of the Farnsworth Petroleum Field

Higley [16] developed a 4D petroleum system model of the Mississippian-Pennsylvanian petroleum system in the Anadarko Basin, and found that present and past oil-migration flow paths in the Texas Panhandle are directed SE-NW, sourced in the deep portion of the Anadarko Basin, where Transformation Ratios are close to 1 [16,75]. Petroleum in Farnsworth Field is likely commingled [75], and sourced from four distinct families of oil (Ordovician-Viola, Woodford, Morrow, and Upper Pennsylvanian). Migration paths may be over 100 km long [16,75]. In the Farnsworth Field, distribution of oil accumulation is controlled by the Morrow incised valley complex.

Basin-scale studies of the Anadarko basin [16] provide insight into the petroleum system of the basin but do not provide detailed insight into petroleum sources and migration paths for our study site. To fill this gap, we developed a 2D petroleum system model of the Farnsworth Field region with Schlumberger PetroMod® (2013) software. We also constructed a burial history curve to understand the time-depth relation of the cap rock (Thirteen Finger Limestone formation).

5.1. Setup of 1D Petroleum System Models

A burial history curve was constructed for well 13-10A (Figures 13 and 14, Table 2). Sixteen stratigraphic units and basement were included as a unit in the 1D petroleum system model. Each unit was assigned lithofacies information (Table 2), thickness, age of deposition, and, if appropriate, age of erosion. Each unit was also assigned its role in the petroleum system (source, reservoir, seal, underburden, or overburden). Four source rocks were included: the Woodford shale, lower Morrowan shale, Upper Morrowan shale, and Thirteen Finger Limestone. Each source rock was assigned a hydrogen index (HI), total organic carbon percentage (TOC), and a kinetics model. Lithofacies were generalized based on SWP well log and core data; TOC and HI values are from SWP core analyses.

Geologic ages of the different formations (Table 2) are based on Higley et al. [62]. All lithologies were generalized based on the dominant lithology present within a model layer. Erosion amounts and timing for the pre-Pennsylvanian unconformity were based on interpretations of the unconformity being regional [21]. Well log data of well 13-10A did not reach basement depths because operators were not targeting intervals below the Morrowan Formation. In order to include the complete Anadarko Basin's burial history, the modeled wells were extended to basement using 2D seismic data. The top of the basement in the seismic data has a distinctive seismic character change from sub-parallel laterally continuous reflectors to chaotic discontinuous reflectors. Basement depths were obtained by performing a seismic well-tie, and used a velocity model for the depth conversion, as described above. Once the seismic data were depth-converted, depth estimates for the Woodford Shale, Cambrian-Devonian, and basement were added to the one-dimensional and two-dimensional models. The Woodford Shale depth, thickness (40 ft), and initial TOC (~1.5–2%) were approximated from Higley et al. [62]

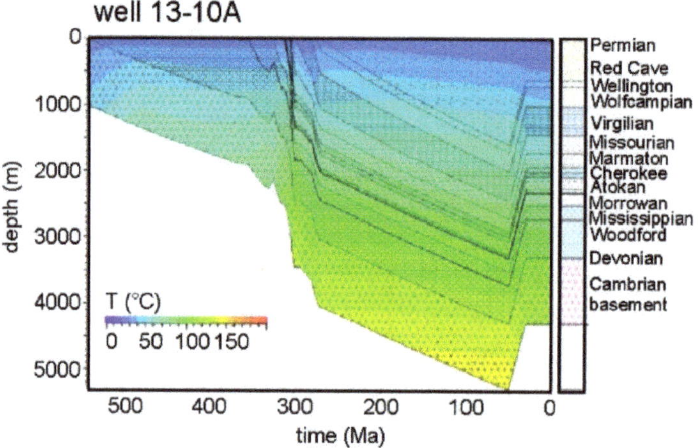

Figure 13. Burial and temperature history for well 13-10A. Before onset of the Laramide uplift, the source rocks from which hydrocarbons in Farnsworth Field are sourced (Table 1; Woodford, Morrowan, and Atokan) reached the oil/gas windows.

We constrained the burial history curve and basin thermal history with vitrinite reflectance, production data, and Rock-Eval pyrolysis data [60]. Schlumberger PetroMod® software was used to predict temperatures in the well, and we calibrated those temperatures with thermal maturity indicators. As an example, the well 13-10A 1D petroleum system model predicts a Morrowan reservoir temperature of 74 °C and a maximum reservoir temperature of 104 °C that occurred around 50 Ma (Figure 13). Hinds [76] documented the actual reservoir temperature of Farnsworth at time of discovery as 75.5 °C (168 F), similar to our predicted value. The resulting burial history curve is shown in Figure 13.

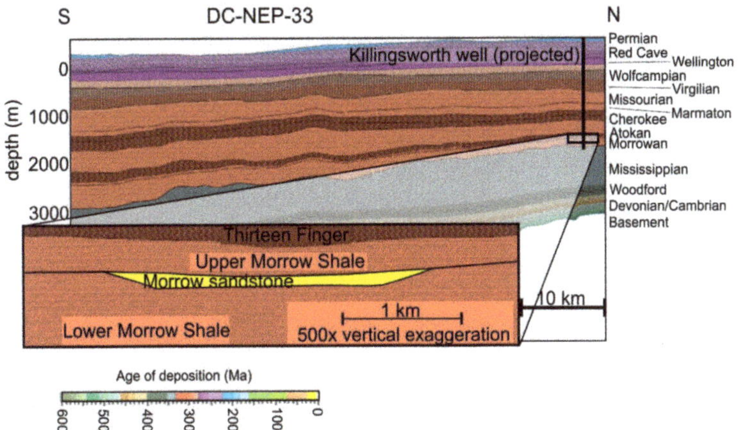

Figure 14. Details of seismic line DC-NEP-33 (location in Figure 2) with a Morrowan sandstone layer (yellow) that is encased in shale (light red). The projected location of the Killingsworth well is marked. Geometry of the sand body is based on well logs and average reservoir dimensions (see text for discussion). Colors correspond to the age of deposition, except for the Morrow sandstone, which is of Morrowan age.

5.2. Results of 1D Petroleum System Modeling

Slow subsidence with intermittent periods of uplift characterized the pre- Late-Pennsylvanian history of the western Anadarko Basin (Figure 13). A phase of relatively rapid subsidence occurred during the Late Pennsylvanian- Early Permian, when the Anadarko Basin subsidence accelerated as part of foreland basin development [22,27,43,62,77]. Subsidence continued until the start of the Laramide orogeny. Temperatures increased to about 130 °C in the deepest part of the column (Figure 13) until the onset of the Laramide deformation. In agreement with previous studies [16], our 1D model predicts that the Woodford Shale, Morrowan Formations, and the Thirteen Finger Limestone are currently in the oil and gas windows.

5.3. 2D Petroleum System Modeling Setup

Interpreted seismic line DC-NEP-33 (north and south, Figure 4) provides the depths and geometries for the layers in our 2D petroleum system model. The 2D petroleum system model gives insight into petroleum migration and paleo-leakage pathways, and allows for more detailed analysis of fluid accumulations and compositions in the region than the 1D analysis. Only thicker sedimentary sequences or packages (hundreds of meters) were included in the petroleum system models, except around the Morrowan reservoir of interest where layers decrease to thicknesses of tens of meters. Two erosional events were modeled: the pre-Pennsylvanian subsurface unconformity and Laramide uplift and erosion (Table 2). The pre-Pennsylvanian unconformity is present across the NE Texas Panhandle [21,46]. In our study area, the unconformity associated with the Laramide uplift is at the surface, where 1–3 km of Mesozoic and Permian deposits have been eroded as the basin was tilted eastward in the Cenozoic [54,60].

The Booker Field (location in Figure 1) well logs indicate a reservoir sandstone facies that is not resolvable on the 2D legacy seismic lines; this was included in the 2D model as it is an important reservoir rock. Dimensions of this sand body are estimated from SWP well logs and literature [44,60]. This sandstone body is the active injection interval for CO_2-EOR operations at Booker Field. Figure 14 shows formations and corresponding depositional ages of the 2D model.

5.4. Results of 2D Petroleum System Modeling

The 2D petroleum system models predict hydrocarbon accumulation and fluid properties (Figure 15). The Behar et al. [78] type II compositional kinetic model was selected for the Thirteen Finger Limestone and Upper Morrow shale; the Behar et al. [78] type III compositional kinetic model was selected for the lower Morrow shale, and the Lewan and Ruble [79] Woodford shale hydrous pyrolysis kinetics model was used for the Woodford shale (Table 2). We selected the hybrid Darcy/flowpath model to describe petroleum flow in Schlumberger PetroMod® software.

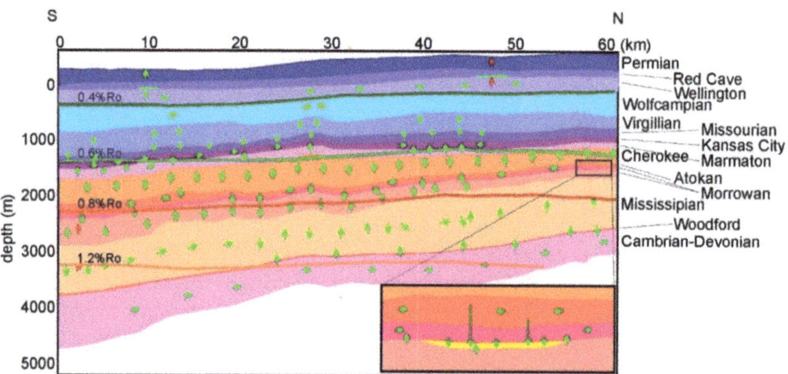

Figure 15. Vitrinite reflectance and liquid (green) and gas (red) migration pathways predicted by Schlumberger Petromod® software. The Morrowan formation is divided into an Upper and Lower Morrow shale. The black box is enlarged in the inset; the reservoir sandstone formation (yellow) is charged by the Upper and Lower Morrowan shales, the Woodford, and the Thirteen Finger Limestone.

Modeled temperatures in the Morrowan formations increase southward along line DC-NEP-33 from 70 °C to 105 °C. The model predicts that the CO_2 reservoir is presently at 74 °C at the locations of Farnsworth and Booker Fields. The Thirteen Finger Formation temperature ranges from 67.5 °C to 87 °C. This southward increase in temperature is reflected in the basin's maturation; maturation within a formation generally increases southward (Figure 15). In our modeled transect, the Woodford Shale is most mature, and its predicted vitrinite reflectance is between 1–1.3% R_o; vitrinite reflectance of the Atokan Thirteen Finger is between 0.65–0.8% R_o. In our models, only the deep Woodford Shale is in the gas window. This formation has a Transformation Ratio of 100% in the south, and 42% in the far-north part of the section. The Thirteen Finger's Transformation Ratio ranges from 30.5–6%, the Upper Morrow Shale from 40–7%, and the lower Morrow Shale from 33–1.3% (south to north; Transformation Ratios are lower in the north). Even though the Transformation Ratio of the lower Morrow Shale is high, it generates low volumes of hydrocarbons in our model because the organic carbon content is low.

In our model, hydrocarbons were expelled from the Woodford Shale from c. 301 Ma in the southern part of the section and around 206 Ma in the northern part. The lower Morrow Shale is the next deepest potential source layer but does not indicate hydrocarbon expulsion onset anywhere. This is explained by its low hydrogen index of 10 (Table 2). The expulsion onset for the Upper Morrow Shale is at 270 Ma in the south and 125 Ma in the north. The Thirteen Finger has expulsion onsets at 299 Ma and 260 Ma in the south and north, respectively. Hydrocarbon migration pathways (Figure 15) are generally vertical and up-dip northward, with some exceptions; primary migration is locally downward, and the Morrowan sandstone reservoir is charged by the Upper Morrow shale, the Woodford (through the Lower Morrow shale), and the Thirteen Finger Limestone (Figure 15).

Table 2. 1D Petroleum system model input, well 13-10A. HI is Hydrogen Index, TOC is Total Organic Carbon.

Layer Name	Top (m)	Base (m)	Deposited from-to (Ma)	Erosion (Ma) [Amount]	Lithology	HI	TOC	Kinetics
Permian-Cenozoic	0	600	30	50–30 Ma [1000 m]	Sandstone (subarkose, clay rich)	-	-	-
Red Cave	600	712	275–270	-	Organic lean siltstone	-	-	-
Wellington	712	1006	280–275	-	Sandstone	-	-	-
Wolfcampian	1006	1472	299–280	-	Dolomite	-	-	-
Virgilian	1472	1731	303–299	-	Shale	-	-	-
Missourian	1731	1932	304–303	-	Shale	-	-	-
Kansas City	1932	2021	305–304	-	Limestone (shaly)	-	-	-
Marmaton	2021	2094	305.3–305	-	Limestone (shaly)	-	-	-
Cherokee	2094	2279	310–305.3	-	Shale	-	-	-
Thirteen Finger	2279	2320	311–310	-	Limestone (shaly)	355	9.18	[78] TII
Upper Morrow shale	2320	2339	311–310	-	Shale (organic rich)	57	3.62	[78] TII
Morrow B	2339	2350	314–313.7	-	Sandstone, subarkose	-	-	-
Lower Morrow shale	2350	2533	324–314.5	314.5–314 [15 m]	Shale (organic rich)	10	1.1	[78] TIII
Mississippian	2533	2745	354–330	330–324 [150]	Limestone (organic rich)	-	-	-
Woodford	2745	2758	369–354	-	Shale (organic rich)	300	1.8	[79]
Cambrian-Devonian	2758	3320	542–369	-	Limestone	-	-	-
Basement	3320	-	-	-	Granite (>1000 Ma)	-	-	-

Migration in the Thirteen Finger Limestone began around 299 Ma along the section. Migration paths contain both vertical and lateral pathways as hydrocarbons meet resistance from low permeability strata (Figure 15). In some locations, variations in formation depths cause local migration paths to diverge from the overall trend of south to north upward migration. The charging mechanism for the Morrowan sandstone is downward migration from the overlaying organic rich shales in the Thirteen Finger, Upper Morrowan shales from the deeper basin, and possibly Woodford shale through upward migration. In two locations (at ~9 km in the section, and at ~47 km in the section), the model predicts a breakthrough and upward migration of hydrocarbons toward the surface. Matrix-based CO_2 migration risk through the caprocks is assumed to be low. The caprocks entrapped large accumulations of hydrocarbons for millions of years, and modeling [60] shows that permeability decreased and entry pressures increased since oil migration and charging began through the Cenozoic uplift event.

We used Schlumberger PetroMod® software to analyze the fluid components and saturation. PetroMod divides groups of carbon chains (i.e., C_1, $C_{2–5}$, $C_{6–14}$, and C_{15+}) differently from how the Farnsworth Field 1956 oil components are grouped. In order to compare the two, both the Farnsworth Field and PetroMod results were re-grouped as follows; C_1, $C_{2–5}$, C_{6+}. C_{7+} (Farnsworth Field) was grouped with $C_{6–14}$ and C_{15+} (PetroMod); as a result, some of the carbon chains' resolution was lost in this process. Results indicate inter-reservoir variability of saturation and composition. The model predicts that the southern part of reservoir has a saturation of oil (S_o) of ~73% and the northern part has a S_o of ~64%. Reports indicate that the average S_o for Booker Field is 70% and for Farnsworth Field around 69% [80]. Hinds' [76] analysis at time of discovery shows an API of 38° for Farnsworth Field. PetroMod predicts no gas accumulation within the reservoir, which is in agreement with observations at time of discovery. The compositional predictions show largest inter-reservoir variability in the higher order carbon chains. The overall compositional trend includes ~40 mol% from C_{1-5} chains and ~60 mol% from C_{6-15+} chains throughout the Morrowan sandstone reservoir.

In summary, petroleum migration paths are, like the well-to-well flow paths, generally up-dip, with local exceptions. Along the modeled line, migration paths are predicted to have reached the surface in two locations. The sandstone reservoir was charged by the Upper Morrow shale, Lower Morrow shale, the Woodford, and the Thirteen Finger Limestone.

6. Conclusions

CO_2 leakage from sequestration reservoirs may occur via geologic (structural, sedimentary, igneous) pathways as well as via (abandoned) wellbores. This study analyzed the geologic migration pathways in Farnsworth Oil Field, northern Texas. Although faults have been reported previously in the northwest Anadarko Basin, we found no direct evidence for tectonic faults in the reservoir or caprock in Farnsworth Field. Analysis of 2D legacy and 3D seismic datasets do reveal depth and thickness variations of the Morrow B reservoir rock; our interpretation is that they are related to erosional events and paleo-topography, including karst formation ad erosion of the underlying Hunton Formation. No igneous or sedimentary chimneys have been detected in Farnsworth Field.

Combining the 3D seismic data interpretations with results from tracer experiments provides a mechanism to understand inter-well flow patterns and to predict flow directions of injected CO_2. Tracer study analysis suggests that inter-well flow is generally up-dip or horizontal within the Morrow B. Flow patterns are affected by depth variations in the Morrow B and erosional features that may prohibit south-southeast ward flow crossing the center of Farnsworth Field where the depth of the Morrow B changes due to erosion of the underlying Hunton Formation. Here, the impermeable caprock of the Morrow B- the Thirteen Finger- might prohibit southward flow.

1D and 2D Petroleum system models were developed to understand the petroleum system and petroleum migration pathways in the Farnsworth Field area. Four petroleum source rocks were modeled in the northwest Anadarko Basin: the Woodford, Lower- and Upper Morrow shale, and the Thirteen Finger Limestone. The models predict a basin-ward increase in temperature and maturation; at the location of Farnsworth Field, the model predicted CO_2 reservoir is at present 74 °C, which is in excellent agreement with the measured temperature at the time of discovery. In our modeled transect, the most basin-ward location of the Woodford Shale is in the gas window; all other source rocks are in the oil window. Woodford shale began to expulse hydrocarbons around 301 Ma in the southern part of the section and around 206 Ma in the northern part; migration in the other source rocks showed a similar temporal-spatial relation, with a more recent onset. In the northwestern Anadarko Basin, petroleum migration was generally up-dip with local exceptions; the Morrow B sandstone was likely charged by both formations below and overlaying the reservoir rock. Along the modeled transect were several locations where petroleum may have escaped to the surface. Our modeling shows that matrix-based CO_2 migration risk through the caprocks is low.

Based on these analyses, vertical CO_2 migration from the reservoir formation in Farnsworth Field via geologic pathways seems unlikely; higher-quality geophysical datasets than presently available should be analyzed to confirm this finding. Abandoned and aged wells remain a risk for CO_2 escape from the reservoir formation and deserve further research.

Author Contributions: Conceptualization, J.v.W., N.H., P.R., M.M. and E.G.; writing—original draft J.v.W., E.G., P.R., N.H.; writing—review and editing, G.A. All authors have read and agreed to the published version of the manuscript.

Funding: Funding for this project is provided by the U.S. Department of Energy's (DOE) National Energy Technology Laboratory (NETL) through the Southwest Partnership on Carbon Sequestration (SWP) under Award No. DE-FC26-05NT42591.

Acknowledgments: We thank two anonymous reviewers, and Richard Esser, Jim Evans, Peter Mozley, and Paige Csozki for suggestions on and contributions to an earlier draft of this manuscript.

Conflicts of Interest: The authors declare no conflict of interest.

References

1. Bachu, S. Sequestration of CO_2 in geological media: Criteria and approach for site selection in response to climate change. *Energy Convers. Manag.* **2000**, *41*, 953–970. [CrossRef]
2. Holloway, S. Storage of fossil fuel-derived carbon dioxide beneath the surface of the Earth. *Annu. Rev. Energy Environ.* **2001**, *26*, 145–166. [CrossRef]
3. Lackner, K.S. A guide to CO_2 sequestration. *Science* **2003**, *300*, 1677–1678. [CrossRef] [PubMed]
4. Morgan, A.; Grigg, R.; Ampomah, W. A gate-to-gate life cycle assessment for the CO2-EOR operations at Farnsworth Unit (FWU). *Energies* **2021**, *14*, 2499. [CrossRef]

5.   Bachu, S.; Gunter, W.D.; Perkins, E.H. Aquifer disposal of CO_2: Hydrodynamic and mineral trapping. *Energy Convers. Manag.* **1994**, *35*, 269–279. [CrossRef]

6.   Davis, H.G.; Northcutt, R.A. The greater Anadarko Basin: An overview of petroleum exploration and development. In *Anadarko Basin Symposium*; Johnson, K.S., Ed.; 1988; Volume 90, pp. 13–24.

7.   Metz, B.; Davidson, O.; De Coninck, H.; Loos, M.; Meyer, L. *IPCC Special Report on Carbon Dioxide Capture and Storage: Prepared by Working Group III of the Intergovernmental Panel on Climate Change*; Cambridge University Press: Cambridge, NY, USA, 2005.

8.   Scherer, G.W.; Celia, M.A.; Prevost, J.H.; Bachu, S.; Bruant, R.G.; Duguid, A.; Fuller, R.; Gasda, S.E.; Radonjic, M.; Vichit-Vadakan, W. Leakage of CO_2 through abandoned wells: Role of corrosion of cement. In *Carbon Dioxide Capture for Storage in Deep Geologic Formations*; Elsevier: Amsterdam, The Netherlands, 2005; pp. 827–848.

9.   Carey, J.W.; Wigand, M.; Chipera, S.J.; WoldeGabriel, G.; Pawar, R.; Lichtner, P.C.; Wehner, S.C.; Raines, M.A.; Guthrie, G.D. Analysis and performance of oil well cement with 30 years of CO_2 exposure from the SACROC Unit, West Texas, USA. *Int. J. Greenh. Gas Control* **2007**, *1*, 75–85. [CrossRef]

10.   Carey, J.W.; Svec, R.; Griff, R.; Lichtner, P.C.; Zhang, J.; Crow, W. Wellbore integrity and CO_2-brine flow along the casing-cement microannulus. *Energy Procedia* **2009**, *1*, 3609–3615. [CrossRef]

11.   Sminchack, J.; Gupta, N.; Byrer, C.; Bergman, P. *Issues Related to Seismic Activity Induced by the Injection of CO_2 in Deep Saline Aquifers (No. DOE/NETL-2001/1144)*; National Energy Technology Laboratory: Pittsburgh, PA, USA; Morgantown, WV, USA, 2001.

12.   Streit, J.E.; Watson, M.N. Estimating rates of potential CO_2 loss from geological storage sites for risk and uncertainty analysis. *Greenh. Gas Control Technol.* **2005**, *7*, 1309–1314.

13.   Cartwright, J.; Huuse, M.; Aplin, A. Seal bypass systems. *AAPG Bull.* **2007**, *91*, 1141–1166. [CrossRef]

14.   Chadwick, R.A.; Williams, G.A.; Noy, D.J. CO_2 storage: Setting a simple bound on potential leakage through the overburden in the North Sea Basin. *Energy Procedia* **2017**, *114*, 4411–4423. [CrossRef]

15.   Parker, R.L. Farnsworth Morrow Oil Field. *Panhand. Geonews* **1956**, *4*, 5–12.

16.   Higley, D.K. 4D petroleum system model of the Mississippian system in the Anadarko Basin Province, Oklahoma, Kansas, Texas, and Colorado, U.S.A. *Mt. Geol.* **2013**, *50*, 81–98.

17.   Perry, W.J., Jr. *Tectonic Evolution of the Anadarko Basin Region, Oklahoma*; US Geological Survey: Reston, VA, USA, 1989; p. 19.

18.   Johnson, K.S. (Ed.) Geological evolution of the Anadarko basin. In *Anadarko Basin Symposium, Oklahoma Geological Survey Circular*; Johnson, K.S. (Ed.) Oklahoma Geological Survey : Toulsa, OK, USA, 1988; Volume 90, pp. 3–12.

19.   Eddleman, M.W. Tectonics and geologic history of the Texas and Oklahoma Panhandles. In *Oil and Gas Fields of the Texas and Oklahoma Panhandles*; Panhandle (Texas) Geological Society: Amarillo, TX, USA, 1961; pp. 61–68. [CrossRef]

20.   Ham, W.E.; Denison, R.E.; Merritt, C.A. Basement rocks and structural evolution of southern Oklahoma. *Okla. Geol. Surv. Bull.* **1964**, *95*, 302.

21.   Forgotson, J.M.; Statler, A.T.; David, M. Influence of regional tectonics and local structure on deposition of Morrow Formation in western Anadarko Basin. *AAPG Bull.* **1966**, *50*, 518–532.

22.   Evans, J.L. Major structural and stratigraphic features of the Anadarko Basin. In *Pennsylvanian Sandstones of the Mid-Continent*; Hyne, N.J., Ed.; Tulsa Geological Society: Tulsa, OK, USA, 1979; Volume 1, pp. 97–113.

23.   Brewer, J.A.; Good, R.; Oliver, J.E.; Brown, L.D.; Kaufman, S. COCORP profiling across the Southern Oklahoma aulacogen: Overthrusting of the Wichita Mountains and compression within the Anadarko Basin. *Geology* **1983**, *11*, 109–114. [CrossRef]

24.   Ball, M.M.; Henry, M.E.; Frezon, S.E. *Petroleum Geology of the Anadarko Basin Region, Province (115), Kansas, Oklahoma, and Texas*; U.S. Geological Survey Open-File Report; U.S. Geological Survey: Reston, VA, USA, 1991.

25.   Keller, G.R.; Stephenson, R.A.; Hatcher, R.D.; Carlson, M.P.; McBride, J.H. The southern Oklahoma and Dniepr-Donets aulacogens: A comparative analysis. *Mem.-Geol. Soc. Am.* **2007**, *200*, 127.

26.   Tave, M.; Gurrola, H. Lithospheric structure of the Southern Oklahoma Aulacogen and surrounding region as determined from broadband seismology and gravity. *Geol. Soc. Am.* **2013**, *45*, 4.

27.   Feinstein, S. Subsidence and thermal history of Southern Oklahoma aulacogen- Implications for petroleum exploration. *AAPG Bull.* **1981**, *65*, 2531–2533.

28.   Huffman, G.G. Pre-Desmoinesian isopachous and paleogeologic studies in central Mid-Continent region. *AAPG Bull.* **1959**, *43*, 2541–2574.

29.   Jones, C.L., Jr. *An Isopach, Structural, and Paleogeologic Study of Pre-Desmoinesian Units in North Central Oklahoma*; Oklahoma City Geological Society, The Shale Shaker Digest III: Oklahoma City, OK, USA, 1961; pp. 216–233.

30.   McCaskill, J.G., Jr. *Multiple Stratigraphic Indicators of Major Strike-SLip Movement along the Eola Fault, Subsurface Arbuckle Mountains, Oklahoma*; Oklahoma City Geological Society, The Shale Shaker: Oklahoma City, OK, USA, 1998; Volume 48, pp. 119–133.

31.   Rottmann, K. *Isopach Map of Woodford Shale in Oklahoma and Texas Panhandle*; Oklahoma Geological Survey: Oklahoma City, OK, USA, 2000.

32.   Fritz, R.D.; Medlock, P.; Kuykendall, M.J.; Wilson, J.L. The geology of the Arbuckle Group in the midcontinent: Sequence stratigraphy, reservoir development, and the potential for hydrocarbon exploration. In *The Great American Carbonate Bank: The Geology and Economic Resources of the Cambrian—Ordovician Sauk Megasequence of Laurentia*; Derby, J.R., Fritz, R.D., Longacre, S.A., Morgan, W.A., Sternbach, C.A., Eds.; American Association of Petroleum Geologists: Tulsa, OK, USA, 2012; pp. 203–273. [CrossRef]

33. Eriksson, M.E.; Leslie, S.A.; Bergman, C.F. Jawed polychaetes from the upper Sylvan shale (Upper Ordovician), Oklahoma, USA. *J. Paleontol.* **2005**, *79*, 486–496. [CrossRef]

34. Amati, L.; Westrop, S.R. Sedimentary facies and trilobite biofacies along an Ordovician shelf to basin gradient, Viola Group, south-central Oklahoma. *Palaios* **2006**, *21*, 516–529. [CrossRef]

35. Pearson, O.J.; Miller, J.J. Tectonic and structural evolution of the Anadarko Basin and structural interpretation and modeling of a composite regional 2D seismic line. In *Petroleum Systems and Assessment of Undiscovered Oil and Gas in the Anadarko Basin province, Colorado, Kansa, Oklahoma, and Texas—USGS Province 58*; Higley, D.K., Ed.; Digital Data Series DDS-69-EE; U.S. Geological Survey: Reston, VA, USA, 2014. [CrossRef]

36. Frezon, S.E.; Jordan, L. Oklahoma. In *Paleotectonic Investigations of the Mississippian Systems in the United States*; Craig, L.C., Varnes, K.L., Eds.; USGS: Reston, VA, USA, 1979; Volume 1010, pp. 147–159.

37. Craig, L.C.; Conner, C.W. Paleotectonic investigations of the Mississippian System in the United States. In *History of the Mississippian System, an Interpretive Summary*; Craig, L.C., Varnes, K.L., Eds.; USGS: Reston, VA, USA, 1979; Volume 101, pp. 371–406.

38. Mapel, W.J.; Johnson, R.B.; Backman, G.O.; Varnes, K.L. Southern Midcontinent and southern Rocky Mountains region. In *Introduction and Regional Analysis of the Mississippian System, Part 1 of Paleotectonic Investigations of the Mississippian Systems in the United States*; Craig, L.C., Varnes, K.L., Eds.; USGS: Reston, VA, USA, 1979; Volume 1010-J, pp. 161–187.

39. Suriamin, F.; Pranter, M.J. Lithofacies, depositional, and diagenetic controls on the reservoir quality of the Mississippian mixed siliciclastic-carbonate system, eastern Anadarko Basin, Oklahoma, USA. *Interpretation* **2021**, *9*, 1–72. [CrossRef]

40. Johnson, K.S.; Amsden, T.W.; Denison, R.E.; Dutton, S.P.; Goldstein, A.G.; Rascoe, B., Jr.; Sutherland, P.K.; Thompson, D.M. Southern Midcontinent region. In *Sedimentary Cover–North American Craton, U.S.*; Sloss, L.L., Ed.; Geological Society of America: Boulder, CO, USA, 1988; pp. 307–359.

41. Cather, M.; Rose-Coss, D.; Gallagher, S.; Trujillo, N.; Cather, S.; Hollingworth, R.S.; Mozley, P.; Leary, R.J. Deposition, diagenesis, and sequence stratigraphy of the Pennsylvanian Morrowan and Atokan intervals at Farnsworth Unit. *Energies* **2021**, *14*, 1057. [CrossRef]

42. Lyday, J.R. Atokan (Pennsylvanian) Berlin Field: Genesis of recycled detrital dolomite reservoir, deep Anadarko Basin, Oklahoma. *AAPG Bull.* **1985**, *69*, 1931–1949. [CrossRef]

43. LoCricchio, E. Granite wash play overview, Anadarko basin: Stratigraphic framework and controls on Pennsylvanian Granite wash production, Anadarko Basin, Texas and Oklahoma. In Proceedings of the Granite Wash and Pennsylvanian Sand Forum, Oklahoma City, OK, USA, 25 September 2014.

44. Swanson, D. Deltaic Deposits in the Pennsylvanian upper Morrow Formation in the Anadarko Basin, in Pennsylvanian Sandstones of the Mid-Continent. Tulsa Geological Society: Tulsa, OK, USA, 1979; Volume 1, pp. 115–168.

45. Krystinik, L.F.; Blakeney, B.A. Sedimentology of the upper Morrow Formation in eastern Colorado and western Kansas. In *Morrow Sandstones of Southeast Colorado and Adjacent Areas*; Sonnenberg, S.A., Shannon, L.T., Rader, K., von Drehle, W.F., Martin, G.W., Eds.; Rocky Mountain Association of Geologists: Denver, CO, USA, 1990; pp. 37–50.

46. Blakeney, B.A.; Krystinik, L.F.; Downey, A.A. Reservoir heterogeneity in Morrow valley fills, Stateline trend: Implications for reservoir management and field expansion. In *Morrow Sandstones of Southeast Colorado and Adjacent Area*; Sonnenberg, S.A., Shannon, L.T., Rader, K., von Drehle, W.F., Martin, G.W., Eds.; Rocky Mountain Association of Geologists: Denver, CO, USA, 1990; pp. 131–142.

47. Sonnenberg, S.A.; Shannon, L.T.; Rader, K.; Von Drehle, W.F. Regional structure and stratigraphy of the Morrowan Series, southeastern Colorado and adjacent areas. In *Morrow Sandstones of Southeast Colorado and Adjacent Area*; Sonnenberg, S.A., Shannon, L.T., Rader, K., von Drehle, W.F., Martin, G.W., Eds.; Rocky Mountain Association of Geologists: Denver, CO, USA, 1990; pp. 1–8.

48. Al-Shaieb, Z.; Puckette, J.; Abdalla, A. Influence of sea-level fluctuation on reservoir quality of the upper Morrowan sandstones, northwestern shelf of the Anadarko Basin. In *Sequence Stratigraphy of the Midcontinent*; Hyne, N.J., Ed.; Tulsa Geological Society: Tulsa, OK, USA, 1995; Volume 4, pp. 249–268.

49. McKay, R.H.; Noah, J.T. Integrated perspective of the depositional environment and reservoir geometry, characterization, and performance of the Upper Morrow Buckhaults Sandstone in the Farnsworth Unit, Ochiltree County, Texas. *Okla. Geol. Surv. Circ.* **1996**, *98*, 101–114.

50. Puckette, J.; Abdalla, A.; Rice, A.; Al-Shaieb, Z. The upper Morrow reservoirs: Complex fluvio-deltaic depositional systems. In *Deltaic Reservoirs in the Southern Midcontinent*; Johnson, K.S., Ed.; 1996; Volume 98, pp. 47–84.

51. Bowen, D.W.; Weimer, P. Regional sequence stratigraphic setting and reservoir geology of Morrow incised-valley sandstones (lower Pennsylvanian), eastern Colorado and western Kansas. *AAPG Bull.* **2003**, *87*, 781–815. [CrossRef]

52. Bowen, D.W.; Weimer, P. Reservoir geology of Nicholas and Liverpool cemetery fields (lower Pennsylvanian), Stanton county, Kansas, and their significance to the regional interpretation of the Morrow Formation incised-valleyfill systems in eastern Colorado and western Kansas. *AAPG Bull.* **2004**, *88*, 47–70. [CrossRef]

53. Puckette, J.; Al-Shaieb, Z.; Van Evera, E. Sequence stratigraphy, lithofacies, and reservoir quality, upper Morrow sandstones, northwestern shelf, Anadarko Basin. In *Morrow and Springer in the Southern Midcontinent*; Andrews, R.D., Ed.; 2008; Volume 111, pp. 81–97.

54. Gallagher, S.R. Depositional and Diagenetic Controls on Reservoir Heterogeneity: Upper Morrow Sandstone, Farnsworth Unit, Ochiltree County, Texas. Master's Thesis, New Mexico Institute of Mining and Technology, Socorro, NM, USA, 2014; p. 214.
55. Lee, Y.; Deming, D. Overpressures in the Anadarko basin, southwestern Oklahoma: Static or dynamic. *AAPG Bull.* **2002**, *86*, 145–160.
56. Sorenson, R.P. A dynamic model for the Permian Panhandle and Hugoton fields, western Anadarko basin. *AAPG Bull.* **2005**, *89*, 921–938. [CrossRef]
57. Rose-Coss, D. Geologic Characterization of the Morrowan and Atokan Formation: Implications for CCUS. Master's Thesis, New Mexico Institute of Mining and Technology, Socorro, NM, USA, 2017; p. 251.
58. Hentz, T.F. *Depositional, Structural and Sequence Framework of the Gas-Bearing Cleveland Formation (Upper Pennsylvanian): Western Anadarko Basin, Texas Panhandle*; Bureau of Economic Geology, University of Texas at Austin: Austin, TX, USA, 1994; p. 213.
59. Marsh, S.; Holland, A. Comprehensive fault database and interpretive fault map of Oklahoma. In *Oklahoma Geological Survey Open-File Report OF2-2016*; The University of Oklahoma: Norman, OK, USA, 2016.
60. Holloway, S.; Holland, A.; Keller, G.R. *Oklahoma Fault Database Contributions from the Oil and Gas Industry*; Oklahoma Geological Survey Open-File Report OF1-2016; Oklahoma Geological Survey: Toulsa, OK, USA, 2016.
61. Gragg, E. Petroleum System Modeling in the Northwest Anadarko Basin: Implications for Carbon Storage. Master's Thesis, New Mexico Institute of Mining and Technology, Socorro, NM, USA, 2016; p. 149.
62. Totten, R.B. General geology and historical development, Texas and Oklahoma panhandles. *AAPG Bull.* **1956**, *40*, 1945–1967.
63. Higley, D.K.; Cook, T.A.; Pawlewicz, M.J. Petroleum systems and assessment of undiscovered oil and gas in the Anadarko Basin Province, Colorado, Kansas, Oklahoma, and Texas—Woodford Shale Assessment Units. In *Petroleum Systems and Assessment of Undiscovered Oil and Gas in the Anadarko Basin Province, Colorado, Kansas, Oklahoma, and Texas—USGS Province 58*; USGS Digital Data Series DDS-69-EE; Higley, D.K., Ed.; Digital Data Series DDS-69-EE; U.S. Geological Survey: Reston, VA, USA, 2014.
64. Hutton, A. Geophysical Modeling and Structural Interpretation of a 3D Reflection Seismic Survey in Farnsworth Unit, TX. Master's Thesis, New Mexico Institute of Mining and Technology, Socorro, NM, USA, 2015; p. 85.
65. White, M.D.; Esser, R.P.; McPherson, B.P.; Balch, R.S.; Liu, N.; Rose, P.E.; Garcia, L.; Ampomah, W. Interpretation of tracer experiments on inverted five-spot well-patterns within the western half of the farnsworth unit oil field. *Energy Procedia* **2017**, *114*, 7070–7095. [CrossRef]
66. Chopra, S.; Marfurt, K.J. *Seismic Attributes for Prospect Identification and Reservoir Characterization*; SEG Geophysical Developments Series; SEG: Tulsa, OK, USA, 2007.
67. Chopra, S.; Kumar, R.; Marfurt, K.J. Seismic discontinuity attributes and Sobel filtering. In Proceedings of the SEG Denver 2014 Annual Meeting, Denver, CO, USA, 26–31 October 2014; pp. 1624–1628.
68. Trujillo, N. Influence of Lithology and Diagenesis on Mechanical and SEALING properties of the Thirteen Finger limestone and Upper Morrow Shale, Farnsworth Unit, Ochiltree County, Texas. Master's Thesis, New Mexico Institute of mining and Technology, Socorro, NM, USA, 2017; p. 156.
69. Rose, P.E.; Benoit, W.R.; Kilbourn, P.M. The application of the polyaromatic sulfonates as tracers in geothermal reservoirs. *Geothermics* **2001**, *30*, 617–640. [CrossRef]
70. Rose, P.E.; Capuno, V.; Peh, A.; Kilbourn, P.M.; Kasteler, C. The use of the naphthalene sulfonates as tracers in high temperature geothermal systems. In Proceedings of the 23rd PNOC Geothermal Conference, Manila, Philippines, 14 March 2002.
71. Rose, P.E.; Johnson, S.D.; Kilbourn, P.M.; Kasteler, C. Tracer testing at Dixie Valley, Nevada using 1-naphthalene sulfonate and 2,6-naphthalene disulfonate. In Proceedings of the Twenty-Seventh Workshop on Geothermal Reservoir Engineering, Stanford, CA, USA, 27–29 January 2002.
72. Sanjuan, B.; Pinault, J.L.; Rose, P.E.; Gérard, A.; Brach, M.; Braibant, G.; Crouzet, C.; Foucher, J.C.; Gautier, A.; Touzelet, S. Tracer testing of the geothermal heat exchanger at Soultz-sous-Forêts (France) between 2000 and 2005. *Geothermics* **2001**, *35*, 622–653. [CrossRef]
73. Mountain, B.W.; Winick, J.A. The thermal stability of the naphthalene sulfonic and naphthalene disulfonic acids under geothermal conditions: Experimental results and a field-based example. In Proceedings of the New Zealand Geothermal Workshop, Auckland, New Zealand, 19–21 November 2012.
74. Nimmo, J.R.; Perkins, K.S.; Rose, P.E.; Rousseau, J.P.; Orr, B.R.; Twining, B.V.; Anderson, S.R. Kilometer-scale rapid transport of naphthalene sulfonate tracer in the unsaturated zone at the Idaho National Engineering and Environmental Laboratory. *Vadose Zone J.* **2002**, *1*, 89–101. [CrossRef]
75. Beserra, T. Oil classification and exploration opportunity in the Hugoton Embayment, western Kansas, and Las Animas Arch, eastern Colorado. *AAPG Search Discov.* **2008**, *10146*, 21.
76. Hinds, R.F. *Reservoir Fluid Study, Farnsworth Field*; Core Laboratories, Inc.: Amsterdam, The Netherlands, 1956.
77. Hobbs, N.F.; van Wijk, J.W.; Leary, R.; Axen, G.J. Late Paleozoic evolution of the Anadarko Basin: Implications for Laurentian tectonics and the assembly of Pangea. *Tectonics* **2021**. revised.

78. Behar, F.; Vandenbroucke, M.; Tang, Y.; Marquis, F.; Espitalie, J. Thermal cracking of kerogen in open and closed systems: Determination of kinetic parameters and stoichiometric coefficients for oil and gas generation. *Org. Geochem.* **1997**, *26*, 321–339. [CrossRef]
79. Lewan, M.D.; Ruble, T.E. Comparison of petroleum generation kinetics by isothermal hydrous and nonisothermal open-system pyrolysis. *Org. Geochem.* **2002**, *33*, 1457–1475. [CrossRef]
80. Munson, T.W. Depositional, diagenetic, and production history of the Upper Morrowan Buckhaults Sandstone, Farnsworth Field, Ochiltree County Texas. Master's Thesis, West Texas University, Canyon, TX, USA, 1988; p. 117.

Article

Multiscale Assessment of Caprock Integrity for Geologic Carbon Storage in the Pennsylvanian Farnsworth Unit, Texas, USA

Natasha Trujillo [1,†], Dylan Rose-Coss [1,‡], Jason E. Heath [2], Thomas A. Dewers [3,*], William Ampomah [4], Peter S. Mozley [5] and Martha Cather [4]

[1] New Mexico Institute of Mining and Technology, Earth and Environmental Science Department, Socorro, NM 87801, USA; natasha.trujillo@pxd.com (N.T.); dylanh@state.nm.us (D.R.-C.)

[2] Sandia National Laboratories, Geomechanics Department, Albuquerque, NM 87185, USA; jeheath@sandia.gov

[3] Sandia National Laboratories, Nuclear Waste Disposal Research and Analysis, Albuquerque, NM 87185, USA

[4] New Mexico Institute of Mining and Technology, Petroleum Resource Recovery Center, Socorro, NM 87801, USA; William.Ampomah@nmt.edu (W.A.); Martha.Cather@nmt.edu (M.C.)

[5] New Mexico Institute of Mining and Technology, Associate Vice President for Academic Affairs and Earth and Environmental Science Department, Socorro, NM 87801, USA; Peter.Mozley@nmt.edu

* Correspondence: tdewers@sandia.gov; Tel.: +1-505-845-0631

† Present address: Pioneer Natural Resources Company, Midland, TX 75039, USA.

‡ Present address: New Mexico Oil Conservation Division, Santa Fe, NM 87505, USA.

Citation: Trujillo, N.; Rose-Coss, D.; Heath, J.E.; Dewers, T.A.; Ampomah, W.; Mozley, P.S.; Cather, M. Multiscale Assessment of Caprock Integrity for Geologic Carbon Storage in the Pennsylvanian Farnsworth Unit, Texas, USA. *Energies* **2021**, *14*, 5824. https://doi.org/10.3390/en14185824

Academic Editor: Alireza Nouri

Received: 2 February 2021
Accepted: 10 June 2021
Published: 15 September 2021

Publisher's Note: MDPI stays neutral with regard to jurisdictional claims in published maps and institutional affiliations.

Abstract: Leakage pathways through caprock lithologies for underground storage of CO_2 and/or enhanced oil recovery (EOR) include intrusion into nano-pore mudstones, flow within fractures and faults, and larger-scale sedimentary heterogeneity (e.g., stacked channel deposits). To assess multiscale sealing integrity of the caprock system that overlies the Morrow B sandstone reservoir, Farnsworth Unit (FWU), Texas, USA, we combine pore-to-core observations, laboratory testing, well logging results, and noble gas analysis. A cluster analysis combining gamma ray, compressional slowness, and other logs was combined with caliper responses and triaxial rock mechanics testing to define eleven lithologic classes across the upper Morrow shale and Thirteen Finger limestone caprock units, with estimations of dynamic elastic moduli and fracture breakdown pressures (minimum horizontal stress gradients) for each class. Mercury porosimetry determinations of CO_2 column heights in sealing formations yield values exceeding reservoir height. Noble gas profiles provide a "geologic time-integrated" assessment of fluid flow across the reservoir-caprock system, with Morrow B reservoir measurements consistent with decades-long EOR water-flooding, and upper Morrow shale and lower Thirteen Finger limestone values being consistent with long-term geohydrologic isolation. Together, these data suggest an excellent sealing capacity for the FWU and provide limits for injection pressure increases accompanying carbon storage activities.

Keywords: carbon sequestration; caprock integrity; noble gas migration; seal by-pass

1. Introduction

The Site Characterization program of the Southwest Regional Partnership on Carbon Sequestration (SWP) assesses the integrity of caprock formations that immediately overlie the Pennsylvanian Morrow B sandstone, which is the target reservoir for a combined carbon capture, utilization, and storage (CCUS) project involving enhanced subsurface oil recovery (EOR)–CO_2 storage at the Farnsworth Unit (FWU), Texas, USA. Caprock integrity is the ability of generally low permeability and high capillary entry pressure formations overlying a reservoir—typically referred to as caprocks—to impede movement of fluids from the reservoir below and thus display sealing. Capillary sealing behavior arises from the nanoscale pore throats and interfacial fluid properties of the wetting phase in the caprock (e.g., brine) and the non-wetting phase of the reservoir (e.g., CO_2 or hydrocarbons). However, caprocks may contain "seal-bypass systems", which are features and/or processes that allow reservoir fluids to move out of the reservoir [1], while wettability of reservoir and caprock can be altered by carbon storage and/or enhanced hydrocarbon recovery procedures [2]. Seal-bypass systems may include discontinuous coverage of the sealing lithologies over the reservoir, natural or induced fracture networks, faults, permeable injectites (i.e., structures formed by sediment injection) or other geologic pipe structures, and man-made intrusions such as leaky wellbores [1]. Implicit in the concept of caprock sealing behavior is a timescale of interest, which for geologic CO_2 storage is 100 s to 1000 s of years.

This chapter presents a novel caprock integrity study that focuses on evaluating geologic sealing behavior for capillary, fluid flow, and mechanical properties at different spatial and temporal scales. Specifically we assess: 1. Pore-scale capillary sealing and microstructure; 2. local seal bypass mechanisms; 3. seal regional lateral continuity across reservoir scales; 4. mechanical integrity of the reservoir-caprock package to fluid pressure perturbation; and 5. reservoir-scale hydrologic isolation over relevant time scales. This requires a multiscale approach using a variety of techniques to cover the range of length and time scales, processes, and features involved in caprock integrity (Figure 1). To these ends we examine well log- to sub-core scale heterogeneity of the reservoir and sealing lithologies of the Morrow B sandstone lithologies, the upper Morrow shale top seal, and the overlying Thirteen Finger limestone secondary sealing lithologies. Capillary heterogeneity is examined using mercury porosimetry. We examine evidence for existing fractures and faults that could serve as seal bypass systems under present day stress orientations, as well as geomechanical constraints on induced seal bypass features associated with CCUS and EOR activities within the FWU. The lateral continuity of sealing units in the FWU is assessed via subsurface mapping. The large-scale sealing capacity of these lithologies as have occurred over the geologic time is assessed using noble gas measurements collected from fresh core. Formation-scale features examining the scale of the entire FWU and regional stratigraphic architecture using seismic methods have been discussed by Rose-Coss et al. [3,4] and Ampomah et al. [5]. Together this data set provides a unique time-integrated assessment of caprock integrity over engineering to geologic time- and length-scales relevant to CCUS.

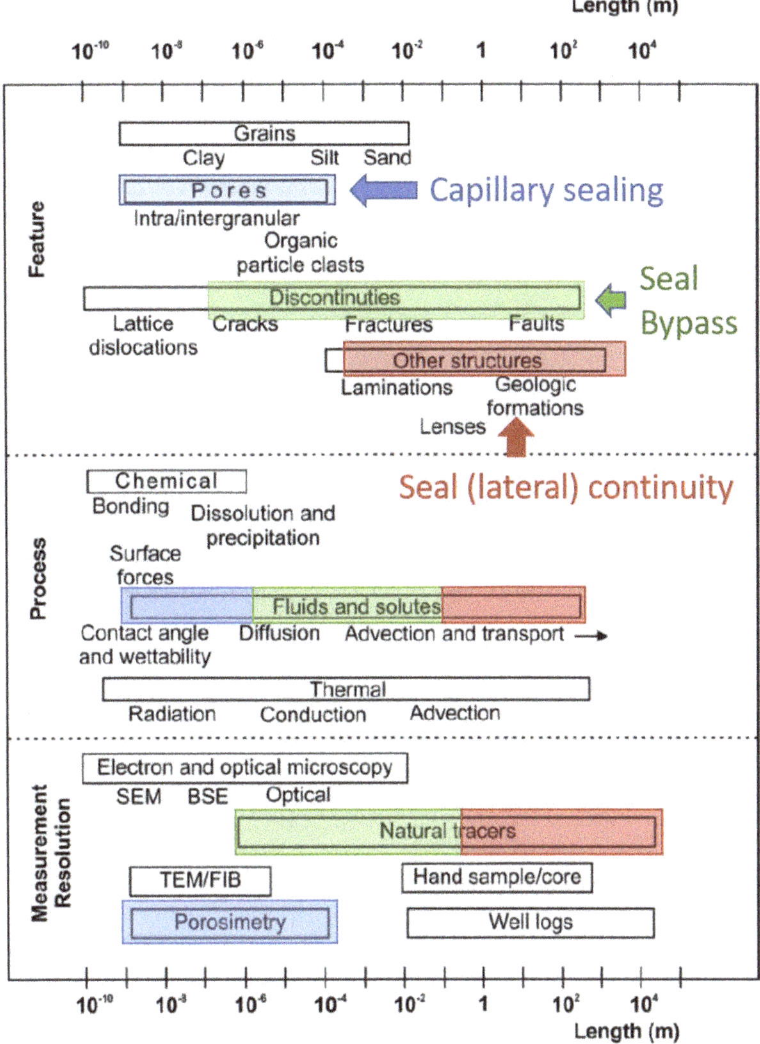

Figure 1. Features, processes, and measurement resolution relevant to the assessment of caprock integrity (adapted from the original Figure 36 in [6] and the modification thereof in Figure 1.1 in [7]).

2. Site Location and Geologic Setting

2.1. Unit History and General Geology

The FWU is located in Ochiltree county, Texas, USA (Figure 2), with the nearby Arkalon Ethanol Plant and Agrium Fertilizer Plant supplying anthropogenic CO_2 for enhanced oil recovery in the field. Production and injection at FWU occur strictly within the Pennsylvanian Morrow B sandstone (Figure 3; [5]). "Morrow" is an operational name that refers to a sequence of alternating mudstone and sandstone intervals deposited during the Morrowan period of the late Pennsylvanian [8]. The Morrow B delineates the first sandstone package deposited below the Atokan Thirteen Finger limestone [9,10].

The primary caprock intervals at FWU are comprised of the upper Morrow shale and the Thirteen Finger limestone (both operational names/units in the FWU; Figure 3). The Thirteen Finger limestone is an informal name for a series of predominantly carbonate cementstone intervals that are intercalated with black carbonaceous mudstone deposited during the Atokan period of the late Pennsylvanian (Figure 3; see also Trujillo [10] and Rose-Coss [9]).

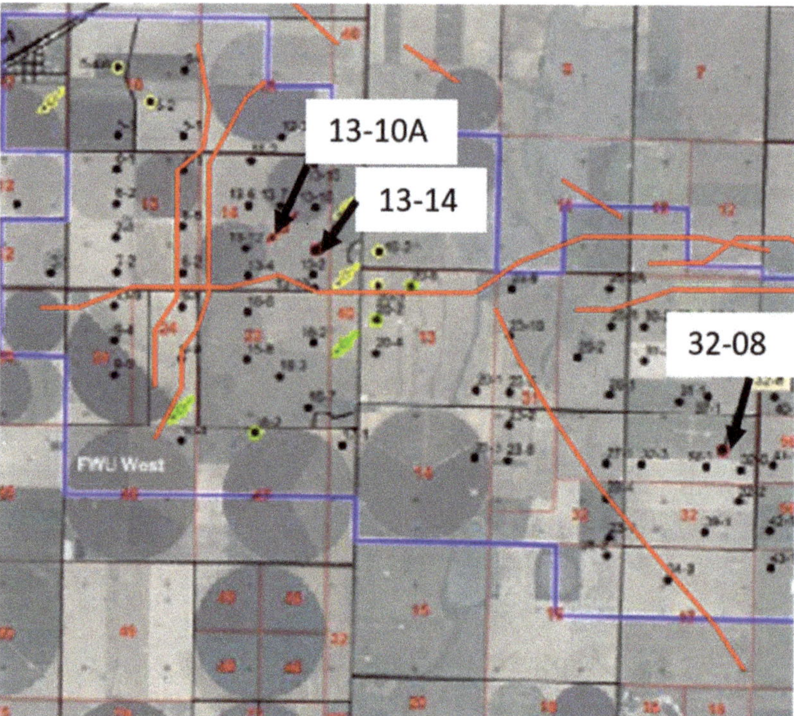

Figure 2. Locations of wells in the Farnsworth Unit (outlined in blue) used in assessing local and field-wide caprock integrity. Red lines in show locations of inferred faults (modified from Balch and McPherson [11] and Hutton [12]).

Both the Morrowan and Atokan intervals are common throughout the Texas and Oklahoma panhandles, southeastern Colorado, and western Kansas. Overlying stratigraphy includes Late Pennsylvanian through the Middle Permian shales and limestones, with lesser amounts of dolomite, sandstone, and evaporites [8,12–14].

2.2. Tectonic Setting

The FWU sits on the northwest shelf of the Anadarko basin in the Texas Panhandle. From the FWU, the basin plunges to the southeast where it reaches depths of over 40,000 ft (12,192 m) adjacent to the Amarillo-Wichita Uplift [15,16]. Maximum rates of subsidence occurred during Morrowan to Atokan times [14–17]. Positive features which might have influenced deposition within the region include the Ancestral Rockies to the north, the Central Kansas uplift to the north-east, and the Wichita-Amarillo uplift to the south [17,18]. The structural grain of the basin was inherited from the Precambrian to Cambrian failed arm of a triple junction known as the Southern Oklahoma Aulacogen [15,16]. The region was then tectonically quiet until the beginning of the Chesterian-Morrowan, when the Wichita-Amarillo uplift and the ancestral Rockies formed as a result of the northeast-directed basement-involved thrust faulting associated with collision between the North American and Gondwanan plates [14,15,19,20]. Fault movement within the Wichita-Amarillo uplift is characterized by the vertical block movement and left-lateral strike slip movement. The vertical fault movement began in the Chesterian and then continued predominantly at the end of the Morrowan into the Atokan time period [17]. This period of faulting created normal faults with a down-to-the-south displacement parallel to the axis of shear associated with the Amarillo uplift. The left-lateral strike slip movement occurred afterwards in the late-to-post Atokan and is expressed as anticlinal horst blocks and synformal grabens diverging at intersections from the main shear zones [17]. Tectonic activity slowed after the Atokan and the region was quiescent by the end of the Pennsylvanian. Local uplifts and associated basins combined with climate variations at time of deposition set the stage for the stratigraphic variations seen in the core, and especially evident in the Thirteen Finger limestone.

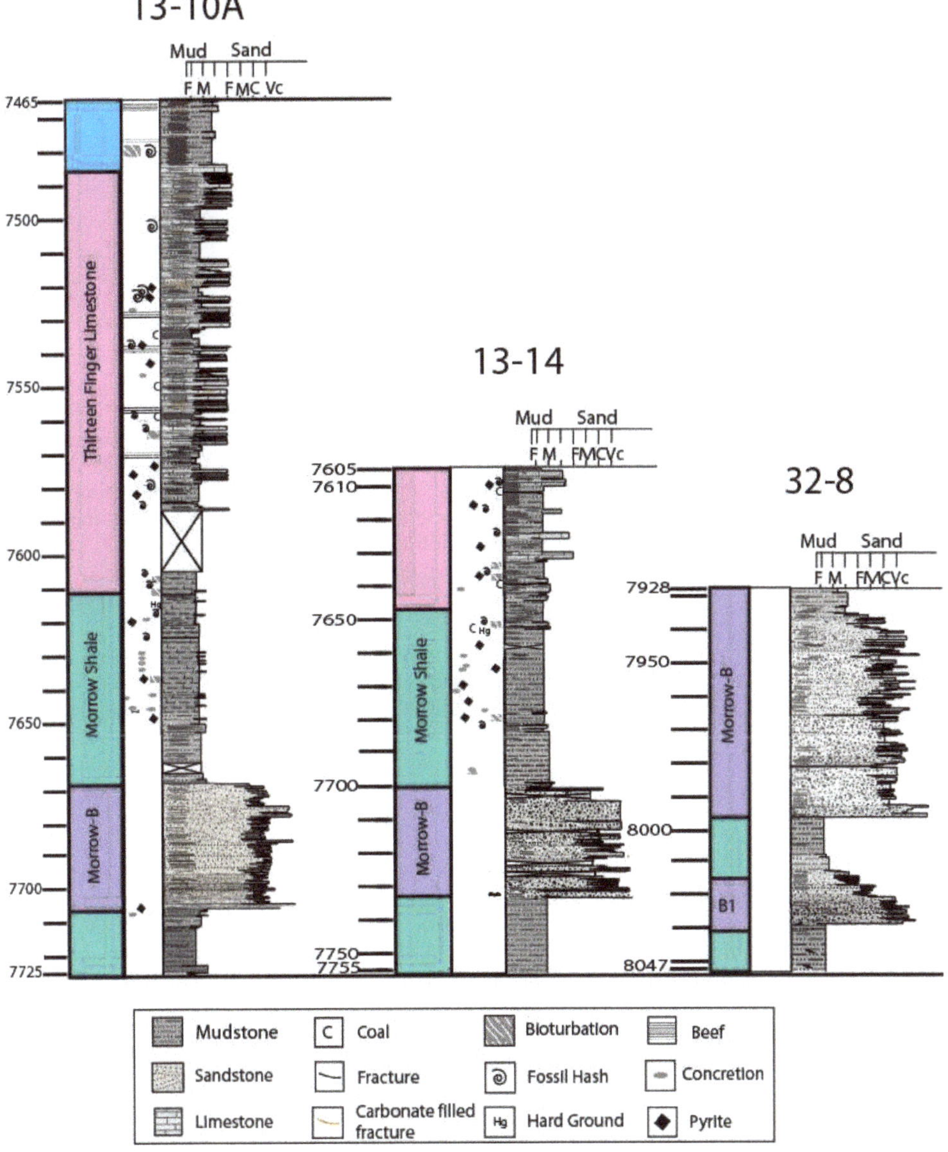

Figure 3. Stratigraphic columns of three SWP characterization wells (modified from Rose-Coss [9]).

3. Materials and Methods

3.1. Coring Program, Petrologic Description, and Well Log Analysis

The coring program was designed and implemented in 2013 and 2014 by the SWP and former operator Chaparral Energy LLC, which targeted the primary Morrow B sandstone reservoir and the overlying caprocks, the upper Morrow shale, and the Thirteen Finger limestone. The coring program included core analysis plans to support major SWP project objectives and/or research topics on CO_2 storage capacity, injectivity, and plume extent; storage permanence; and injection- and/or production-induced reservoir damage, and included a suite of petrophysical, petrological, geomechanical, and geochemical testing.

Schlumberger ran a large suite of wire-line tools for caprock and reservoir characterization, and wellbore integrity assessment in cooperation with the SWP, which had personnel in the field to observe drilling and coring of Well 13-10 A and 13-14, and to assist with core preservation and core handling. Terra Tek, now a Schlumberger company, performed core handling in the field and initially housed the core for initial characterization and sampling. Initial core reviews were performed by Sandia National Laboratories (SNL), New Mexico Tech (NMT), and Chaparral Energy to choose sample locations for petrologic, petrophysical, geomechanical, and geochemical analysis to be performed by Terra Tek, SNL, and NMT. SNL and NMT coauthors submitted formal plans to Terra Tek, which included sampling and/or analysis for thin sections (of the caprocks and reservoir rocks and of fractures), relative permeability and capillary pressure, routine plug analysis, mercury porosimetry, Routine Core Plug (RCPA), and Tight Rock Analysis (TRA) (both by Terra Tek), X-ray diffraction, geochemical analyses including pyrolysis and vitrinite reflectance, and geomechanical testing.

To help quantify heterogeneity and guide sampling densities for laboratory testing, the multi-well Heterogeneous Rock Analysis (HRA; [21]) was performed by Schlumberger using proprietary methods. For HRA of FWU reservoir and caprock lithologies, results of gamma ray, deep resistivity, bulk density, neutron porosity, and compressional slowness logs were combined with caliper responses to make a preliminary assessment of rock classes, which resulted in determining eleven separate rock unit classes: Two for the reservoir lithology (Morrow B) and nine for the caprock lithologies (upper and lower Morrow shale and Thirteen Finger limestone). More discussion on core descriptions and core photographs are found in Rose-Coss [9] and Trujillo [10].

3.2. Petrologic Characterization

Petrologic methods involved standard optical thin-section petrography and backscattered electron microscopy conducted at both SNL and NMT, using methods described by Rose-Coss [9] and Trujillo [10] and facilities at the New Mexico Bureau of Geology and Mineral Resources (NMBGMR) and New Mexico Tech. We report here on some backscattered imaging results, with additional details and petrography given by Rose-Coss [9] and Trujillo [10].

3.3. Petrophysics

Detailed descriptions of methods used in the report analysis of the coal and organic-rich shales (performed by Terra Tek) are given by Rose-Coss [9]. Descriptions of Terra Tek tight rock analysis and pressure pulse decay methods for permeability of caprock lithology core plugs are given in Trujillo [10]. Intrusion-extrusion mercury porosimetry was performed on core plugs by Poro-Technology, a Micromeritics company, using a Micromeritics AutoPore IV 9500 Series porosimeter. Core plugs were oriented either vertically or horizontally (i.e., parallel or perpendicular to the long axis of the core). The core plugs were approximately 0.9-inch (2.3 cm) diameter by 0.9-inch (2.3 cm) long and were jacketed with epoxy for directional intrusion. Poro-Technology made closure corrections accounting for volumes of mercury injected that did not penetrate into the pore space prior to the pressure achieving the mercury entry pressure of the pore space. Breakthrough pressure or the pressure at which a non-wetting phase penetrates a rock through the connected pore space [22], was estimated for core plugs using methods of Dewhurst et al. [23]. Breakthrough pressures were converted from the mercury-air system to a CO_2-water system and to CO_2 column heights using the methods of Ingram et al. [24]. We use an interfacial tension value of 484 mN/m for the mercury-air-rock system, and a contact angle of 140°. We assumed a geothermal gradient of 25 °C/km and a hydrostatic pressure gradient of 0.0098 MPa/m to estimate the density of CO_2 and water at the depths of the core plugs. Interfacial tension values for the water-CO_2 system assumed zero ionic strength and used the methods of Heath et al. [25]. Contact angles for the water-CO_2-mineral system were estimated from Iglauer et al. [26] for quartz, calcite, and mica, resulting in a range of 10 to 57°.

3.4. Geomechanics

A series of rock mechanical tests were performed on rock core sampled and tested at Terra Tek's laboratories in Salt Lake City, Utah, using standard techniques. These include Brazil tension (or cylinder splitting) tests, unconfined compression tests, and triaxial compression tests. These were used to extract static elastic properties, rock unconfined and triaxial strength, and tensile strength information from samples from all three SWP characterization wells in the FWU. A standard Mohr-circle analysis was used to delineate failure envelopes for sampled lithologies.

3.5. Fracture Analysis

Under the guidance of coauthors, Terra Tek performed a detailed analysis of macro-scopic fractures on nearly 270 ft of continuous whole core from Well 13-10A. The fracture descriptions focused on identifying fracture types based on morphologic characteristics and intensity. As the core was not oriented, Terra Tek drew an arbitrary "North" line on the core to enable the measurement of relative orientation of measured fractures. Fracture attributes measured include fracture strike and dip relative to this North line, general fracture type, type of mineral fill, type of oil stain, fracture porosity, fracture spacing, and intensity of fractures for each cored interval. Fracture classes include those induced from drilling or coring versus natural fractures that may or may not exhibit shear, extension or mineralization. Terra Tek analysis included tabulation of fracture types by depth and stereo plots of relative fracture orientations by fracture type.

3.6. Preservation of Fresh Core and Noble Gas Analysis

Core preservation for noble and other pore fluid gases followed procedures found in Osenbrück et al. [27]. Especially, designed canisters were built from high-vacuum service equipment to seal samples against atmospheric contamination. Sub-samples of core were weighted and sealed in canisters immediately in the field after the core was retrieved to the Earth. A purging and vacuum pump-down process evacuated atmospheric noble gases from the canisters using methods described by Heath [7]. Helium, neon, and argon isotopes were analyzed at the University of Utah Dissolved and Noble Gas Laboratory in Salt Lake City, Utah, USA. After the transfer of gases into a purification line, the analysis followed the methods described by Hendry et al. [28].

4. Results

As stated in the introduction, assessing caprock integrity for EOR-CCUS involves a multiscale examination of the ability of a caprock lithology or set of lithologies to sustain emplacement of a body of CO_2 for a given time. For CCUS, this may be 100 s or 1000 s of years. One aspect for CCUS that is favorable for use of CO_2 for oil recovery and storage is the fact that the same caprock invoked for CO_2 uses the same caprock involved in oil and gas storage over the geologic time. We know from the long history of subsurface engineering at FWU that storage under EOR conditions is favorable for CO_2 containment. We need to build confidence that injection and emplacement conditions under CCUS best practices does nothing to threaten the integrity of sealing potential of caprock lithologies.

To this end, we examine the integrity of the FWU upper Morrow shale and Thirteen Finger limestone from five perspectives: 1. Characterizing the heterogeneity within the caprock lithologies from well logging and core properties; 2. characterizing the capillary heterogeneity of unfractured lithologies and calculating abilities to sustain a capillary seal to CO_2 of a given volume or column height; 3. evaluating the potential of existing fractures and faults to serve as seal bypass features, as well as assessing the potential for creation of new fractures or reactivating old ones under injection-perturbed stress conditions; 4. characterizing the physical, stratigraphic continuity, and consistency/heterogeneity in the caprock lithologies over the reservoir extent; and 5. assessing the sealing capacity over the geologic time using noble gas distributions. Together, these five topics constitute a multi-length and -time scale assessment of caprock integrity for CCUS at the FWU.

4.1. Heterogeneity at "Well Log"- to Core-Scales

Well logging has been the stalwart of petroleum exploration in determining rock heterogeneity and its application for CCUS, which is of special import for the sealing potential in that core recovery of delicate mudstone lithofacies, is difficult. A workflow for the caprock analysis begins with well logging, proceeds to core description and analysis (if the core is available), and then to subsampling for laboratory analysis. In this section, we utilize the Terra Tek HRA to categorize the FWU reservoir and caprock into eleven distinct lithofacies, which are mapped onto core descriptions and form a basis for subsequent core plug sampling and analysis including mercury porosimetry and mechanical testing.

4.1.1. Lithofacies Interpretations of Caprock Units

Figure 3 shows stratigraphic columns of each core obtained from the three characterization wells at Farnsworth (the 13-10A, 13-14, and 32-8), depicting the extent of mud and sand in the clastic mixture, including fractured zones, depositional fabrics (i.e., carbonate hardgrounds, burrows, and coal cleats), and diagenetic features (i.e., carbonate "beef", concretions, and mineralized fractures) that could exert positive or negative influences on caprock integrity. Details about these features are found with the accompanying core descriptions in [9,10].

The upper Morrow shale is a marine mudstone that directly overlies the Morrow B sandstone reservoir (Figure 3) and thus serves as the primary caprock. It is composed of three common mudstone lithofacies (Table 1) including the black laminated mudstone (blM), calcareous mudstone.

The (cM) and green bioturbated mudstone (gbM) as determined by [9]. The lower portions of the upper Morrow shale consist of the gbM facies, which is transitional from the sands of the Morrow B reservoir. The green bioturbated mudstone (gbM) lithology is a slightly fossiliferous, organic-rich, slightly calcareous mudstone, that contains scattered quartz, feldspar, muscovite, and calcareous fossil-hash silt. The middle portion of the upper Morrow shale consists of the blM facies, which is interpreted by [9] to be deposited under anoxic conditions, consisting of fissile, slightly fossiliferous organic-rich mudstone. This facies gradually transitions upward into the cM facies, a more friable and calcareous mudstone that contains several hardgrounds (i.e., cemented paleo sea-floor surfaces) that are found to be laterally continuous through the FWU [9,10]. The variable degree of cementation in the cM facies imparts a heterogeneity to the geomechanical response, as discussed later.

The overlying Thirteen Finger limestone was deposited in a marine environment that underwent several cycles of transgression and regression during deposition [29] and consists mostly of black carbonaceous mudstone (bcM) lithofacies alternating with limestone layers [9,10]. On inspection, the limestone layers consist of diagenetic carbonate cement or diagenetically enhanced carbonate content and thus are denoted as cementstone lithofacies (cC; [9,10]). Fossil hash concentrations and pyrite nodules occur in varying amounts throughout the mudstone lithology, and there are some coal seams of varying thickness. The Thirteen Finger limestone is a widely distributed formation with a distinct wireline log signature with recognized open and healed vertical fractures that may provide a permeable network, especially towards the top of the unit [29,30].

Table 1. FWU caprock lithofacies descriptions (after Rose-Coss, 2017 [9]), TOC analysis, and assigned color for HRA rock classification (Terra Tek).

Facies and Description	Sedimentary Features	TOC%	HRA Color
Thirteen Finger limestone			
(cC) Carbonate cementstone, micritic dolomite, and dedolomite, grey to white	Well indurated, smooth, sparse cemented fractures, abrupt to gradational bounding surfaces	2.3	Light Blue
(bcM-a) Well indurated black carbonaceous mudstone and siltstone, locally calcareous	pyrite nodules, fossil hash, bioturbation, bedding-parallel fibrous calcite veins ("beef" of Cobold, 2013)	0.44–10.7	**Black**
(bcM-b) Fairly well indurated carbonaceous black mudstone	Floating sand grains, abundant to moderate burrowing	0.44–10.7	Purple
(bcM-c) Black mudstone and coal	Coal with thin layers of mudstone	0.44–10.7	Olive
(bcM-d) Black to grey laminated mudstone with silt partings	Locally dolomitic, fossiliferous, organic rich partings (plant fragments)	0.44–10.7	Orange
(bcM-e) Poorly indurated black carbonaceous mudstone	Locally dolomitic, fossiliferous	0.44–10.7	Grey
Morrow shale			
(cM) Calcareous mudstone, brown to grey, green laminated to massive, broken and bioturbated sections, friable	Slightly fossiliferous, laminations, bioturbation, coal	0.44–10.7	Brown
(blM) Black, laminated mudstone	Low angle to planar laminations, concretions, fossil hash	0.53–2.7	Red
(gbM) Friable, bioturbated mudstone, olive to gray, laminated to massive, friable	Low angle to planar weak laminations, low to moderate bioturbation, abundant microfossils	0.30–1.0	Yellow

The caprock lithofacies in Table 1 are mapped onto the Terra Tek HRA classification scheme as described in [31] denoted by the colors in the last column of Table 1. Note that the Terra Tek HRA procedure recognizes the variability in the mechanical integrity of the bcM lithofacies of the Thirteen Finger limestone, and we have denoted this by additional labels (i.e., bcM-a, bcM-b, etc.). Figure 4 summarizes the results of the HRA performed by Terra Tek from the well log variability in Well 13-10A [31]. The HRA facies designations are shown by the color strip down to the middle of the figure, and one can discern the Morrow B sandstone (dark blue), lower Morrow shale (red), upper Morrow shale (yellow, red, and brown going from deep to shallow), and the Thirteen Finger limestone, with multiple alternating bands of color. The lithologic breaks in the Thirteen Finger limestone are easily discernable from gamma and density logs, for example, a close examination of the density logs and gamma ray "kicks" reveals the storied thirteen shale members of the Thirteen Finger limestone alternating with the carbonate layers (designated by the light blue color bands). Variability in the mechanical integrity within the shale layers of the Thirteen Finger limestone is manifested in five distinguishable subunits of the bcM lithology, shown by separate colors. Later, we distinguish these layers in terms of mechanical behavior, and, for example, moving from elastically stiff to compliant layers we would have the HRA color designations Black > Purple > Olive > Orange > Grey. HRA logs for Wells 13-14 and 32-8 are given by Figures S1 and S2 in the Supplementary Materials accompanying this paper. The HRA analysis is commonly performed in the realm of unconventional "shale" reservoirs [21,32,33] and is a means to better understand the mudstone lithological variability given the poor core recovery and general difficulty in obtaining core-plug samples for standard analyses, which can lead to a sampling bias and over-representation of the strengths of these materials.

A mapping of HRA color designations and facies designations from core logging performed by Rose-Coss [9] is facilitated by petrographic observations from thin sections prepared from the core obtained from the 13-10A, 13-14, and 32-8 Wells. These were performed by Steve Cather of the NMBGMR (Personal Communication, 2017) and are summarized in Tables S1–S3 in the Supplementary Materials. From this, we determine that the Orange HRA refers to Bcm facies with more abundant carbonate, the Olive designation refers to coal layers and the black mudstone above and below that encapsulates them, and the grey Bcm lithofacies is the least indurated, and as such has the least values of dynamic and static elastic moduli (shown later). Black and purple HRA units are the most indurated of the bcM facies.

The caprock lithologies would classify as silt- (quartz, carbonate, organic matter) bearing clay-rich mudstone and clay-dominated mudstone (Figure 5A,B; classification of Macquaker and Adams [34]). From thin section observation [9,10], the mudstones contain variable amounts of organic matter, quartz, and macro and microfossils. Authigenic pyrite, calcite, and dolomite are common. Many of the cementstone layers contain fractures, which are commonly filled with carbonate cement [10]. The limestone is somewhat unusual, in that it is dominated by diagenetic carbonate (Figure 5C,D). The limestone locally contains a significant biogenic carbonate [10]. Thus, most of the limestones in the Thirteen Finger limestone are more properly classified as cementstones. This interpretation is supported by the obvious occurrence of concretions in the core (Figure 5D), which are similar in character to the limestone beds.

Although most of our attention is devoted to the caprock, we refer to lithologies in the Morrow B sandstone reservoir for comparison purposes. Lithologic descriptors associated with the color codes for the Morrow B sandstones are far simpler and relate to the hydrologic flow units discussed by [2,4], carrying a dark blue or green HRA designation as shown in Figure 4 above and Figures S1 and S2 in the Supplementary Materials.

4.1.2. Porosity and Permeability of Sealing Lithofacies

With the HRA mapped onto core descriptions, representative core plugs of the eleven lithofacies were analyzed for porosity, water and oil saturation, and gas permeability via Terra Tek's RCA at depth intervals of approximately 3 ft if core plugs were attainable, and these are mostly Morrow B (reservoir) lithologies. Data for Well 13-10A are given in the Appendix of Rasmussen et al. [2], data for the other two Wells (13-14 and 32-8) are given in Table S4 in the Supplementary Materials, color coded by the HRA rock class. Here, we present the porosity and permeability of the mudstone and limestone lithofacies that were recoverable in the coring and represent the sealing lithologies for CO_2 containment at FWU. The relatively poor core recovery of mudstones is evident as a sampling bias with more core plug data evident from the Morrow B sandstone and the Thirteen Finger limestone lithologies.

These results are summarized in Table S5 for all three characterization wells and are mapped to both the caprock facies designation of [9] and the HRA color unit. In Figure 6, we plot the total porosity and permeability by depth for mudstone and limestone members of the Morrow B sandstone (depth range shown in purple), the overlying upper Morrow shale (depth range shown in green), and of the Thirteen Finger limestone (depth range shown in pink). Although there is well-to-well variability, in general, the Morrow shale mudstones have higher porosity and slightly higher permeability than mudstone and limestone in the other formations. Porosities and permeabilities of the hydrologic flow units in the Morrow B sandstone are considerably higher, with porosity values ranging largely from 15 to 20%, and permeability ranging from 10 to 1000 mD (orders of magnitude higher than the mudstone facies of the caprock units at FWU shown in Figure 6; see Figure 2 in [2]).

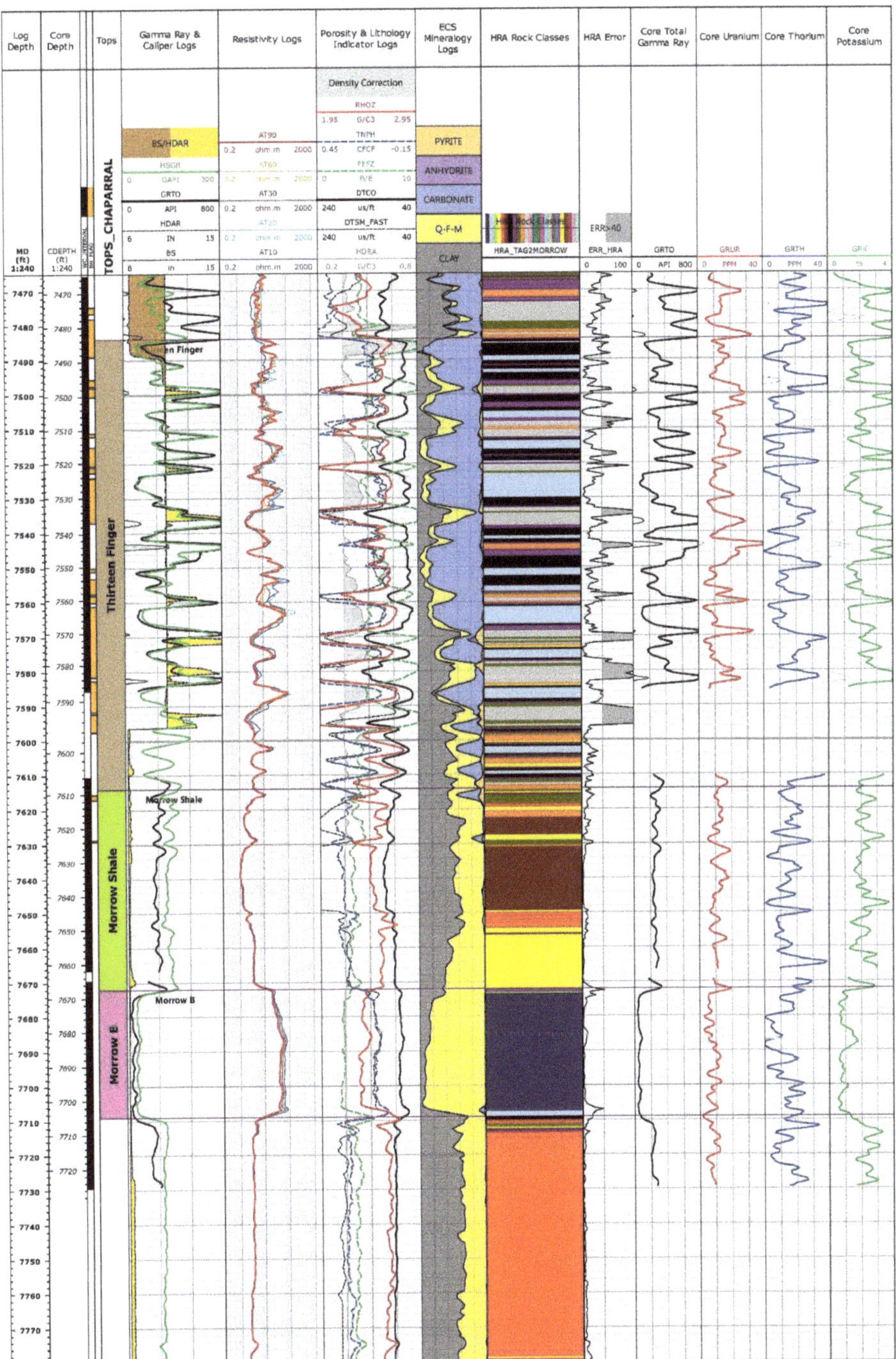

Figure 4. Well 13-10A heterogeneous rock analysis log showing the color by the rock unit and depth. Depths are given in feet and converted to meters by multiplying by 0.3022. Note that the logging depth and coring depth contain discrepancies associated with the coring procedures.

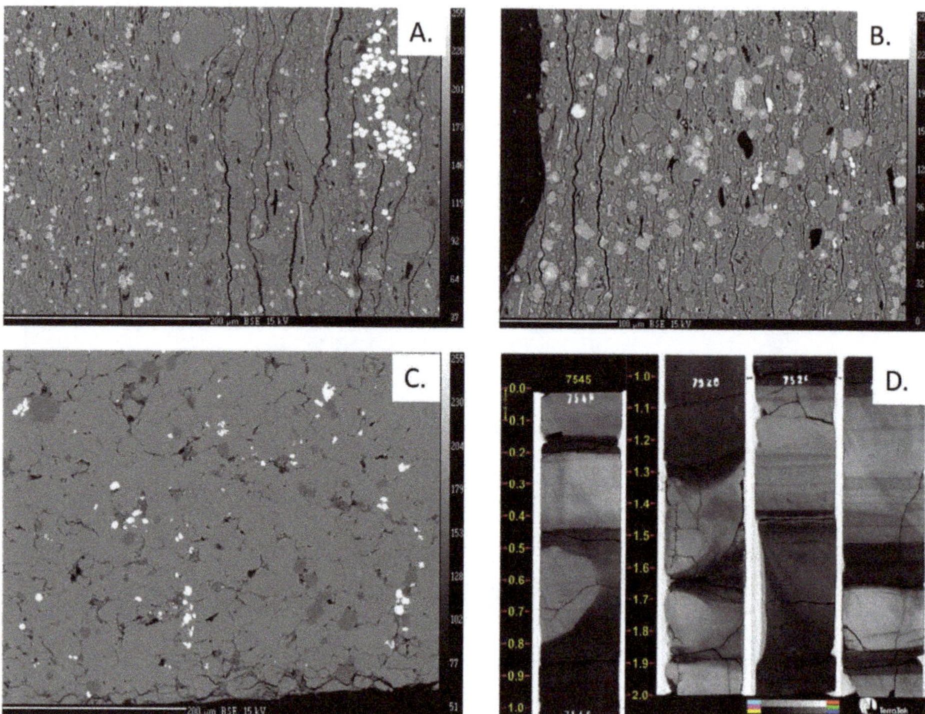

Figure 5. (A) Back-scattered electron image of a silt (carbonate, organic matter) bearing clay-rich mudstone in the Morrow shale. Stratigraphic-up is to the left, with siltier rich laminae on the left and clay-rich laminae on the right. Black portions are organic matter, and white features are pyrite framboids. Well 13-10A, 7633.76 ft. **(B)** Backscattered electron image of silt (quartz, calcite, organic matter) bearing clay-rich mudstone in the Morrow shale. Stratigraphic-up is to the left. Black particles are organic matter, irregular dark gray particles are quartz, light gray rhombs are ankerite, white grains are pyrite. Well 13-10A, 7632.6 ft. **(C)** Backscattered electron image showing calcite (light gray), dolomite (dark gray), and pyrite (white) cemented mudstone (cementstone) in the Thirteen Finger limestone, Well 13-10A, 7540.65 ft. **(D)** Core photographs showing the variable geometry of limestones in the Thirteen Finger limestone. Some limestones are laterally continuous throughout the width of the core, whereas other are laterally discontinuous (concretionary). Well 13-10A, depths indicated on the photo. See [9,10]) for additional images and descriptions.

4.2. Capillary Heterogeneity and Sealing Capacity

The sealing capacity in the context of CCUS is the CO_2 column height that is retained by the capillarity of a water wet rock. Here, we estimate CO_2 column heights (using calculated pore throat diameters and breakthrough pressures) for the different reservoirs and caprock lithologies using MICP analyses from core plug samples, which is summarized in Table S6 in the Supplementary Materials. The pore throat size distributions for the Morrow B sandstone, upper Morrow shale, and the Thirteen Finger limestone for the west and east side of the FWU are compared in Figure S3 and the accompanying text in the Supplementary Materials. There is a clear difference in the reservoir and sealing lithologies across the FWU as one would expect. The Morrow B sandstone pore throat diameters show a broad range over five orders of magnitude (0.0003 to 1000 μm), whereas the upper Morrow shale and Thirteen Finger limestone show a much narrower range over two orders of magnitude (~0.003–0.1 μm).

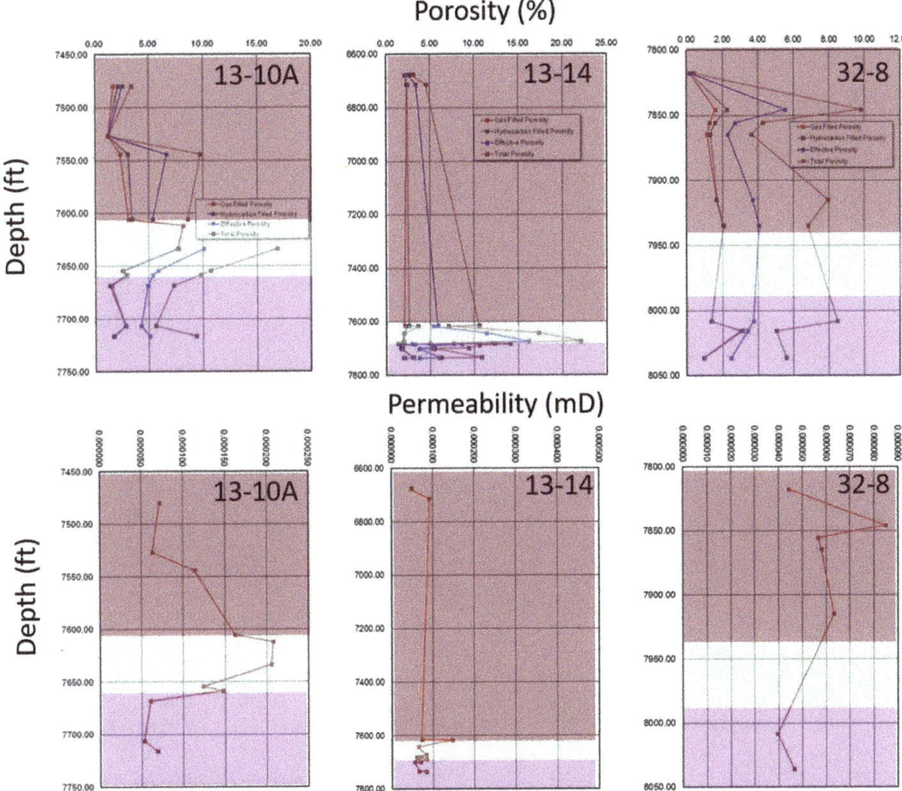

Figure 6. Porosity and permeability of Farnsworth mudstone lithofacies within the Morrow B sandstone (purple), Morrow shale (green), and Thirteen Finger limestone (pink) for Wells 13-10A, 13-14, and 32-8. Color schemes and depths correspond to those in Figure 3. Data were determined from the Terra Tek tight rock analysis (TRA) methodology.

The mercury porosimetry results were used to calculate the CO_2 column heights for the caprock and reservoir formations using the standard methods [7] and are shown by depth for the three characterization wells in Figure 7. The CO_2 column heights for the upper Morrow shale and the Thirteen Finger limestone range from 1000 to 10,000 m (3280–32,808 ft). The cementstone lithology in the Thirteen Finger limestone has 11,000 m of CO_2 column height, with an average of 9000 m (29,527 ft). Two of the cementstone samples reached the upper limit of pressure 60,000 psi (414 MPa), that the MICP instrument is capable of sustaining, with no observable intrusion of mercury.

The mudstone lithologies within the upper Morrow shale and Thirteen Finger limestone have an average CO_2 column height of 2900 m (9514 ft), with a range of 1000 to 10,000 m (3280–32,808 ft). Not unexpectedly, the caprock CO_2 column height values are 1-to-2 orders of magnitude larger than the sandstone reservoir values and exceed the reservoir thickness, suggesting an excellent caprock integrity. The CO_2 column heights for the Morrow B sandstone reservoir ranged from about 1 to 100 m (3.3–328 ft). These results would suggest that the Morrow shale and Thirteen Finger limestone caprock should provide an excellent capillary sealing for CO_2 for CCUS operations in the FWU. Note that these results refer to the lithologic properties of the rock matrices themselves. In order to further examine the question of caprock integrity, we need to examine in detail the potential of various seal by-pass features both in the form of existing natural fractures and in the potential for inducing fractures during injection and operation phases of CCUS-EOR.

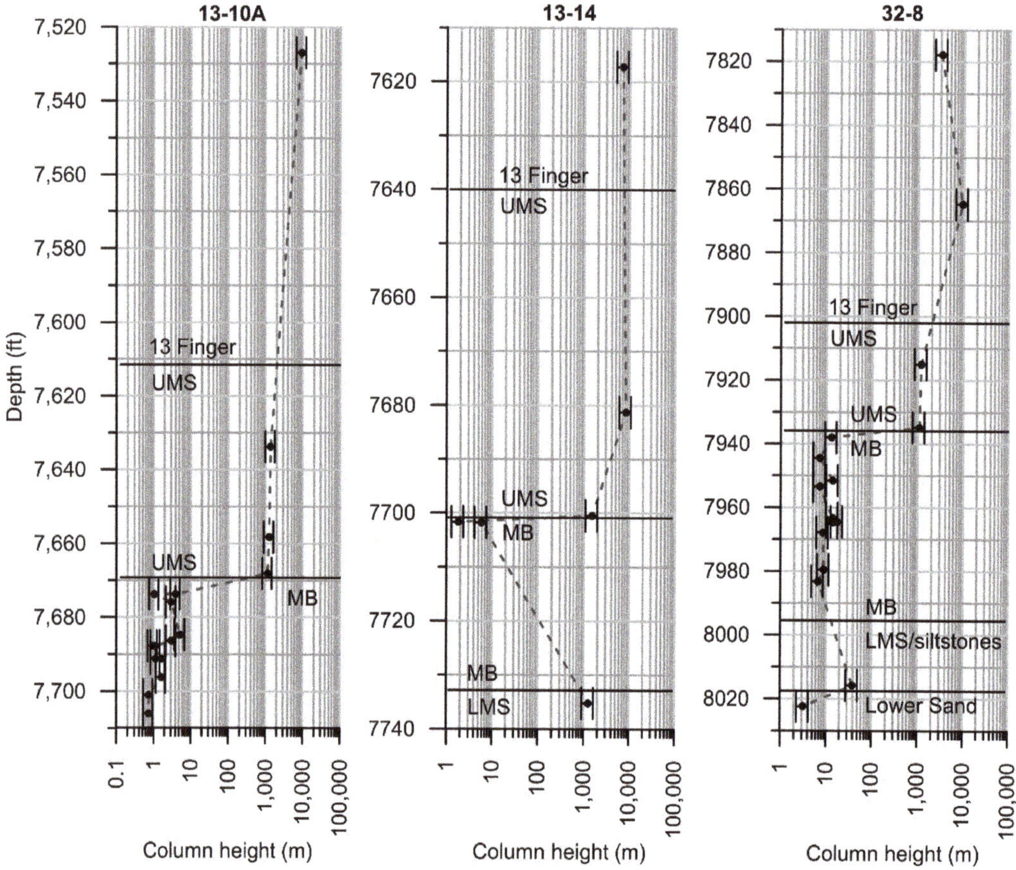

Figure 7. CO_2 column heights for Wells 13-10A, 13-14, and 32-8 determined from the MICP analysis. "MB" denotes the Morrow B sandstone; "UMS" denotes the upper Morrow shale, and "13 Finger" denotes the Thirteen Finger limestone.

4.3. Seal Bypass Potential and Mechanical Integrity

4.3.1. Natural and Induced Fractures

Natural fractures in mudstone lithologies can impact fluid-flow, fracture permeability, and mechanical strength of the rock [35], all critical aspects for caprock integrity. To understand how natural fractures impact the ability of FWU caprock lithologies to prevent CO_2 leakage, we need to characterize fracture apertures and density, as well as orientation spatially and in reference to the current in situ stress orientations. Orientation aspects are especially relevant as: 1. Fractures are generally strength-limiting at above-core length scales, and the increase in pore pressure can induce slip and permeability increases for suitably oriented fractures [36,37]; 2. fracture orientation with respect to the in situ local stress tensor affects aperture width and thus permeability; and 3. fractures induced by fluid injection, i.e., hydrofractures, propagate in directions dictated by the local stress tensor. It is also important to distinguish fractures induced by coring, as these are not indicative of the state of fracturing in the subsurface, but additionally can aid in determining orientations of principal stresses in the subsurface as described below.

Fractures in FWU caprock were described via the fracture class type, orientation, fracture dip, type of mineral fill, fracture porosity, fracture spacing, and intensity for Wells 13-14 and 32-8 [31]. For the Well 13-14 core, a detailed analysis of macroscopic fractures was conducted on nearly 270 ft of continuous whole core material, approximately 123 ft of which contain significant fractures. For our purpose here, we focus on fracture data from Well 13-14 in the interval corresponding to the upper Morrow shale as it is the primary caprock lithology, and Well 13-14, being in the western portion of the FWU, is where the CO_2 injection is being monitored by the SWP. Although the 13-14 core is unoriented, the coring-induced fracture orientations allow an estimation of core orientation with respect to principal stress orientations in the subsurface at FWU.

There are four types of fracture classes identified in the Well 13-14 core that include: Drilling or coring induced fractures; open low angle shear fractures; high-angle partially open fractures that heal through a carbonate interval; and filled fractures, generally a low angle. Drilling-induced fractures are the most abundant. Mineralized fractures are rare, but the most common mineral in-fill recognized is calcite. Rose diagrams indicating relative fracture orientation of all observed fractures in the Morrow shale were created for each fracture class and shown in Figure 8. The natural fractures in 13-14 show a similar orientation to the induced fractures and may indicate that the timing of the natural fractures to be more recent, i.e., formed under current stress orientations. According to the Snee and Zoback [38] stress map of Texas (see also [10]), the FWU should be located in a transitional stress state between a normal faulting regime and a strike-slip faulting regime where the maximum horizontal stress (S_H) is slightly less but approaching the vertical stress S_v in magnitude (i.e., $S_H \sim S_v > S_h$ where S_h is the minimum horizontal stress). The orientations of maximum horizontal stress in the Texas Panhandle, determined from horizontal breakouts, trend from SE-NW to EW, which would be the expected orientations of hydrofracture propagation, as well as open fractures (which open in a direction perpendicular to the least horizontal compressive stress direction).

If these orientations are indeed characteristic of natural fracture orientations at FWU, and if we understand the orientations of the principal stress directions, we can then determine critical dip directions for existing fractures that might be induced to slip upon pore pressure increases associated with injection. This is beyond the scope of the present chapter but will be addressed in a later work. However, the coincidence of the coring induced fractures and the natural fractures would suggest that these fractures would be of a critical orientation for slip (i.e., shear fracture) associated with fluid-injection induced overpressure. What works well for caprock integrity, however, is the relative rare occurrence of fractures overall in the FWU caprock lithologies as represented by core analysis.

4.3.2. Static and Dynamic Geomechanical Behavior

It is important to understand the limiting strength of the shallow crust posed by existing fractures [37], and as well it is necessary to understand the heterogeneity in matrix rock mechanical properties. Static rock mechanics properties concern poro-elastic deformation, yielding, and ultimate rock strength and failure. A typical suite of rock mechanics tests that permit parameterization of constitutive models includes an unconfined compression (UCS) test (a right cylinder of rock is exposed to an axial load with no confining load applied to the round surface of the cylinder), several triaxial (TXC) tests at different confining pressures (an axial load is applied to the long axis of the cylinder with a constant confining pressure applied to the cylinder sides), and a hydrostatic test in which the rock cylinder is subject to a constant applied pressure, with or without the separately controlled pore pressure. A deformable jacket surrounding the cylinder keeps the pore and confining systems separate. These tests allow us to examine the variability of static elastic properties (i.e., Young's Modulus and Poisson's Ratio, which can either be used to represent an elastically isotropic medium or be directionally dependent, which is a simplified means to assess elastic anisotropy, for example, with respect to the primary bedding direction), yielding behavior (involving inelastic processes such as microfracture growth and coalescence or pore collapse), and failure (involving complete loss of cohesion of a deforming rock generally by a through-going shear fracture).

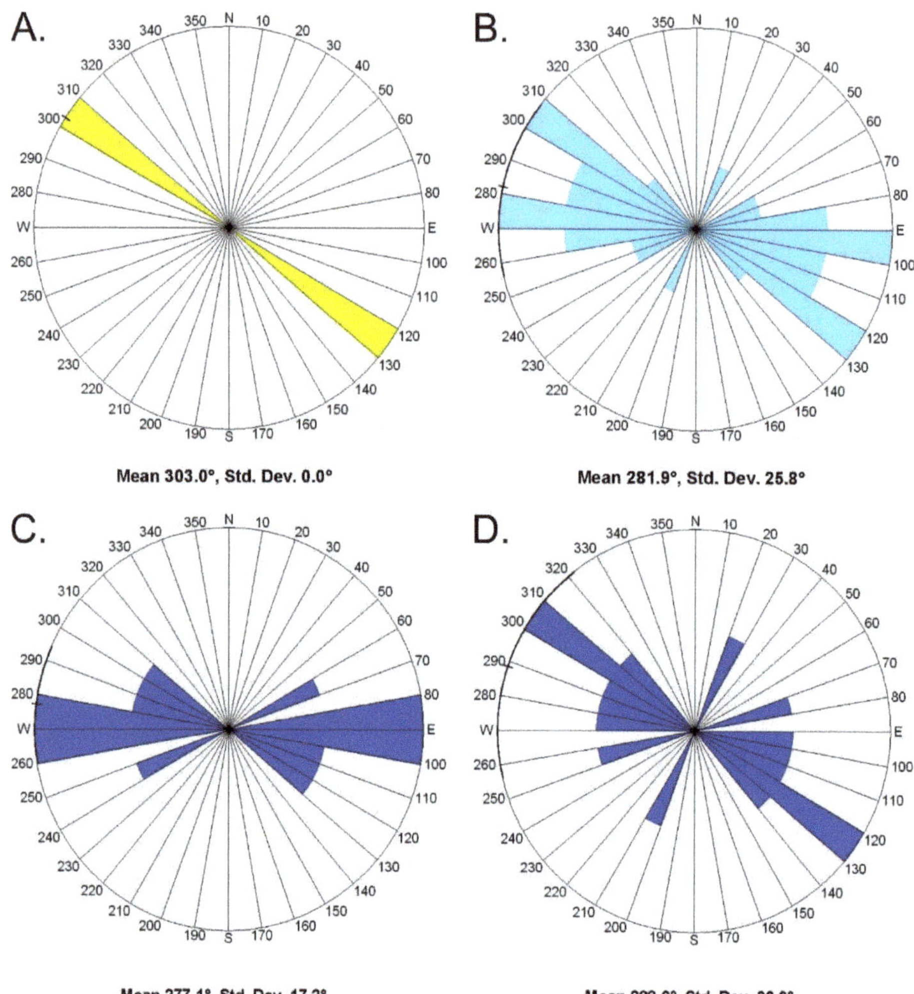

Figure 8. Strike rose diagrams for fractures observed in the Well 13-14 core for the Morrow shale (7660–7722 ft). (**A**) Coring-induced fractures; (**B**) all natural fractures; (**C**) open fractures; and (**D**) partially or totally mineralized fractures. The coincidence in directions between coring induced fractures and natural fractures existing at FWU suggests that the stress conditions resulting in the natural fractures were of similar orientation, and that natural fractures occurring in the FWU caprocks would be kept open under current subsurface stress conditions.

An extensive suite of geomechanical properties have been assembled from the Terra Tek testing program for the FWU characterization wells [31]. One valuable aspect of the data set is the degree to which it maps properties on rock units based on Terra Tek's HRA methodology for sampling and testing. We focus here on elastic properties, failure envelopes, and fracture gradients as they concern sealing lithology responses to fluid injection.

Static and dynamic elastic properties are represented as continuous profiles that combine well logging, Terra Tek's HRA analysis, and laboratory measurements on the core that is used for calibration. These are presented in Figures S4–S6 in the Supplementary Materials. The results for the eleven HRA rock classes for well 13-10A are summarized in Figure 9.

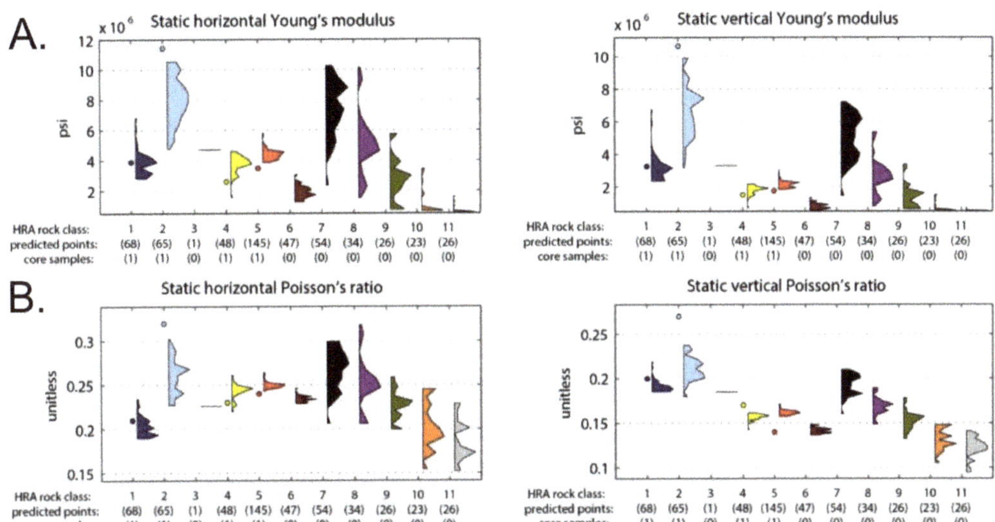

Figure 9. Static Young's Modulus (**A**) and Poisson's Ratio (**B**) determined by combining ranges interpreted from well log static rock compression tests illustrating the contrast in elastic properties of the Morrow B sandstone reservoir and the Morrow shale and Thirteen Finger Limestone caprock for Well 13-10A. The left-hand figures show values for samples oriented horizontally, and the right figures show values for samples oriented vertically. In general, horizontal values are slightly higher than vertical values, reflecting an anisotropy likely derived from depositional fabrics.

In general, horizontal values are slightly higher than vertical values, reflecting an anisotropy likely derived from depositional fabrics. The highest values (with light blue) are found in the Thirteen Finger limestone cementstones, the next highest correspond to the black color (Bcm facies in the Thirteen Finger limestone), whereas the lowest values (orange and grey) are weaker Bcm lithofacies in the Thirteen Finger limestone. Morrow B sandstone lithofacies (dark blue) and Morrow shale (yellow, red, and brown) lithofacies are intermediate in value. In general, this is not an unexpected degree of elastic heterogeneity, but could influence how the caprock responds to a reservoir-scale increase in pore pressure associated with CCUS activities.

Of greater interest for caprock integrity is the failure behavior of the suite of relevant rock types, and these are presented as failure envelopes in Figure S7 in Supplementary Materials. Due to core recovery issues, the triaxial testing results are limited to the more competent lithofacies, whereas for caprock integrity, the weaker lithofacies which were not recovered or were otherwise damaged during coring are of greater interest as these lithologies would represent the greater threat to sealing integrity from fluid injection and over-pressuring. However, the results of the triaxial testing were useful as calibrations for well log data in a proprietary approach by Terra Tek to estimate fracture gradients. Figure S7 shows Coulomb failure envelopes for Thirteen Finger limestone and Morrow shale caprock and Morrow B sandstones created by combining UCS and triaxial test data for these rock types. As discussed by Trujillo [10], Morrow B sandstones are much weaker with much lower cohesion than the mudstone or limestone lithofacies in the upper Morrow shale or the Thirteen Finger limestone.

4.3.3. Fracture Gradients in Farnsworth Reservoir and Sealing Lithologies

We bracketed the orientation of the three principal stresses within the FWU from regional observations [38], which may correspond to directions of open and induced fractures in the upper Morrow shale. There is a vertical maximum principal stress, a minimum horizontal stress in the orientation of 350° (North-South), and a maximum horizontal stress orientation at about 270° (East-West). A more difficult determination is an assessment of magnitude of the principal stresses (vertical/overburden stress, S_v, the maximum horizontal stress, S_H, and the minimum horizontal stress, S_h), and addressing this is beyond the scope of this paper. However, based on the rock mechanics testing data and the Terra Tek analysis (summarized in Figures S4–S6 in the Supplementary Materials), we can bracket the minimum horizontal stress gradient required to exceed rock strength and propagate fractures. These estimates are typically used to limit the extent of injection-associated pore fluid overpressures so as to not damage formations during injection and production activities. For CCUS in the FWU, these values are important to consider. Based on the in situ stress orientations in the FWU, any fractures created in the weaker reservoir lithology would be oriented vertically (for joints or tension fractures) or steeply dipping (in the case of shear fractures), given the orientations of the least principal stresses.

Estimates of the minimum horizontal stress gradient in FWU caprock and reservoir lithologies are determined from the Terra Tek suite of testing shown in Figures S4–S6 and are summarized in Figure 10 for all three characterization wells. The lowest values are in the well 13-14, but in general values cluster around 0.70 to 0.80 psi/ft (1 psi/ft is equivalent to 0.0226 MPa/m). Although some of the lowest values are in the caprock lithologies, so are the highest values, with the Morrow B sandstones falling generally below 0.75 psi/ft. Pore pressure gradients range from 0.43 psi/ft in the Thirteen Finger limestone and increase to 0.585 psi/ft in the upper Morrow shale [10], while an estimated vertical stress gradient is 1.1 psi/ft. With a fracture gradient of 0.7 psi/ft at a depth of 7700 ft (2347 m) in the reservoir, the maximum pore pressure that could be attained within the reservoir without fracture is ~5390 psi or ~37 MPa. At the caprock-reservoir interface the fracture gradient is 0.85 psi/ft at a depth of 7668 ft (2337 m), the maximum pore pressure that could be attained within the caprock without fracture is ~6518 psi (~44.9 MPa). Injection-induced pressures by the CO_2 in the western portion of the field are on the order of 5000 psi (34.5 MPa), which is below the maximum level. The fracture gradients indicate that the Morrow B sandstone reservoir is weaker than the overlying lithologies, so any fracture initiated around CCUS injection wells in the FWU should not propagate into the overlying sealing units.

4.4. Seal Continuity across the Farnsworth Unit

Isopach maps of the upper Morrow shale and Thirteen Finger limestone across the FWU were prepared from formation tops and bottoms after data compiled by [9] and are shown in Figure S8 in the Supplementary Materials, along with the total caprock thickness. The minimum thickness for the upper Morrow shale occurs in the middle of the FWU at ~42 ft (12.8 m). The minimum thickness of the Thirteen Finger limestone occurs in the western portion of the FWU at around 60 ft (18.3 m). In general, the caprock thickness ranges from 240 ft (73.2) in the eastern portion to ~120 ft (36.6 m) in the western portion of the FWU. The lateral caprock continuity easily suggests that seal integrity would be anticipated along the mapped extent of the FWU.

4.5. Seal Integrity Inferred from Noble Gas Analyses

Measurement of naturally occurring noble gases in a vertical profile from the preserved fresh core in the reservoir and caprock units complements the other caprock integrity analyses as the pattern of noble gas isotopic content is the direct in situ integrated result of the driving forces and transport properties through the reservoir and the caprock. Thus, the noble gas isotopic profile reflects the original infiltration of groundwater with atmospheric noble gas contents and the addition of subsurface geogenic noble gases, which are affected by transport via advection and/or diffusion and potentially more recent reservoir activities since FWU water flooding began in the 1950s. Noble gas profiles that reflect diffusion-dominated transport are expected for high sealing quality caprock, whereas caprock with seal bypass systems (i.e., an interconnected fracture network) may result in an advective noble gas profile [7].

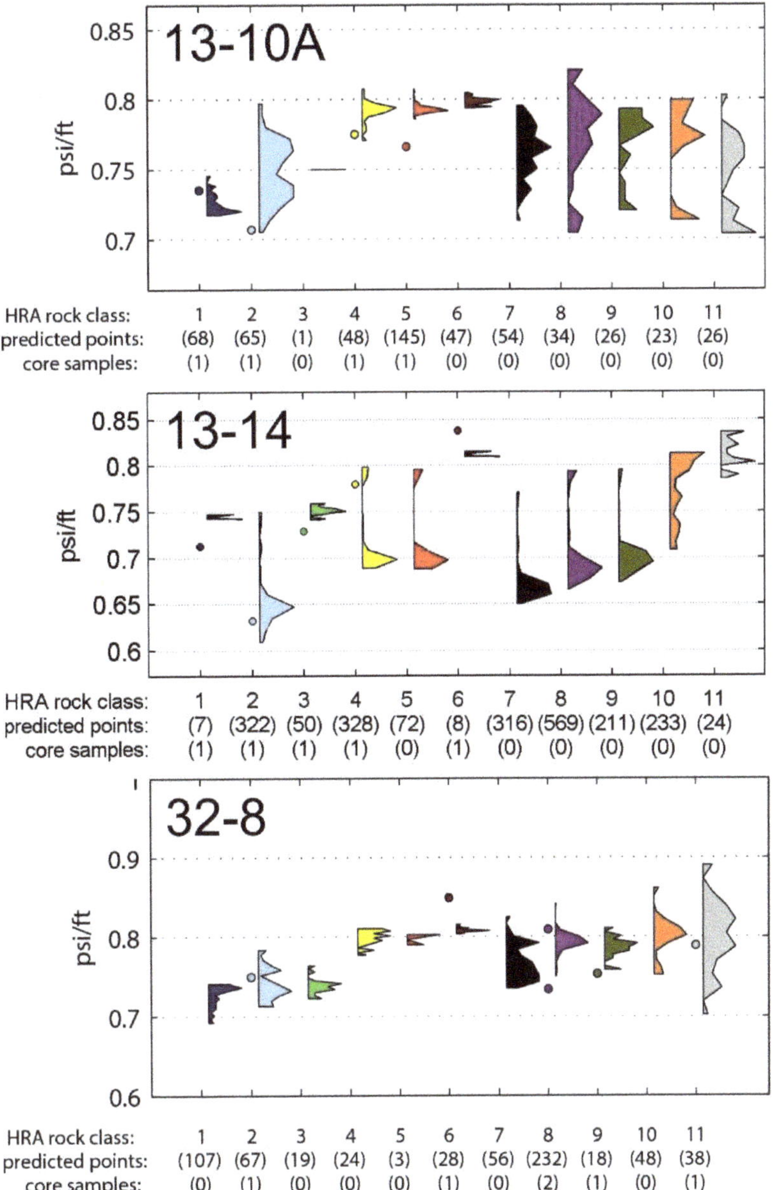

Figure 10. Estimated minimum horizontal stress (fracture) gradients by rock class for each of the three characterization wells. In general, the Morrow B sandstone rock classes (light and dark blue) have lower values indicating they are more easily fractured. Propagating vertical fractures associated with injection-induced hydrofracturing would likely not propagate into the overlying caprock formations.

The context for interpreting noble gas data measured in this study is that atmospherically sourced isotopes are ^{20}Ne and ^{36}Ar, whereas geogenic isotopes sourced in the subsurface are ^{3}He, ^{4}He, ^{40}Ar, and ^{22}Ne [39]. Isotopic ratios for atmospheric, crustal, and mantle reservoirs are well known and used to identify sources of fluids in petroleum systems [40]. The measured ^{3}He over ^{4}He ratio (R) from FWU samples, normalized by that same ratio for the atmosphere (R_a), have values of ~0.02 for most measurements within the Thirteen Finger limestone and the upper Morrow shale (Figure 11, see the left-most column; values are listed in Table S7 in the Supplementary Materials). These values of ~0.02 are consistent with crustal fluids—the crustal end-member R/R_a value is 0.02, which represents the dominant production of ^{4}He from U and Th decay [40]. Deviations from values of ~0.02 occur within the Morrow B sandstone and near the top of the Thirteen Finger Formation (Figure 11).

The Morrow B sandstone deviations from 0.02 R/R_a are most likely caused by the long-term water flooding in the Farnsworth Unit, which initially used groundwater sourced from the Miocene Ogallala Formation in the Texas, USA, Panhandle that probably had R/R_a values at closer to 1.00 than the older caprock fluids (younger groundwaters will probably have values closer to one than the older crustal fluids; see [40]). The R/R_a value greater than 1.00 in the Morrow B sandstone may be an artifact of the laboratory analysis. Mantle-sourced fluids have R/Ra values much greater than one [40], but this is unlikely as a source due to the rest of the samples being less than one. Complex phase partitioning between groundwater, oil, and any gas phase may also lead to ratios greater than one. Ratios of ^{4}He/^{20}Ne, normalized by the atmospheric value, are several orders of magnitude greater than one for most samples, thus indicating high helium concentrations due to the long-term production of ^{4}He from U and Th in the sealing lithologies with low permeability and low effective diffusivity. The very distinct change in R/R_a from the Morrow B into the upper Morrow shale indicates that the upper Morrow shale is a good seal at least at that contact measured by the coring.

Due to the high amount of methane and helium in many samples as observed during the laboratory analysis, sample splitting was necessary, which affected the reliability of the argon values. Thus, argon values were not reported by the laboratory for several samples (Table S7). The argon isotopic values that are available also reflect some processes that introduce fluids into the system that may have been in equilibrium with the atmosphere (Figure 11). A sample within the upper Morrow shale has a relatively low light-to-heavy Ar isotopic ratio, which is expected as ^{40}K within the formation would produce ^{40}Ar.

The deviations from 0.02 R/R_a for the Thirteen Finger limestone may be due to the improper sealing or leakage of the preservation canisters, as such leakage would bring the values closer to 1.00, which may be likely for the sample at depth 7515 ft (2290.57 m) as its R/R_a is 0.47, and its neon and argon isotope ratios are close to the atmospheric values (Figure 11). However, the adjacent sample at depth 7502 (2286.01 m) also has a relatively high R/R_a of 0.034 and its neon and argon ratios are slightly shifted from the atmospheric equilibrium values. Thus, it is possible that some process occurs near the top of the Thirteen Finger limestone to introduce a minor atmospheric source of (meteoric) fluids or is otherwise fractionating the noble gases. We speculate that the observed natural fractures may permit fluid movement that has larger R/R_a than the crustal values of ~0.02 of the rest of the Thirteen Finger limestone and upper Morrow shale samples. This would suggest that although the upper Morrow shale and lower Thirteen Finger limestone appear to have been isolated from the surrounding fluid movement (and fluid contamination associated with decades-long water flooding operations at FWU), the upper portions of the Thirteen Finger limestone may have been infiltrated by fluids in contact with atmospheric noble gas isotopic values.

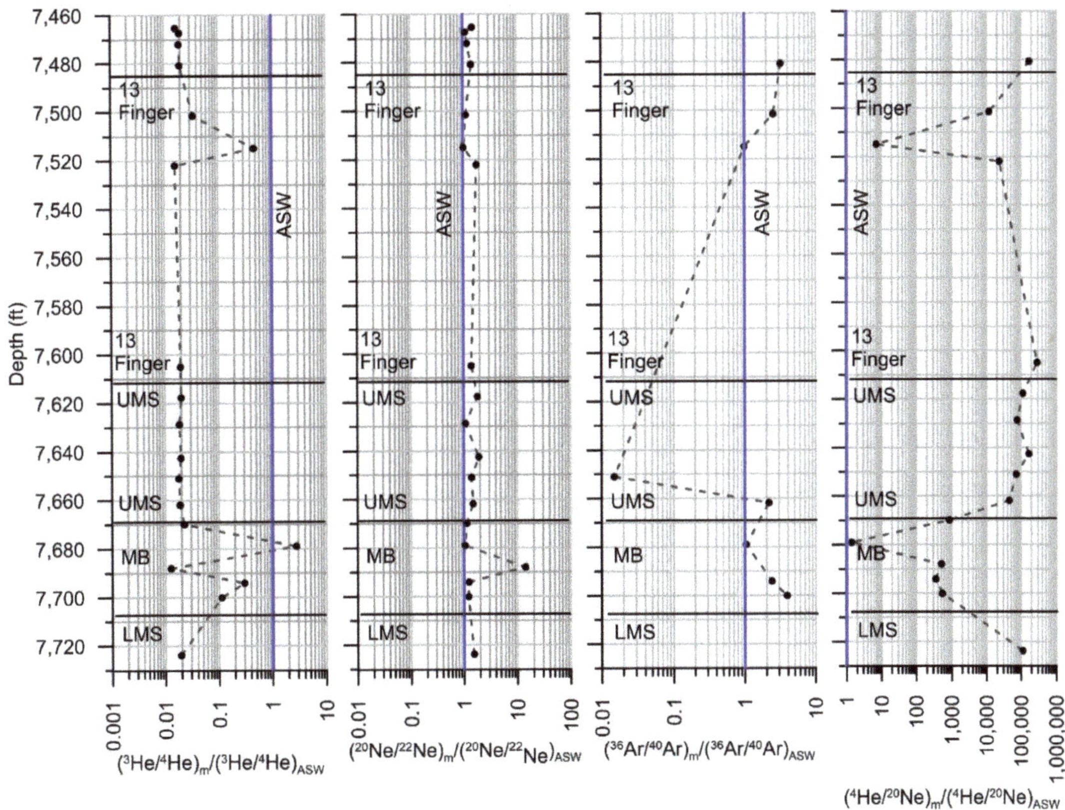

Figure 11. Noble gas ratios versus depth for Well 13-10A. The subscript *m* stands for the ratio measured for the sample, whereas the subscript *ASW* stands for the air saturated water value for the ratio.

5. Discussion

Direct formation-scale assessment of caprock integrity is difficult in a large part from the heterogeneity at all scales. Wire-line logging and seismic techniques may lack resolution to identify potential seal-by systems (e.g., such as connected fracture networks that are below seismic resolution). Assessments of capillary heterogeneity and geomechanical behavior from small-scale core-plug measurements may not be representative of heterogeneity across formation scales and can reflect a sampling bias from the core recovery. However, both methods are used to infer large-scale behavior, which may include modeling to integrate to the reservoir and caprock properties made at different locations and different scales [41]. To build confidence in and understanding of caprock integrity at FWU, this study approaches caprock integrity by systematically assessing processes that govern the sealing quality at different scales following the framework of Figure 1. Thus, the processes are examined from nanoscale capillary-wettability controls to mechanical properties and the full caprock-reservoir system behavior via sedimentological-stratigraphic evaluation and large-scale in situ noble gas transport. Small-scale sealing integrity is confirmed by MICP measurements of high breakthrough pressures and very high estimated CO_2 column heights for the upper Morrow shale and the Thirteen Finger limestone. Mechanical properties from core measurements indicate that, as the Morrow B sandstone is relatively weak, fractures that may be induced from CO_2 injection activities would probably not propagate into the upper Morrow shale. The naturally occurring in situ noble gas isotopic profile builds confidence that water flooding and production reservoir operations prior to coring of Well 13-10A did not damage a high-quality sealing caprock as the helium ratios are quite different across the Morrow B sandstone and upper Morrow shale contact. The thickness and lateral continuity of the upper Morrow shale and Thirteen Finger limestone further strengthen the argument for a high-quality sealing caprock system at FWU.

6. Conclusions

Our main conclusions from this study are drawn from the five topics in the Results Section and are as follows:

- A cluster analysis of rock heterogeneity based on well log and other analyses by Terra Tek shows that the caprock lithologies at FWU can be grouped into nine separate units. This classification forms the basis for subsampling and core plug analysis (when possible), as well as determinations of static and dynamic rock mechanics properties and fracture gradients.
- MICP results show that sealing units in the Morrow shale and Thirteen Finger limestone units should provide excellent sealing capacity for storage of CO_2 in the Morrow B Sandstone injection unit, as calculated CO_2 column heights exceed the thickness of the Morrow B. Cementstones in the Thirteen Finger limestone have anomalously high sealing potential and strength, and the ability of these thin bands of tight carbonate to serve as seals by themselves would be limited only by their lateral continuity.
- An assessment of fracture gradient across the reservoir and caprock lithologies is drawn from the HRA, well log analysis, and laboratory triaxial testing. The range of fracture gradients show that formations should be able to support a relatively large injection-induced overpressure to around 7 or so MPa (~a thousand psi) over hydrostatic pressure values at the depths of interest. Existing fractures are apparently rare in the FWU caprocks but may have orientations that suggest creation under existing stress conditions at FWU, given the similarity in orientations to coring-induced fractures. Failure analyses show that the Morrow B sands are weaker than overlying lithologies, so that any fracture initiation around the injection well would not be expected to propagate into the overlying sealing units.
- Caprock units including the upper Morrow shale and Thirteen Finger limestone show sufficient thickness and lateral continuity across the FWU suggesting good sealing potential, barring any significant seal by-pass features.
- The noble gas analysis from fresh core shows that the caprock lithologies show no degree of leakage from historical water and CO_2 flooding in the FWU, whereas the Morrow B sandstone shows an impact from historical EOR activities. The upper Thirteen Finger limestone contains noble gas values that are consistent with invasion by meteoric waters, whereas the middle and lower Thirteen Finger limestone and upper Morrow B shale contain values that suggest hydrologic isolation.

Together, these analyses conducted at different scales strongly suggest an excellent sealing capacity for the Morrow shale and Thirteen Finger limestone lithologies. This is ascertained from the high degree of capillary sealing, the low potential for seal bypass, and the large regional extent of the caprock lithologies in the FWU.

Supplementary Materials: The following are available online at https://www.mdpi.com/article/10.3390/en14185824/s1, Tables S1–S3: Caprock thin section petrologic descriptors with Schlumberger Heterogeneous Rock Classification (HRA) assigned unit color; Table S4: Summary of petrophysical properties (Terra Tek routine core analysis); Table S5: Terra Tek tight rock analysis for mudstone and limestone lithofacies in the Morrow shale and Thirteen Finger limestone; Table S6: Mercury intrusion data and associated analysis by rock class for the 13-10A, 13-14, and 32-8 characterization wells; Table S7: Results of noble gas analysis, including the sample mass and supplementary estimates of wet bulk density, total porosity based on laboratory analysis on fresh core samples or well log interpretation; Figure S1: Well 13-14 HRA summary and accompanying well logs used in the analysis; prepared by Schlumberger; Figure S2: Well 32-8 HRA summary and accompanying well logs used in the analysis, prepared by Schlumberger; Figure S3: Pore throat diameter in microns vs. incremental non-wetting phase saturation by formation and well location, from mercury porosimetry measurements; Figure S4: Well 13-10A interpolated static and dynamic rock mechanics properties based on Schlumberger/Terra Tek HRA log analysis and lab experiments (shown as red and black dots); Figure S5: Well 13-14 interpolated static and dynamic rock mechanics properties based on Schlumberger/Terra Tek HRA log analysis and lab experiments (shown as red and black dots); Figure S6: Well 32-8 interpolated static and dynamic rock mechanics properties based on Schlumberger/Terra Tek HRA log analysis and lab experiments (shown as red and black dots); Figure S7: Coulomb failure envelopes for representative caprock lithologies from triaxial core testing; Figure S8: Isopach maps of caprock lithologies in the Farnsworth Unit. A. Morrow shale; B. Thirteen Finger limestone; C. Total caprock thickness (after Rose-Coss 2017, [9]).

Author Contributions: Conceptualization, M.C., J.E.H. and P.S.M.; methodology, J.E.H., P.S.M. and T.A.D.; formal analysis, N.T., D.R.-C. and J.E.H.; investigation, N.T. and D.R.-C.; data curation, J.E.H.; writing—original draft preparation, N.T., D.R.-C. and P.S.M.; writing—review and editing, T.A.D., J.E.H. and M.C.; supervision, M.C., P.S.M. and J.E.H.; project administration, M.C. and W.A.; funding acquisition, M.C. and W.A. All authors have read and agreed to the published version of the manuscript.

Funding: Funding for this project was provided by the US Department of Energy's (DOE) National Energy Technology Laboratory (NETL) through the Southwest Regional Partnership on Carbon Sequestration under award no. DE-FC26-05NT42591.

Data Availability Statement: Additional supporting data for this study are found on the US DOE's National Technology Laboratory Energy Data eXchange website (EDX) at: Heath, J.; Dewers, T.; Cather, M. Core analysis e-reports produced for the Southwest Regional Partnership on Carbon Sequestration for Wells 13-10A, 13-14, and 32-8 of the Farnsworth Unit, Texas. **2021**, https://edx.netl.doe.gov/dataset/core-analysis-e-reports-for-the-swp (accessed on 8 August 2021), doi:10.18141/1798503.

Acknowledgments: Much of the analysis in this paper (not including the noble gas data and mercury porosimetry) is based on data from two New Mexico Tech Masters Theses by N. Trujillo and D. Rose-Coss. We thank Steve Cather for his lithologic interpretations found in Tables S1 through S3 in the Supplementary Material. We thank Joseph Hall, a geologist formerly of Chaparral Energy LLC, for assistance in designing the coring program, his geologic expertise of the Farnsworth Unit, and his assistance with characterizing the core. Wayne Rowe of Schlumberger managed well logging efforts for this project. Christopher Gillespie of Terra Tek, formerly a Schlumberger company, managed the myriad geomechanical, geochemical, and petrophysical measurements provided by Terra Tek for this project. Sandia National Laboratories is a multi-mission laboratory managed and operated by National Technology and Engineering Solutions of Sandia, LLC, a wholly owned subsidiary of Honeywell International Inc., for the US Department of Energy's National Nuclear Security Administration under contract DE-NA0003525. This paper describes objective technical results and analysis. Any subjective views or opinions that might be expressed in the paper do not necessarily represent the views of the US Department of Energy or the United States Government. The core from which the samples were collected for this project are housed at the Subsurface Data and Core Libraries of the New Mexico Bureau, Socorro, New Mexico, USA.

Conflicts of Interest: The authors declare no conflict of interest.

References

1. Cartwright, J.; Huuse, M.; Aplin, A. Seal bypass systems. *AAPG Bull.* **2007**, *91*, 1141–1166. [CrossRef]
2. Rasmussen, L.; Fan, T.; Rinehart, A.; Luhmann, A.; Ampomah, W.; Dewers, T.; Heath, J.; Cather, M.; Grigg, R. Carbon storage and enhanced oil recovery in Pennsylvanian Morrow Formation clastic reservoirs: Controls on oil-brine and oil-CO_2 relative permeability from diagenetic heterogeneity and evolving wettability. *Energies* **2019**, *12*, 3663. [CrossRef]
3. Rose-Coss, D.; Ampomah, W.; Hutton, A.C.; Gragg, E.; Mozley, P.; Balch, R.S.; Grigg, R. Geologic characterization for CO_2-EOR Simulation: A Case Study of the Farnsworth Unit, Anadarko Basin, Texas. *AAPG Search Discov.* **2015**, *80484*, 29.
4. Ross-Coss, D.; Ampomah, W.; Cather, M.; Balch, R.S.; Mozley, P.; Rasmussen, L. An improved approach for sandstone reservoir characterization. In Proceedings of the SPE Western Regional Meeting, Anchorage, AK, USA, 23–26 May 2016; Society of Petroleum Engineers: Richardson, TX, USA, 2016.

5. Ampomah, W.; Balch, R.S.; Grigg, R.B.; Will, R.; Dai, Z.; White, M.D. Farnsworth Field CO_2-EOR project: Performance Case History. In Proceedings of the SPE Improved Oil Recovery Conference, Tulsa, OK, USA, 11–13 April 2016; Society of Petroleum Engineers: Richardson, TX, USA, 2016.
6. U.S. Department of Energy; Office of Basic Energy Sciences. Basic Research Needs for Geosciences: Facilitating 21st Century Energy Systems. Report from the Workshop Held 21–23 February 2007. Available online: https://www.osti.gov/servlets/purl/935430 (accessed on 8 August 2021).
7. Heath, J. Multi-Scale Petrography and Fluid Dynamics of Caprocks Associated with Geologic CO_2 Storage. Ph.D. Dissertation, New Mexico Institute of Mining and Technology, Socorro, NM, USA, 2010; 410p.
8. Puckette, J.; Al-Shaieb, Z.; Van Evera, E. Sequence Stratigraphy, Lithofacies, and Reservoir Quality, Upper Morrow Sandstones, Northwestern Shelf, Anadarko Basin. In *Morrow and Springer in the Southern Midcontinent, 2005 Symposium*; Andrews, R.D., Ed.; Oklahoma Geological Survey Circular, Oklahoma Geological Survey: Norman, OK, USA, 2008; pp. 81–97.
9. Rose-Coss, D. A Refined Depositional Sequence Stratigraphic and Structural Model for the Reservoir and Caprock Intervals at the Farnsworth Unit, Ochiltree County, TX. Master's Thesis, New Mexico Institute of Mining and Technology, Socorro, NM, USA, 2017; 258p.
10. Trujillo, N. Influence of Lithology and Diagenesis on Mechanical and Sealing Properties of the Thirteen Finger Limestone and Upper Morrow Shale, Farnsworth Unit, Ochiltree County, Texas. Master's Thesis, New Mexico Institute of Mining and Technology, Socorro, NM, USA, 2017; 83p.
11. Balch, R.; McPherson, B. Phase III Demonstration: Farnsworth Unit, Mastering the Subsurface through Technology Innovation and Collaboration. In Proceedings of the Carbon Storage and Oil and Natural Gas Technologies Review Meeting, Pittsburgh, PA, USA, 16–18 August 2016.
12. Hutton, A.C. Geophysical Modeling and Structural Interpretation of a 3D Reflection Seismic Survey in Farnesworth Unit, TX. Master's Thesis, New Mexico Institute of Mining and Technology, Socorro, NM, USA, 2015; 101p.
13. Ross, C.A.; Ross, J.P. Late Paleozoic transgressive-regressive deposition. *Soc. Econ. Paleontol. Mineral. Spec. Publ.* **1988**, *42*, 227–247.
14. Higley, D.K. Compiler, 2014, Petroleum systems and assessment of undiscovered oil and gas in the Anadarko Basin Province, Colorado, Kansas, Oklahoma, and Texas—USGS Province 58: U.S. Geological Survey Digital Data Series DDS–69–EE, 327 p. 8 pls. [CrossRef]
15. Perry, W.J. *Tectonic Evolution of the Anadarko Basin Region, Oklahoma*; U.S. Geological Survey Bulletin: Norman, OK, USA, 1989; Volume 1866, p. 19.
16. Cather, M.; Rose-Coss, D.; Gallagher, S.; Trujillo, N.; Cather, S.; Hollingworth, S.; Mozley, P.; Leary, R. Deposition, diagenesis, and sequence stratigraphy of the Pennsylvanian Morrowan and Atokan Intervals at Farnsworth Unit. *Energies* **2021**, *14*, 1057. [CrossRef]
17. Evans, J.L. Major Structural and Stratigraphic Features of the Anadarko Basin. In *Pennsylvanian Sandstones of the Mid-Continent*; Hyne, N.J., Ed.; Tulsa Geological Society, Special Publication: Tulsa, OK, USA, 1979; Volume 1, pp. 97–113.
18. Munson, T.W. Depositional, Diagenetic, and Production History of the Upper Morrowan Buckhaults Sandstone, Farnsworth Field, Ochiltree County Texas. Master's Thesis, West Texas University, Canyon, TX, USA, 1988; 117p.
19. Brewer, J.A.; Good, R.; Oliver, J.E.; Brown, L.D.; Kaufman, S. COCORP profiling across the Southern Oklahoma Aulacogen: Overthrusting of the Wichita Mountains and compression within the Anadarko Basin. *Geology* **1983**, *11*, 109–114. [CrossRef]
20. Ye, H.; Royden, L.; Burchfiel, C.; Schuepbach, M. Late Paleozoic deformation of interior North America: The greater ancestral Rocky Mountains. *AAPG Bull.* **1996**, *80*, 1397–1432.
21. Petriello, J.; Marino, S.; Suarez-Rivera, R.; Handwerger, D.A.; Herring, S.; Woodruff, W.; Stevens, K. Integration of Quantitative Rock Classification with Core-Based Geologic Studies: Improved Regional-Scale Modeling and Efficient Exploration of Tight Shale Plays. In *167048-MS, Proceedings of the SPE Unconventional Resources Conference and Exhibition-Asia Pacific, Brisbane, Australia, 11–13 November 2013*; Society of Petroleum Engineers: Richardson, TX, USA, 2013.
22. Dullien, F.A.L. *Porous Media: Fluid Transport and Pore Structure*, 2nd ed.; Academic Press: San Diego, CA, USA, 1992; 574p.
23. Dewhurst, D.N.; Jones, R.M.; Raven, M.D. Microstructural and petrophysical characterization of Muderong Shale: Application to top seal risking. *Pet. Geosci.* **2002**, *8*, 371–383. [CrossRef]
24. Ingram, G.M.; Urai, J.L.; Naylor, M.A. Sealing processes and top seal assessment. *Nor. Pet. Soc. Spec. Publ.* **1997**, *7*, 165–174.
25. Heath, J.E.; Dewers, T.A.; McPherson, B.J.O.L.; Petrusak, R.; Chidsey, T., Jr.; Rinehart, A.; Mozley, P. Pore networks in continental and marine mudstones: Characteristics and controls on sealing behavior. *Geosphere* **2011**, *7*, 429–454. [CrossRef]
26. Iglauer, S.; Salamah, A.; Sarmadivaleh, M.; Liu, K.; Phan, C. Contamination of silica surfaces: Impact on water-CO_2-quartz and glass contact angle measurements. *Int. J. Greenh. Gas Control* **2014**, *22*, 325–328. [CrossRef]
27. Osenbrück, K.; Lippmann, J.; Sonntag, C. Dating very old pore waters in impermeable rocks by noble gas isotopes. *Geochim. Cosmochim. Acta* **1998**, *62*, 3041–3045. [CrossRef]
28. Hendry, M.J.; Kotzer, T.G.; Solomon, D.K. Sources of radiogenic helium in a clay till aquitard and its use to evaluate the timing of geologic events. *Geochim. Cosmochim. Acta* **2005**, *69*, 475–483. [CrossRef]
29. Sonnenberg, S.A.; Shannon, L.T.; Rader, K.; Von Drehle, W.F.; Martin, G.W. Regional Structure and Stratigraphy of the Morrowan Series, Southeast Colorado and Adsjacent Areas. In *Morrow Sandstones of Southeast Colorado and Adjacent Areas*; Rocky Mountain Association of Geologists Guidebook: Denver, CO, USA, 1990; pp. 37–50.

30. Bagley, M.E.; Puckette, J.; Boardman, D.; Watney, W.L. Lithofacies and reservoir assessment for the Thirteen Finger Limestone, Hugoton Embayment, Kansas. *AAPG Search Discov.* **2001**, *50513*. Available online: https://www.searchanddiscovery.com/pdfz/documents/2011/50513bagley/ndx_bagley.pdf.html (accessed on 9 February 2021).

31. Heath, J.; Dewers, T.; Cather, M. Core analysis e-reports produced for the Southwest Regional Partnership on Carbon Sequestration for Wells 13-10A, 13-14, and 32-8 of the Farnsworth Unit, Texas. 2021. Available online: https://edx.netl.doe.gov/dataset/core-analysis-e-reports-for-the-swp (accessed on 8 August 2021). [CrossRef]

32. Suarez-Rivera, R.; Handwerger, D.; Rodriguez-Herrera, A.; Herring, S.; Stevens, K.; Vaaland Dahl, G.; Borgos, H.; Marino, S.; Paddock, D. Development of a heterogeneous Earth model in unconventional reservoirs for early assessment of reservoir potential, ARMA 13-667. In Proceedings of the 47th US Rock Mechanics/Geomechanics Symposium, San Francisco, CA, USA, 23–26 June 2013.

33. Suarez-Rivera, R.; Chertov, M.; Willberg, D.; Green, S.; Keller, J. Understanding Permeability Measurements in Tight Shales Promotes Enhanced Determination of Reservoir Quality. In *SPE 162816-MS, Proceedings of the CURC 2012, Calgary, AB, Canada, 30 October–1 November 2012*; Society of Petroleum Engineers: Richardson, TX, USA, 2013.

34. Macquaker, J.H.S.; Adams, A.E. Maximizing information from fine-grained sedimentary rocks: An inclusive nomenclature for mudstones. *J. Sediment. Res.* **2003**, *73*, 735–744. [CrossRef]

35. Dehandschutter, B.; Vandycke, S.; Sintubin, M.; Vandenberghe, N.; Wouters, L. Brittle fractures and ductile shear bands in argillaceous sediment: Inferences from Oligocene Boom Clay (Belgium). *J. Struct. Geol.* **2005**, *27*, 1095–1112. [CrossRef]

36. Lucier, A.; Zoback, M.; Gupta, N.; Ramakrishnan, T.S. Geomechanical aspects of CO_2 sequestration in a deep saline reservoir in the Ohio River Valley region. *Environ. Geosci.* **2006**, *13*, 85–103. [CrossRef]

37. Lucier, A.; Zoback, M. Assessing the economic feasibility of regional deep saline aquifer CO_2 injection and storage: A geomechanics-based workflow applied to the Rose Run sandstone in Eastern Ohio, USA. *Int. J. Greenh. Gas Control* **2008**, *2*, 230–247. [CrossRef]

38. Snee, J.E.L.; Zoback, M.D. State of stress in Texas: Implications for induced seismicity. *Geophys. Res. Lett.* **2016**, *43*, 10208–10214. [CrossRef]

39. Ballentine, C.J.; Burgess, R.; Marty, B. Tracing Fluid origin, transport and interaction in the crust. In *Noble Gases in Geochemistry and Cosmochemistry: Reviews in Mineralogy and Geochemistry*; Porcelli, D., Ballentine, C.J., Wieler, R., Eds.; Mineralogical Society of America: Washington, DC, USA, 2002; Volume 47, pp. 539–614.

40. Prinzhofer, A. Noble Gas in Oil and Gas Accumulations. In *The Noble Gases as Geochemical Tracers, Advances in Isotope Geochemistry*; Burnard, P., Ed.; Springer: New York, NY, USA, 2013; pp. 225–247.

41. Xiao, T.; Xu, H.; Moodie, N.; Esser, R.; Jia, W.; Zheng, L.; Rutqvist, J.; McPherson, B. Chemical-mechanical impacts of CO_2 intrusion into heterogeneous caprock. *Water Resour. Res.* **2020**, *56*, e2020WR027193. [CrossRef]

Article

Probabilistic Assessment and Uncertainty Analysis of CO_2 Storage Capacity of the Morrow B Sandstone—Farnsworth Field Unit

Jonathan Asante [1,*], William Ampomah [1,*], Dylan Rose-Coss [2], Martha Cather [1] and Robert Balch [1]

[1] Department of Petroleum and Natural Gas Engineering, Faculty of Petroleum Engineering, New Mexico Tech Campus, Socorro, NM 87801, USA; Martha.Cather@nmt.edu (M.C.); robert.balch@nmt.edu (R.B.)

[2] New Mexico Oil Conservation Division, Santa Fe, NM 87801, USA; DylanH.Rose-Coss@state.nm.us

[*] Correspondence: jonathan.asante@student.nmt.edu (J.A.); William.ampomah@nmt.edu (W.A.)

Abstract: This paper presents probabilistic methods to estimate the quantity of carbon dioxide (CO_2) that can be stored in a mature oil reservoir and analyzes the uncertainties associated with the estimation. This work uses data from the Farnsworth Field Unit (FWU), Ochiltree County, Texas, which is currently undergoing a tertiary recovery process. The input parameters are determined from seismic, core, and fluid analyses. The results of the estimation of the CO_2 storage capacity of the reservoir are presented with both expectation curve and log probability plot. The expectation curve provides a range of possible outcomes such as the P90, P50, and P10. The deterministic value is calculated as the statistical mean of the storage capacity. The coefficient of variation and the uncertainty index, P10/P90, is used to analyze the overall uncertainty of the estimations. A relative impact plot is developed to analyze the sensitivity of the input parameters towards the total uncertainty and compared with Monte Carlo. In comparison to the Monte Carlo method, the results are practically the same. The probabilistic technique presented in this paper can be applied in different geological settings as well as other engineering applications.

Keywords: carbon dioxide storage; storage efficiency factor; probabilistic; expectation curve; Monte Carlo

Citation: Asante, J.; Ampomah, W.; Rose-Coss, D.; Cather, M.; Balch, R. Probabilistic Assessment and Uncertainty Analysis of CO_2 Storage Capacity of the Morrow B Sandstone—Farnsworth Field Unit. *Energies* **2021**, *14*, 7765. https://doi.org/10.3390/en14227765

Academic Editor: Alireza Nouri

Received: 2 September 2021
Accepted: 9 November 2021
Published: 19 November 2021

Publisher's Note: MDPI stays neutral with regard to jurisdictional claims in published maps and institutional affiliations.

1. Introduction

In recent decades, global warming as a result of the greenhouse gas effect has become the forefront of every discussion worldwide [1–4], and it signifies how much this unabated issue is of concern to the inhabitants of the world. To address this issue, this study focuses on the ways in which the greenhouse gas effect can be reduced by capturing, utilizing, and storing anthropogenic CO_2 gas in geologic formations.

From the literature, CO_2 can be stored in deep saline formations, unmineable coal beds, and oil and gas depleted reservoirs [5–8]. In analyzing these geologic formations, coal beds that may serve as good candidates for geological storage of CO_2 are those in which the coal is unlikely to be mined in the future and which have sufficient permeability. This option for CO_2 storage is still under the demonstration phase [9]. CO_2 storage in hydrocarbon-bearing reservoirs and deep saline formations is normally required to be at a depth exceeding about 2652 ft [9]. At this depth, the ambient temperature and pressure will usually cause the CO_2 to exist in the liquid or supercritical state. At these conditions, the density of the CO_2 varies from 50% to 80% of that of water [9].

The liquid CO_2 behaves as some crude oils do and tends to rise due to buoyant forces. The presence of a high-quality cap rock is required to impede the upward migration of the liquid CO_2 and ensure that it remains trapped underground. When the CO_2 is injected into the reservoir, it compresses and fills the void spaces available by partly displacing the in-situ fluids. In a hydrocarbon depleted reservoir, the displacement of in situ fluids results in improved oil recovery and also creates a large volume of space for CO_2 storage. Generally, potential storage volumes are estimated to range from a few percent to over 30% of the rock bulk volume [9]. Upon injection of the CO_2 into the underground storage, the fraction of the CO_2 that would be stored depends on both the physical and geochemical trapping mechanisms. The physical trapping mechanism includes a cap rock that prevents the upwards movement of the liquid CO_2 and capillary forces that keep the liquid CO_2 in the tiny pore spaces.

Nonetheless, porosity can remain open below the caprock, and this allows for the lateral migration of the CO_2. For such conditions, additional trapping mechanisms are necessary for continual CO_2 entrapment. The geochemical trapping involves the reaction of the injection of CO_2 with in-situ fluids and rock. The CO_2 dissolves in the formation water, which becomes denser and sinks further into the formation. Also, injection of CO_2 increases the reservoir pressure and, in the presence of formation water, there is the tendency for hydrate formation to occur at low temperatures [10]. The formation of hydrates can impact the CO_2 storage by reducing the capacity through flow blockage or improve the storage capacity through the trapping of the CO_2 in an ice-like solid. Hence, a fraction of the injected CO_2 is being stored over millions of years.

Methods of estimating the storage capacity for different types of storage can vary greatly. Deriving an estimate for a saline aquifer can be complex because as many as four trapping mechanisms may be present at different times or rates, or may all be operating simultaneously [6]. Estimating capacity for an oil or gas reservoir may be easier, partly because of the greater availability of data acquired during exploration and production. The quantification of the CO_2 storage capacity is based on the fundamental assumption that the injected CO_2 will fully occupy the void spaces created by the produced oil. This assumption holds for reservoirs not in hydrodynamic contact with an aquifer or that have not undergone flooding during secondary and tertiary recovery [6]. Based on this assumption, the prospective CO_2 storage estimation for any reservoir may be assessed using two techniques: the production and volumetric approaches.

The production-based method utilizes the estimated ultimate recovery (EUR) of a reservoir and assumes CO_2 can replace its equivalence. Using the production approach to estimate CO_2 storage capacity for a geologic formation is not a common procedure. A few researchers have suggested methodologies to be used for the production approach. Among them is Frailey [11], who has developed models analogous to decline curve analysis (DCA) and mass balance (for gas reservoirs), used in the petroleum industry, to estimate storage capacity for a saline aquifer formation. The DCA model is valid for an exponential decline under pseudo-steady state conditions. The production-based approach is associated with several uncertainties. The amount of CO_2 injected is not a direct replacement for the amount of hydrocarbon produced. The differences in molecular size, shapes, chemistry, and adsorption properties between injected CO_2 and methane (CH_4), a major component of produced gas, also contribute to the uncertainty [12].

The use of the volumetric approach to quantify the storage capacity of a geologic formation is common. Many, including the United States Department of Energy (USDOE), have suggested a formulation to estimate the storage capacity for different geologic formations. The volumetric approach for the quantification of the CO_2 storage capacity is based on the methodology used in the industry to calculate the oil initially in place (OIIP) or gas initially in place (GIIP) [7]. It requires the product of the area, net thickness, average effective porosity, formation volume factor, and in situ CO_2 density [7]. Frailey [11] also presented another approach using compressibility to estimate the storage capacity of a geologic formation. Injecting CO_2 into an underground reservoir causes the in-situ fluid (water) to compress further into the micropores as the pore pressure increases, and this, in turn, increases the effective pore volume. The addition of these volume changes is the extra volume that the CO_2 can occupy. This is only valid for a closed reservoir. In comparison, the uncertainty of the volumetric approach stems from the original estimation of the rock volume and the pore space available for CO_2 storage.

The quantification of reservoir properties is not a straightforward approach due to uncertainty accompanied by the vague and imprecise knowledge gained from the interpretation of data acquired from the formation [13]. The reservoir is not uncertain; the uncertainty lies in the ability to fully describe and understand the heterogeneity of the reservoir [14]. A reservoir is influenced by many complex geological processes, such as movements due to plate tectonics, alteration of fluid properties due to uplifts or burials, and precipitation and dissolution of a variety of minerals due to various diagenetic processes, and all contribute to the uncertainty associated with the quantification of a reservoir properties. Hence, to quantify a reservoir using a deterministic method will not yield a good result since there is a range of possible outcomes. This is due to the uncertainties associated with the imprecise data obtained from the technology employed in revealing the information of the reservoir. In fact, the probabilistic method is better in quantifying a reservoir since its model incorporates a range of values and generates a range of possible outcomes. Additionally, this method also provides a viable technique to analyze the uncertainty [15].

The main goal of reservoir uncertainty analysis is to quantify and reduce the total uncertainty to help yield better output. The process requires identifying the factor which contributes most greatly to the total uncertainty. Uncertainty about an estimate needs to be minimized to be the most useful in decision making. Achieving this does not come easily, as the process of reducing uncertainty can be expensive. To improve the outcome of a project to enhance decision-making at a minimal cost, Zee Ma [14] suggests these questions should be addressed: How much data is available? How much data can be available? Moreover, how much data is needed? Addressing these questions improves efficiency while preventing additional costs from data acquisition because uncertainty will always exist regardless of the methodology [15].

The storage efficiency factor is one of the most sensitive parameters which affect the total mass storage of the CO_2 and a poor estimation of this can cause the estimation of reservoir storage capacity to be grossly inaccurate. The storage efficiency factor was first introduced in the regional scale assessment of the storage capacity in the United States and Europe in 2007 [16]. Since then, various authors have delved into the subject using different approaches. Bachu [16] reviewed and analyzed the gaps of different methodologies presented by various authors on the storage efficiency factor. The efficiency factor depends on the rock and fluid properties, CO_2 storage operation, and regulatory constraints. Below are some reviews which illustrate different approaches to estimate the storage efficiency factor.

Brennan [17] determined the residual storage efficiency using pressure, temperature gradients, depth ranges, irreducible water and gas saturation, and relative permeability between CO_2 and the existing pore fluids. The residual efficiency outcomes were mapped against the reservoir depth to generate an efficiency gradient at respective depths. Although this is a promising approach to determine the efficiency factor since it is sensitive to pore geometry, it is computationally expensive.

Park et al. [18] also determined the efficiency factor by flooding a core plug with water and displacing the water with $ScCO_2$. Although the experiment was conducted under reservoir temperature and pressure, the distilled water used is subjective to true output representation of the efficiency factor affected by total dissolved solids (TDS) [17]. Also, Rasmussen et al. [19] show that aged cores better represent the wettability of the reservoir than cleaned cores; hence, this method will not reflect the storage efficiency factor.

For this paper, the storage efficiency factor will be estimated using saturation differences from the relative permeability curve of the Morrow B reservoir as a function of the hydraulic flow unit (HFU).

The main objective of this paper is to improve our existing assessment of the amount of CO_2 that can be stored in the Morrow B sandstone at the Farnsworth Field Unit (FWU) in Ochiltree County, Texas. To achieve the objective, this study will focus on the quantification of the CO_2 storage capacity of the Morrow B reservoir and analyzing the uncertainties associated with it. The estimation and computations conducted are based on data obtained from the Farnsworth Unit located in Texas. Ampomah et al. [20] compared and demonstrated that among the probabilistic techniques—first-order, parametric, and Monte Carlo simulation methods— the parametric method, which is an analytical procedure, can be generated with ease within a spreadsheet and also can be used as a substitute for the Monte Carlo simulation with a minute to no difference in the output. Hence, the parametric method will be used to quantify the storage capacity of the Morrow B reservoir. This paper will greatly contribute to the scope of carbon capture, utilization, and storage (CCUS) research and industrial projects.

2. Methods

2.1. Theory

This section presents the mathematical formulations employed to estimate the CO_2 storage capacity successfully. Although there are many analytical procedures used in the estimation of reserves, the volumetric approach is generally employed in the petroleum industry to estimate reserves. However, the input parameters for the volumetric equation are underlain with many uncertainties, which are dependent on the geologic setting and quality of geologic and engineering data available [21]. Hence, it is expedient not to depend on deterministic computations that provide a single best estimate but rather probabilistic computations, which provide a range of outcomes that reflect the input parameters' underlying uncertainties.

The statistical distribution of input parameters used in mathematical models is an important facet of probabilistic analysis. In the volumetric approach of reserves estimation, the probability density function (PDF) for input parameters can be a normal distribution, triangular, and/or lognormal. Chen [22] presented a new approach for generating statistical input distributions such as mean and standard deviation, from an existing data set. Normal or Gaussian distribution is bell-shaped and symmetric to the mean. The normal distribution is mostly defined by the mean and standard deviation of the input parameter $X \sim N\ (m, s^2)$.

The PDF equation for a random variable X represented by normal distribution is shown in Equation (1). The mean, median, and mode are generally equal for normal distributions due to their symmetrical nature. In situations where the amount of data is limited, the triangular distribution is mostly employed. A triangular distribution is represented by a minimum (a), mean (m), and a maximum (b) to indicate a random variable X ($X \sim$ triangular (a, m, b)). A PDF equation to specify a triangular distribution of a random variable X is shown in Equation (2). A random function is lognormally distributed when the logarithms of the values of X are normally distributed. The lognormal distribution is mostly asymmetric and represented by $\ln(X) \sim N\ (\alpha, \beta^2)$. For a lognormal distribution, the degree of skewness increases with an increase in standard deviation. For a minute standard deviation, the lognormal distribution behaves like that of normal distribution. Equation (3) illustrates a PDF of the lognormal distribution. Normal and lognormal distributions are used in this paper.

$$f(x) = \frac{1}{\sqrt{2\pi s}} \exp\left[-\frac{1}{2}\left(\frac{x-m}{s}\right)^2\right] \tag{1}$$

$$f(x) = \begin{cases} \frac{2(x-a)}{(b-a)(m-a)} & \text{where } a < x < m \\ \frac{2(b-x)}{(b-a)(b-m)} & \text{where } m \leq x < b \end{cases} \tag{2}$$

$$f(x) = \frac{1}{x\sqrt{2\pi\beta}} \exp\left[-\frac{1}{2}\left(\frac{In(x)-\alpha}{\beta}\right)^2\right] \tag{3}$$

The volumetric equation proposed by USDOE [8] for the determination of CO_2 storage for an oil and gas depleted reservoir was altered and used for this project. Equation (4) shows the altered equation utilized for this project.

$$GCO_2 = A \times h_{net} \times \phi \times S_{hi} \; X \; \rho_{CO2} \; X \; E_{oil/gas} \qquad (4)$$

where GCO_2 = mass CO_2 storage capacity, MMtons; A = area, acre; h_{net} = net thickness, ft; ϕ = porosity, fraction; S_{hi} = hydrocarbon saturation, fraction; and $E_{oil/gas}$ = CO_2 storage efficiency factor.

The efficiency factor was estimated by using the saturation differences of water, oil, and gas. The fluids' saturation was determined from water–oil and oil–gas relative permeability curves from the various HFUs of the Farnsworth Field examined by Rasmussen et al. [19]. Relative permeability is a function of pore geometry, as well as mineralogy of the rocks in and surrounding pores, and rock properties that influence pore geometry also influence relative permeability [23]. Hence, wettability, fluid distribution, and fluid saturation history also affect relative permeability. This makes the use of relative permeability an even better method to determine the efficiency factor since such factors also influence the efficiency factor.

Figure 1 illustrates both water–oil and oil–gas relative permeability curves with the water and oil being the wetting phase, respectively. To determine the efficiency factor, it was assumed for the oil–gas relative permeability curve that the water saturation in the reservoir rock does not exceed its irreducible value. That is, water is present but immobile; it only reduces the void spaces of the reservoir. Equation (5) was used in estimating the efficiency factor.

$$E = \frac{1 - Swc - Sorg - Sgc}{1 - Swc - Sor} \qquad (5)$$

where E = efficiency factor; Swc = the connate water saturation; $Sorg$ = the residual oil saturation in the presence of gas; Sgc = the critical gas saturation; and Sor = the residual oil saturation (before gas injection). From Equation (5), the efficiency factor for each of the HFUs was determined and the mean and standard deviation were computed for use in the parametric method.

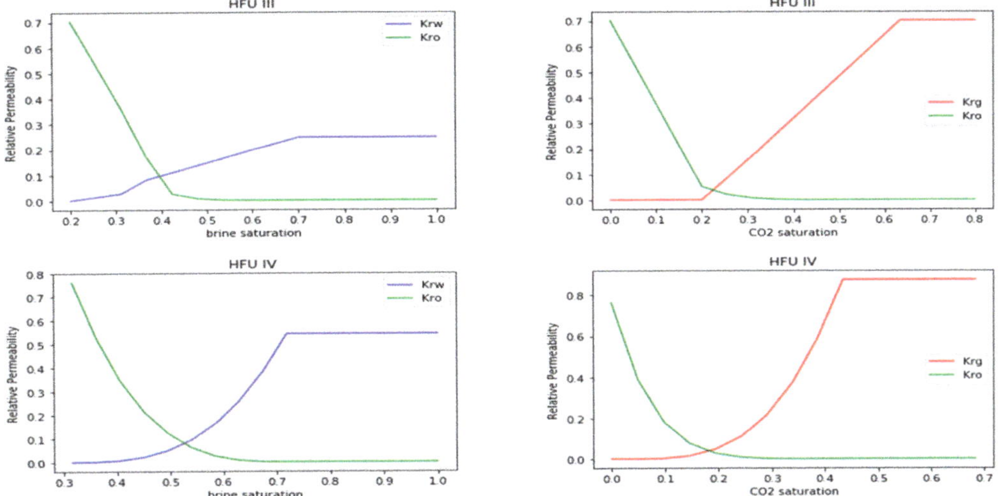

Figure 1. *Cont.*

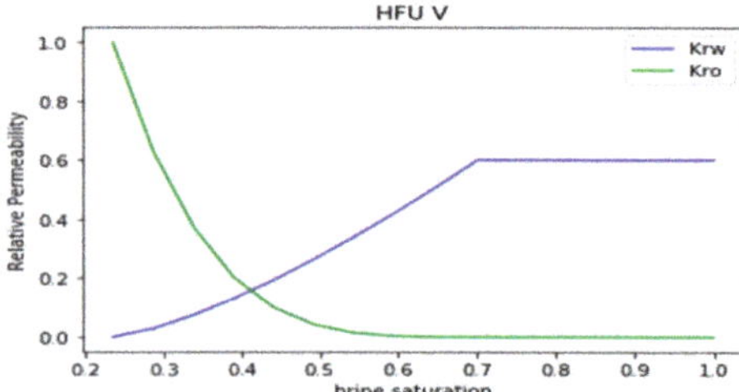
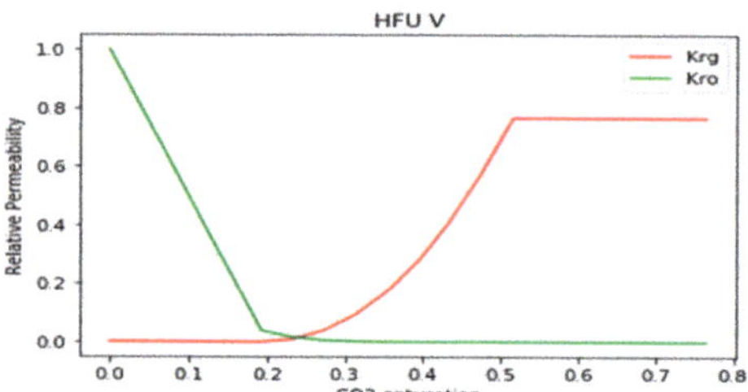

Figure 1. Relative permeability curves of water(brine)-oil (**left**) and gas (CO_2)-oil (**right**) created as a function of HFU of the Farnsworth Field Unit. HFU III-V were considered for both water-oil and oil-gas relative permeability curves. HFU III-V represents intermediate to highest permeability of Morrow B sandstone at a given porosity.

2.2. Parametric Method

The parametric method is an analytical procedure that uses common statistical information such as mean (m_j) and standard deviation (s_j) and easily quantifies uncertainty. The mathematical procedures are illustrated in Equations (6)–(11). The first step in parametric uncertainty quantification is to transform the mean (m_j) and standard deviation (s_j) from the original space to the lognormal mean (α_j) and lognormal standard deviation (β_j) as shown in Equations (7) and (8), respectively. The output lognormal mean (α_Ω) and lognormal variance ($\beta_\Omega{}^2$) of hydrocarbon initially in place (HCIIP) is computed by summing input lognormal mean (α_j) and lognormal variance ($\beta_j{}^2$) 8s illustrated in Equations (8) and (9), respectively. Equation (14) is used to compute the relative impact of individual input. A flow chart illustrating the parametric method is shown in Figure 2.

$$V_j = \frac{s_j}{m_j}, \text{ where } V \text{ is the coefficient of variation} \tag{6}$$

$$\beta^2{}_j = In\left(1 + V^2{}_j\right), \text{ where } V^2 \text{ is computed by equation} \tag{7}$$

$$\alpha_j = In(m_j) - 0.5 \times \beta^2{}_j \tag{8}$$

$$\alpha_\Omega = \sum \alpha_j \tag{9}$$

$$\beta_\Omega^2 = \sum \beta_j^2 \tag{10}$$

$$\text{Relative impact, } \mu_j = \frac{\beta_j^2}{\sum \beta_j^2} \tag{11}$$

A lognormal output of mean (α_Ω) and standard deviation (β_Ω) are used to generate various statistical measurements. These statistical outputs are categorized into three main divisions that include measures of location, measures of variability, and measures of shape. For example, P90, P50, P10, variance, and P10-to-P90 ratio are shown in Equations (12)–(16), respectively [24].

$$P90(90\% \text{ probability at least this value}) = \exp(\alpha - 1.2816\beta) \tag{12}$$

$$P50(50\% \text{ probability at least this value}) = \exp(\alpha) \tag{13}$$

$$P10(10\% \text{ probability at least this value}) = \exp(\alpha + 1.2816\beta) \tag{14}$$

$$\text{Variance, } s^2 = \left[\exp\left(2\alpha + \beta^2\right)\right]\left[\exp\left(\beta^2\right) - 1\right] \tag{15}$$

$$P10 - to - P90 \text{ ratio (uncertainty measure)} = \exp(2 \times 1.2816\beta) \tag{16}$$

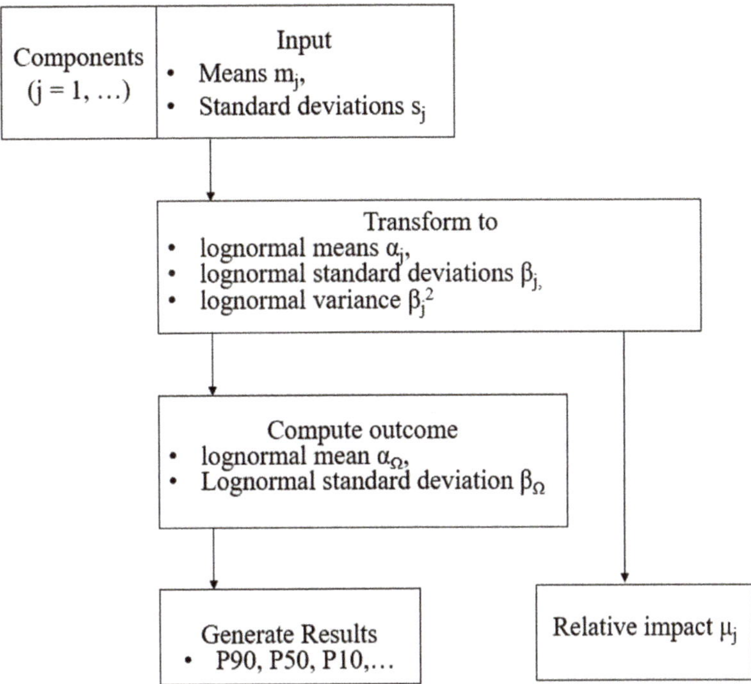

Figure 2. A flow chart showing a parametric procedure in probabilistic reserves estimation.

2.3. Plotting Tools

In this work, two approaches were utilized to present the probabilistic distribution (expectation plots) of HCIIP estimation. They were the "manual" approach and the Microsoft Excel built-in function. The manual approach is generated with Equation (17). The z_c in the equation can be interpreted as the number of lognormal standard deviations. It is obvious from the equation that one needs only the lognormal mean (α) and lognormal standard deviation (β). The expectation curve becomes the exceedance probabilities (at various confidence levels) vs. Pc (GCO_2).

$$P_c = \exp(\alpha + z_c \beta) = P50 \times \exp(z_c \beta) \tag{17}$$

Microsoft Excel has two built-in functions that can be utilized to generate expectation curves (e.g., GCO_2) at various confidence levels. These are LOGNORM.INV (x, lognormal mean, lognormal standard deviation) and LOGNORM.INV (p, lognormal mean, lognormal standard deviation). All input variables (x, α, β) > 0. The "p" is classified as a cumulative probability ($0 \leq p \leq 1$). A plot of exceedance probabilities ($1 - p$) vs. x is the expectation curve from these two functions. Outcomes from parametric studies can also be illustrated by using a "log probability plot". This straight-line plot elaborates the relationships between key confidence levels.

3. Farnsworth Field Unit Static Model

The SWP has developed several generational geological models over the years, revolving around the acquisition of a new dataset. Ampomah et al. [25] presented the second-generation geological model, which was based on preliminary seismic interpretations. In this work, reservoir properties such as porosity and permeability distributions were modeled with geostatistical methods using data from core and well logs. In 2016,

Ampomah et al. [26] presented an improved geological model that modeled the field permeability distribution by developing correlations that separated the east and west sides of the field, which appear to have different behaviors. Detailed geological descriptions for FWU have been presented in several publications [1,25–32]. Ross-Coss et al. [33] presented an updated geological model for the SWP project, which significantly improved the site characterization efforts at FWU. This work developed a hydraulic flow unit (HFU) workflow based on pore throat aperture. The approach delineated the target reservoir into eight geological units. The analysis showed that diagenetic processes highly influenced reservoir property distribution as compared to depositional processes. The HFU methodology has been the basis for subsequent model improvements within the Morrow B Sand at FWU.

Balch et al. [34] presented the latest static model, which forms the basis for this work. This model extends from the vertical extent of the model to units above and below the reservoir and cap rock. There are 14 geologic horizons. The improved static model has a grid distribution of 189 × 179 × 106 with a total 3D grid of 3.6 million cells. Each grid block has a dimension of 100 × 100 ft. Figure 3 shows the geological zones used for the structural framework. This model does not include any previously mapped faults as ongoing work has thrown existence of such faults into doubt. Reservoir properties for this latest static model have been mapped from the Thirteen Finger Limestone (one of the caprocks) to the base of the Morrow B sandstone. The property modeling workflow applied to each formation depended on the data available and the formation characteristics. Integration methods included artificial neural network facies identification from well logs and cores, spatial variogram analysis, discrete and continuous distributions, and co-simulation with elastic inversion properties. Spatial variograms from 3D seismic acoustic impedance were used to condition property modeling within the upper layers due to limited well-log information. The prominent caprock layers within this upper section include the Thirteen Finger Limestone and Morrow Shale.

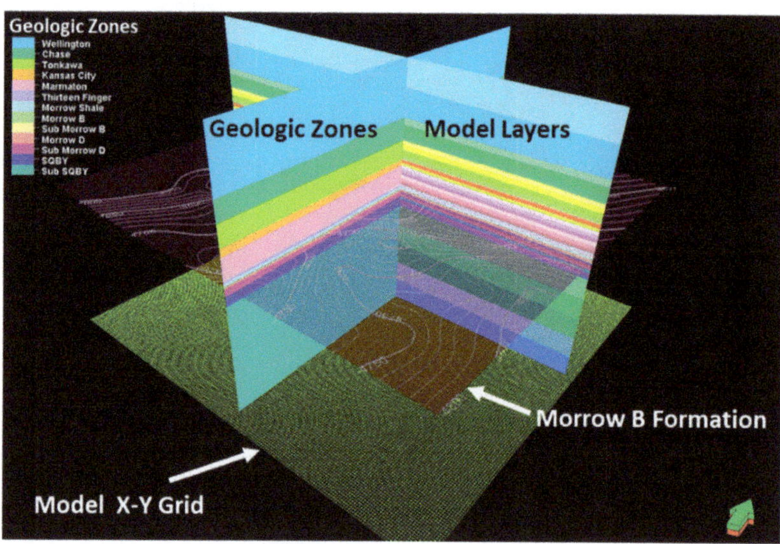

Figure 3. Geologic model framework elements.

The Morrow B formation was the target reservoir; it ranges in depth from 7550 ft to 7950 ft with an average dip of less than one degree. It has been interpreted as a fluvial incised-valley deposit [29,30]. The thickness maps were used to estimate formation volume and architecture. The Morrow B has an average thickness of 22 ft on the west side of the field and 52 ft thick on the east side, with a mean thickness for the field of 24.47 ft and a standard deviation of 11.7 ft [33]. Figure 4 shows the net thickness map for FWU within the field unit boundary. The thickest sands are restricted to the middle of the field, whereas the sands thin along the periphery. Thicker accumulations occur in discreet areas such as to the west of well 13-10A and to the east of well 32-8. The established HFU methodology based on the Winland R35 was used to model porosity and permeability distribution within the Morrow B Sand. Fifty-one wells with core porosity and permeability were utilized in the HFU workflow, which quantified the heterogeneity within the target zone resulting in eight (8) distinct delineations.

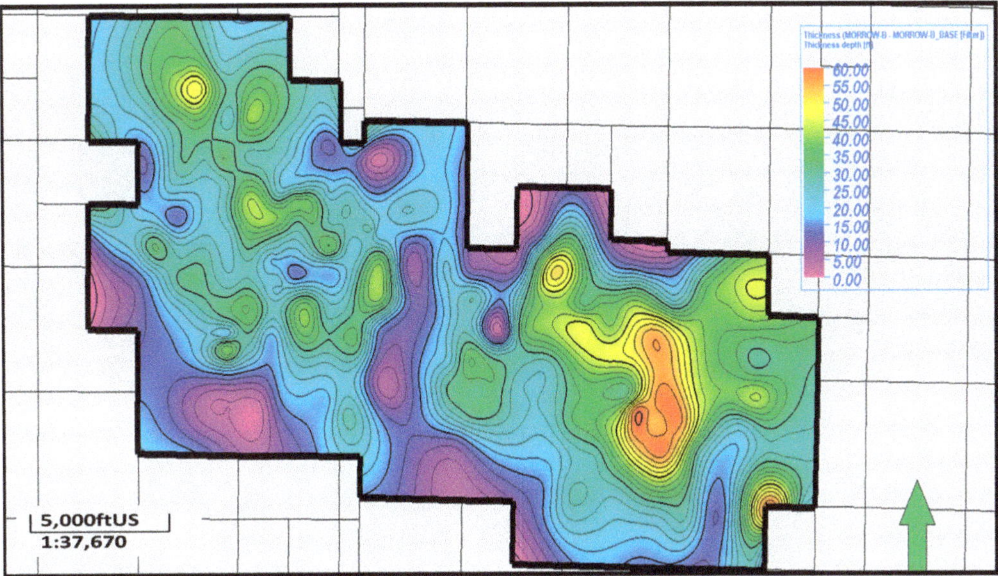

Figure 4. Net sand thickness within Morrow B reservoir based on well top information. The mean of the net sand is 24.47 ft with a standard deviation of 11.65 ft.

Figure 5 illustrates how the depth and GR log for well 13-10A correspond to bedding and core description. Subsequent columns illustrate log, core, and thin section porosity along with log and core permeability. Porosity and permeability values were used to compute the LogR35 log. Thin section mineralogy and pore type summation columns were created based on point count data. Type thin sections for HFU1, HFU3, and HFU8 were placed in relative order to display where within the reservoir those particular HFUs were located, and a depiction of the pore type associated with that HFU. Thin Section 1 in Figure 5 is representative of HFU3. Here, intergranular space is almost filled with siderite cement and detrital clay. However, there is a large amount of intragranular macroporosity associated with grain dissolution. This hydraulic unit's relatively high value for porosity does not correspond to a high permeability value because of the lower likelihood of intragranular space creating interconnected networks. From the thin section derived mineralogy and porosity classes, it becomes apparent that the degree of cementation and the degree of intergranular porosity development are primary controls on porosity and permeability. Figure 6 shows a plot of porosity and log permeability to depict the delineation within the reservoir based on the HFU. Figure 7 shows the 3D distribution of porosity of the target Morrow B reservoir. Figure 8 shows modeled permeability based on the porosity–permeability relationship.

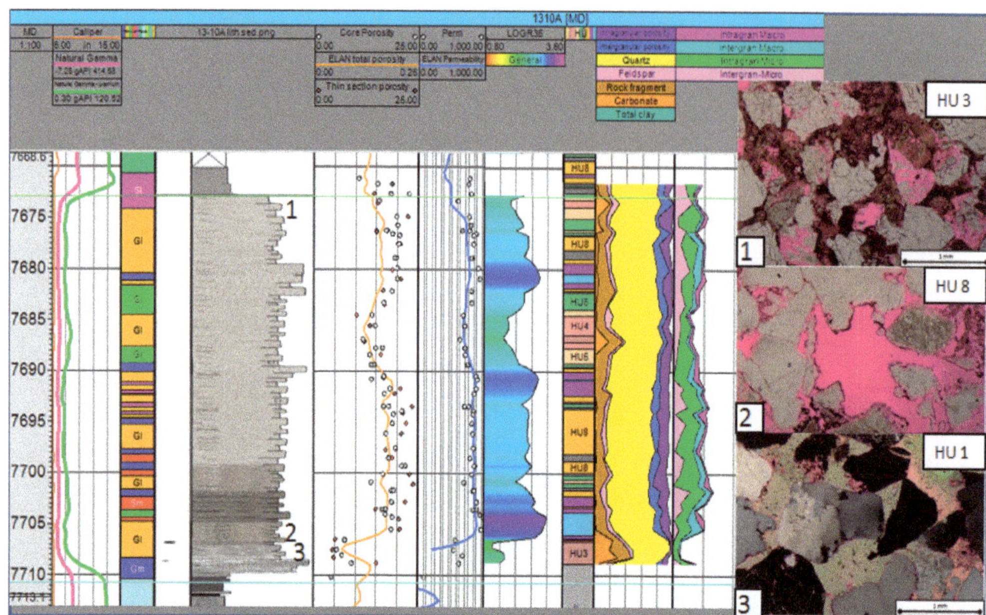

Figure 5. Well section displaying reservoir interval for well 13-10A on west side of field. From left to right columns are depth (MD), gamma, bedding, sedimentary features, core description, porosity, permeability, LogR35, HFU categories, thin section derived mineralogy, thin section pore type, and thin section photos displaying scale bar and HFU (designated HU in this figure). Number on thin section image corresponds to location in core. Pink coloration in thin sections is pink-dyed epoxy in pore space.

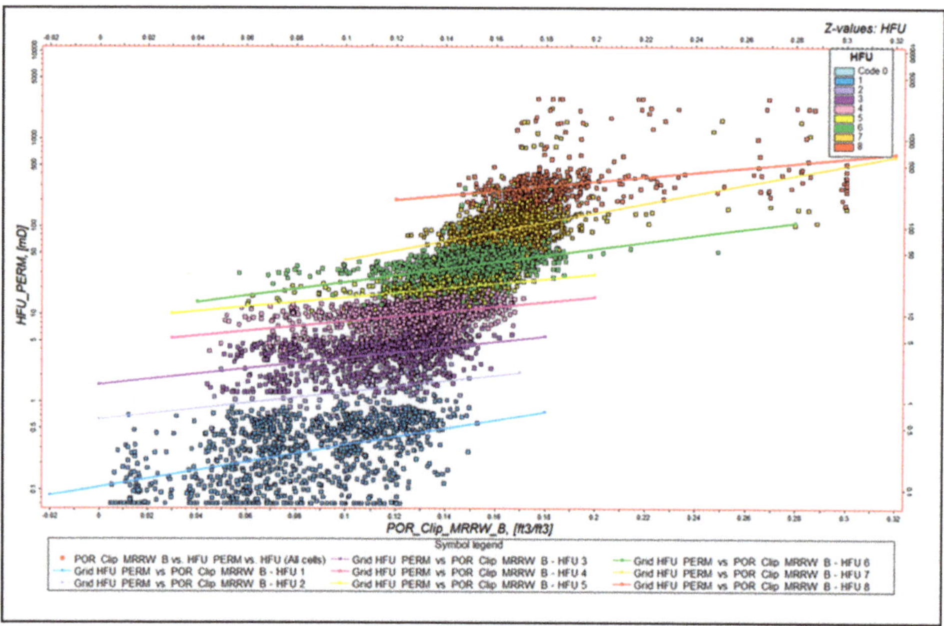

Figure 6. Porosity versus permeability for the 51 cored wells, separated by pore throat size into hydraulic flow units.

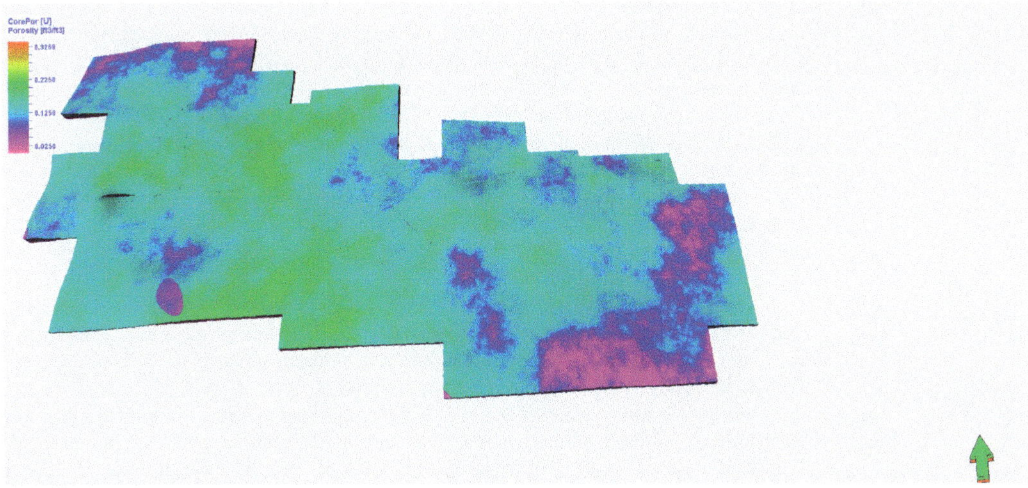

Figure 7. Porosity distribution for top of Morrow B created using Gaussian random function simulation co-kriged using the volumes computed from the HFU property.

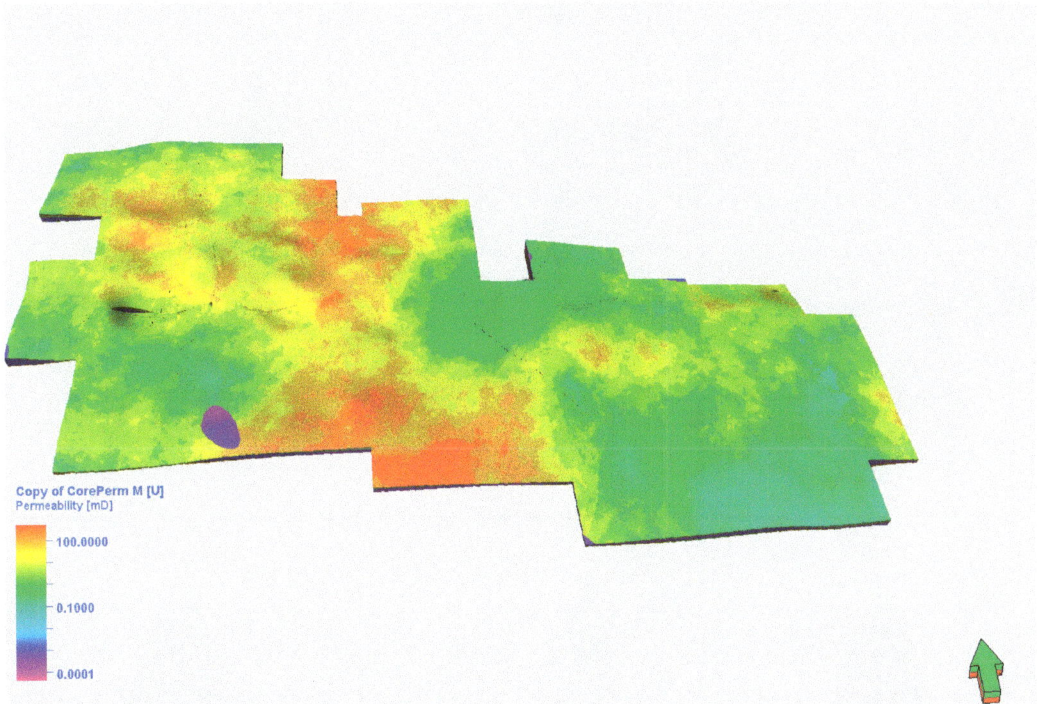

Figure 8. Permeability distribution created by propagating permeability as a function of porosity defined by hydraulic units. Defining permeability in this way ensured that trends followed those predicted by the HFUs and did not simply assign high permeability values to high porosity values.

Volumetric Analysis

Initial fluid saturations for FWU were measured from special core analysis. The mean of oil saturation was 69%, with a standard deviation of 3.45%. The initial reservoir pressure and the temperature were 2217.7 psia at a datum of 4900 ft subsea and 168°F, respectively. The original bubble point pressure was 2073.7 psia, and a miscibility mean pressure range of about 4200–4500 psia was also observed. The reservoir was slightly undersaturated at the time of discovery. The supercritical CO_2 (ScCO_2) density was calculated using the Span and Wagner equation of state [35]. The mean and standard deviation of the ScCO_2 was estimated to be 47.7 lbm/ft^3 and 0.47 lbm/ft^3, respectively. Based on the relative permeability curve of the FWU in conjunction with Equation (5), the mean efficiency factor was 69.9%, with a standard deviation of 22.6%.

The several volumetric reserves' computational parameters mentioned in the above paragraphs, including area, net thickness, porosity, fluid saturation, ScCO_2 density, and storage efficiency factor, were used in analyzing the uncertainty of storage capacity potential for the Morrow B reservoir using the parametric methods. A summary of input parameters is presented in Table 1. All input parameters were assumed to be lognormal distributed, although porosity, saturation, and ScCO_2 density depicted close to normal distribution due to their minute standard deviations, and these are illustrated in Figure 9.

Table 1. Mean and standard deviation of the input parameters used for the analysis.

Parameters	Units	Mean	Standard Deviation
		mj	*sj*
Area, A	Acre	12,652.500	3795.750
Net thickness, hn	ft	24.470	11.650
Porosity, Φ	Frac.	0.141	0.020
Initial oil saturation, Soi	Frac.	0.690	0.036
Efficiency factor	Frac.	0.699	0.226
Density of CO_2	lbm/ft^3	47.700	0.470

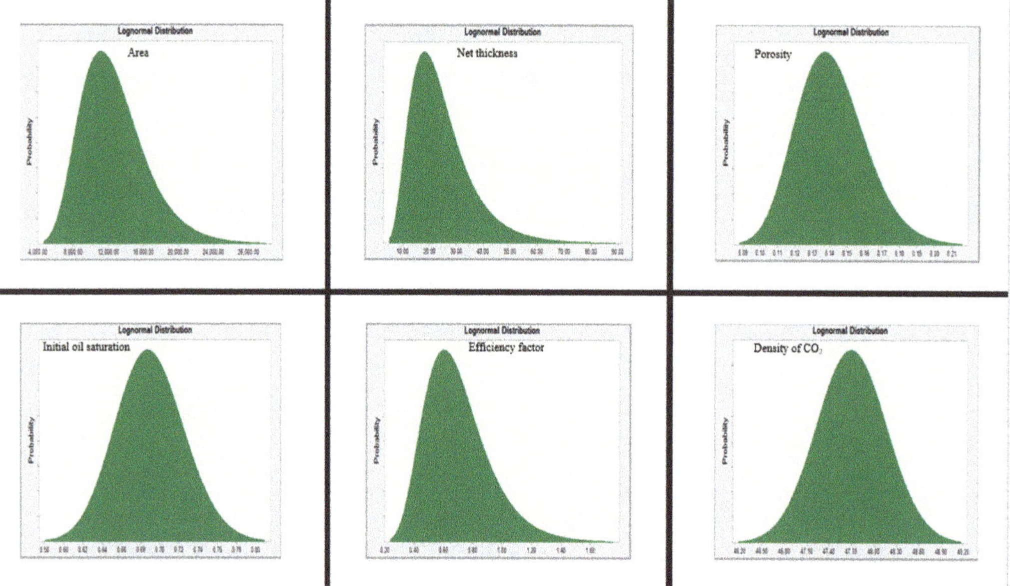

Figure 9. The probability density function distribution of the input parameters generated by the Monte Carlo simulation.

4. Results and Discussion

The main objective was to quantify the storage capacity of CO_2 in the Farnsworth Field Unit using the parametric method and to analyze the contribution of uncertainties of the input parameters to the total uncertainty. The Monte Carlo simulation was also presented to validate the results generated by the parametric method. This paper included the presentation of a new static model of the field and presented a different methodology in the estimation storage efficiency factor.

Most studies and simulations of the FWU have been focused on the west side of the field, since this is the part of the field where the CO EOR project has been conducted. The new model presented in this paper examines the CO_2 storage capacity of both the west and east sides of the FWU, as the east side may eventually have interest for CO_2 EOR or carbon storage. In 2010, the average pressure recorded on the east side was about 4700 psia [27]. The high pressure was a result of a successful waterflood and significantly demonstrates that minimum miscibility pressure could be achieved if CO_2 is injected into this side of the field.

Figure 9 shows the distribution of the input parameters of the FWU. Distribution in nature was assumed to follow the logarithmic distribution. When the deviation of a group of variables is small, the values are closer to each other. Then, the distribution of the variables in the group is normal. However, when the deviation of the variables is large, that is, the values are farther from each other, then the distribution of the variables in the group is skewed. Now, from the outputted distribution of the input parameters, the porosity and the initial oil saturation were normally distributed, but the area, net thickness, and efficiency factor were highly dispersed, which further confirms the high heterogeneity of the formation.

The storage efficiency factor is one of the most uncertain parameters using the volumetric approach in estimating the storage capacity of a geologic formation. It is influenced by both rock and fluid properties. Studies show that the FWU is highly heterogeneous [19,27], and also, due to the alternating water and gas injection, the wettability of the Morrow B reservoir has changed to intermediate wet or mixed wet [19]. The estimation of the storage efficiency factor using the relative permeability curve as a function of HFU better represents the changes that have taken place in the reservoir due to productivity-enhancing techniques. HFUs III-V were considered in the estimation of the storage efficiency factor, as these represent the intermediate to highest permeability of the Morrow B sandstone at a given porosity (Figure 9). The relative permeability generated from these core samples was measured at an increasing order of pressure values from 3000 psi to 4000 psi, which mimics the miscibility pressure of the reservoir. From Equation (5), the denominator represents the theoretical space available for CO_2 while the numerator represents the actual space available for CO_2. Assuming a constant irreducible water saturation (Swc), the storage efficiency factor increases with reducing irreducible oil saturation after CO_2 injection (Sorg) and critical gas saturation (Sgr).

The volumetric approach suggested by the USDOE [8] for oil and gas depleted reservoirs was altered and used to estimate the storage capacity of the Morrow B reservoir. The volumetric analysis uses reservoir rock and reservoir fluid properties to calculate hydrocarbon initially in place and the portion that can be recovered. Estimations of reserves using this approach always result in uncertainties with the output. Based on the degree of uncertainty, reserves are classified as proven (1P or P90), probable (2P or P50), and possible (3P or P10) [36]. As a result, the probabilistic approach was used, which gives details of the entire range of possible outcomes of the estimates instead of a single value like the deterministic approach. Below is a statistical measure illustrating the quantities of the storage capacity of the supercritical CO_2 (GCO2) in the Farnsworth Unit (Morrow B reservoir).

From Table 2, P90 of both the parametric and the Monte Carlo simulation methods yielded about 8 MMtons, indicating a 90% probability that the estimated GCO2 will be at least 8 MMtons or 10% confidence that the estimated GCO2 will be less than 8 MMtons. The P50, which equals about 18 MMtons, is the best estimate in terms of the probabilistic approach, and it signifies a 50% probability that the estimated GCO2 will be at least 18 MMtons. The P10 is the most unlikely estimate in comparison to the P90 and P50. The

P10 for both methods is about 41 MMtons, and it shows a 90% probability that the GCO_2 will be less than 41 MMtons. The deterministic value is computed as the mean, which is about 22 MMtons for both methods. There is a 37.6% probability that at least the estimated GCO_2 will be equal to the deterministic value.

Table 2. Summary of the statistical measure of the storage capacity distribution.

Various Statistical Measures of Storage Distribution	Units	Parametric	Monte Carlo
P90 (90% probability at least this value), $P90 = \exp(\alpha - 1.2816\,\beta)$	MMtons	7.81	7.68
P50 (50% probability at least this value), $P50(m_G) = \exp(\alpha)$	MMtons	17.79	17.63
Mean (arithmetic), $m = \exp(\alpha + 0.5\,\beta^2)$	MMtons	21.87	21.63
P10 (10% probability at least this value), $P10 = \exp(\alpha + 1.2816\,\beta)$	MMtons	40.52	40.58
Variance, $s^2 = [\exp(2\alpha + \beta^2)][\exp(\beta^2) - 1]$	MMtons	244,029,852.89	234,984,702.95
Standard deviation, $s = (\text{variance})^{1/2}$	MMtons	15.62	15.33
Coefficient of variation, $V = [\exp(\beta^2) - 1]^{1/2}$		0.71	0.71
P10-to-P90 ratio, $P10/P90 = \exp(2 \times 1.2816\,\beta)$		5.19	5.28
Skewness, $S = 3V + V^3$		2.51	2.30
Closeness of mean-P50 $[(\text{Mean} - P50)/P50]$, $C = \exp(0.5\,\beta^2) - 1$		0.23	0.23

Figure 10 shows the expectation curve, which illustrates the probabilistic distribution of the CO_2 storage estimation of the Morrow B reservoir. The expectation curve shows the outcome of the estimated storage capacity at different confidence levels. The P99 and P1 on the expectation curve can serve as a good practical boundary for estimating the storage capacity of the Morrow B sandstone. The expectation curve of both the parametric and the Monte Carlo simulation yielded practically the same results. However, 10,000 simulation runs were needed for the Monte Carlo simulation to generate these outputs. Also, for the Monte Carlo simulation, prior knowledge of the results should be known to serve as a guide for the number of simulations runs. Most of the time, the mean value is used. From the expectation curve, it can be deduced that the probability that the GCO_2 will lie between 7.81 and 40.52 MMtons is 80%.

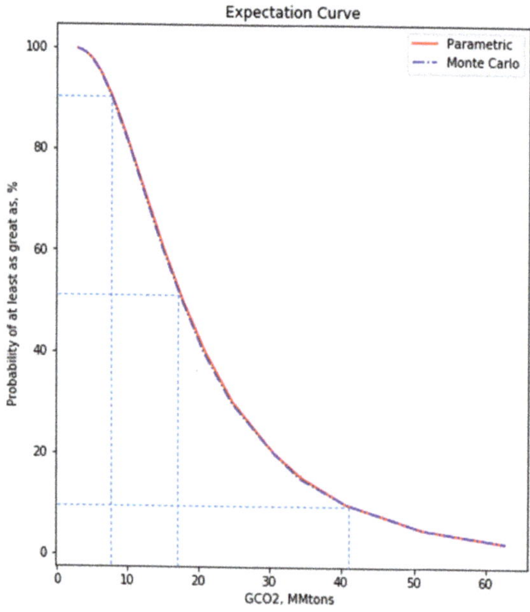

Figure 10. Expectation curve illustrating the CO_2 storage capacity at a different confidence level.

Figure 11 illustrates a log probability plot (P90-P50-P10 straight line). From the log probability plot, P90, P50, and P10 correspond to -1.2816, 0, and 1.2816 on the number of lognormal standard deviations axis, respectively. Hence, in terms of the probability distribution, it is in reverse order to the expectation curve. That is, the P90 comes first, followed by the P50, then the P10 in ascending order. This P90-P50-P10 straight line is similar to the expectation curve because all other statistical measures can be found on this straight line. Hence, the P90-P50-P10 straight line displays the entire range of the estimated distribution.

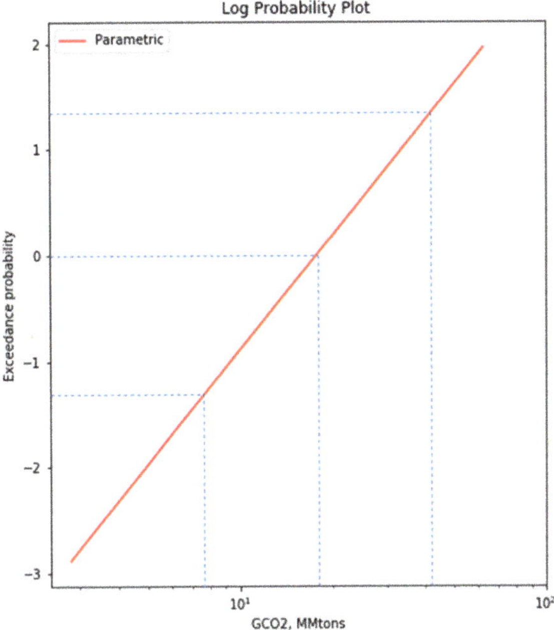

Figure 11. Log probability plot showing the probabilistic distribution of the storage capacity of the Morrow B reservoir.

The standard deviation allocated to each of the input parameters indicates the uncertainty with the mean value. The degree of the uncertainty value stems from how difficult it is to determine that parameter. For instance, from Table 1, the standard deviation of the net thickness is about half of its mean value, which signifies how uncertain and difficult it is to accurately determine the exact value. However, the standard deviation of the $ScCO_2$ density is insignificant as compared to its mean value. This is valid because the mean value is established using different sophisticated pressure equipment in reading the miscibility pressure through repeated experiments. These are also observed from Figure 9; the net thickness is positively skewed while the S_cCO_2 density seems symmetrical.

From the expectation curve, the mean $\neq$ median $\neq$ mode and signifies the skewness of the output probability density function. The expectation curve provides a good way of visualizing the total uncertainty. The narrower the curve or the closer the curve is to the vertical axis, the less is the total uncertainty. The log probability plot (P90-P50-P10 straight line) also provides a way of analyzing the uncertainty. The steeper the P90-P50-P10 straight line, the lesser the uncertainty to the total outcome.

The overall uncertainty of the Morrow B has a coefficient of variation of 71%, which translated into a standard deviation of about 16 MMtons. The total uncertainty of output increases with the product of input parameters. Hence, the total uncertainty is a result of the degree of uncertainty of the individual input parameters. A relative impact plot was constructed to analyze the uncertainty of the individual input parameters towards the total uncertainty.

From the relative impact plot showing in Figures 12 and 13, the net thickness contributed the most to the total uncertainty of about 50%, followed by the efficiency factor, which contributed about 25% to the total uncertainty for both the parametric and Monte Carlo simulation. The net thickness, efficiency factor, and area together contributed about 95% to the total uncertainty. Relatively, the contribution of the efficiency factor to the total uncertainty was less considering the different parameters which affect it. However, this was expected considering how long the FWU has been produced; there were enough data to estimate the efficiency factor.

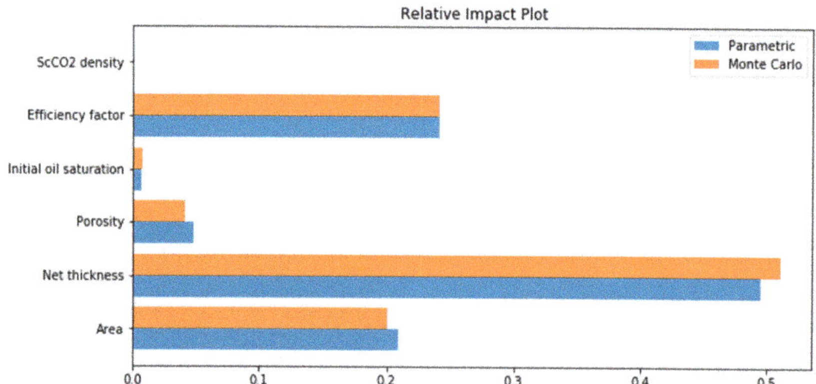

Figure 12. Relative impact plot illustrating the contribution of the input parameters to the total uncertainty. This plot also compares the sensitivity analysis of both the parametric method and the Monte Carlo simulation.

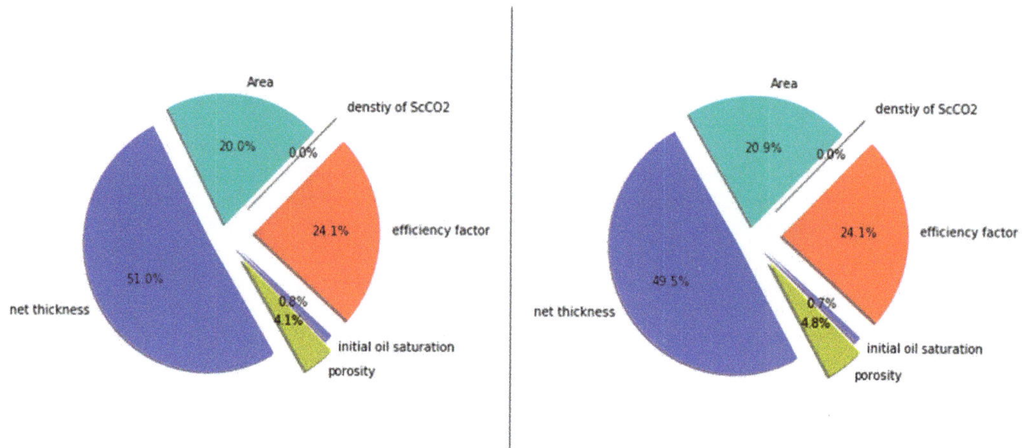

Figure 13. Pie chart showcasing impact of the input parameters to the total uncertainty in percentages. The left and right charts represent the results generated by the Monte Carlo simulation and the parametric method, respectively.

5. Concluding Remarks

This work utilized an approach using the relative permeability curve as a function of the hydrologic flow unit to determine the storage efficiency factor and employed the parametric method to estimate the storage capacity of the Morrow B reservoir. It also shows sensitivity analysis of the input parameters towards the total uncertainty.

From the new static model presented in this paper, it appears that the eastern side of the field has sufficient storage capacity to make CCUS a feasible proposition, given the right economic conditions.

The use of relative permeability curves to estimate storage efficiency factors is an effective and feasible approach. It is concluded that for constant irreducible water saturation, the storage efficiency factor increases with the reduction of irreducible oil saturation after CO_2 injection (Sorg) and critical gas saturation (Sgc).

From the probabilistic output generated by both techniques, the parametric method results show that at least 7.81 MMtons can be stored, 17.79 MMtons of CO_2 can probably be stored, and it may be possible to store as much as 40.52 MMtons of CO_2 in the Morrow B reservoir. The results outputted by the Monte Carlo simulation show similar results; 7.68 MMtons is at least proven to be stored, 17.63 MMtons can probably be stored, and it may be possible to store as much as 40.58 MMtons of CO_2 in the Morrow B reservoir. The deterministic value, which is the single best estimate, was determined from the parametric method and Monte Carlo simulation to be 21.87 and 21.63 MMtons, respectively. From the relative impact plot, the net thickness, storage efficiency factor, and area contributed about 95% to the total uncertainty for both techniques. To significantly improve the estimation of the storage capacity of the Morrow B reservoir, this percentage needs to be reduced drastically. The net thickness contributed the most to the total uncertainty, and this should be the top-most priority to reduce the total uncertainty and better estimate the storage capacity.

The probabilistic approach (parametric method) was successfully used to estimate the storage capacity. Comparing both the parametric method and the Monte Carlo simulation, the results were practically the same, although 10,000 simulation runs were used, and this illustrates how computationally expensive the Monte Carlo simulation is. The parametric method assumes input variables to be lognormally distributed. It is an analytical procedure and can be performed in a simple spreadsheet application. This technique can be applied in all disciplines that seek to quantify the feasibility of a project while analyzing and quantifying total uncertainty. This technique may also assist management in decision-making procedures by helping them to arrive at a viable conclusion.

Author Contributions: Conceptualization, W.A.; methodology, W.A., J.A. and D.R.-C.; software, J.A. and W.A.; validation, R.B., M.C. and D.R.-C.; formal analysis, J.A., W.A. and D.R.-C.; writing—original draft preparation, J.A., W.A. and D.R.-C.; writing—review and editing, M.C. and R.B.; visualization, J.A., W.A. and D.R.-C.; supervision, W.A.; funding acquisition, R.B. All authors have read and agreed to the published version of the manuscript.

Funding: Funding for this project was provided by the U.S. Department of Energy's (DOE) National Energy Technology Laboratory (NETL) through the Southwest Regional Partnership on Carbon Sequestration (SWP) under Award No. DE-FC26-05NT42591.

Institutional Review Board Statement: Not applicable.

Informed Consent Statement: Not applicable.

Data Availability Statement: Not applicable.

Conflicts of Interest: The authors declare no conflict of interest.

References

1. Morgan, A.; Grigg, R.; Ampomah, W. A Gate-to-Gate Life Cycle Assessment for the CO_2-EOR Operations at Farnsworth Unit (FWU). *Energies* **2021**, *14*, 2499. [CrossRef]
2. Ruttinger, A.W.; Kannangara, M.; Shadbahr, J.; De Luna, P.; Bensebaa, F. How CO_2-to-Diesel Technology Could Help Reach Net-Zero Emissions Targets: A Canadian Case Study. *Energies* **2021**, *14*, 6957. [CrossRef]
3. García-Mariaca, A.; Llera-Sastresa, E. Review on Carbon Capture in ICE Driven Transport. *Energies* **2021**, *14*, 6865. [CrossRef]

4. Shreyash, N.; Sonker, M.; Bajpai, S.; Tiwary, S.K.; Khan, M.A.; Raj, S.; Sharma, T.; Biswas, S. The Review of Carbon Capture-Storage Technologies and Developing Fuel Cells for Enhancing Utilization. *Energies* **2021**, *14*, 4978. [CrossRef]

5. Pingping, S.; Xinwei, L.; Qiujie, L. Methodology for estimation of CO_2 storage capacity in reservoirs. *Pet. Explor. Dev.* **2009**, *36*, 216–220. [CrossRef]

6. Bachu, S.; Bonijoly, D.; Bradshaw, J.; Burruss, R.; Holloway, S.; Christensen, N.P.; Mathiassen, O.M. CO_2 storage capacity estimation: Methodology and gaps. *Int. J. Greenh. Gas Control* **2007**, *1*, 430–443. [CrossRef]

7. Goodman, A.; Hakala, A.; Bromhal, G.; Deel, D.; Rodosta, T.; Frailey, S.; Small, M.; Allen, D.; Romanov, V.; Fazio, J.; et al. U.S. DOE methodology for the development of geologic storage potential for carbon dioxide at the national and regional scale. *Int. J. Greenh. Gas Control* **2011**, *5*, 952–965. [CrossRef]

8. Hall, G. *The US 2012 Carbon Utilization and Storage Atlas*; US Department of Energy: Washington, DC, USA, 2012; p. 130.

9. IPCC. IPCC Special Report: Carbon Dioxide Capture and Storage. *CMAJ* **2004**, *171*, 1327. [CrossRef]

10. Rossi, F.; Gambelli, A.M. Thermodynamic phase equilibrium of single-guest hydrate and formation data of hydrate in presence of chemical additives: A review. *Fluid Phase Equilibria* **2021**, *536*, 112958. [CrossRef]

11. Frailey, S.M. Methods for estimating CO_2 storage in saline reservoirs. *Energy Procedia* **2009**, *1*, 2769–2776. [CrossRef]

12. Levine, J.S.; Fukai, I.; Soeder, D.; Bromhal, G.; Dilmore, R.M.; Guthrie, G.D.; Rodosta, T.; Sanguinito, S.; Frailey, S.; Gorecki, C.; et al. U.S. DOE NETL methodology for estimating the prospective CO_2 storage resource of shales at the national and regional scale. *Int. J. Greenh. Gas Control* **2016**, *51*, 81–94. [CrossRef]

13. Pyrcz, M.J.; White, C.D. Uncertainty in reservoir modeling. *Interpretation* **2015**, *3*, SQ7–SQ19. [CrossRef]

14. Ma, Y.Z.; La Pointe, P.R. Uncertainty Analysis in Reservoir Characterization and Management: How much should we know about what we don't know? *AAPG Memoir* **2011**, *96*, 1–15. [CrossRef]

15. Ziegel, E.R.; Deutsch, C.V.; Journel, A.G. Geostatistical Software Library and User's Guide. *Technometrics* **1998**, *40*, 357. [CrossRef]

16. Bachu, S. Review of CO_2 storage efficiency in deep saline aquifers. *Int. J. Greenh. Gas Control.* **2015**, *40*, 188–202. [CrossRef]

17. Brennan, S.T. The U.S. Geological Survey carbon dioxide storage efficiency value methodology: Results and observations. *Energy Procedia* **2014**, *63*, 5123–5129. [CrossRef]

18. Park, J.; Yang, M.; Kim, S.; Lee, M.; Wang, S. Estimates of scCO$_2$ Storage and Sealing Capacity of the Janggi Basin in Korea Based on Laboratory Scale Experiments. *Minerals* **2019**, *9*, 515. [CrossRef]

19. Rasmussen, L.; Fan, T.; Rinehart, A.; Luhmann, A.; Ampomah, W.; Dewers, T.; Heath, J.; Cather, M.; Grigg, R. Carbon Storage and Enhanced Oil Recovery in Pennsylvanian Morrow Formation Clastic Reservoirs: Controls on Oil–Brine and Oil–CO_2 Relative Permeability from Diagenetic Heterogeneity and Evolving Wettability. *Energies* **2019**, *12*, 3663. [CrossRef]

20. Ampomah, W.; Balch, R.S.; Chen, H.-Y.; Gunda, D.; Cather, M. Probabilistic reserves assessment and evaluation of sandstone reservoir in the Anadarko Basin. In Proceedings of the SPE/IAEE Hydrocarbon Economics and Evaluation Symposium, Houston, TX, USA, 17–18 May 2016; Volume 16. [CrossRef]

21. Cronquist, C. *Estimation and Classification of Reserves of Crude Oil, Natural Gas and Condensate*; Society of Petroleum Engineers: Richardson, TX, USA, 2001.

22. Chen, H.-Y. Engineering Reservoir Mechanics. Ph.D. Thesis, New Mexico Tech, Socorro, NM, USA, 2012.

23. Morgan, J.; Gordon, D. Influence of Pore Geometry on Water-Oil Relative Permeability. *J. Pet. Technol.* **1970**, *22*, 1199–1208. [CrossRef]

24. Capen, E.C. Probabilistic reserves! Here at last? *SPE Reserv. Eval. Eng.* **2001**, *4*, 387–394. [CrossRef]

25. Ampomah, W.; Balch, R.S.; Grigg, R.B. Analysis of Upscaling Algorithms in Heterogeneous Reservoirs with Different Recovery Processes. In Proceedings of the SPE Production and Operations Symposium, Oklahoma City, OK, USA, 1–5 March 2015.

26. Ampomah, W.; Balch, R.S.; Ross-Coss, D.; Hutton, A.; Cather, M.; Will, R.A. An Integrated Approach for Characterizing a Sandstone Reservoir in the Anadarko Basin. In Proceedings of the Offshore Technology Conference, Houston, TX, USA, 2–5 May 2016.

27. Ampomah, W.; Balch, R.S.; Grigg, R.B.; Will, R.; Dai, Z.; White, M.D. Farnsworth Field CO_2-EOR Project: Performance Case History. In Proceedings of the SPE Improved Oil Recovery Conference, Tulsa, OK, USA, 11 April 2016. OnePetro.

28. Ampomah, W.; Balch, R.S.; Grigg, R.B.; Dai, Z.; Pan, F. Compositional Simulation of CO_2 Storage Capacity in Depleted Oil Reservoirs. In Proceedings of the Carbon Management Technology Conference, Sugar Land, TX, USA, 17–19 November 2015.

29. Hutton, A.C. Geophysical Modeling and Structural Interpretation of a 3D Reflection Seismic Survey in Farnsworth Unit, TX. Master's Thesis, New Mexico Institute of Mining and Technology, Socorro, NM, USA, 2015.

30. Gallagher, S.R. Depositional and Diagenetic Controls on Reservoir Heterogeneity: Upper Morrow Sandstone, Farnsworth Unit, Ochiltree County, Texas. Ph.D. Thesis, New Mexico Institute of Mining and Technology, Socorro, NM, USA, 2014.

31. Czoski, P. Geologic Characterization of the Morrow B Reservoir in Farnsworth Unit, TX Using 3D VSP Seismic, Seismic Attributes, and Well Logs. Master's Thesis, New Mexico Institute of Mining and Technology, Socorro, NM, USA, 2014; p. 101.

32. Munson, T.W. Depositional, diagenetic, and production history of the Upper Morrowan Buckhaults Sandstone, Farnsworth Field, Ochiltree County Texas. *Shale Shak. Dig.* **1994**, *40*, 2–20.

33. Ross-Coss, D.; Ampomah, W.; Cather, M.; Balch, R.S.; Mozley, P.; Rasmussen, L. An Improved Approach for Sandstone Reservoir Characterization. In Proceedings of the SPE Western Regional Meeting, Anchorage, AL, USA, 23–26 May 2016.

34. Balch, R.; McPherson, B.; Will, R.A.; Ampomah, W. Recent Developments in Modeling: Farnsworth Texas, CO 2 EOR Carbon Sequestration Project. In Proceedings of the 15th Greenhouse Gas Control Technologies Conference, Abu Dhabi, UAE, 15–18 March 2021; Volume 2.
35. Span, R.; Wagner, W. A New Equation of State for Carbon Dioxide Covering the Fluid Region from the Triple-Point Temperature to 1100 K at Pressures up to 800 MPa. *J. Phys. Chem. Ref. Data* **1996**, *25*, 1509–1596. [CrossRef]
36. Etherington, J.; Ritter, J. The 2007 SPE/WPC/AAPG/SPEE Petroleum Resources Management System (PRMS). *J. Can. Pet. Technol.* **2008**, *47*. [CrossRef]

Article

Relative Permeability: A Critical Parameter in Numerical Simulations of Multiphase Flow in Porous Media

Nathan Moodie [1], William Ampomah [2], Wei Jia [1] and Brian McPherson [1,*]

[1] Carbon Science & Engineering Research Group, University of Utah, Salt Lake City, UT 84112, USA; nathan.moodie@m.cc.utah.edu (N.M.); wei.jia@utah.edu (W.J.)
[2] New Mexico Tech, Socorro, NM 87801, USA; william.ampomah@nmt.edu
* Correspondence: b.j.mcpherson@utah.edu; Tel.: +1-801-558-4043

Abstract: Effective multiphase flow and transport simulations are a critical tool for screening, selection, and operation of geological CO_2 storage sites. The relative permeability curve assumed for these simulations can introduce a large source of uncertainty. It significantly impacts forecasts of all aspects of the reservoir simulation, from CO_2 trapping efficiency and phase behavior to volumes of oil, water, and gas produced. Careful consideration must be given to this relationship, so a primary goal of this study is to evaluate the impacts on CO_2-EOR model forecasts of a wide range of relevant relative permeability curves, from near linear to highly curved. The Farnsworth Unit (FWU) is an active CO_2-EOR operation in the Texas Panhandle and the location of our study site. The Morrow 'B' Sandstone, a clastic formation composed of medium to coarse sands, is the target storage formation. Results indicate that uncertainty in the relative permeability curve can impart a significant impact on model predictions. Therefore, selecting an appropriate relative permeability curve for the reservoir of interest is critical for CO_2-EOR model design. If measured laboratory relative permeability data are not available, it must be considered as a significant source of uncertainty.

Keywords: relative permeability; geologic carbon storage; multi-phase flow simulation

Citation: Moodie, N.; Ampomah, W.; Jia, W.; McPherson, B. Relative Permeability: A Critical Parameter in Numerical Simulations of Multiphase Flow in Porous Media. *Energies* **2021**, *14*, 2370. https://doi.org/10.3390/en14092370

Academic Editor: Jianchao Cai

Received: 18 February 2021
Accepted: 19 April 2021
Published: 22 April 2021

Publisher's Note: MDPI stays neutral with regard to jurisdictional claims in published maps and institutional affiliations.

1. Introduction

With the prospects of climate change looming and an ever-increasing demand for power generation and heavy industry, reducing greenhouse gas emissions from large point-source emitters, such as coal and natural gas power plants or fertilizer operations, has become paramount. Geologic carbon storage (GCS) is one potential path for emission reductions. Carbon dioxide is captured from the large point-source emitters, compressed into a supercritical state and injected into a suitable storage formation such as a depleted oil and gas reservoir or a deep saline aquifer [1–5]. Mature oil fields undergoing CO_2-Enhanced Oil Recovery (CO_2-EOR) are another promising option for GCS that also offer an economic benefit. The incremental recovery for CO_2-EOR operations can produce an additional 7–23% of the oil in place while simultaneously storing roughly 40% of the CO_2 injected [6].

Multiphase flow and transport simulation that can characterize CO_2 effects in oil and water are an integral part of designing GCS projects for oil and gas fields. These simulations are used in project design, permitting, forecasting oil production and storage capacity, and quantifying possible site risks. Understanding the uncertainty in the simulation model inputs and their impact on performance and predictions is critical for project success. The permeability distribution in a reservoir is probably the biggest source of uncertainty, followed closely by relative permeability. Three-phase relative permeability has significant impacts on fluid flow and storage capacity, yet is poorly understood and often generalized. Laboratory measurements have historically been focused on measuring two- and three-phase relative permeability curves for oil and gas (CH_4) reservoirs [7–9]. From these data, empirical models have been developed which promise broad applicability, including use in GCS numerical simulations [7–14]. For this study, we utilized an empirical formula for relative permeability developed by Corey [10] to described oil and gas flow in porous media. This empirical formula was to create relative permeability curves for CO_2-EOR numerical simulations at our field site [15,16].

It is important to understand that when measuring three-phase relative permeability (gas/oil/water), experimental methods generally fall into one of two categories. The first general method measures pairs' two-phase relative permeability, gas/oil and oil/water, and then uses an empirical combination model such as Stone II or the Baker model to calculate the three-phase relative permeability [7,13,14,17]. In the second general method, all three phases are measured concurrently to create a true three-phase relationship for the fluids of interest [8,17,18]. Generally, numerical simulation codes do not leverage three-phase relative permeability data. Simulators such as STOMP use a two-phase empirical model along with critical parameters like the residual wetting and non-wetting phase saturations and a curve parameter to calculate the relative permeability from the fluid saturations [19]. Alternatively, other numerical simulators, such as the Eclipse® numerical simulator used in this study, leverage table data in the form of two-phase saturation versus relative permeability tabular data coupled with linear interpolation between data points [20]. These methods then use a combination model to calculate the three-phase relative permeability. Either method requires the input data to be in the form of a pair of two-phase relative permeability versus saturation curves, one for the gas/oil pair and one for the oil/water pair. In this study, we leveraged the functionality in Eclipse® to use tabular data by generating a suite of plausible curves using the Corey's Curve empirical formula, then importing those curves into the simulation model.

The choice of relative permeability curve can be a significant source of model uncertainty. This uncertainty can come from the uncertainty inherent in laboratory measurements or the general lack of relative permeability curves for most GCS candidate sites and formations. At the time of this study, the target reservoir at our study site had a single pair of binary relative permeability curves that were derived from a laboratory study [21,22]. Studies quantifying the model uncertainty related to the relative permeability curve on numerical simulation forecasts are rare, and for our study site do not exist. Therefore, a primary goal of this study is to evaluate how uncertainty in the relative permeability curves impacts CO_2-EOR model forecasts.

2. Study Site: Farnsworth Unit, Texas

The study site is the Farnsworth Unit, an active CO_2-EOR site since 2010 located in the Anadarko Basin of northern Texas. The target formation is in the Upper Morrow sequence called the Morrow 'B' Sandstone. The field has produced 27 billion cubic feet of gas and more than 1.9 million barrels of oil from the Morrow 'B' Sandstone, a fluvial valley-fill sandstone [22–24]. The formation is at a depth of between 7550 and 7950 feet and is a series of connected sandstone bodies dipping to the West at less than one degree [21,23,25]. The sandstone sequences occur at the base of the 'B' unit and are up to 44 ft thick consisting of medium to coarse sand and conglomerates that are thought to be deposited as a series of fluvial point-bars that are all connected [23,24,26]. Laboratory measurements on core samples and well log analysis indicate a mean porosity of 14% to 17% and average permeabilities that range from 27 mD to 140 mD [24,25,27] The older work done by Bolyard (1989) indicated higher porosity and permeability while the newer work done by Rose-Coss et al. (2016) indicated lower average porosity and permeability [24,26].

3. Numerical Model Development

3.1. FWU Geological Model

A geological model of the Farnsworth Unit was developed by the Southwest Regional Partnership on Carbon Sequestration (SWP) using the full suite of petrophysical data collected during the ongoing characterization effort including SEM, XRD, seismic data sets, well logs, core samples, and thin sections [26–29]. The full geological model represents the whole oil field at Farnsworth. Petrophysical properties of porosity and permeability were populated across the domain by stochastic algorithm constrained by well logs [16]. A subset of the domain centered on the west half of the field was used for this study. This is the current active injection and production area and primary focus of the SWP research.

3.2. Relative Permeability

For this study, 17 different relative permeability curves were constructed that represent a possible set of curves which may apply to the reservoir. The goal is to study a wide range of relative permeability curves with the assumption that as the relative permeability curve becomes more linear, the saturation end-points and the relative permeability end-points become larger, representing a transition from low fluid mobility to high fluid mobility. The residual phase saturation, maximum relative permeability, and the curve shape were varied to create a range of curves that bracket the parameter space from near linear to highly curved. As of the time of this study, there was only a single relative permeability curve measured for the Morrow Sandstone [21,22]. This lack of measured data necessitates that a wide range of input parameters be examined to understand the influence relative permeability has on a CO_2-EOR operation. Figure 1 highlights a representative selection of curves that transition from highly curved to near linear. The remaining relative permeability curves used in the study are shown in Figures 2, A1 and A2.

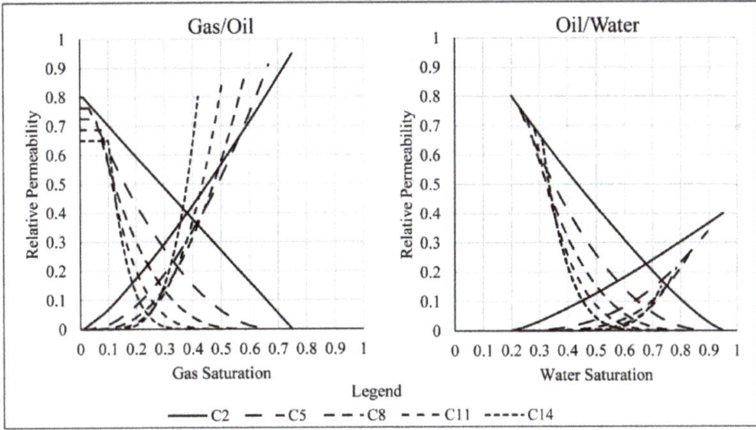

Figure 1. A representative selection of relative permeability curves used in this study. All curves used follow a similar trend of high residual saturation and highly curved to low residual saturation and near linear curve. Only those shown here vary the maximum relative permeability in addition to the other two variables.

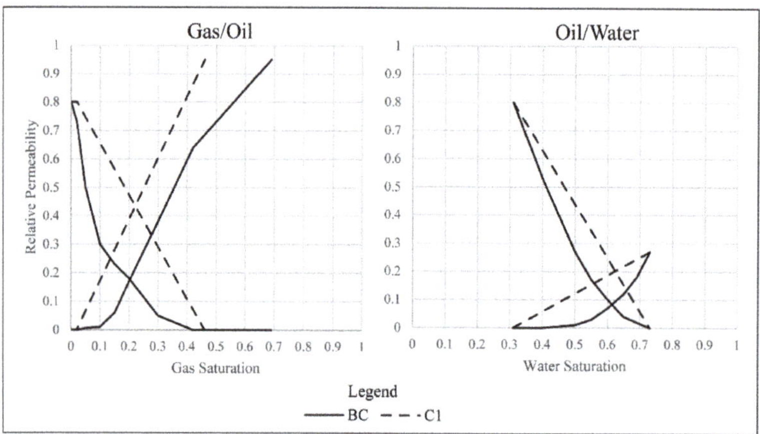

Figure 2. Plot shows the Base Case relative permeability curve from a UNOCAL study (solid lines) and the C1 relative permeability curve that represents the most linear curve used in the study.

In this study we used the Eclipse® numerical simulation software for all the simulations. A key feature that we leveraged is its ability to use lookup tables with linear interpolation between points instead of empirical formulas to calculate the relative permeability. This effectively decouples the empirical formula from the numerical simulator, allowing much greater flexibility in specifying the saturation and relative permeability endpoints as well as the degree of curvature. Any empirical formula could have been chosen to create the curves.

We chose to use a modified version of the Corey's Curve formula to create all the Eclipse® lookup tables used except for the Base Case, discussed below. This relationship provides great flexibility for developing a wide variety of curves from highly non-linear to linear with the added ability to modify the residual phase saturation and maximum relative permeability endpoints. The Corey's Curve formula allows the creation of a wide range of relative permeability curves based on three sets of inputs, residual saturation, maximum relative permeability, and lambda (the curve parameter). The gas and non-aqueous liquid (oil) curve are described by Equations (1) and (2), where S_t is the liquid saturation defined as the non-aqueous liquid (oil) saturation plus the residual aqueous liquid saturation, S_{gr} is the residual gas saturation, S_{tr} is the residual liquid saturation defined as the residual oil saturation plus the residual water saturation, λ defines the shape of the curve, k_{rg}max is the maximum gas relative permeability, k_{rn}max is the maximum non-aqueous liquid (oil) relative permeability, and $k_{rg}(S_t)$ and $k_{rn}(S_t)$ are the gas and non-aqueous liquid relative permeability at liquid saturation S_t [19].

$$k_{rg}(S_t) = k_{rg}max(1 - (S_t - S_{tr})/(1 - S_{tr} - S_{gr}))^\lambda \tag{1}$$

$$k_{rn}(S_t) = k_{rn}max((S_t - S_{tr})/(1 - S_{tr} - S_{gr}))^\lambda \tag{2}$$

The water and oil curves are described by Equations (3) and (4), where S_l is the aqueous saturation, S_{lr} is the residual aqueous saturation, S_{nr} is the residual non-aqueous (oil) saturation, λ defines the shape of the curve, k_{rl}max is the maximum aqueous liquid relative permeability, k_{rn}max is the maximum non-aqueous liquid relative permeability, and $k_{rl}(S_l)$ and $k_{rn}(S_l)$ are the aqueous and non-aqueous liquid relative permeability at aqueous liquid saturation S_l [19]. Tables 1 and 2 have the parameters (residual phase saturation, maximum phase relative permeability, and lambda) used to calculate all the relative permeability curves in this study, including the subset shown in Figure 1.

$$k_{rl}(S_l) = k_{rl}max((S_l - S_{lr})/(1 - S_{lr} - S_{nr}))^\lambda \tag{3}$$

$$k_{rn}(S_l) = k_{rn}\max(1 - (S_l - S_{lr})/(1 - S_{lr} - S_{nr}))^{\lambda} \tag{4}$$

Table 1. Phase residual saturation and curve parameter lambda used in Equations (1)–(4) to create the relative permeability curves used in this study.

Model	S_{gr}	S_{tr}	S_{lr}	S_{nr}	Lambda(G/O) Oil	Gas	Lambda(O/W) Oil	Water
C1	0.020	0.540	0.310	0.270	1.000	1.250	1.250	1.250
C2	0.01	0.25	0.2	0.05	1	1.25	1.25	1.25
C3	0.01	0.25	0.2	0.05	1	1.25	1.25	1.25
C4	0.07	0.2	0.2175	0.0875	1	1.25	1.25	1.25
C5	0.0325	0.3325	0.2275	0.105	1.875	1.8125	2.0625	2.4375
C6	0.0325	0.3325	0.2275	0.105	1.875	1.8125	2.0625	2.4375
C7	0.13	0.1433	0.2533	0.12	1.875	1.8125	2.0625	2.4375
C8	0.055	0.415	0.255	0.16	2.75	2.375	2.875	3.625
C9	0.055	0.415	0.255	0.16	2.75	2.375	2.875	3.625
C10	0.08	0.1	0.15	0.09	2.75	2.375	2.875	3.625
C11	0.0775	0.4975	0.2825	0.215	3.625	2.9375	3.6875	4.8125
C12	0.0775	0.4975	0.2825	0.215	3.625	2.9375	3.6875	4.8125
C13	0.0733	0.26	0.3	0.153	3.625	2.9375	3.6875	4.8125
C14	0.1	0.58	0.31	0.27	4.5	3.5	4.5	6
C15	0.1	0.58	0.31	0.27	4.5	3.5	4.5	6
C16	0.12	0.13	0.215	0.15	4.5	3.5	4.5	6

Table 2. Maximum relative permeability values used in Equations (1)–(4) to create the relative permeability curves used in this study.

Model	$k_{rn}(S_t)$Max	$k_{rg}(S_t)$Max	$k_{rl}(S_l)$Max	$k_{rn}(S_l)$Max
C1	0.8	0.95	0.27	0.8
C2	0.8	0.95	0.4	0.8
C3	0.8	0.95	0.27	0.8
C4	0.8	0.95	0.27	0.8
C5	0.7625	0.9125	0.3375	0.7625
C6	0.8	0.95	0.27	0.8
C7	0.8	0.95	0.27	0.8
C8	0.725	0.875	0.275	0.725
C9	0.8	0.95	0.27	0.8
C10	0.8	0.95	0.27	0.8
C11	0.6875	0.8375	0.2125	0.6875
C12	0.8	0.95	0.27	0.8
C13	0.8	0.95	0.27	0.8
C14	0.65	0.8	0.15	0.65
C15	0.8	0.95	0.27	0.8
C16	0.8	0.95	0.27	0.8

The Base Case model permutation uses a relative permeability curve (BC in Figure 2) developed by UNOCAL for a simulation study focused on the efficacy of water and CO_2 flooding at the Farnsworth Unit [21,22]. The UNOCAL study did not provide a capillary pressure curve for the Morrow 'B' Sandstone. For this site, it is initially believed that capillary pressure had a negligible effect on phase movement and was thus ignored in the previous models to aid in simplicity and computational speed. For consistency with previous work and to reduce extra variables that may influence the simulation results capillary pressure was not included [16,25]. We plan to study the influence that the addition of capillary pressure may have on model forecasts to determine if this assumption is valid.

3.3. Farnsworth Units Model Domain

The Farnsworth Unit is divided into an east and west half that appear to be hydraulically split [22,23]. The west half is the site of most production and injection operations historically and are where current and future CO_2-EOR operations are occurring [22,23]. A detailed geological model encompassing the Morrow 'B' Sandstone in the west half of the reservoir was the basis of our simulation model [26]. We up-scaled this geologic model with over 26 million cells to a simulation model consisting of 33,756 active cells. Along with the grid geometry, we up-scaled the permeability and the porosity (Figure 3). This yielded a mean permeability of 39 mD with a standard deviation of 54 mD and a mean porosity of 14% with a standard deviation of 3%, in line with the original geological model values of 13.6% porosity and 27 mD permeability. We assumed the sealing formation, the Morrow Shale, made a no-flow boundary on all sides as well as the top and bottom of the reservoir, so only the reservoir interval is included in the simulation model. See Moodie et al. (2019) for more details on the model domain description [30].

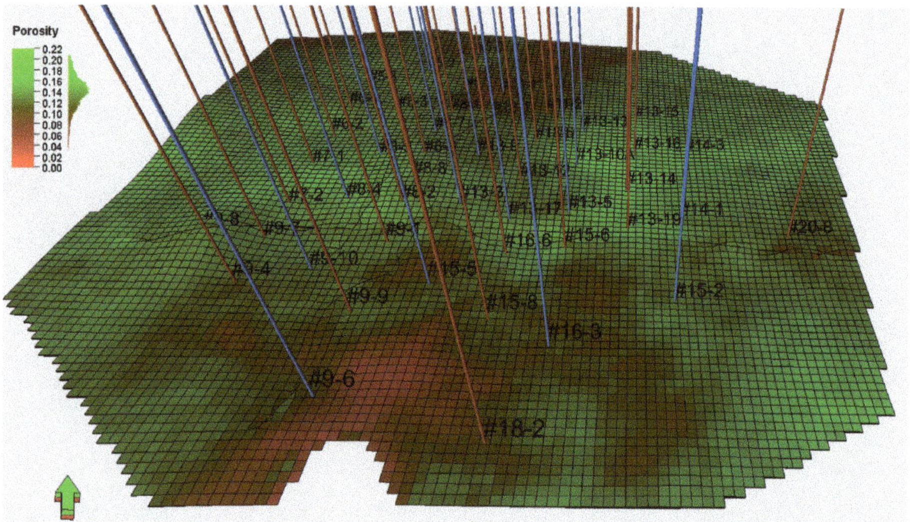

Figure 3. The dynamic model domain showing the porosity distribution for the top of the model with the injection wells in blue and production wells in red.

The initial conditions were derived from the results of the history-matched primary production and water-flooding modeling study done previously by the SWP and include the oil saturation and oil component distribution, the water saturation and the pressure [16]. The simulation is initialized with no gas phase. All methane (CH_4) and other light volatiles are dissolved in the oil. A compositional fluid model is specified for this study based on a fluid properties report from initial exploration, then refined by fluid modeling [31,32].

3.4. Well Operations Schedule and Model Fit

The Farnsworth Unit operates under a water alternating gas (WAG) injection scheme. The schedule used in this study mimics current practices and future plans. Figure 3 indicates the location of the injection and production wells used in the model. The production and injection schedule are broken into two main operation periods. The first period extends from 1 December 2010 to 31 July 2016 and uses historical monthly injection and production data to define well rates. The second period extends from 1 August 2016 to 1 January 2036 and models potential future operations through to the end of the field's lifetime. During this second period, the CO_2 injection volume gradually decreases until it is completely reliant on recycled CO_2 to meet targets. See Moodie et al. (2019) for a detailed breakdown of the operations schedule [30].

To assess the model performance, we perform a regression analysis comparing the historical oil production from December 2010 to January 2016 to the Base Case simulation model. A R^2 value of 0.94 indicates a reasonable fit and comparing this data to a history-matched model developed by Ampomah et al. (2016) [16]. Reviewing Figure 4 indicates a strong correlation between the historical data and the model data, with the Base Case (dashed line), the history-matched model of Ampomah et al. (2016) (solid black line), and the FWU historical data (open circles) all plotted [16].

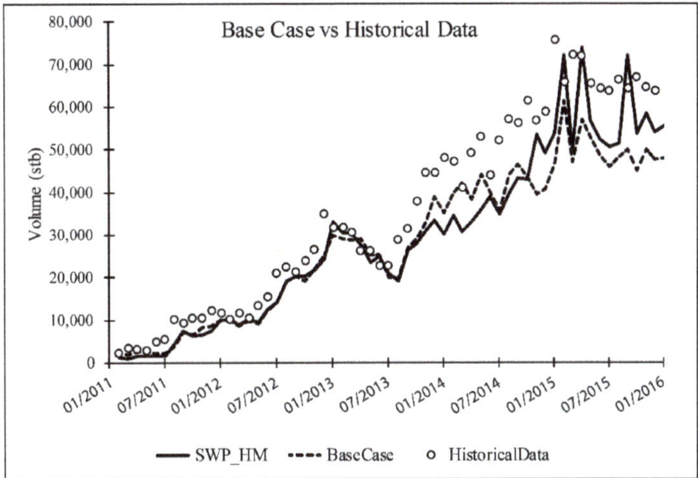

Figure 4. Plot shows the historical monthly oil production for the Farnsworth Units (black open circles), Base Case model forecasts (black dashed line), and history-matched model forecasts (SWP_HM) (solid black line). Both numerical models show a good data fit to the historical production data.

4. Discussion

Results indicate that the shape and endpoints of the relative permeability curve impart a significant influence on model forecasts. Generally, the more non-linear the curve, the more CO_2 that is predicted to be stored, and the less oil, water, and gas is predicted to be produced. However, the saturation endpoints and the assumed maximum relative permeability have a significant impact on this trend, with some of the more non-linear curves (C11 & C12) predicting more oil production than the most linear curve (C1).

4.1. Carbon Dioxide Storage

The total amount of stored CO_2 forecasted in this modeling study is between 2.8 million tons and 3.2 million tons by the end of the simulation. The largest fraction of the CO_2 is dissolved in the oil phase, between 1.2 and 2.4 million tons. The cases using highly non-linear relative permeability curves with narrow saturation ranges (C11 through

C16) had overall a larger portion of the CO_2 dissolved in the oil phase while the more linear curves with broader saturation ranges (C1 through C7) have almost a million tons less CO_2 dissolved in the oil phase but significantly more CO_2 in the supercritical phase. This indicates that the CO_2 in the supercritical phase has an inverse relationship to the CO_2 dissolved in the oil phase with respect to the changes in the relative permeability curve (Figure 5). As the curve becomes more non-linear, the fluid mobility decreases, leading to a decrease in the volume of CO_2 dissolved in the oil and a corresponding increase in the supercritical CO_2. The highest fluid mobility curve (model C1) predicts an almost even CO_2 distribution between the supercritical phase (46%) and oil phase (44%); while the lowest fluid mobility curves (C14 to C16) predict a significant difference in CO_2 phase distribution, 10% in the supercritical phase and 82% dissolved in the oil phase for model C14, see Table 3 for stored CO_2 mass distribution. The CO_2 mass dissolved in the water phase is mostly unaffected by changes in the relative permeability curve, varying by only 2% across all model permutations, from 8% to 10% of the total CO_2 stored. The Base Case (BC) model predicts the largest mass of CO_2 stored in the reservoir and the phase distribution matches the models with relative permeability curves that describe intermediate fluid mobility. This indicates that the Base Case relative permeability curve describes an intermediate fluid mobility relationship, close to what C8 and C9 models predict.

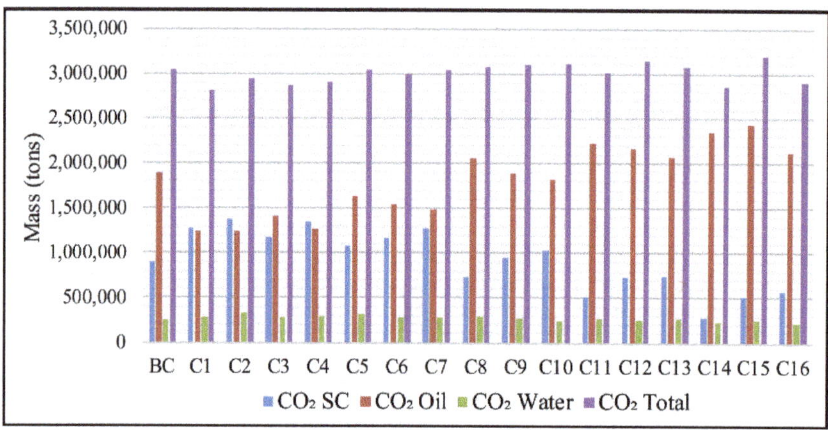

Figure 5. This chart shows CO_2 storage for each of the model permutations. 'CO_2 SC' is the supercritical CO_2, 'CO_2 Oil' is the CO_2 and CH_4 that is dissolved in the oil phase, 'CO_2 Water' is the CO_2 dissolved in the aqueous phase, and 'CO_2 Total' is the total amount of gas, both CO_2 and CH_4, in the reservoir.

Table 3. The total mass of CO_2 stored in each phase; supercritical, dissolved in the oil phase, dissolved in the aqueous phase. These data are the totals for the end of the simulation.

Model	Supercritical CO_2 (Tons)	(%)	CO_2 in Oil (Tons)	(%)	CO_2 in Water (Tons)	(%)	Total (Tons)
BC	8.98×10^5	30%	1.88×10^6	62%	2.61×10^5	9%	3.04×10^6
C1	1.28×10^6	46%	1.24×10^6	44%	2.79×10^5	10%	2.80×10^6
C2	1.37×10^6	47%	1.24×10^6	42%	3.24×10^5	11%	2.94×10^6
C3	1.18×10^6	41%	1.41×10^6	49%	2.80×10^5	10%	2.87×10^6
C4	1.35×10^6	46%	1.27×10^6	44%	2.89×10^5	10%	2.90×10^6
C5	1.08×10^6	36%	1.63×10^6	54%	3.18×10^5	10%	3.03×10^6
C6	1.16×10^6	39%	1.54×10^6	52%	2.81×10^5	9%	2.99×10^6
C7	1.28×10^6	42%	1.48×10^6	49%	2.77×10^5	9%	3.04×10^6
C8	7.30×10^5	24%	2.05×10^6	67%	2.97×10^5	10%	3.08×10^6
C9	9.46×10^5	31%	1.89×10^6	61%	2.66×10^5	9%	3.10×10^6
C10	1.03×10^6	33%	1.83×10^6	59%	2.50×10^5	8%	3.11×10^6
C11	5.14×10^5	17%	2.22×10^6	74%	2.75×10^5	9%	3.01×10^6
C12	7.27×10^5	23%	2.17×10^6	69%	2.53×10^5	8%	3.15×10^6
C13	7.48×10^5	24%	2.06×10^6	67%	2.69×10^5	9%	3.08×10^6
C14	2.84×10^5	10%	2.34×10^6	82%	2.28×10^5	8%	2.85×10^6
C15	5.08×10^5	16%	2.43×10^6	76%	2.52×10^5	8%	3.19×10^6
C16	5.77×10^5	20%	2.11×10^6	73%	2.22×10^5	8%	2.91×10^6

4.2. Oil Production

Oil production does not follow the same trend as the CO_2 storage. Models C1 through C9 indicate a declining oil production as fluid mobility described by the relative permeability curves decreases, except models C4 and C7 (Table 4). The relative permeability curve used in models C11, C12, and C14 describes a lower fluid mobility condition but predicts similar oil production as the highest fluid mobility models (C1 and C2); whereas the lowest fluid mobility relative permeability models predict the lowest oil production, as would be expected. A possible reason for the high oil production show in models C11, C12, and C14 is that the oil saturation falls within a range on the relative permeability curve that promotes the oil phase mobility, the mid-range of the relative permeability curve. Oil saturations are between 40% and 60% in the active production and injection areas giving relative permeability ranges of 0.23 to 0.35. Within the same area, the relative permeabilities to water never get above 0.1, promoting oil mobility over water mobility.

Table 4. Table shows the forecasted total oil produced for each simulation case and the magnitude of difference when compared to the Base Case (BC) forecasts.

Model	Oil Produced (STB)	
	Total	**Delta vs. BC Model**
BC	2.16×10^7	0
C1	2.68×10^7	5.14×10^6
C2	2.57×10^7	4.08×10^6
C3	2.32×10^7	1.59×10^6
C4	2.77×10^7	6.03×10^6
C5	2.20×10^7	3.56×10^5
C6	2.04×10^7	-1.23×10^6
C7	2.60×10^7	4.39×10^6
C8	1.89×10^7	-2.77×10^6
C9	1.77×10^7	-3.96×10^6
C10	2.30×10^7	1.38×10^6
C11	2.63×10^7	4.69×10^6
C12	2.90×10^7	7.40×10^6
C13	2.26×10^7	9.84×10^5
C14	2.60×10^7	4.33×10^6
C15	1.51×10^7	-6.54×10^6
C16	1.75×10^7	-4.17×10^6

The Base Case (BC) model's oil production falls within the middle of the range predicted by this study, similar to models C6, C10, C13 (Table 4). This indicates that the Base Case relative permeability curve was measured from an area of the field that exhibits medium fluid mobility when compared to the range of fluid mobilities predicted by the synthetic relative permeability curves.

4.3. Pressure

The influence of relative permeability on the pressure field was highly time dependent. Figure 6 indicates that during the first phase of injection when historical data are used to control the injection rates (2010 to 2017), there is less difference in pressure between the relative permeability curves tested. During the predictive phase of the injection schedule (2017 to 2036), the difference across all the relative permeability curves tested increased to 27% by the end of injection. This variation in pressure increases significantly as the proportion of recycled CO_2 in the injection stream increases. An inflection point in all of the models on 1 January 2024 marks the point when CO_2 availability for injection becomes tied to production volumes and hence the fluid movement within the reservoir. On 1 January 2030, there is another inflection point that marks when new CO_2 to the model is stopped and only recycled CO_2 is available to the injection wells.

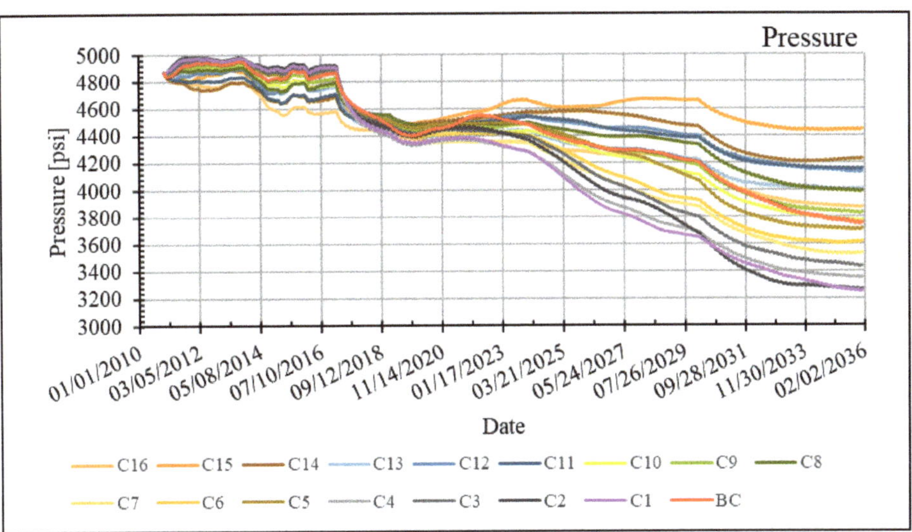

Figure 6. Pressure across the whole domain through time. The bold red line is the Base Case results.

The relative permeability curves appear to have a smaller, but still significant, impact on the average reservoir pressure when there is an unlimited source of CO_2 for the injection wells to meet their targets. Once the fluid mobility within the reservoir influences the volume of CO_2 available to meet injection targets, the impact of the relative permeability curves becomes much more pronounced. Curves that restrict the fluid movement through a high degree of curvature and narrow saturation range, such as C15 and C16, have lower oil production, less CO_2 present in the reservoir, and a higher average pressure, while the curves that promote fluid movement through more linear curvature and a wider saturation range (C1 and C2) have higher oil production, more CO_2 in the reservoir, and a lower average reservoir pressure.

5. Conclusions

Results of this study indicate that small variations in the shape of the relative permeability curve have a significant impact on the model forecasts. While all models predicted nearly the same total CO_2 stored in the reservoir, the phase it is stored as (supercritical vs. dissolved in oil vs. dissolved in water) is greatly influenced by the relative permeability curve. Relative permeability curves that describe low fluid mobility predict most of the CO_2 is stored in the oil phase with very little in the supercritical phase. The relative permeability curves that describe the highest overall fluid mobility predicts that there is an even distribution of CO_2 in the supercritical phase and the oil phase, allowing the CO_2 to migrate faster and in greater quantities to the production wells, leading to lower amounts of total stored CO_2. The higher the mobility, the more contact between the CO_2 plume and the oil, increasing the amount of CO_2 that is dissolved in the oil phase and thereby increasing production when compared to the relative permeability curves that predict low overall fluid mobility. The reduction in the oil's viscosity due to CO_2 dissolution allows it greater fluid mobility and may be why there is an increase in oil production when compared to some of the relative permeability curves that predict low fluid mobility. This may also account for why there is less CO_2 stored in the oil phase with the relative permeability curves that describe high fluid mobility.

The findings of this study indicate that the relative permeability curve is a critical parameter that must be given careful consideration when designing multiphase flow models. It is essential to understand and quantify the uncertainty in the relative permeability curve.

If measured laboratory relative permeability data are not available or limited for the study domain, the relative permeability curve should be considered a significant source of model uncertainty and accounted for as part of the simulation effort.

Author Contributions: Conceptualization, N.M. and B.M.; Data curation, N.M.; Formal analysis, N.M.; Funding acquisition, B.M.; Investigation, N.M.; Methodology, N.M., W.J. and B.M.; Project administration, B.M.; Software, N.M.; Supervision, B.M.; Validation, W.A.; Visualization, N.M.; Writing—original draft, N.M.; Writing—review & editing, N.M., W.A., W.J. and B.M. All authors have read and agreed to the published version of the manuscript.

Funding: Funding for this study was provided by the Southwest Regional Partnership on Carbon Sequestration (SWP) under Award No. DE-FC26-05NT42591.

Institutional Review Board Statement: Not applicable.

Informed Consent Statement: Not applicable.

Conflicts of Interest: The authors declare no conflict of interest. The funders had no role in the design of the study; in the collection, analyses, or interpretation of data; in the writing of the manuscript, or in the decision to publish the results.

Appendix A

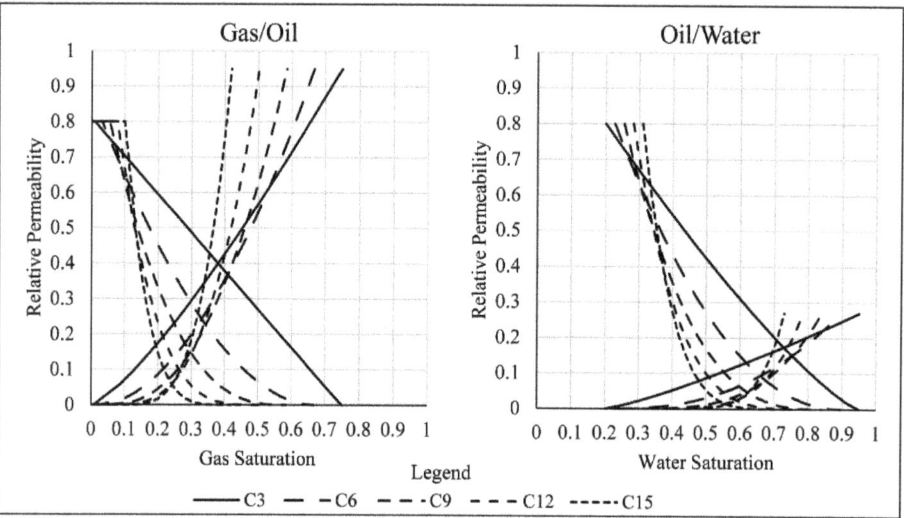

Figure A1. The C3, C6, C9, C12, and C15 relative permeability curves. The saturation endpoints and curvature are varied, while the relative permeability end points remain constant.

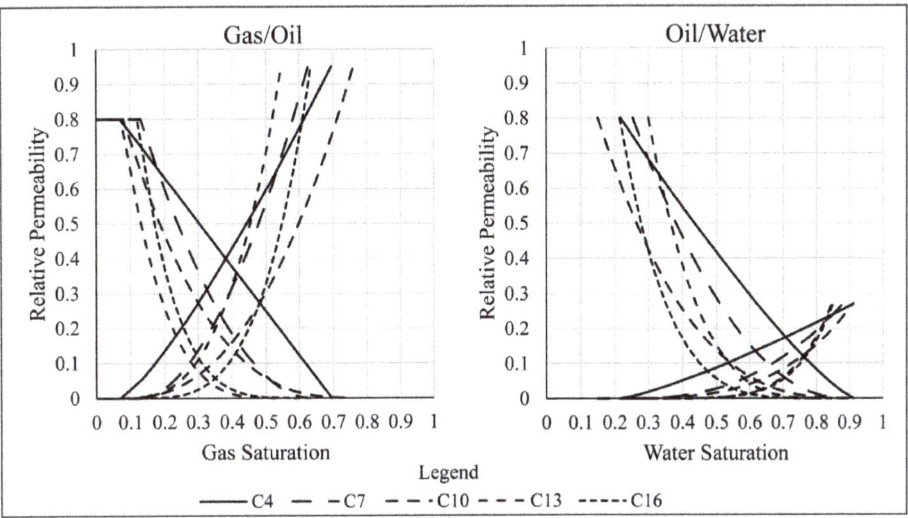

Figure A2. The C4, C7, C10, C13, and C16 relative permeability curves. The saturation endpoints and curvature are varied and the relative permeability endpoints are fixed.

References

1. Yang, C.; Dai, Z.; Romanak, K.D.; Hovorka, S.D.; Treviño, R.H. Inverse Modeling of Water-Rock-CO2 Batch Experiments: Potential Impacts on Groundwater Resources at Carbon Sequestration Sites. *Environ. Sci. Technol.* **2014**, *48*, 2798–2806. [CrossRef] [PubMed]
2. Dai, Z.; Middleton, R.; Viswanathan, H.; Fessenden-Rahn, J.; Bauman, J.; Pawar, R.; Lee, S.-Y.; McPherson, B. An Integrated Framework for Optimizing CO2 Sequestration and Enhanced Oil Recovery. *Environ. Sci. Technol. Lett.* **2014**, *1*, 49–54. [CrossRef]
3. Bachu, S. Identification of oil reservoirs suitable for CO2-EOR and CO2 storage (CCUS) using reserves databases, with application to Alberta, Canada. *Int. J. Greenh. Gas Control.* **2016**, *44*, 152–165. [CrossRef]
4. Bacon, D.H.; Qafoku, N.P.; Dai, Z.; Keating, E.H.; Brown, C.F. Modeling the impact of carbon dioxide leakage into an unconfined, oxidizing carbonate aquifer. *Int. J. Greenh. Gas Control.* **2016**, *44*, 290–299. [CrossRef]
5. Dai, Z.; Stauffer, P.H.; Carey, J.W.; Middleton, R.S.; Lu, Z.; Jacobs, J.F.; Hnottavange-Telleen, K.; Spangler, L.H. Pre-site Characterization Risk Analysis for Commercial-Scale Carbon Sequestration. *Environ. Sci. Technol.* **2014**, *48*, 3908–3915. [CrossRef] [PubMed]
6. Metz, B.; Davidson, O.; de Coninck, H.X.; Loos, M.; Meyer, L. *Carbon Dioxide Capture and Storage: Special Report of the Intergovernmental Panel on Climate Change*; Cambridge University Press: Cambridge, UK, 2005.
7. Dietrich, J.K.; Bondor, P.L. Three-Phase Oil Relative Permeability Models. *Soc. Pet. Eng.* **1976**, *6044*, 1976.
8. Oak, M.; Baker, L.; Thomas, D. Three-Phase Relative Permeability of Berea Sandstone. *J. Pet. Technol.* **1990**, *42*, 1054–1061. [CrossRef]
9. Saraf, D.; Batycky, J.; Jackson, C.; Fisher, D. An Experimental Investigation of Three-Phase Flow of Water-Oil-Gas Mixtures Through Water-Wet Sandstones. *Soc. Pet. Eng.* **1982**, *10761*. [CrossRef]
10. Corey, A.T. The interrelation between gas and oil relative permeabilities. *Prod. Mon.* **1954**, *19*, 38–41.
11. Land, C.S. Calculation of Imbibition Relative Permeability for Two- and Three-Phase Flow From Rock Properties. *Soc. Pet. Eng. J.* **1968**, *8*, 149–156. [CrossRef]
12. Stone, H. Probability Model for Estimating Three-Phase Relative Permeability. *J. Pet. Technol.* **1970**, *22*, 214–218. [CrossRef]
13. Stone, H. Estimation of Three-Phase Relative Permeability And Residual Oil Data. *J. Can. Pet. Technol.* **1973**, *12*, 53–61. [CrossRef]
14. Baker, L.E. Three-Phase Relative Permeability Corrections. In Proceedings of the Enhanced Oil Recovery Symposium, Tulsa, Oklahoma, 16–21 April 1988; Society of Petroleum Engineers: Tulsa, Oklahoma, 1988.
15. Moodie, N.; McPherson, B.; Lee, S.-Y.; Mandalaparty, P. Fundamental Analysis of the Impacts Relative Permeability has on CO2 Saturation Distribution and Phase Behavior. *Transp. Porous Media* **2014**, *108*, 233–255. [CrossRef]
16. Ampomah, W.; Balch, R.S.; Grigg, R.B.; Will, R.; Dai, Z.; White, M.D. Farnsworth Field CO2-EOR Project: Performance Case History. In Proceedings of the SPE Improved Oil Recovery Conference, Tulsa, OK, USA, 11–13 April 2016.
17. Dullien, F.A.L. *Porus Media: Fluid Transport and Pore Structure, Second*; Academic Press, Inc.: San Diego, CA, USA, 1992.
18. Sarem, A. Three-Phase Relative Permeability Measurements by Unsteady-State Method. *Soc. Pet. Eng. J.* **1966**, *6*, 199–205. [CrossRef]

19. PNNL. *STOMP—Subsurface Transport Over Multiple Phases*; Pacific Northwest National Laboratory: Richland, WA, USA, 2015; Available online: http://stomp.pnnl.gov/ (accessed on 1 June 2015).
20. Schlumberger. *Eclipse: Technical Description, Eclipse Reservoir Simulation Software*; Schlumberger Ltd.: Houston, TX, USA, 2010.
21. May, R.S. *A Numerical Simulation Study of the Farnsworth Unit Waterflood (West Side)*; UNOCAL Corporation: El Segundo, CA, USA, 1987.
22. May, R.S. *A Simulation Study for the Evaluation of Tertiary Oil Recovery by Co2 Injection in the Farnsworth Unit*; UNOCAL Corporation: El Segundo, CA, USA, 1988.
23. Munson, T.W. Depositional, Diagenetic, and Production History of the Upper Morrowan Buckhaults Sandstone, Farnsworth Field, Ochiltree County Texas. 1994. Available online: https://www.osti.gov/pages/servlets/purl/1639065 (accessed on 21 April 2021).
24. Bolyard, D.W. Upper Morrowan B sandstone reservoir, Flank field, Baca County, Colorado. In *Search for the Subtle Trap, Hydrocarbon Exploration in Mature Basins*; West Texas Geological Society: Midland, TX, USA, 1989; pp. 255–268.
25. Ampomah, W.; Balch, R.S.; Grigg, R.B.; Dai, Z.; Pan, F. Compositional Simulation of CO2 Storage Capacity in Depleted Oil Reservoirs. In Proceedings of the Carbon Management Technology Conference, Sugar Land, TX, USA, 17–19 November 2015.
26. Ross-Coss, D.; Ampomah, W.; Cather, M.; Balch, R.S.; Mozley, P.; Rasmussen, L. An Improved Approach for Sandstone Reservoir Characterization. In Proceedings of the SPE Western Regional Meeting, Anchorage, AK, USA, 23–26 May 2016; p. SPE-180375. [CrossRef]
27. Czoski, P. *Geologic Characterization of the Morrow B Reservoir in Farnsworth Unit, TX Using 3D VSP Seismic, Seismic Attributes, and Well Logs*; New Mexico Institute of Mining and Technology, Department of Earth and Environmental Science, Geophysics: Socorro, NM, USA, 2014.
28. Hutton, A.C. *Geophysical Modeling and Structural Interpretation of a 3D Reflection Seismic Survey in Farnsworth Unit, TX*; New Mexico Institute of Mining and Technology: Socorro, NM, USA, 2015.
29. Cather, S.M.; Cather, M.C. Comparative petrography and paragenesis of Pennsylvanian (Upper Morrow) sandstones from the Farnsworth Unit 13-10A, 13-14, and 32-8 wells, Ochiltree County, Texas. *PRRC Rep.* **2016**, 1–16. Available online: https://www.mdpi.com/1996-1073/12/19/3663/htm (accessed on 21 April 2021).
30. Moodie, N.; Ampomah, W.; Jia, W.; Heath, J.; McPherson, B. Assignment and calibration of relative permeability by hydrostratigraphic units for multiphase flow analysis, case study: CO2-EOR operations at the Farnsworth Unit, Texas. *Int. J. Greenh. Gas Control.* **2019**, *81*, 103–114. [CrossRef]
31. Hinds, R.F. *Reservoir Fluid Study, Messall No. Well, Farnsworth (Upper Morrow) Field*; Core Laboratories, Inc.: Amsterdam, The Netherlands, 1956.
32. Ampomah, W.; Gunda, D. *Farnsworth Field Reservoir Fluid Analysis*; New Mexico Tech; Petroleum Recovery Research Center: Socorro, NM, USA, 2015.

Article

Practical CO_2—WAG Field Operational Designs Using Hybrid Numerical-Machine-Learning Approaches

Qian Sun [1,2], William Ampomah [1,*], Junyu You [1,3], Martha Cather [1] and Robert Balch [1]

[1] Petroleum Recovery Research Center, New Mexico Institute of Mining and Technology, Socorro, NM 87801, USA; sunqian.psu@vip.126.com (Q.S.); junyu.you@student.nmt.edu (J.Y.); martha.cather@nmt.edu (M.C.); robert.balch@nmt.edu (R.B.)

[2] School of Energy Resources, China University of Geoscience, Beijing 100083, China

[3] School of Petroleum and Natural Gas Engineering, Chongqing University of Science and Technology, Chongqing 401331, China

* Correspondence: william.ampomah@nmt.edu

Abstract: Machine-learning technologies have exhibited robust competences in solving many petroleum engineering problems. The accurate predictivity and fast computational speed enable a large volume of time-consuming engineering processes such as history-matching and field development optimization. The Southwest Regional Partnership on Carbon Sequestration (SWP) project desires rigorous history-matching and multi-objective optimization processes, which fits the superiorities of the machine-learning approaches. Although the machine-learning proxy models are trained and validated before imposing to solve practical problems, the error margin would essentially introduce uncertainties to the results. In this paper, a hybrid numerical machine-learning workflow solving various optimization problems is presented. By coupling the expert machine-learning proxies with a global optimizer, the workflow successfully solves the history-matching and CO_2 water alternative gas (WAG) design problem with low computational overheads. The history-matching work considers the heterogeneities of multiphase relative characteristics, and the CO_2-WAG injection design takes multiple techno-economic objective functions into accounts. This work trained an expert response surface, a support vector machine, and a multi-layer neural network as proxy models to effectively learn the high-dimensional nonlinear data structure. The proposed workflow suggests revisiting the high-fidelity numerical simulator for validation purposes. The experience gained from this work would provide valuable guiding insights to similar CO_2 enhanced oil recovery (EOR) projects.

Keywords: multi-objective optimization; CO_2-WAG; machine learning; numerical modeling; hybrid workflows

Citation: Sun, Q.; Ampomah, W.; You, J.; Cather, M.; Balch, R. Practical CO_2—WAG Field Operational Designs Using Hybrid Numerical-Machine-Learning Approaches. *Energies* **2021**, *14*, 1055. https://doi.org/10.3390/en14041055

Academic Editor: Valentina Colla

Received: 21 January 2021
Accepted: 12 February 2021
Published: 17 February 2021

Publisher's Note: MDPI stays neutral with regard to jurisdictional claims in published maps and institutional affiliations.

1. Introduction

The Southwest Regional Partnership on Carbon Sequestration (SWP) project focuses on the design and monitoring of a field-scale CO_2 enhanced oil recovery (EOR) process in the Farnsworth Unit (FWU) located in the Anadarko Basin, Texas. From 2010 to 2014, 16.82 billion standard cubic feet of anthropogenic CO_2 was injected into the Morrow-B sand [1]. the CO_2 utilized in this project is captured by the Arkalon Ethanol Plant and the Agrium Fertilizer Plant locating in Liberal, Kansas, and Borger, Texas, respectively [2]. According to Munson [3], the original oil (OOIP) and gas in place (OGIP) are approximately 120 million (MM) barrels and 41.48 billion standard cubic feet (SCF), respectively. The field development was initiated in 1955 and the waterflood started in 1963. Starting from the end of 2010, the FWU field has been undergoing a water alternative CO_2 injection process to extract the residual oil in place.

The Morrow B sandstone formation is located at a depth interval between 7550 ft and 7950 ft. The formation has an average dip angle of <1° [4] and was deposited in the late Pennsylvanian by a fluvial system in an incised valley [5]. The average net pay thickness is 22 ft. The initial reservoir and bubble point pressure are 2203 psi and 2059 psi, respectively [4]. The reservoir temperature at a depth of 7900 ft is measured to be 168°F. The reservoir has a mean porosity of 14.6% and a mean permeability of 58 md [6]. The west half of the field is considered highly permeable and porous when compared with the east half [3]. The reservoir was initially undersaturated with gas–oil solution ratio of 345 SCF/stock tank barrel (STB), and the oil saturation and formation volume factor were characterized as 69% and 1.192 res bbl/STB, respectively [6].

Reservoir engineers have successfully structured many different versions of numerical simulation models to investigate the fluid flow transportation dynamics in Morrow-B formation and monitor the long-term fate of the CO_2 plume [7]. A numerical compositional model coupled with geological, geophysical, and engineering data was reported in previous work [8]. The initial SWP's FWU geological static model was first presented in 2015 [9] and was employed by many related works [10–12]. Investigations on the hydraulic flow unit (HFU) was performed to characterize the heterogeneous petrophysical properties [13]. A facies model was developed to populate the petrophysical property distributions with assistance from the HFU. There are numerous history-matching studies imposed on reservoir models considering the field historical injection and production data in the primary, secondary, and tertiary recovery period [14]. The history-matched model can be used to assess various CO_2 water-alternative-gas (WAG) forecasting scenarios and more importantly, optimize the project design strategies.

The machine-learning technologies exhibit strong competences to solve a large spectrum of petroleum engineering problems, including sweet spot identification [15], history matching [16], fluid property characterization [17], and field development strategy optimization [18]. In the field of reservoir simulation, the machine-learning models comprehend the fluid transportation dynamics in porous media via learning the data structure presented by a knowledge base instead of solving the partial differential equations using numerical and analytical methods [19]. The training of the machine-learning model takes advantages of the knowledge base containing field and synthetic data samples. To validate the machine-learning model, rigorous blind testing applications must be imposed to investigate the generalization capability. The blind testing performances of the machine-learning models used in this work were less than 1%. Notably, the computational overhead using the machine-learning model could be reduced by several orders of magnitudes more than using the high-fidelity numerical simulator [20]. Literature surveys indicate that machine-learning models have been successfully developed to simulate production performance of gas condensate reservoirs [21], shale gas reservoirs [22], coalbed methane reservoirs [23], enhanced oil recovery processes [24], etc. Moreover, for certain field specified problems, the machine-learning models can also be trained utilizing seismic, well-log, and production data, which is competent to assess the production performance within the area covered by the seismic survey [25]. In this case, the machine-learning model can make reliable predictions without the presence of hydrodynamic (permeability, relative permeability, etc.) and petrophysical (fluid composition, pressure-volume-temperature (PVT) data, etc.) data. Moreover, recent research efforts make attempts to develop machine-learning models as an alternative to the equation of state and conduct flash calculations in the compositional simulations [26]. Such well-trained machine-learning models (also known as expert systems) can be employed as "proxies" of the high-fidelity numerical simulator to generate a large volume of realizations in the computational demanding processes.

The SWP project would demand the use of machine-learning methods to solve the history-matching and CO_2-WAG injection optimization problems. As a carbon capture, utilization, and sequestration (CCUS) project, the field operator considers not only the oil recovery, but also the volume of CO_2 sequestration, both of which determine the project economics from different perspectives. When more objective functions are included, the computational overhead of the optimization study would be even heavier. In this article, a robust machine-learning assisted multi-objective optimization workflow is presented. Expert proxy models, a global optimization algorithm, and numerical models are effectively coupled in the workflows to generate fast and high-quality reservoir engineering analysis. The discussion of this paper starts from the description of the reservoir model. Then the machine-learning proxies employed by this work are summarized. Afterwards, the optimization algorithms and treatment of the multi-objective problem are presented. By coupling the machine-learning proxy and the optimization protocols, two workflows are structured to solve the history-matching and CO_2-WAG design optimization problems. Last but not least, we employ case studies by imposing the proposed workflow on the FWU field.

2. Reservoir Modeling

A reservoir model is the most fundamental element of the proposed workflow. The establishment of the numerical simulation model assembles geological, petrophysical, and field historical data. Reservoir engineers have structured many versions of numerical simulation models for the FWU field considering hydrodynamic, geo-mechanical, and geochemical mechanisms of the fluid transportation in Morrow-B formation. The field scale numerical models are validated via rigorous history-matching processes using the historical injection and production data and employed to build forecasting scenarios. In addition, sector models around the well 13–10A pattern are utilized to validate the optimization workflows.

2.1. Hydraulic Flow Unit

The geological model takes advantage of the petrophysical properties interpolated from the laboratory measurements using the hydraulic flow unit (HFU). Morrow-B sandstone exhibits strong heterogeneity divergences from the diagenetic processes, which leads to various pore structures, multiphase flow characteristics, and wettability. Moreover, rock and fluid property variations could evolve progressively after CO_2-WAG injection starts. The porosity–permeability relationship can be identified using the HFU to classify the sedimentologic and diagenetic heterogeneity of the Morrow-B formation.

Morrow B sandstone was first classified into five porosity facies and eight subfaces; and eight HFUs were identified to characterize hydrogeologic heterogeneities using the well log and petrophysical data collected from various scientific wells [13]. Followed up research works suggest lumping HFU 5 to 8 into an identical group due to the similar flow features observed from core-flooding experiments [27].

The geological model continuously utilizes the permeability–porosity relationship developed from the HFUs. When a reservoir simulation model is structured, grids are assigned with different relative permeability curves based on the HFU characteristics. In Figure 1, the HFU distributions at different simulation layers of the reservoir model are displayed.

2.2. Updated Geological Model

As with the advancement of the reservoir characteristic analysis on the FWU field, the original simulation model is restructured using an updated outer boundary and property population protocol. As shown in Figure 2, the permeability and porosity distributions of the updated geological mode are displayed in Figure 2a,c using a 100 ft by 100 ft mesh grid system. To implement the updated geological model, the history-matching process must be revisited from the primary and secondary stages. Due to the long production period (from 1956 to 2010), completing one simulation case takes more than 12,000 s of central processing unit (CPU) time, which makes the computational cost of the history-matching process prohibitively expensive. With such a running speed, even preparing the dataset to train the proxy model becomes unrealistic. Thus, the upscale of the original model becomes necessary. Figure 2b,d show the upscaled permeability and porosity distribution using a 200 ft by 200 ft mesh grid system. Figure 3 illustrates the comparison of the simulation results regarding the oil and water production using the original and upscaled reservoir model with an average disparity of less than 0.8%. The use of the upscaled model significantly reduces the computational time to less than 300 s of CPU time, which is 40 times faster than the original model. With the help of upscaling, the history-matching of the primary and secondary recovery periods is done using the coarser grid first and then validated by the finer grids.

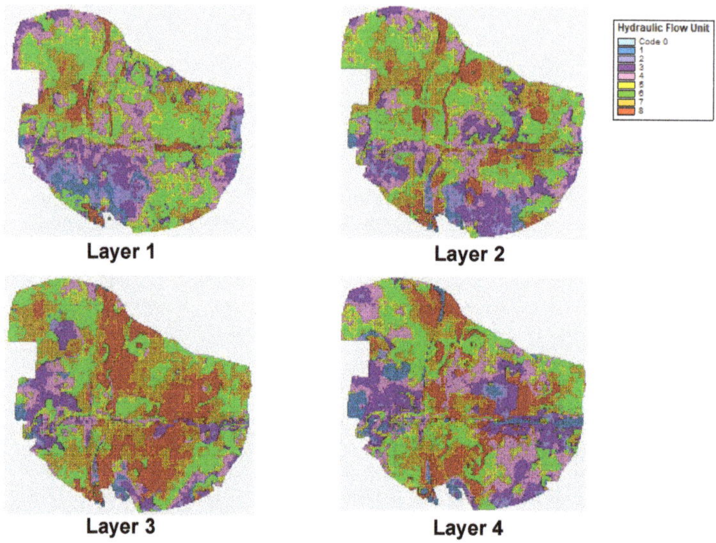

Figure 1. HFU (hydraulic flow unit) distributions in different layers of Morrow-B formation.

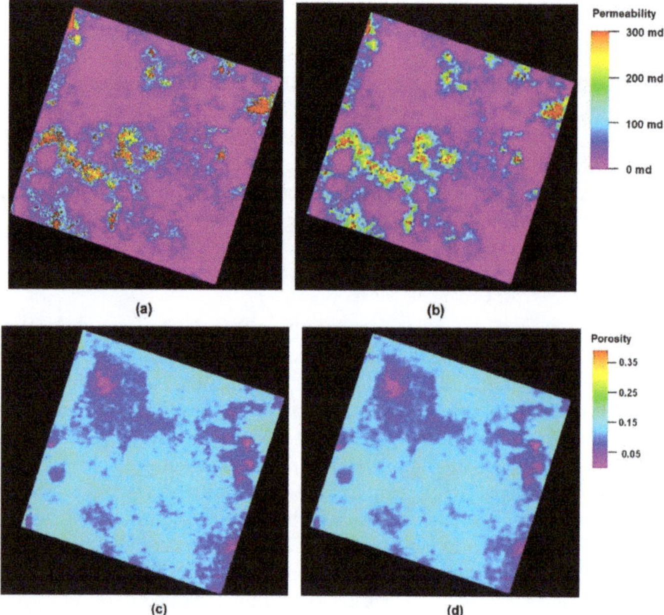

Figure 2. Updated geological model and the upscaled grid displayed as (**a**) permeability in a finer grid, (**b**) permeability in a coarse grid, (**c**) porosity in a finer grid, and (**d**) porosity in a coarse grid.

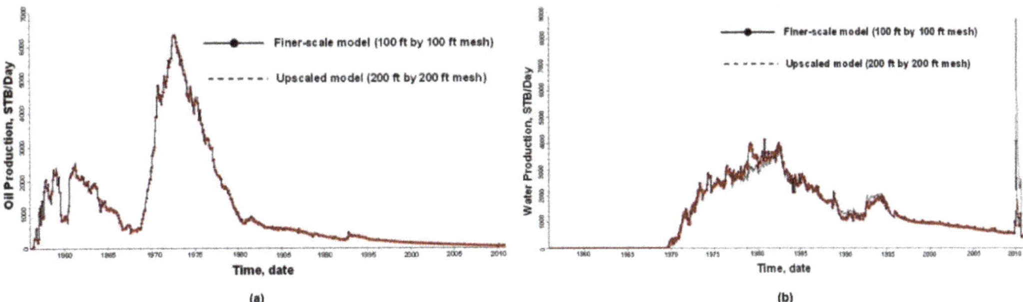

Figure 3. Comparison of the results generated by the finer and upscaled (coarser) grid simulation models in terms of (**a**) oil production and (**b**) water production.

2.3. Injection Pattern Model

A sector model focusing on the Well 13–10A pattern was structured to simulate the smaller area. As shown in Figure 4, the sector model is discretized using a 60 by 60 Cartesian grid system (mesh size is 26.7 ft by 26.7 ft) and simulates 1/8 of the five-spot injection pattern. The property distribution is established using the data collected from 13-A well. This work employs the sector model to generate datasets for the testing purposes of the optimization workflow, since the computational cost of completing one simulation is low. Moreover, the injection pattern model is suitable to investigate multiple mechanisms near the scientific well.

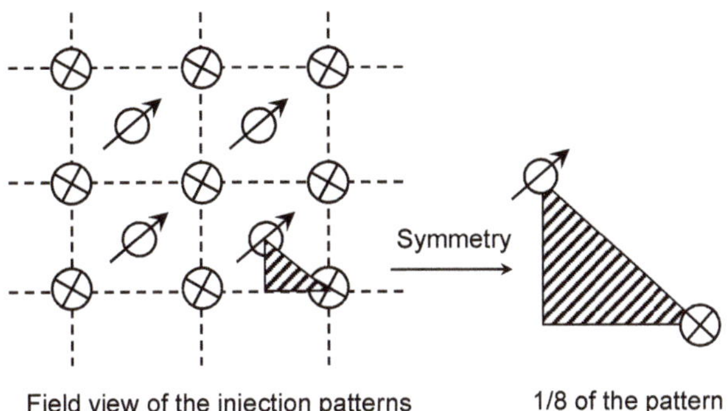

Figure 4. Illustration of the injection pattern-based model.

The aforementioned simulation models played a crucial role in this work in terms of comprehending the fluid flow dynamics in the Morrow-B sands, and more importantly, generating the desired data to develop the machine-learning proxy models. It is worth stressing that the interaction between the high-fidelity numerical and machine-learning protocols is critical for the long-term development and updating of the workflow.

3. Machine-Learning Proxies

In this work, the history-matching and CO_2 injection design processes employed many machine-learning technologies such as response surface models (RSM), multi-layer neural networks (MLNN), and support vector machines (SVM). These expert machine-learning models (also called "proxies") are developed to learn from the numerical simulation models and investigate the fluid transportation mechanisms via a data-driven perspective. Due to their high computational efficacies and robust generalization competence, machine-learning models have been successfully developed and act as regression tool to evaluate the field performance in terms of the fluid recovery, injection performance, pressure responses, and economic assessments.

In the history-matching applications, the proxy models correlate the uncertain hydro-dynamical properties with the fluid production/injection and pressure data, which assists the evaluation of the fitting error during the tuning process. There are many generations of reservoir simulation models evolving as more geological, petrophysical, and field operational data become available. The initial history matching works can be done without the help of machine-learning technologies because the earlier version of the reservoir model is simpler in terms of the reservoir characteristics. When the new geological model arrives, the permeability distributions are updated and more importantly, various relative permeability curves are assigned to the grids based on the HFU indices. Consequentially, the number of uncertain parameters becomes considerably larger, and so does the required volume of simulation realizations. Therefore, the use of machine-learning technologies becomes quite necessary to history-match the current version of the simulation model.

After a history-matched model was structured, another class of proxy model was developed to learn the data structure presented between the CO_2-WAG injection strategies and the forecasted project's techno-economic responses. The development of the proxy models includes the training process and a rigorous blind testing protocol to ensure the prediction accuracy. The following discussions summarize the machine-learning models developed in the history-matching and optimization studies of the SWP CO_2-WAG projects.

3.1. Response Surface Models

The response surface model (RSM) is one of the most classic nonlinear regression algorithms. The prediction model can be expressed via Equation (1):

$$y = b_0 + \sum_{i=1}^{m} b_i x_i + \sum_{i=1}^{m} \sum_{j \geq i}^{m} b_{ij} x_i x_j + \sum_{i=1}^{m} b_{ii} x_i^2 \tag{1}$$

where b_0, b_i, b_{ii}, and b_{ij} are the regression coefficients; x_i is the ith input feature; m is the total number of input features; and y is the output feature. To avoid overfitting issues, the higher than quadratic order terms are usually omitted. Gradient decent and least square methods [28] can be used to determine the regression coefficient. The RSM model is the earliest machine-learning model used in the CO_2-WAG optimization study of the SWP project. Four RSM proxies are developed to predict oil and water productions, CO_2 in place, and production volumes in the CO_2-WAG injection period of the project [6]. The R^2 value observed by comparing the results predicted by proxy and high-fidelity numerical models is close to one with relative error values of less than 0.1%, which confirms the validity of the RSM proxies.

3.2. Multi-Layer Neural Networks

Multi-layer neural networks (MLNN) model a class of artificial neural network (ANN) model that is inspired by the signal transportation process of biological neural units. Typically, an MLNN model is composed of an input, an output, and serval hidden layers. Artificial neurons are included in these layers to store, transform, and transport data signals. The numerical values connecting the neurons in adjacent layers are "weights." An MLNN model transfers the signal via Equation (2):

$$a_i^n = f_n \left(\sum_j w_{i,j}^n a_j^{n-1} + b_i^n \right) \tag{2}$$

where the notation n numbers the current layer ($n - 1$ is the previous layer) of the layer, a is the artificial neuron, and w and b are the weight and the bias, respectively. The index notations i and j number the layer and neuron, respectively. Notably, f_n is a transfer function constraining the signal within the range of [0, 1] or [−1, 1], which enables the MLNN model to deal with the nonlinearity of the dataset. The sigmoid and linear functions are commonly imposed on the hidden and output layers, respectively. The training process of the MLNN model essentially minimizes the error function depicted by Equation (3):

$$E(w) = \frac{1}{2}||o - t||^2 \tag{3}$$

where w refers to the weight vector, and o and t are the prediction and training target, respectively. Thus, the training epoch of the MLNN iteratively updates w to minimize the error function until a prescribed stopping criterion is achieved. Many training algorithms such as the scaled conjugate gradient and the Levenberg–Marquardt and Bayesian regularization [29] methods are widely used in training MLNN models with complex topologies.

There are many MLNN applications in the SWP project: In the machine-learning-assisted history-matching process, MLNN models are trained to predict the field production responses with varying reservoir properties. Two representative blind test cases are shown in Figure 5. The proxy model is competent to predict the oil and water productions when the hydrodynamical properties are changed, with overall blind testing errors at 0.5% and 9.87%, respectively. As shown in Figure 5, the satisfactory blind testing performance indicates that the ANN models have been well trained and could be used to evaluate the history-matching error in the proposed workflow. Notably, such results have not been history-matched yet.

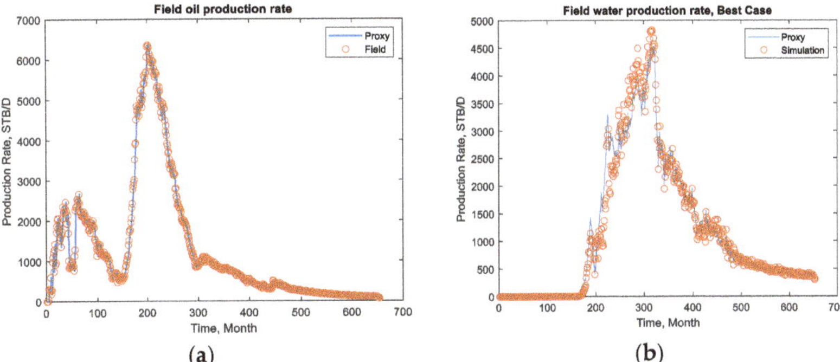

Figure 5. Blind testing performance of multi-layer neural networks (MLNN) proxies in the history-matching applications of (**a**) oil production and (**b**) water production.

The other class of MLNN models are trained to forecast the long-term field development responses in terms of hydrocarbon production, CO_2 sequestration volume, and project economic assessments considering the CO_2-WAG design parameters as input [18]. The CO_2-WAG operational parameters include the durations of CO_2 and water injection cycles, the water injection rate, and production well specifications. In this work, the error statistics of blind testing applications for an injection pattern and the field case applications, with overall mean errors observed as 0.71% and 1.00%, respectively. Therefore, the MLNN model can be employed as a surrogate model to evaluate the techno-economic objective functions considered in the design of CO_2-WAG projects.

In the SWP project, MLNN models have been successfully trained and act as robust regression tools. The high computational speed enables many time-consuming studies such as global sensitivity analysis, history-matching, multi-objective optimization, etc.

3.3. Support Vector Machines

The support vector machine (SVM) is a robust regression model employed in this work. A linear SVM regression model is an extension of a linear regression model that aims at [30]:

$$\min\left(\frac{1}{2}||\omega||^2\right) \tag{4}$$

$$\text{subject to} \begin{cases} y - \langle \omega, x \rangle - b \le \varepsilon \\ \langle \omega, x \rangle + b - y \le \varepsilon \end{cases} \tag{5}$$

The model is suitable to make predictions for problems with n input and 1 output, where $x \in \mathcal{R}^n$ and $y \in \mathcal{R}$. Note that in Equations (4) and (5), ω is the coefficient vector and b is the intercept. The flatness of the model measures how small the norm of ω could be. ε is the error tolerance of the regression problem. The operator $\langle x, \omega \rangle$ indicates the dot product of vectors x and ω. To extend the applicability of SVM model to solve real problems, the soft margin is introduced to Equation (6) via a slake variable ξ:

$$\min\left[\frac{1}{2}||\omega||^2 + C \sum_{i=1}^{m}(\xi_i + \xi_i^*)\right] \tag{6}$$

$$\text{subject to} \begin{cases} y_i - \langle \omega, x_i \rangle - b \le \varepsilon + \xi_i \\ \langle \omega, x_i \rangle + b - y_i \le \varepsilon + \xi_i^* \\ \xi_i, \xi_i^* \ge 0, \; i = 1, \ldots, m \end{cases}$$

where C is a constant to balance the flatness and the tolerance degree of the deviations larger than ε. A loss function depicted in Equation (7) is used:

$$|\xi|_\varepsilon = \begin{cases} 0 \text{ if } |\xi| \le \varepsilon \\ |\xi| - \varepsilon \text{ otherwise} \end{cases} \tag{7}$$

The modern SVM models are typically written via a dual optimization form, which can be expressed as follows:

$$\text{maximize} \begin{cases} -\frac{1}{2} \sum_{i,j=1}^{m} (\alpha_i - \alpha_i^*)(\alpha_j - \alpha_j^*)\langle x_i, x_j \rangle \\ -\varepsilon \sum_{i=1}^{m} (\alpha_i + \alpha_i^*) + \sum_{i=1}^{m} y_i(\alpha_i - \alpha_i^*) \end{cases} \tag{8}$$

$$\text{subject to} \sum_{i=1}^{m}(\alpha_i - \alpha_i^*) = 0; \alpha_i, \alpha_i^* \in [0, C]$$

where α and α^* are Lagrange multipliers.

For nonlinear problems, kernel functions are employed to map the original training patterns into an implicit feature space so that the linear SVM regression can be used. In this case, Equation (8) is modified as Equation (9):

$$\text{maximize} \begin{cases} -\frac{1}{2} \sum_{i,j=1}^{m} (\alpha_i - \alpha_i^*)(\alpha_j - \alpha_j^*)k(x_i, x_j) \\ -\varepsilon \sum_{i=1}^{m} (\alpha_i + \alpha_i^*) + \sum_{i=1}^{m} y_i(\alpha_i - \alpha_i^*) \end{cases} \tag{9}$$

where $k(x_i, x_j)$ is a kernel. In this work the Gaussian kernel written in Equation (10) was employed:

$$k(x_i, x_j) = \exp\left(-\frac{||x_i - x_j^2||}{2\sigma^2}\right) \tag{10}$$

Similar to the MLNN model, the training of an SVM model needs to identify three hyperparameters, including the constant C, magnitude of ε, and the kernel scale factor δ. In this work, the SVM models were developed to act as proxy models to predict the oil production (Proxy-Oil) and CO_2 storage volume (Proxy-CO_2). The cross validation results indicate that the relative error of the Proxy-Oil and Proxy-CO_2 comparing to the high-fidelity numerical simulator to be 0.06% and 0.01%, respectively [31].

By developing the machine-learning proxies, the following experiences can be summarized:

1. The RSM is more suitable for problems with a smaller input dimension and single output parameter. Compared to MLNN and SVM models, the training overhead is much lighter.
2. The MLNN model exhibits robust generalization capability for problems with large input and output dimensions. However, the degree of freedom of the hyperparameter is more than that of RSM and SVM models. Thus, more computational costs are required to obtain an optimum model with optimum prediction performance.
3. The SVM model is more suitable for problems with strong nonlinearity and a large input dimension. The number of hyperparameters to be tuned is smaller than in the MLNN model. However, it cannot make a prediction for more than one output variable.

All of the machine-learning proxies play vital roles in the optimization study by enabling various computational expensive processes, which require giant amounts of assessments on the objective functions.

4. Optimization Protocols

In this work, the metaheuristic and stochastic optimization protocols such as particle swarm optimization and genetic algorithms were employed to optimize various technical and economic objective functions. Compared to the conventional gradient-based optimization methods, such optimization protocols are not constrained by the continuity and differentiability of the objective function. Therefore, they are more suitable for solving problems with complex and implicit objective functions. However, the global optimization technologies used in this work would demand the establishment of populations with a certain volume of samples. Thus, totally relying on the high-fidelity numerical simulators would make the computational overhead prohibitively heavy. This section briefly goes through the critical strategies of global technologies included in this work.

4.1. Objective Functions and Constraints
Economic Objective Functions

The project net present value (NPV) is a major economic consideration in the optimization study. A general definition of NPV can be depicted as Equation (11):

$$NPV = \text{CAPEX} + \sum_{i=1}^{m} \frac{(q_{oi} \times Oil\ price) \times (1 - tax\ rate) - C_i}{(1 + interest\ rate)^i} \tag{11}$$

In Equation (11), $CAPEX$ is the project capital expenditure, m is the total number of counted timesteps of the project, q_{oi} is the cumulative oil production of the ith timestep, and C_i is the operational cost. In this work, the NPV definition was modified to adapt to the CO_2-WAG field operation as Equation (12):

$$NPV = CAPEX + \sum_{i=1}^{n} \frac{(q_o \times Oil\ price + q_{co2,store} \times r_{co2,credit} - C_i)}{(1 + interest\ rate)^i} \tag{12}$$

where the $CAPEX$ = 10 million dollars (USD), oil price is 50 USD/bbl, interest rate is 5%, the CO_2 storage credit ($r_{co2,credit}$) is 45 USD/metric ton. The cost term C_i includes water and CO_2 injection cost, CO_2 purchase rate, and produced water treatment cost, which can be expressed by Equation (13):

$$C_i = q_{w,inj} \times r_w - q_{w,pro} \times r_{w,pro} - q_{inj,\,co2} \times r_{co2} - q_{co2,p} \times r_{co2,p} \tag{13}$$

where the water injection cost (r_w) is 1.03 USD/STB, the produced water treatment cost ($r_{w,pro}$) is 0.64 USD/STB, the CO_2 purchase cost is 1.72 USD/thousand(M) SCF, CO_2 injection cost is 0.85 USD/MSCF. The variables $q_{w,inj}$ and $q_{w,pro}$ are water injection and production rates in STB/day, respectively; $q_{inj,CO2}$ and $q_{CO2,p}$ are CO_2 injection and purchasing rates in MSCF/day, respectively.

The evaluation of the NPV value requires the oil/CO_2/water production and CO_2/water injection data reported by the numerical simulation model, which is determined by the CO_2-WAG design strategies. When the NPV is considered one of the objective functions in the optimization workflow, a large volume of realizations is required to find the optimum CO_2-WAG injection protocol to maximize the NPV. In this work, the calculation of the NPV was done by the proxy models with a low computational cost.

4.2. Technical Objective Functions

History matching error: The history-matching error is a significant objective function to measure the misfit of the numerical model results with the field historical data. In this work, the square error was used to account for the differences between the results generated by the simulator with the field historical measurement for oil, water, gas production data, water, gas injection data, and pressure data via Equation (14) through Equation (19), respectively.

$$E_o = \sum_i^m \left[(O_o - T_o)_i^2 \right] \tag{14}$$

$$E_{wp} = \sum_i^m \left[(O_{wp} - T_{wp})_i^2 \right] \tag{15}$$

$$E_{gp} = \sum_i^m \left[(O_{gp} - T_{gp})_i^2 \right] \tag{16}$$

$$E_{wi} = \sum_i^m \left[(O_{wi} - T_{wi})_i^2 \right] \tag{17}$$

$$E_{gi} = \sum_i^m \left[(O_{gi} - T_{gi})_i^2 \right] \tag{18}$$

$$E_p = \sum_i^m \left[(O_p - T_p)_i^2 \right] \tag{19}$$

where O and T represent the model prediction and historical data, respectively, and m is the total number of timesteps of the available field data. Notably, in this work, O was generated by machine-learning proxy models to reduce the computational overhead.

CO_2 storage volume: For a CCUS project, the CO_2 storage volume is a critical technical objective function considered in the optimization study, which can be defined as [6]:

$$CO_2 \text{ storage} = CO_2 \text{ purchased} - CO_2 \text{ produced} + \text{Recycle} \tag{20}$$

The CO_2 storage volume plays a crucial role in the project economic perspective by bringing considerable tax allowance to the project. Since the SWP project has rigorous plans to strategically purchase a certain volume of CO_2 based on the surface operational facility capacities, maximizing the CO_2 storage volume is essentially minimizing the CO_2 production volume in the CO_2-WAG process.

Oil recovery: The cumulative oil production (which could be normalized by the residual oil in place as the recovery factor) in the forecasting period is another technical objective function to be maximized in the design of the CO_2-WAG project, which also decides the major income of the project NPV.

Physical and Engineering Constraints

The optimization protocols design the CO_2-WAG operational criteria from the algorithmic perspective. Without imposing proper physical engineering and physical constraints, the optimization results would be unrealistic. In this work, the following constraints were imposed on the optimization workflow to ensure the engineering practicability:

1. The history-matching study specifies the oil production, CO_2, and water injection rates, and considers the CO_2 and water production the primary objective functions. The constraints imposed on the history-matching work is that the average pressure must be below 5400 psi.
2. The CO_2-WAG optimization is constrained by an average reservoir pressure range of [3700, 5400] psi to maintain the miscibility of the sweeping front.

In the optimization process, either for the history-matching or project design purposes, the solutions in the population pool would be screened by the constraints to sustain the engineering applicability. To meet such a requirement, proxy models need to be developed to predict the average reservoir pressure based on various history-matching and CO_2-WAG design scenarios.

4.3. Treatment of Multiple-Objective Optimizations

The optimization studies in this work considered more than one objective function, which are called multi-objective optimization problems (MOO). For instance, the history-matching work needs to minimize the error functions defined by Equation (14) through Equation (19) simultaneously, and the CO_2-WAG design problem aims at maximizing the NPV, oil recovery, and CO_2 storage volume in the meantime. There were two different ways employed in this work to treat the MOO problems:

Aggregated method: An aggregated objective function can be defined by Equation (21):

$$f_a = \sum_{i=1}^{m} (w_i \times f_i) \tag{21}$$

where f_i refers to various objective functions, w_i is the associated weights applied to the functions, and f_a is the aggregated objective function. In this way, the MOO is converted to a single-objective optimization study. In this work, the weighted method was used to solve the history matching problem by defining the history matching error as Equation (22):

$$f_{HME} = E_o + E_{gp} + E_{wp} + E_{wi} + E_{gi} \tag{22}$$

where the weights imposed on the error terms are identical. In addition, the optimization of the CO_2-WAG design uses the aggregated objective function depicted in Equation (23) [6]:

$$f = w_1 \times \text{Cumulative Oil Production} + w_2 \times CO_2 \text{ Storage} + w_3 \times \text{NPV} \tag{23}$$

The weight factors can be justified based on the operational preference.

Pareto optimum theory: An alternative approach to treating the MOO problem is to generate a Pareto front solution by employing the Pareto optimum theory, which is suitable for establishing a solution repository considering the tradeoff relationship amongst multiple objective functions [32].

The Pareto optimum theory defines a vector $\vec{x} = [x_1, x_3, x_3, \cdots x_n]$ with n variables, and a vector $\vec{f}(\vec{x}) = \left[\vec{f}_1(\vec{x}), \vec{f}_2(\vec{x}), \vec{f}_3(\vec{x}), \cdots \vec{f}_m(\vec{x})\right]$ with m objective functions considered in the optimization process. Ω assembles the candidate solutions in the searching domain. A solution $\vec{u} = [u_1, u_3, u_3, \cdots u_m]$ is defined to dominate solution $\vec{v} = [v_1, v_3, v_3, \cdots v_m]$, where $\vec{u}$ and $\vec{v}$ are two objective function vectors, if $\forall i \in \{1, 2, 3, \cdots m\}$ to make $u_i \leq v_i$ and $\exists i \in \{1, 2, 3, \cdots m\}$ to make $u_i < v_i$. This concept can be denoted as $\vec{u} \preceq \vec{v}$. The Pareto optimum set collects the solution vectors as $P^* = \{ x \in \Omega | \exists x' \in \Omega \vec{f}(x') \preceq \vec{f}(x) \}$ and the corresponding objective function ensemble is defined as the Pareto Front: $PF^* = \left\{ \vec{f} = \left[f_1\left(\vec{x}\right), f_2\left(\vec{x}\right), f_3\left(\vec{x}\right), \cdots, f_m\left(\vec{x}\right) \right] | x \in P^* \right\}$.

The Pareto optimum theory can be illustrated by Figure 6, which is a hypothetical problem aiming at minimizing f_1 and f_2. The solutions $\vec{x}_1^*$ and $\vec{x}_2^*$ are two solutions in the Pareto optimum set. $\vec{x}_1^*$ intends to improve objective function f_2 compared to $\vec{x}_2^*$ by sacrificing (increasing) f_1 due to the tradeoff relationship. In addition, $\vec{x}'$ is a dominating solution since solution $\vec{x}_1^*$ exists, which exhibits smaller f_1 and f_2 at the same time.

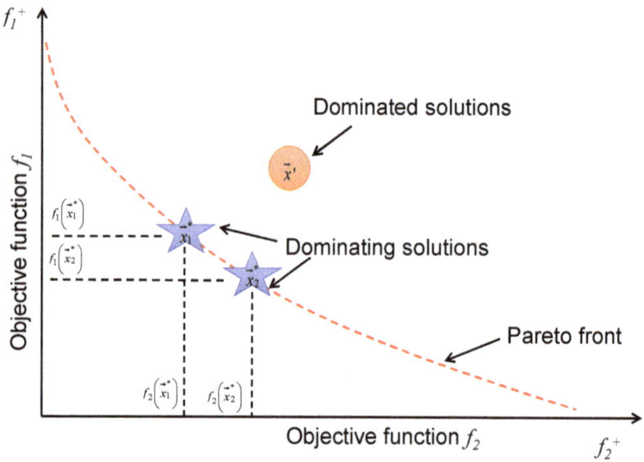

Figure 6. Pareto front, dominated, and dominating solutions.

It should be emphasized that regardless of the optimization algorithm, the application of the Pareto front theory by itself is a computational expensive process since it ranks a large volume of solutions regarding multiple objective functions. The CO_2-EOR project is a good candidate upon which to impose the Pareto front theory since the incremental oil production would yield more produced CO_2, which reduces the volume of carbon sequestration. More importantly, considering the project NPV, the impacts of the tax allowance and the oil production benefits need to be comprehensively investigated. More importantly, the Pareto front solution suggests the operators have more flexibility to design the CO_2-WAG processes under various techno-economic conditions. However, such an optimization protocol is not suitable for the history-matching problems. The reason is that history-matching processes shoot for a solution that minimizes all the objective functions simultaneously. A solution fitting the oil production history with large water production error could be a dominating solution in the Pareto front, but it cannot be considered in the history-matching study. Therefore, the majority of the Pareto optimum set would be abandoned.

4.4. Optimization Algorithms

Genetic algorithm (GA): The GA refers to a family of computational models derived from Darwin's theory of biological evolution. The idea is one of the natural selection organization principles for optimizing the individuals of populations. GAs mimic natural selection to optimize more successfully. Problems are solved by an evolutionary process resulting in a best solution (fittest survivor). GAs do not search via gradients; the searching is done via sampling by stochastic operators rather than by deterministic rules. Figure 7 briefly illustrates the workflow of a GA optimizer. The application of a GA includes the following operators:

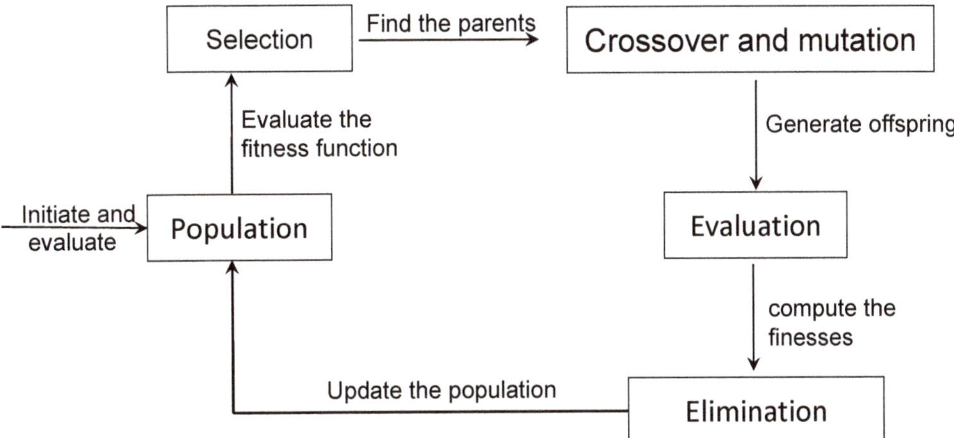

Figure 7. Optimization workflow using genetic algorithm (GA).

Encoding: The very first step is to convert the solution of a problem, typically composed of vectors of control parameters into a chromosome-type data structure.

Selection: The selection operator is employed to find individuals that exhibit good fitness functions and become candidates as parents to generated offspring. The most used selection operators include tournament and roulette wheel selection.

Reproducing: Once the parents are selected, the next step is to make them reproduce offspring via crossover. The crossover is a GA operator to generate new chromosomes based on the individuals of previous generations. A certain crossover rate needs to be specified so that a certain volume of selected individuals will be used to generate offspring. Typically, the crossover rate ranges from 60% to 95%.

Mutation: Mutation is a GA operator that makes the search range wider. What mutation does is change a gene of a chromosome completely without any reason. In a genetic algorithm, the mutation rate is extremely low, typically less than 1%.

Decoding: At the end of a genetic algorithm iteration, the chromosomes need to be converted back to real number solutions.

The GA is one of the first trials in the SWP project and optimizes multiple techno-economic objective functions for the CO_2-WAG injection process [6].

Particle swarm optimization (PSO): PSO is a nature-inspired evolutionary and stochastic optimization technique to solve computationally hard optimization problems. It is a fast technology based on the movement and intelligence of swarms. PSO was built by mimicking the working mechanisms of biological swarm migrations. In a PSO application, particles communicate directly or indirectly with one another via searching directions. During the iteration process, the particles update their position according to their previous experience and the experience of their neighbors. The process of PSO optimization includes the following three steps [33]:

1. Evaluate the fitness by the proxy model.
2. Calculate the velocity term v using Equation (24):

$$v_i(k+1) = wv_i(k) + c_1r_1[\bar{x}_i(k) - x_i(k)] + c_2r_2[g(k) - x_i(k)] \tag{24}$$

where k is the iteration level, $\bar{x}$ is the local best, and g is the global best.

3. Update the particle position via Equation (25):

$$x_i(k+1) = x_i(k) + v_i(k+1) \tag{25}$$

Compared to GA, the optimization process of PSO does not include any encoding/decoding procedure, which accelerates the convergence. However, the PSO can be easily trapped by the local minima when a complex objective function is considered. Figure 8 is the optimization workflow using a PSO algorithm.

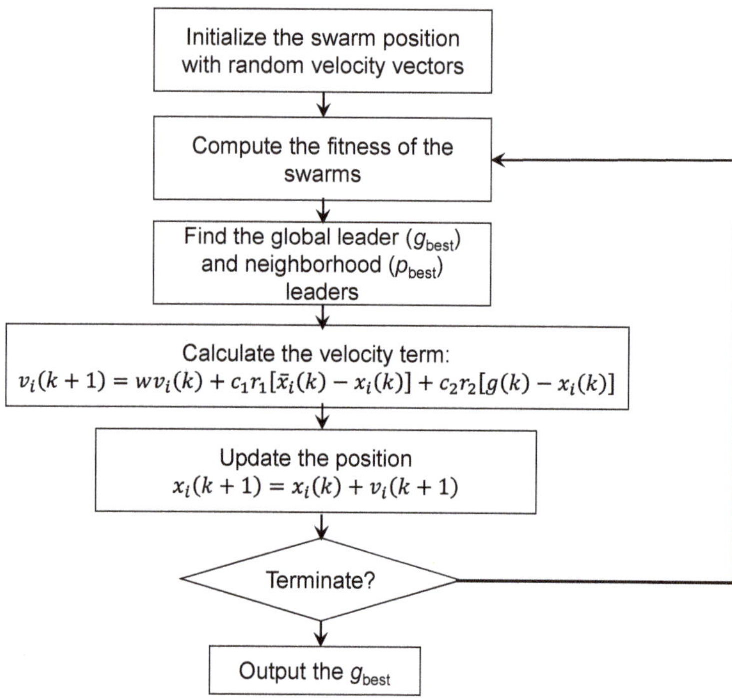

Figure 8. Optimization workflow using particle swarm optimization (PSO).

Multi-objective particle swarm optimization (MOPSO): Compared to the single-objective PSO, MOPSO introduces a "repository (REP)" to restore the nondominated solutions of each iteration. The size of the repository can be prescribed from 100 to 250 particles. When the repository is full, the dominated solution in REP needs to be replaced by the nondominated solutions appearing in the next generation.

A "hypercube" is introduced in MOPSO to quantify the "neighbor" (also known as the searching domain) of the particles. Hypercubes are a subdivision of the objective function domain separated by hyperlines that are uniformly distributed in each objective function domain. We illustrated a hypercube (Figure 9) of a problem with two objective functions of project NPV and oil recovery. The dots represent the particles of certain MOPSO iterations. It is worth emphasizing that the hyperlines need to be justified after the fitness of population (POP) is updated in each iteration due to the upper and lower limits of the objective function domains possibly being changed.

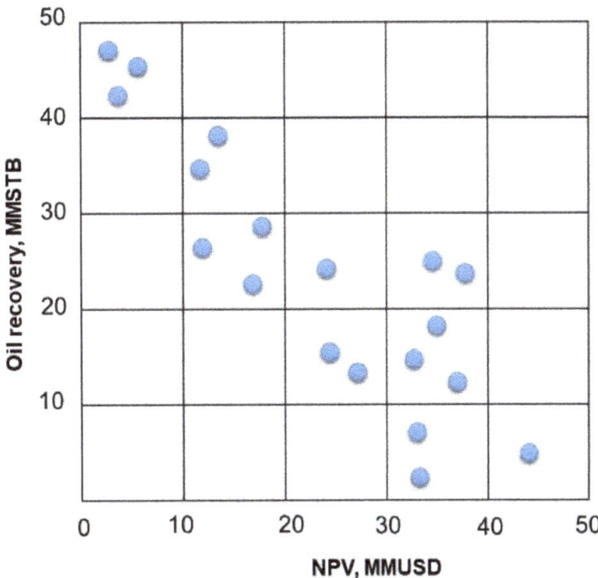

Figure 9. Hypercube illustration of a problem with two objective functions.

Similar to the single objective PSO, the MOPSO iteration also includes three stages [34]:

Stage 1: Initialization: The initial population (*POP*) is generated and the velocity (*VEL*) of each particle is set to 0. The objective functions of all the particles are computed and their fitness is assessed. Then, the initial solution repository (*REP*) is structured using the nondominated solutions in the *POP*. Then the initial personal best (*PBEST*) of the particles is set to be the initial fitness.

Stage 2: Velocity calculation: Compute the movement velocity of the particles using Equation (26):

$$VEL[i] = 0.4 \times VEL[i] + C_1(PBEST[i] - POP[i]) + C_2(REP[h] - POP[i]) \qquad (26)$$

where C_1 and C_2 are two weight factors within [0, 1]. Based on the hypercube structure in the solution repository, a global best *REP* [h] is randomly picked from the grid where each particle is located.

Stage 3: Update: The particles in the population are updated via Equation (27):

$$POP[i] = POP[i] + VEL[i] \qquad (27)$$

Then, the fitness of the particles is re-evaluated based on the updated population. At this stage, the *REP* needs to be updated by removing the dominating solutions. The hypercube structure and the *PBEST* would also change accordingly.

Current research works still focus on quantifying the convergence criteria of MOPSO. One of the most broadly accepted opinions to determine the convergence of MOPSO is that the iteration can be terminated when none of the new particles can dominate any of the particles involved in *REP*. Figure 10 shows the workflow of the MOPSO process. MOPSO is proved to be a robust algorithm that finds the repository of the close to the true Pareto front by many searches.

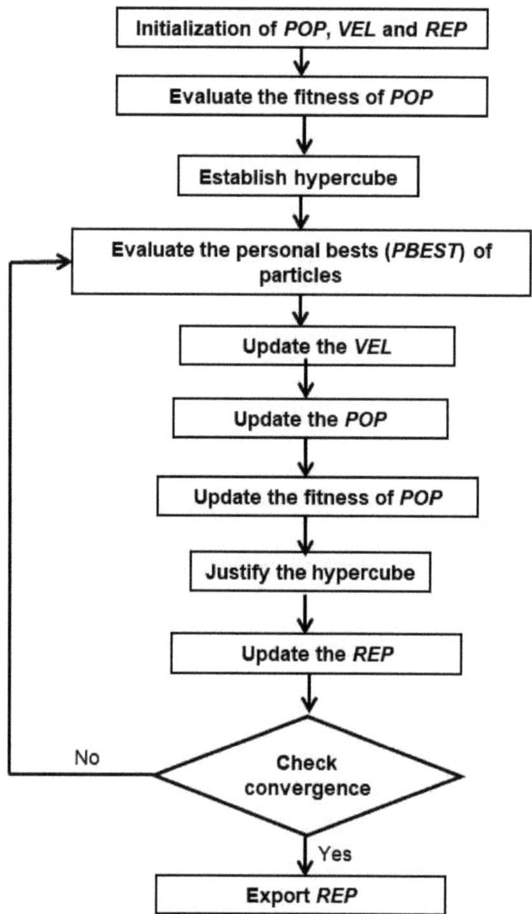

Figure 10. Workflow of multi-objective particle swarm optimization (MOPSO).

5. Structuring the Hybrid Numerical Machine-Learning Workflow

In the previous sections, we introduced the machine-learning proxy models and optimization protocols. The coupling of them would build robust machine-learning assisted optimization protocols for various engineering purposes in the CO_2-WAG injection project. In this paper, a machine-learning assisted history-matching and a multi-objective optimization workflow is presented.

5.1. History-Matching Workflow

Due to the high dimension and considerable computational overhead, history-matching cannot be completed by totally relying on the high-fidelity numerical model. Therefore, machine-learning models are needed to assist the history-matching work. Using the prepared numerical simulation runs as the knowledge base, a series of ANN models can be successfully trained to predict the oil, water, and gas production considering the uncertain reservoir properties as inputs. The importance of using the proxy models is that one can conduct a large volume of numerical simulation realizations with little computational cost. The trained proxy models are coupled with PSO to minimize the history-matching error function. The error function is defined as the summation of the square differences between the field historical data and the model prediction.

In the history-matching process, the oil, gas, and water production rates were considered primary unknowns to calculate the error function. Considering the extensive number of tuning parameters, the results of the history-matching model could exhibit strong non-uniqueness, which means that there is more than one combination of the 62 parameters that could generate a similar level of history-matching error. After an optimal solution is obtained, the history matching solution found by the machine-learning workflow is revisited by the high-fidelity numerical reservoir simulator, which confirms the matching quality. In Figure 11, the machine-learning assisted history-matching workflow is displayed, which was successfully imposed to develop history-matched reservoir models for the primary/secondary and CO_2-WAG period.

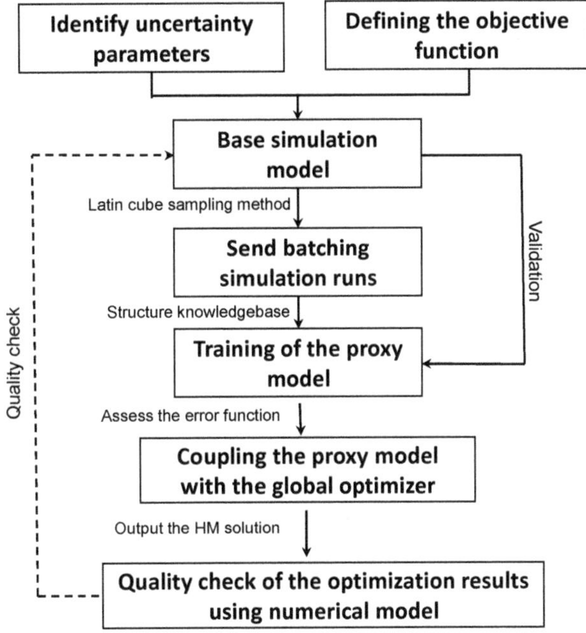

Figure 11. Machine-learning assisted history-matching workflow.

5.2. Multi-Objective Optimization Workflow

Figure 12 illustrate the machine-learning assisted multi-objective optimization workflow coupling the numerical reservoir simulation model, proxy models, and global optimization algorithm.

Data preparation: The original high-fidelity numerical model is utilized to prepare a certain volume of numerical realizations. Those realizations are utilized to prepare the knowledge base that is used to train the proxy model. Considering a blind testing error margin of <5%, the training of an injection pattern base model needs at least 100 simulation runs due to the smaller model size and dimensions. Training for the field scale proxy model may require more than 500 simulation runs.

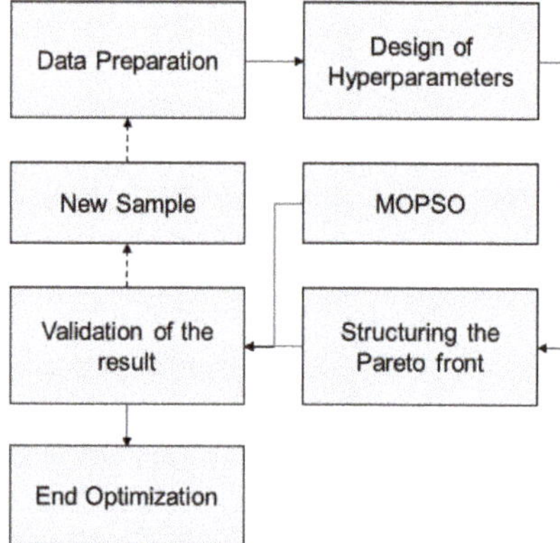

Figure 12. Machine-learning assisted multi-objective optimization workflow.

Design of hyperparameters: It is known that the hyperparameters of the machine-learning models have significant impacts on the prediction performance. The hyperparameters of SVM are optimized using Bayesian optimization and the MLNN topology is designed with a self-adaptive training protocol [35].

Structuring the Pareto front: MOPSO employs the proxy models to generate the Pareto front by considering various t objective functions. The 2- and 3-objective Pareto fronts can be visualized using a two-dimensional curve and a surface, respectively.

Validation of the result: The Pareto front solutions must be validated before advising any operational decisions. The input parameters of the Pareto front solutions revisit the numerical model and compare the between the proxy and simulation results. If large disparities are observed, then some new samples are added to the training database to re-train the proxy. This loop can be continued several times until the error between the simulation results and the proxy predictions is lower than a prescribed error tolerance.

6. Case Studies

6.1. A History-Matching Application

A history-matching application using the proposed machine-learning assisted workflow is presented in this discussion. The objective of this study was to tune the reservoir hydrodynamic properties including the permeability along the x-, y-, and z- directions, and the Corey's relative permeability coefficients. The permeability distributions were tuned by imposing anisotropic multipliers. The reservoir model assigned various three-phase relative permeability curves based on the HFU characterization. There were five different relative permeability sets considered in this work. Notably, HFU 5, 6, 7, and 8 were lumped into one group and shared identical relative permeability data. Considering the permeability multiplier and five different sets of Corey's coefficients, there were 62 parameters involved in the history-matching processes.

The machine-learning assisted tuning process was carried out by the workflow discussed in the previous sections. The Latin cube sampling protocol was used to prepare 100 simulation cases by varying the uncertain hydrodynamic parameters. The dataset was used to train an expert MLNN proxy to predict the field responses based on different hydrodynamic properties. In the machine-learning assisted workflow, the proxy model substituted the high-fidelity numerical simulator to evaluate the history-matching error defined by Equation (22). Figure 13 shows the history-matching results obtained by the machine-learning assisted workflow coupling ANN proxy models and PSO optimizer. It illustrates the fluid production matching quality predicted by the ANN models when the history-matching error was minimized. The average relative matching errors were 0.9%, 42.9%, and 17.2% for oil, water, and gas production rates comparing the real-field historical data, respectively. Although the overall history-matching results obtained by the PSO algorithm indicated a 20.2% relative error, water production exhibited much worse performance compared to that of oil and gas. Thus, a confirmation by revisiting the high-fidelity numerical simulator was quite necessary.

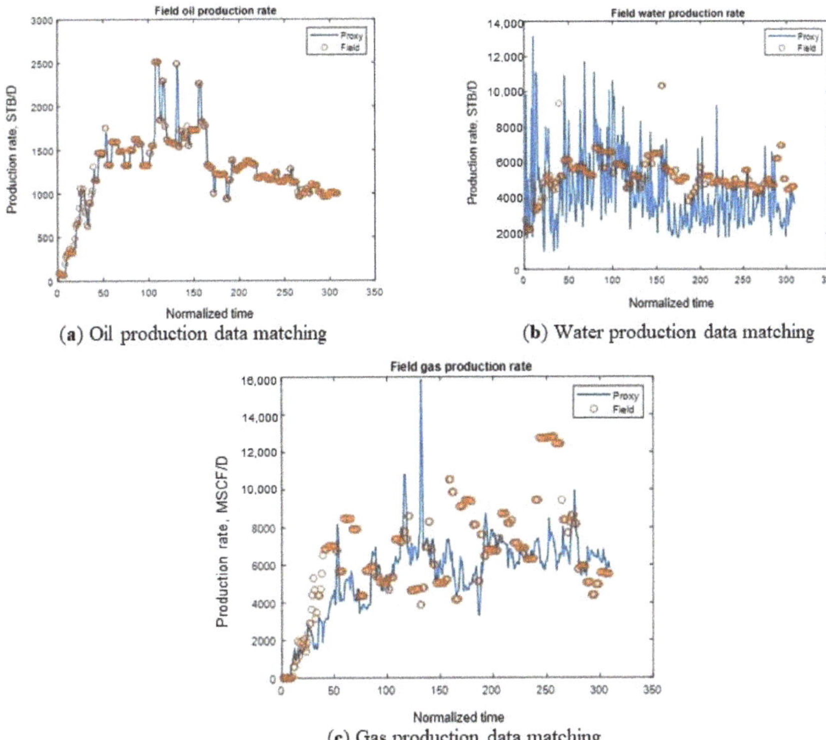

(**a**) Oil production data matching

(**b**) Water production data matching

(**c**) Gas production data matching

Figure 13. History-matching results obtained by the proxy model for (**a**) oil production, (**b**) water production, and (**c**) gas production data.

The major objective of the machine-learning assisted history-matching workflow was to find the combination of reservoir hydrodynamic properties that make the reservoir simulation model predictions agree with the field historical data. When the PSO optimizer minimizes the history-matching error during the iterative processes, the ANN proxies are employed to predict the field responses. Notably, a set of reservoir properties would be obtained after the PSO iteration converges, which is considered the solution of the history-matching process. The history matching solution must revisit the high-fidelity model for the following reasons:

1. Although the proxy models were well-trained, there existed potential error margins. A history-matching solution must feed into the high-fidelity simulator and confirm the matching quality.
2. To structure the forecasting scenarios, a base-case numerical simulation model needed to be established by re-running the high-fidelity simulator using the history matching solution suggested by the machine-learning assisted workflow.

Figure 14 shows the history-matching quality confirmed by the high-fidelity simulator considering the solution found by the machine-learning assisted workflow. Good agreements in terms of oil/gas/water production and the gas injection rate were observed, which indicates that the numerical model was well tuned and effectively characterized the underground hydrodynamic environment. More importantly, the successful history-matching study stressed the robustness of the proxy model by learning the data structure presented by the dataset. The computational cost was significantly reduced by the machine-learning assisted workflow. Preparing the 100 simulation runs took 80 h of CPU time. With the help of the proxy model, the workflow completed more than 600 PSO iterations using a population size of 100 (60,000 realizations in total) within 300 s.

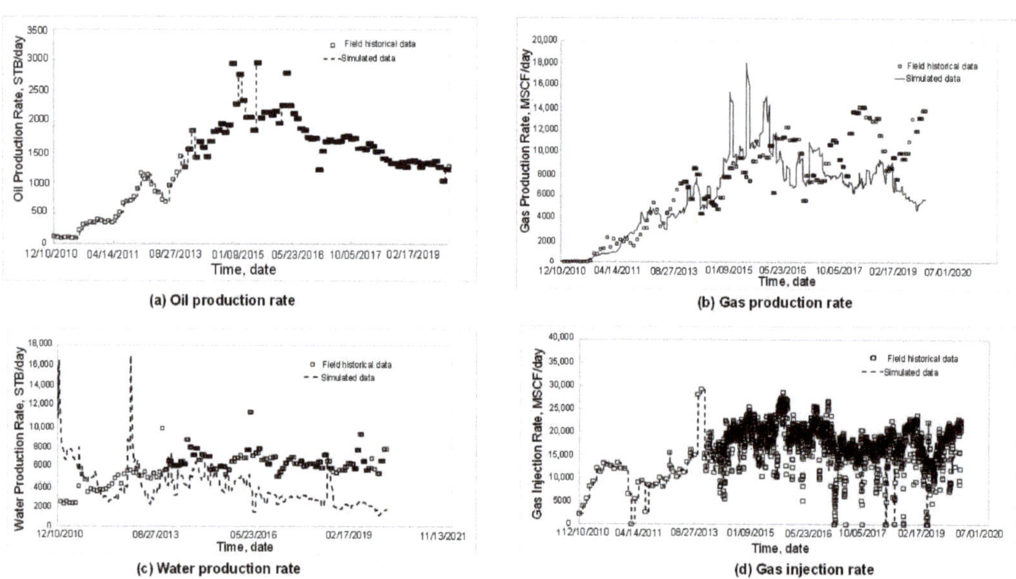

Figure 14. History-matching results confirmed using the high-fidelity numerical model for (**a**) the oil production rate, (**b**) the gas production rate, (**c**) the water production rate, and (**d**) the gas injection rate.

6.2. A Multi-Objective Optimization Application

Another application of the proposed workflow successfully designed the CO_2-WAG injection in the FWU field for the period from January 2020 to January 2038. The CO_2 injection includes the purchased CO_2 and the recycled CO_2 from the produced gas. According to the initial analysis of the simulation results and previous studies [31], as shown in Figure 15, the purchased CO_2 rate varies from 2020 to 2033. At the end of 2033, no more CO_2 is purchased.

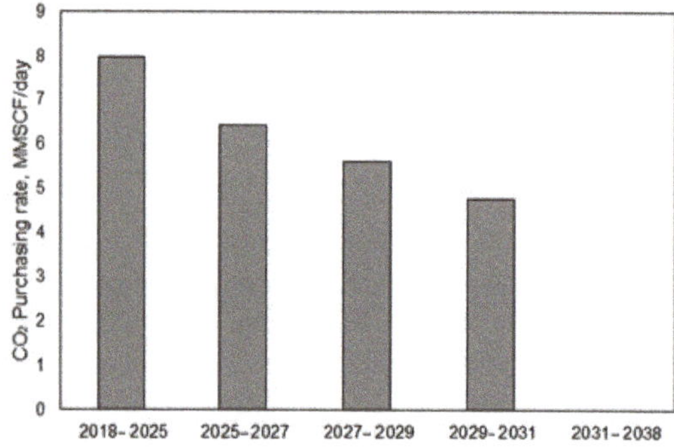

Figure 15. CO_2 purchasing rate.

A base-case forecasting scenario was established by dividing the CO_2-WAG injection patterns into four groups (WAG A, B, C, and D). The entire project timeline was split into eight time periods, as shown in Table 1. There are more wells added to the groups as the project processes.

Table 1. Time period details of the development plan.

Stages	Start	End
1	January 2020	June 2020
2	July 2020	January 2022
3	January 2022	January 2023
4	January 2023	January 2026
5	January 2026	January 2028
6	January 2028	January 2030
7	January 2030	January 2032
8	January 2032	January 2038

The CO_2-WAG injection parameters included the water and gas injection duration, water injection rates, and production well specifications, which vary for each group and time period. Thus, there were 37 design specifications considered in the optimization study. The NPV, oil recovery, and CO_2 storage volume were the objective functions. A base case model was structured using default design parameters suggested by the field operator [2] The incremental oil production, NPV, and CO_2 injection/production volume of the base case model are displayed in Figure 16. The objective functions of the base case model are summarized in Table 2. Note that the CO_2 storage efficacy is the ratio of CO_2 sequestrated to the total purchased volume.

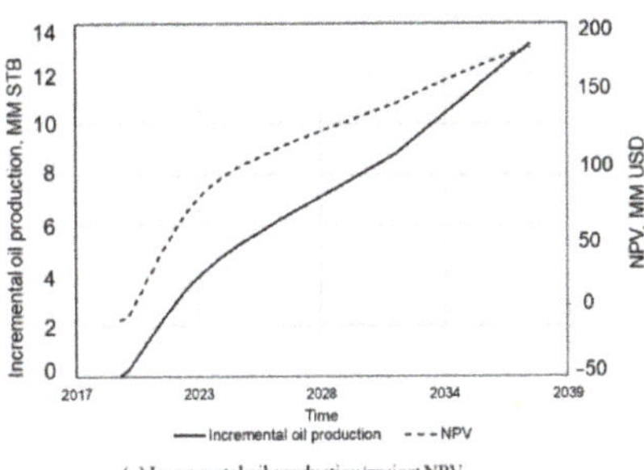

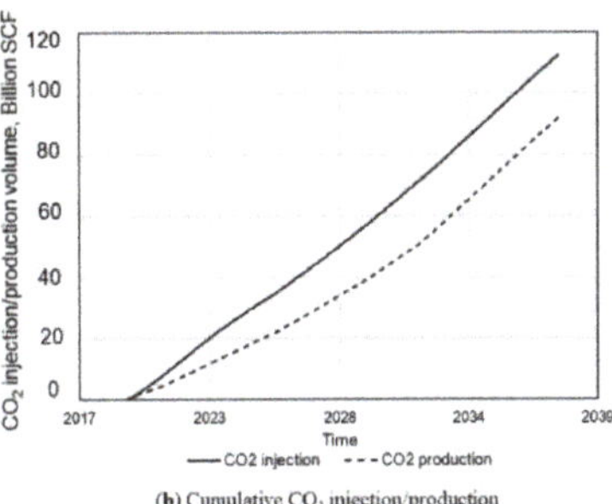

Figure 16. Simulation results of the base case model for (**a**) incremental oil production/project NPV and (**b**) cumulative CO_2 injection/production.

Table 2. Objective functions of the base case model.

Objective Function	Unit	Value
Oil production increments	MM bbl	13.3
CO_2 storage volume increments	MM metric ton	1.06
Project NPV	Million USD	183
Cumulative Oil Production by January 2038	MM STB	16.8
Total CO_2 storage volume by January 2038	MM metric ton	2.36
CO_2 storage efficacy	percentage	81.19%

The cumulative oil recovery was 16.8 MM bbl and the total CO_2 storage was 2.36 million metric tons, which was 81.2% of the total purchased CO_2. The project NPV was USD 183 million. It is worth stressing that the average reservoir pressure of $\geq$4000 psi was imposed as the physical constraint on the multi-objective optimization to sustain a miscible flooding process [6]. The proposed optimization workflow aims at improving all the objective functions simultaneously.

As shown in Figure 17, the Pareto front solution considering three objective functions is illustrated as a space curve in the domain of f_1 (oil recovery), f_2 (CO_2 storage), and f_3 (NPV). The dominating solution sitting on the Pareto front was validated by the numerical simulator. The good agreements can be observed in Figure 17, which confirms the validity of the proxy model and the MOPSO optimization results. Another interesting observation was drawn by plotting the projections of the three-objective Pareto front to three of the orthogonal surfaces. In Figure 18, a strong tradeoff relationship can be observed between the oil recovery and NPV, as well as the oil recovery and CO_2 storage volume, which means that by improving the oil recovery, the NPV and CO_2 storage volume has to be sacrificed. Therefore, the field operator could have a flexible range of choices to design the CO_2-WAG process based on the primary desire of the project. However, due to the tax credit brought by the CO_2 sequestration, the project NPV and CO_2 storage volume increases monotonically.

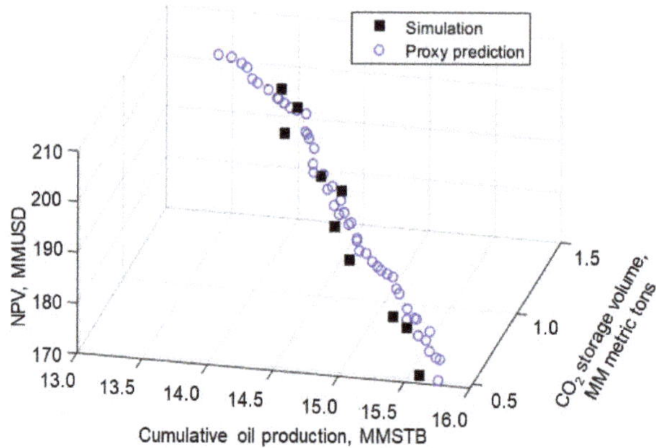

Figure 17. Comparison between the Pareto front and numerical simulation results using some of the optimized input parameters: three-objective.

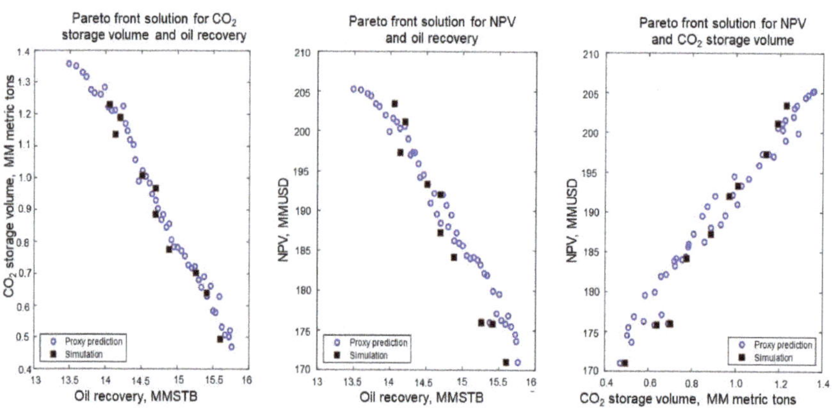

Figure 18. Projection view of the comparison between the Pareto front and numerical simulation results using some of the optimized input parameters: three-objective.

Table 3 summarizes the optimization results using the proposed workflow and the base case values. The base case oil recovery fell out of the range of the Pareto front solution, which means that the base case design scenario was part of the dominating solution. Notably, the range of oil recovery included in the Pareto front was quite narrow (13.3–15.7 MM STB), and the improvement to the base case was 16.0%. The design of CO_2-WAG made a considerable difference for the CO_2 storage volume and NPV. Therefore, the solution yielding the highest NPV and CO_2 storage volume, even with the lowest oil recovery, should be considered with priority. However, that investigation could be altered when the oil price increases, or the tax allowance brought by CO_2 utilization becomes lower.

Table 3. Comparison between the results of the base case and the optimized solutions.

Item	Unit	Range/Value	Base Case
Oil production increments	MM STB	13.3–15.7	13.2
CO_2 storage volume increments	MM metric ton	0.42–1.4	1.06
Project NPV	Million USD	170–205	-
Max cumulative oil production	MM STB	19.3	16.8
Max cumulative CO_2 storage	MM metric ton	2.7	3.63
Max project NPV	Million USD	205	183
CO_2 storage efficacy	percentage	92.90%	81.19%

The field operator collaborating with the SWP project aims to extract hydrocarbon in the Morrow-B formation using CO_2-EOR technologies. The primary goal of the operation is to produce residual oil, and the sequestration of CO_2 during the EOR stage would help improve the project NPV via tax credits. That is why oil recovery, the NPV, and CO_2 storage volume are selected as the objective functions. The key to maximizing the CO_2 storage (minimizing the CO_2 production) during the EOR process is the injection specifications. The miscible flooding processes inject CO_2 gas in continuous or cyclic manners. For a multiphase flow system, the mobility of the gaseous phase relates to the gas saturation. As the gas injection processes, a high gas saturation channel could form from which the injected gas could break through. Such an issue would occur in the continuous CO_2 injection process. In the practical CO_2 injection process, the water and CO_2 may inject in a cyclic manner and avoid the persistent buildup of gas saturation. More importantly, since the pressure and saturation distributions of the system would exhibit strong heterogeneity, the CO_2-WAG design should vary for different injectors. For instance, injectors in the high-water saturation region could take a higher gas injection volume (longer gas injection cycle) and vice versa. Thus, an accurate characterization of the pressure and saturation distribution is also critical for a successful CO_2-WAG design. The deployment of CO_2 foam [36] exhibits more robust mobility control competences. Although these technologies are not used in the Morrow-B CO_2-EOR project, the coupling of chemical slug and CO_2 gas would be beneficial for CO_2 storage and oil recovery under feasible project economic criteria.

7. Conclusive Remarks

This paper presents a comprehensive summary of machine-learning assisted workflows to solve various engineering problems associated with the CO_2-WAG design. By coupling the proxy model and the advanced optimization protocols, the computational overheads of the history-matching and multi-objective optimization processes are significantly reduced. Besides the successful experiences obtained from this work, the following limitations cannot be ignored:

1. The proxy model developed in this work is a field-specified model that only works for the Morrow-B formation. Its implementation in other fields needs a new reservoir simulation model structure and needs to go through the proposed workflow.

2. The CO_2-EOR process only includes the CO_2-WAG process. Other CO_2-EOR technologies such as CO_2 foam, continuous injection, CO_2 huff-n-puff, etc., are not considered.

From this work, the following conclusive statements can be drawn:

1. The selection of the machine-learning algorithm may comprehensively consider the dimension of the problem and the demand of error margin. The RSM, SVM, and MLNN are suitable for different types of datasets and a wise choice of method could essentially enhance the prediction performance of the proxy model.
2. Although machine-learning approaches exhibit many superiorities over the conventional numerical approach, a precise reservoir engineering analysis should take advantage of both. The error margin of the proxy model is the tradeoff for accelerating the computational speed. Thus, validation via the high-fidelity numerical model is necessary before deploying the results in operational practices.
3. The Pareto front optimum protocol provides an alternative way to address multi-objective optimization problems. However, a successful application of the Pareto front optimum solution must be based on the tradeoff relationship between various objective functions.
4. The calculation of the project economic objective functions strongly depends on the tax allowance and crude oil price. Therefore, the operational deployment of the optimum design suggested by the workflow needs to take practical considerations such as crude oil market condition, government policies, etc., into account.

SI-Field Unit Conversion Factor

ft $\times$	0.3048	= m
ft^2 $\times$	0.0929030	= m^2
ft^3 $\times$	0.0283169	= m^3
bbl $\times$	0.1589873	= m^3
psi $\times$	6.894757	= kPa

Author Contributions: Conceptualization, Q.S. and W.A.; methodology, Q.S. and W.A.; software, Q.S. and J.Y.; validation, Q.S., W.A., and J.Y.; formal analysis, M.C. and R.B.; investigation, Q.S. and J.Y.; resources, W.A., M.C., and R.B.; data curation, W.A. and J.Y.; writing—original draft preparation, Q.S. and W.A.; writing—review and editing, Q.S., W.A., and M.C.; visualization, Q.S., J.Y., and W.A.; supervision, W.A.; project administration, M.C., W.A., and R.B.; funding acquisition, W.A. and R.B. All authors have read and agreed to the published version of the manuscript.

Funding: Funding for this project was provided by the U.S. Department of Energy's (DOE) National Energy Technology Laboratory (NETL) through the Southwest Regional Partnership on Carbon Sequestration (SWP) under Award No. DE-FC26-05NT42591.

Institutional Review Board Statement: Not applicable.

Informed Consent Statement: Not applicable.

Data Availability Statement: No new data were created or analyzed in this study. Data sharing is not applicable to this article.

Conflicts of Interest: The authors declare no conflict of interest.

Abbreviations

ANN	artificial neural network
CAPEX	capital expenditure
CCUS	Carbon capture utilization and sequestration
CPU	central processing unit
EOR	enhanced oil recovery
FWU	Farnsworth Unit
GA	genetic algorithm

HFU	hydraulic flow unit
MLNN	multi-layer neural networks
MM	million
MOO	multi-objective optimization
MOPSO	Multi-objective particle swarm optimization
NPV	net present value
OGIP	original gas in place
OOIP	original oil in place
POP	population
PSO	particle swarm optimization algorithm
REP:	repository
res bbl	reservoir barrel
RSM	response surface models
SCF	standard cubic feet
STB	stock tank barrel
SVM	support vector machines
SWP	Southwest Regional Partnership on Carbon Sequestration
VEL	velocity

References

1. You, J.; Ampomah, W.; Kutsienyo, E.J.; Sun, Q.; Balch, R.S.; Aggrey, W.N.; Cather, M. Assessment of Enhanced Oil Recovery and CO_2 Storage Capacity Using Machine Learning and Optimization Framework. In Proceedings of the SPE Europec Featured at 81st EAGE Conference and Exhibition, London, UK, 3–6 June 2019; Society of Petroleum Engineers: Dallas, TX, USA, 2019.
2. Ampomah, W.; Balch, R.; Grigg, R.B.; Cather, M.; Gragg, E.; Will, R.A.; White, M.; Moodie, N.; Dai, Z. Performance assessment of CO_2-enhanced oil recovery and storage in the Morrow reservoir. *Géoméch. Geophys. Geo-Energy Geo-Resour.* **2017**, *3*, 245–263. [CrossRef]
3. Munson, T.W. Depositional, Diagenetic, and Production History of the Upper Morrowan Buckhaults Sandstone, Farnsworth Field, Ochiltree County Texas. *Shale Shak. Dig.* **1994**, *XXXX–XXXXIV*, 2–20.
4. Ampomah, W.; Balch, R.S.; Grigg, R.B.; Will, R.; Dai, Z.; White, M.D. Farnsworth field CO2-EOR project: Performance Case History. In Proceedings of the SPE Improved Oil Recovery Conference, Tulsa, OK, USA, 11–13 April 2006; Society of Petroleum Engineers: Dallas, TX, USA, 2006.
5. Ampomah, W.; Balch, R.S.; Ross-Coss, D.; Hutton, A.; Cather, M.; Will, R.A. An integrated Approach for Characterizing a Sandstone Reservoir in the Anadarko Basin. In Proceedings of the Offshore Technology Conference, Houston, TX, USA, 2–5 May 2016.
6. Ampomah, W.; Balch, R.; Cather, M.; Will, R.; Gunda, D.; Dai, Z.; Soltanian, M. Optimum design of CO_2 storage and oil recovery under geological uncertainty. *Appl. Energy* **2017**, *195*, 80–92. [CrossRef]
7. Ampomah, W.; Balch, R.S.; Grigg, R.B.; McPherson, B.; Will, R.A.; Lee, S.-Y.; Dai, Z.; Pan, F. Co-optimization of CO_2-EOR and storage processes in mature oil reservoirs. *Greenh. Gases Sci. Technol.* **2017**, *7*, 128–142. [CrossRef]
8. Kutsienyo, E.J.; Ampomah, W.; Sun, Q.; Balch, R.S.; You, J.; Aggrey, W.N.; Cather, M. Evaluation of CO_2-EOR Performance and Storage Mechanisms in an Active Partially Depleted Oil Reservoir. In Proceedings of the SPE Europec featured at 81st EAGE Conference and Exhibition, London, UK, 3–6 June 2019; Society of Petroleum Engineers: Dallas, TX, USA, 2019.
9. Ampomah, W.; Balch, R.S.; Grigg, R.B. Analysis of upscaling algorithms in heterogeneous reservoirs with different recovery processes. In Proceedings of the SPE Production and Operations Symposium, Oklahoma City, OK, USA, 1–5 March 2015; Society of Petroleum Engineers: Dallas, TX, USA, 2015.
10. Dai, Z.; Viswanathan, H.; Middleton, R.; Pan, F.; Ampomah, W.; Yang, C.; Jia, W.; Xiao, T.; Lee, S.-Y.; McPherson, B. CO_2 Accounting and Risk Analysis for CO_2 Sequestration at Enhanced Oil Recovery Sites. *Environ. Sci. Technol.* **2016**, *50*, 7546–7554. [CrossRef] [PubMed]
11. Pan, F.; McPherson, B.J.; Dai, Z.; Jia, W.; Lee, S.-Y.; Ampomah, W.; Viswanathan, H.; Esser, R. Uncertainty analysis of carbon sequestration in an active CO_2-EOR field. *Int. J. Greenh. Gas Control* **2016**, *51*, 18–28. [CrossRef]
12. Ahmmed, B.; Appold, M.S.; Fan, T.; McPherson, B.J.O.L.; Grigg, R.B.; White, M.D. Chemical effects of carbon dioxide sequestration in the Upper Morrow Sandstone in the Farnsworth, Texas, hydrocarbon unit. *Environ. Geosci.* **2016**, *23*, 81–93. [CrossRef]
13. Ross-Coss, D.; Ampomah, W.; Cather, M.; Balch, R.S.; Mozley, P.; Rasmussen, L. An Improved Approach for Sandstone Reservoir Characterization. In Proceedings of the SPE Western Regional Meeting, Anchorage, AK, USA, 23–26 May 2016; Society of Petroleum Engineers: Dallas, TX, USA, 2016.
14. You, J.; Ampomah, W.; Sun, Q.; Kutsienyo, E.J.; Balch, R.S.; Dai, Z.; Cather, M.; Zhang, X. Machine learning based co-optimization of carbon dioxide sequestration and oil recovery in CO_2-EOR project. *J. Clean. Prod.* **2020**, *260*, 120866. [CrossRef]
15. Ketineni, S.P.; Ertekin, T.; Anbarci, K.; Sneed, T. Structuring an Integrative Approach for Field Development Planning Using Artificial Intelligence and its Application to an Offshore Oilfield. In Proceedings of the SPE Annual Technical Conference and Exhibition, Houston, TX, USA, 28–30 September 2015; Society of Petroleum Engineers: Dallas, TX, USA, 2015.

16. Zhang, Z.; Jung, H.Y.; Datta-Gupta, A.; Delshad, M. History Matching and Optimal Design of Chemically Enhanced Oil Recovery Using Multi-Objective Optimization. In Proceedings of the SPE Reservoir Simulation Conference, Galveston, TX, USA, 10–11 April 2019; Society of Petroleum Engineers: Dallas, TX, USA, 2019.
17. Zhang, X.; Wu, J.; Coutier-Delgosha, O.; Xiao, H. Recent progress in augmenting turbulence models with physics-informed machine learning. *J. Hydrodyn.* **2019**, *31*, 1153–1158. [CrossRef]
18. You, J.; Ampomah, W.; Sun, Q. Development and application of a machine learning based multi-objective optimization workflow for CO_2-EOR projects. *Fuel* **2020**, *264*, 116758. [CrossRef]
19. Ertekin, T.; Ayala, L.F. *Reservoir Engineering Models: Analytical and Numerical Approaches*, 1st ed.; McGraw-Hill Education: Columbus, OH, USA, 2018; pp. 1–368.
20. Putcha, V.B.; Ertekin, T.A. Fast and Robust Compositional, Multi-Phase, Non-Isothermal Wellbore Hydraulics Model for Vertical Wells. In Proceedings of the SPE Annual Technical Conference and Exhibition, San Antonio, TX, USA, 9–11 October 2017; Society of Petroleum Engineers: Dallas, TX, USA, 2017.
21. Ayala, L.F.; Ertekin, T.; Adewumi, M.A. Study of Gas/Condensate Reservoir Exploitation Using Neurosimulation. *SPE Reserv. Eval. Eng.* **2007**, *10*, 140–149.
22. Panja, P.; Velasco, R.; Pathak, M.; Deo, M. Application of artificial intelligence to forecast hydrocarbon production from shales. *Petroleum* **2018**, *4*, 75–89. [CrossRef]
23. Sumardi, H.R.; Irawan, D. Coalbed Methane Production Parameter Prediction and Uncertainty Analysis of Coalbed Methane Reservoir with Artificial Neural Networks. In Proceedings of the Indonesian Petroleum Association Fortieth Annual Convention and Exhibition, Jakarta, Indonesia, 25–27 May 2016; American Association of Petroleum Geologists: Tulsa, OK, USA, 2016.
24. Parada, C.H.; Ertekin, T.A. New Screening Tool for Improved Oil Recovery Methods Using Artificial Neural Networks. In Proceedings of the SPE Western Regional Meeting, Bakersfield, CA, USA, 21–23 March 2012; Society of Petroleum Engineers: Dallas, TX, USA, 2012.
25. Thararoop, P.; Karpyn, Z.; Gitman, A.; Ertekin, T. Integration of seismic attributes and production data for infill drilling strategies—A virtual intelligence approach. *J. Pet. Sci. Eng.* **2008**, *63*, 43–52.
26. Li, Y.; Zhang, T.; Sun, S. Acceleration of the NVT Flash Calculation for Multicomponent Mixtures Using Deep Neural Network Models. *Ind. Eng. Chem. Res.* **2019**, *58*, 12312–12322. [CrossRef]
27. Rasmussen, L.; Fan, T.; Rinehart, A.; Luhmann, A.; Ampomah, W.; Dewers, T.; Heath, J.; Cather, M.; Grigg, R. Carbon Storage and Enhanced Oil Recovery in Pennsylvanian Morrow Formation Clastic Reservoirs: Controls on Oil–Brine and Oil–CO_2 Relative Permeability from Diagenetic Heterogeneity and Evolving Wettability. *Energies* **2019**, *12*, 3663. [CrossRef]
28. Bishop, C.M. *Pattern Recognition and Machine Learning*, 1st ed.; Springer: New York, NY, USA, 2006; pp. 1–738.
29. Hassoun, M. *Fundamentals of Artificial Neural Networks*; MIT Press: Cambridge, MA, USA, 2003; pp. 1–540.
30. Smola, A.J.; Schölkopf, B. A tutorial on support vector regression. *Stat. Comput.* **2004**, *14*, 199–222. [CrossRef]
31. You, J. Multi-Objective Optimization of Carbon Dioxide Enhanced Oil Recovery Projects Using Machine Learning Algorithms. Ph.D. Thesis, New Mexico Institute of Mining and Technology, Socorro, NM, USA, 2020.
32. Coello, C.; Pulido, G.; Lechuga, M. Handling multiple objectives with particle swarm optimization. *IEEE Trans. Evol. Comput.* **2004**, *8*, 256–279. [CrossRef]
33. Kennedy, J.; Eberhart, R. Particle swarm optimization. In Proceedings of the ICNN'95—International Conference on Neural Networks, Perth, WA, Australia, 27 November–1 December 1995.
34. Sun, Q. The Development of an Artificial-Neural-Network-Based Toolbox for Screening and Optimization of *Enhanced oil* Recovery Projects. Ph.D. Thesis, University Park, The Pennsylvania State University, State College, PA, USA, 2017.
35. Sun, Q.; Ertekin, T. The Development of Artificial-neural-network-based Universal Proxies to Study Steam Assisted Gravity Drainage (SAGD) and Cyclic Steam Stimulation (CSS) Processes. In Proceedings of the SPE Western Regional Meeting, Garden Grove, CA, USA, 27–30 April 2015; Society of Petroleum Engineers: Dallas, TX, USA, 2015.
36. Ibrahim, A.; Emrani, A.; Nasraldin, H. Stabilized CO_2 Foam for EOR Applications. In Proceedings of the Carbon Management Technology Conference, Houston, TX, USA, 17–20 July 2017.

Article

Numerical Modeling of CO$_2$ Sequestration within a Five-Spot Well Pattern in the Morrow B Sandstone of the Farnsworth Hydrocarbon Field: Comparison of the TOUGHREACT, STOMP-EOR, and GEM Simulators

Eusebius J. Kutsienyo [1,*], Martin S. Appold [1], Mark D. White [2] and William Ampomah [3]

1   Department of Geological Sciences, University of Missouri, Columbia, MO 65211, USA; appoldm@missouri.edu
2   Pacific Northwest National Laboratory (PNNL), Richland, WA 99354, USA; mark.white@pnnl.gov
3   Petroleum Recovery Research Center, New Mexico Tech, Socorro, NM 87801, USA; william.ampomah@nmt.edu
*   Correspondence: ek2qt@umsystem.edu; Tel.: +1-575-418-1593

Abstract: The objectives of this study were (1) to assess the fate and impact of CO$_2$ injected into the Morrow B Sandstone in the Farnsworth Unit (FWU) through numerical non-isothermal reactive transport modeling, and (2) to compare the performance of three major reactive solute transport simulators, TOUGHREACT, STOMP-EOR, and GEM, under the same input conditions. The models were based on a quarter of a five-spot well pattern where CO$_2$ was injected on a water-alternating-gas schedule for the first 25 years of the 1000 year simulation. The reservoir pore fluid consisted of water with or without petroleum. The results of the models have numerous broad similarities, such as the pattern of reservoir cooling caused by the injected fluids, a large initial pH drop followed by gradual pH neutralization, the long-term persistence of an immiscible CO$_2$ gas phase, the continuous dissolution of calcite, very small decreases in porosity, and the increasing importance over time of carbonate mineral CO$_2$ sequestration. The models differed in their predicted fluid pressure evolutions; amounts of mineral precipitation and dissolution; and distribution of CO$_2$ among immiscible gas, petroleum, formation water, and carbonate minerals. The results of the study show the usefulness of numerical simulations in identifying broad patterns of behavior associated with CO$_2$ injection, but also point to significant uncertainties in the numerical values of many model output parameters.

Keywords: reactive solute transport; CO$_2$ sequestration; multi-phase fluid flow; Farnsworth Unit; STOMP; GEM; TOUGHREACT

Citation: Kutsienyo, E.J.; Appold, M.S.; White, M.D.; Ampomah, W. Numerical Modeling of CO$_2$ Sequestration within a Five-Spot Well Pattern in the Morrow B Sandstone of the Farnsworth Hydrocarbon Field: Comparison of the TOUGHREACT, STOMP-EOR, and GEM Simulators. *Energies* **2021**, *14*, 5337. https://doi.org/10.3390/en14175337

Academic Editor: Sohrab Zendehboudi

Received: 16 May 2021
Accepted: 21 August 2021
Published: 27 August 2021

Publisher's Note: MDPI stays neutral with regard to jurisdictional claims in published maps and institutional affiliations.

1. Introduction

The Farnsworth Unit (FWU), a hydrocarbon field in northern Texas, USA, has been studied by the Southwest Regional Partnership on Carbon Sequestration (SWP) since 2013 as a test site for commercial-scale CO$_2$ sequestration and enhanced oil recovery (EOR) in a sandstone reservoir [1,2]. Central to assessing the feasibility of CO$_2$ sequestration in the FWU is determining the behavior of the injected CO$_2$, including where and at what rate the CO$_2$ will migrate, how the CO$_2$ will be distributed among the pore fluid phases (i.e., aqueous, gas, and nonaqueous liquid) and minerals, and how the hydraulic properties of the reservoir and the composition of the pore fluids will be changed.

Answering these questions requires the ability to quantify the flow of multiple fluid phases, their transport of solute and heat, and chemical reactions involving the fluid phases and minerals in the reservoir. Several previous SWP studies have attempted to do this using numerical reactive transport modeling. Ahmmed [3] used the TOUGHREACT software [4] to model reservoir behaviors caused by CO$_2$ injection in the immediate vicinity of an individual well and over the full area of the FWU. His model predicted the pH of the formation water to decline to a minimum of 4.7 because of CO_2 injection, causing most of the native minerals in the reservoir to dissolve, except for quartz, kaolinite, and illite. The model predicted hydrodynamic trapping to be the main mechanism of CO_2 sequestration, with ankerite predicted to be the only mineral that sequestered CO_2. The changes in mineral abundance, however, were predicted to be too small to cause much change in the hydraulic properties of the reservoir. Limitations of the model were that it did not include petroleum, only CO_2 was injected into the reservoir rather than water and CO_2 through a water alternating gas (WAG) scheme as actually implemented in the field, the model did not implement the actual regional pressure gradient occurring in the field, and the model was only carried out to 30 years.

Pan et al. [5] used the TOUGHREACT software to evaluate reactive transport in the Morrow B Sandstone resulting from WAG injection over a 5-spot well pattern. Because of the symmetry of the 5-spot well pattern, the model domain consisted of a triangle representing only one-eighth of the total 5-spot pattern area where the injection well and production well were separated by 504 m. Pan et al. [5] incorporated several chloride and sulfate minerals plus muscovite and dawsonite in their model as potential precipitates that were not incorporated in Ahmmed's [3] model. However, except for halite, these minerals were not predicted to precipitate as described in the model by Pan et al. [5]. Besides the relatively small size of their model domain, limitations of the model constructed by Pan et al. [5] were that it did not treat petroleum and used mineral reactive surface areas per unit mass that are significantly larger than indicated by Gallagher's [6] characterization of the Morrow B. Similar to Ahmmed et al. [3], Pan et al. [5] found the native reservoir minerals, quartz, kaolinite, and illite to increase in abundance over time, whereas the other native reservoir minerals dissolved. Like Ahmmed et al. [3], Pan et al. [5] found ankerite to be a mineral sink for injected CO_2, but also magnesite and siderite. Pan et al. [5] found their predicted changes in mineral abundances to cause only minor changes in the hydraulic properties of the reservoir—a maximum increase in porosity and permeability of 2.7 and 8.4%, respectively, occurring close to the injection well.

Khan [7] also used the TOUGHREACT software to model reactive transport in the Morrow B Sandstone as a result of WAG injection but considered a larger model domain than Pan et al. [5] consisting of the western part of the FWU. Like Ahmmed [3] and Pan et al. [5], Khan [7] considered water and CO_2 in his models but not petroleum. Khan's [7] simulations predicted much of the injected CO_2 to leak from the reservoir into the overlying shales or to migrate across the western boundary of the FWU within a few decades. Khan's [7] predicted mineral precipitation and dissolution behaviors resembled those of Ahmmed [3] and Pan et al. [5]. The native reservoir minerals, ankerite, albite, and illite were predicted to dissolve because of CO_2 injection but quartz; kaolinite; smectite; and the carbonate minerals calcite, dolomite, and siderite were predicted to precipitate. However, Khan's [7] predicted porosity changes of order 0.001% were much smaller than those of Pan et al. [5].

Sun et al. [8] used the Computer Modeling Group Green House Gas (CMG-GHG) simulator, GEM [9,10], to model reactive transport of CO_2 injected through a WAG scheme over a model domain consisting of a 5-spot well pattern like that considered by Pan et al. [5]. The total simulation time in Sun et al. [8] was 1000 years, with injection occurring during the first 20 years. In contrast to Ahmmed [3], Pan et al. [5], and Khan [7], Sun et al. [8] predicted most of the injected CO_2 to be sequestered as an immiscible gas phase. Most of the remaining CO_2 in the model of Sun et al. [8] was sequestered through residual trapping. Like Ahmmed [3], Pan et al. [5], and Khan [7], Sun et al. [8] predicted the smallest amount of CO_2 sequestration to occur through mineral trapping. Sun et al. [8] predicted quartz, kaolinite, and siderite to precipitate in their simulations but albite, calcite, chlorite, dolomite, illite, and smectite to dissolve. Sun et al. [8] did not include ankerite or magnesite in their models, two potentially important mineral sinks for CO_2. Sun et al. [8] predicted porosity changes of less than 1% in their models.

White et al. [11] have developed a multi-fluid phase (water–oil–gas) reactive transport simulator called STOMP-EOR, which they applied to study the physical behavior of fluids in the western FWU. Like GEM, STOMP-EOR solves coupled conservation equations for energy, water mass, CO_2 mass, CH_4 mass, and the masses of multiple petroleum components in variably saturated geologic media. Although STOMP-EOR has the capability to compute reactive transport, STOMP-EOR had not yet been used in reactive transport modeling studies of the FWU.

Because the previous SWP investigations of the FWU used three different numerical simulators—TOUGHREACT, GEM, and STOMP-EOR, this raises the question of how consistent the simulators are in terms of their ability to model the same input conditions. Thus, one of the objectives of the present study was to try to answer this question by building identical five-spot pattern models with the same grid design and parameter values for all three simulators and comparing the results. Because TOUGHREACT currently does not have the capability to treat a separate petroleum fluid phase, the comparison of the TOUGHREACT, GEM, and STOMP-EOR models required that water and CO_2 were the only pore fluids. The TOUGHREACT simulator is analogous to the STOMP-CO_2 simulator, both considering only energy conservation, and conservation of water, CO_2, and salt mass. The equations of state differ significantly between STOMP-CO_2 and STOMP-EOR, with CO_2 properties being computed from the Span and Wager [12] equation of state, and cubic equations of state, respectively. A further comparison was made between GEM and STOMP-EOR for identical five-spot pattern models that included water, CO_2, and petroleum pore fluids.

In addition to providing a rigorous comparison of the TOUGHREACT, GEM, and STOMP-EOR simulators, the first of its kind and which will help build confidence in these simulators for future research, the present study also extends the previous reactive transport modeling studies of the FWU in the following ways:

- The studies by Ahmmed [3], Pan et al. [5], and Khan [7] considered only water and CO_2 as the pore fluids and not oil. The present study included a simulation scenario that considered water, CO_2, and oil.
- Although Sun et al. [8] considered water, CO_2, and oil in their reactive transport model, their model had some differences with actual field conditions: (1) They did not use the actual fluid injection temperature in the field, but rather the 75 °C temperature in the reservoir, and (2) they used generic mineral reactive surface areas from Pan et al. [5] and Xu et al. [13] rather than reactive surface areas determined from the field properties of the Morrow B. The present study used the field-based fluid injection temperatures and mineral reactive surface areas.

2. Geological Setting

The Farnsworth Unit is located in the western Anadarko Basin (Figure 1), a structural basin that formed primarily during the Mississippian and Pennsylvanian periods in response to the collision of southeastern North America with Gondwanaland. The Morrow B Sandstone, the main target for hydrocarbon production and CO_2 sequestration in the study, is part of the Upper Morrowan-age (Early Pennsylvanian) stratigraphic succession in the basin. This succession is characterized by alternating intervals of glacially induced marine transgression and regression. During times of marine transgression, increasingly fine-grained clastic sediments were deposited, culminating in the deposition of nearshore and offshore mud. During times of marine regression, streams flowed through the FWU from the northwest, carving channels through the older transgressive sediments. During subsequent marine transgression, the fluvial sediments deposited in these channels were winnowed to form coarse-grained lag deposits and then were buried by fine-grained clastic sediments. Thus, the Morrow B Sandstone consists of relatively narrow channels of coarse sandstone enclosed within fine-grained sediments, creating conditions favorable for stratigraphic CO_2 trapping.

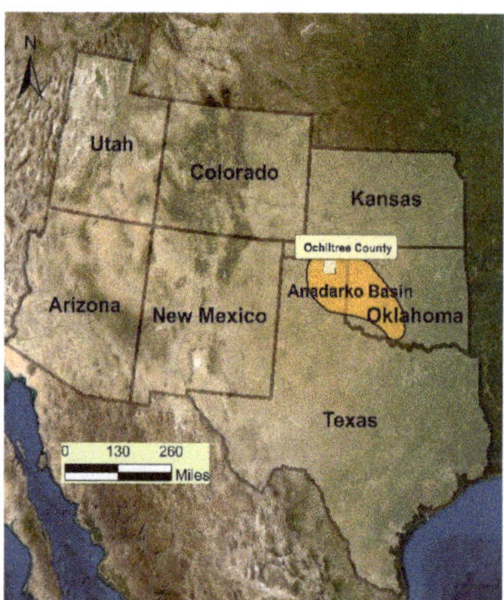

Figure 1. Location of the Farnsworth Unit and the Anadarko basin within the seven-state region of the SWP.

The Morrow B Sandstone is subarkosic, with quartz and albite as the main constituents. The remainder of the Morrow B is composed of minor amounts of chlorite; the carbonate minerals, calcite, siderite, and ankerite; and the clay minerals, smectite, illite, and kaolinite. The Morrow B ranges in thickness from 0 to 16.5 m within the Farnsworth Unit, with an average thickness of 10 m [6,14]. The permeability and porosity of the Morrow B are very heterogeneous, with average values of approximately 48.2 mD and 14.5%, respectively. The Morrow B is overlain by the Morrow Shale and an Atokan-age sequence of low-permeability evaporites and limestone called the Thirteen Finger Limestone (Figure 2; [6,14–17]). Together, the Morrow Shale and Thirteen Finger Limestone act as a caprock for the Morrow B Sandstone [14,15].

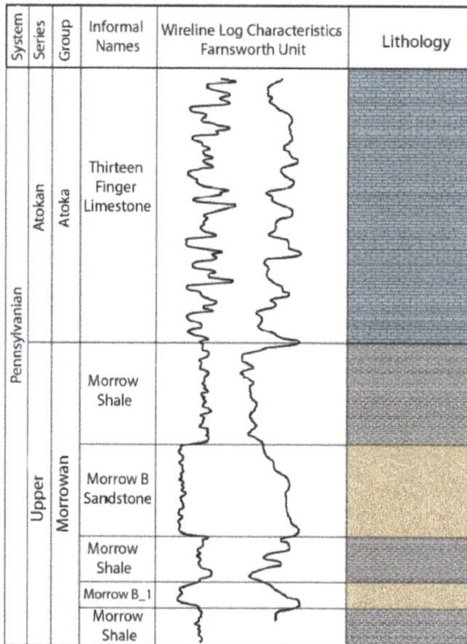

Figure 2. Pennsylvanian stratigraphic sequence of the of Farnsworth Unit. From Gallagher [6].

3. Model Construction and Scenarios

3.1. General Model Characteristics

The field development design within the Farnsworth Unit is a sequence of five-spot well patterns, in which four injection wells are placed at the corners of a square and a production well is placed at the center of the square. The numerical model generated in the present study is based on that five-spot well pattern, specifically on the pattern centered on production well, 13-10A, but because of the symmetry of the well spacing and in the expected model results, the model considers only one-quarter of the full pattern. Thus, the grid employed in the numerical model consisted of a three-dimensional block with an injection well at one corner and a production well at the opposite corner (Figure 3). The numerical grid had horizontal lengths of 504 m each subdivided by 11 equally spaced nodes, and a vertical length of 10 m, coinciding with the thickness of the Morrow B at well 13-10A, subdivided by 4 equally spaced nodes. The small model scale and spatial homogeneity of model parameters within the model domain were chosen to facilitate comparison of results from the three simulators.

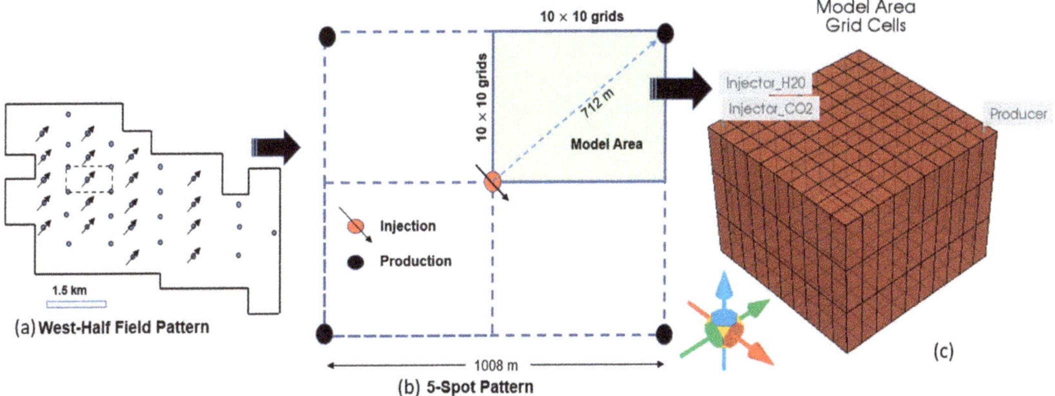

Figure 3. Model spatial domains. (**a**) Areal extent of the FWU. Black circles represent production wells. Black circles with arrows represent injection wells. The dashed rectangle represents the area of the five-spot pattern considered in the present study. (**b**) Plan view of the five-spot well pattern on which the present study was based. The blue shaded rectangle represents the area of the model domain. (**c**) Numerical grid corresponding to the blue shaded area in panel (**b**).

All three numerical simulators solve mass and energy conservation equations for multiphase pressure distribution and fluid flow, solute transport, and heat transport [18–20].

In TOUGHREACT and STOMP-EOR, these equations are solved using the integrated finite difference method. In GEM, these equations are solved using the finite difference method.

Hydrological parameter values used in the models were obtained from studies by Pan et al. [5] and are shown in Table 1.

The initial Morrow B pore water composition used in the models was taken from Ahmmed et al. [21] and is shown in Table 2.

The minerals shown in Table 3 include both the minerals that were initially present in the Morrow B (primary minerals) and new minerals that were expected possibly to precipitate during the model simulations (secondary minerals). The mineral reactive surface areas were obtained from Khan [7], which were calculated from the average radii of mineral grains in the Morrow B reported by Gallagher [6]. Kinetic parameters for mineral precipitation and dissolution were obtained from Palandri and Kharaka [22] and Xu et al. [19]. Mineral precipitation and dissolution reactions initially proceeded according to a neutral pH reaction mechanism because of the initially near-neutral pH of the Morrow B formation water. With the injection of CO_2, the formation water became progressively acidified and mineral precipitation and dissolution reactions then proceeded according to an acidic pH reaction mechanism. Chemical reactions that did not involve minerals were assumed to reach equilibrium within each time step, for example, the intra-aqueous species reactions.

Table 1. Hydrogeologic parameters used in the STOMP-EOR, TOUGHREACT, and GEM simulations.

Matrix compressibility (1/Pa)	4.5×10^{-10}
Diffusion coefficient (m^2/s)	
1. Gas	1.1×10^{-5}
2. Aqueous solutes	2.1×10^{-9}
Rock matrix density (kg/m^3)	2500
Porosity	0.145
Intrinsic Permeability	
1. Horizontal (m^2)	4.74×10^{-14}
2. Vertical Ratio	0.10
Relative Permeability (Corey, 1954 model)	
1. Saturation endpoints	$S_{lr} = 0.3$ and $S_{gr} = 1 \times 10^{-4}$
2. Water/gas endpoints	$Krw = 0.7$ and $Krg = 1$
Capillary Pressure	None
Salt mass fraction in pore water	0.003
Initial aqueous phase saturation	
1. Water–CO_2 models	0.99
2. Water–CO_2–oil models	0.73
Initial gas-phase saturation	0.01
Initial oil-phase saturation (Water-CO_2-oil models)	0.27
Initial field temperature (°C)	75.56
Injection pressure (MPa)	34.47
Injection temperature (°C)	40
Production well screen pressure (MPa)	29.99
WAG Cycle Ratio (Months)	$3:6$
1. Thermal conductivity of saturated rock (W/m K)	2.28
2. Specific heat (J/kg K)	700

Table 2. Concentrations of aqueous component species in the FWU reservoir from well battery AWT4.

Primary Aqueous Species (mol/L):			
Ca^{2+}	8.25×10^{-4}	Ba^{2+}	1.00×10^{-5}
H^+	1.00×10^{-7}	AlO_2^-	2.80×10^{-7}
K^+	1.83×10^{-4}	SO_4^{2-}	1.35×10^{-4}
Mg^{2+}	5.10×10^{-4}	Cl^-	5.90×10^{-2}
Na^+	6.18×10^{-2}	HCO_3^-	1.33×10^{-2}
Fe^{2+}	3.60×10^{-13}	SiO_2	6.69×10^{-4}

Data from Ahmmed et al. [21].

All three numerical simulators use similar kinetic formulations [9,11,13], in which mineral dissolution and precipitation rates are calculated from

$$r_m = \hat{A}_m k_m \left(1 - \frac{Q_m}{K_{eq,\,m}} \right) \tag{1}$$

where r_m is the rate of dissolution or precipitation, $\hat{A}_m$ is the reactive surface area of mineral m, k_m is the rate constant, $K_{eq,\,m}$ is the chemical equilibrium constant, and Q_m is the activity product, expressed as

$$Q_m = \prod_{k=1}^{n_{aq}} a_k^{v_{km}} \tag{2}$$

where n_{aq} is the number of aqueous components, v_{km} is the stoichiometric coefficient of component k in reaction m, and a_k is the activity of the component k.

Table 3. Initial mineral volume fractions, possible secondary mineral phases, reactive surface areas, and kinetic properties at 25 °C.

Minerals	Initial Volume Fraction %	Reactive Surface Area cm^2/g	Neutral pH Mechanism	
			Rates Constant 25 °C [mol m^{-2} s^{-1}]	Activation Energy [kJ mol^{-1}]
Albite	17.973	11.45	1.0×10^{-12}	67.83
Calcite	0.4279	11.07	1.55×10^{-7}	23.50
Clinochlore	0.8559	11.41	1.0×10^{-13}	62.76
Quartz	58.6261	9.80	1.023×10^{-14}	87.70
Illite	0.5991	43.63	1.7×10^{-13}	35.00
Kaolinite	7.018	46.15	1.01×10^{-13}	62.76
Dolomite	0	10.49	6.0×10^{-10}	41.80
Magnesite	0	10	4.57×10^{-10}	23.50
Smectite-ca	0	9.8	1.0×10^{-14}	58.62
Siderite	0	9.8	1.26×10^{-9}	41.80
Ankerite	0	9.84	1.26×10^{-9}	41.80

Mineral volume fraction data from Munson [14] and Gallagher [6].

Reaction rate constants at a temperature of interest are computed from

$$k_m = k_{25}\, exp\left[-\frac{E_a}{R}\left(\frac{1}{T} - \frac{1}{T_0}\right)\right] \tag{3}$$

where E_a is the activation energy and k_{25} is the reaction rate constant at 25 °C.

Other initial conditions are that the model domain had a constant pressure of 30 MPa and a constant temperature of 75 °C.

The boundary conditions for the models constructed for each simulator were the same. All the faces of the model domains had zero fluid flux boundary conditions. The lateral faces of the model domains had zero heat flux boundary conditions. A vertical injection well was placed at one corner of the model domain, and a production well was placed at the opposite corner (Figure 3). The wells were screened over an elevation from 0.0 m to 10.0 m. A WAG scheme was employed at the injection well in which water was injected at a rate of 0.336 kg/s at 40 °C for 90 days, after which CO_2 was injected at a rate of 0.454 kg/s for 180 days at 40 °C. This WAG scheme was employed for the first 25 years of each simulation, after which injection ceased and the simulation was allowed to continue to run to a total time of 1000 years to be able to track long-term effects of CO_2 injection. A constant bottom hole pressure of 30.0 MPa was assigned to the production well. The production and injection rates represent $\frac{1}{4}$ of the average pumping rates in the field from wells 13-9 and 8-4, as the model domains intersect only one-quarter of the perimeter of the wells.

During execution of the model, time step sizes were continuously and automatically adjusted to achieve convergence. In general, time step sizes increased with time as gradients in model parameters diminished.

3.2. Model Scenario 1: Injection of CO_2 into a Saline Water Aquifer

Pore fluids in the Morrow B Sandstone consist of water, petroleum, and methane. Thus, to model as robustly as possible the behavior of CO_2 injected into the Morrow B and to assess the CO_2 sequestration capacity of the Morrow B, all three pore fluids should be treated. This in fact was the objective of the present study's second model scenario described below. However, another objective of the present study was to assess the consistency of major reactive solute transport simulators with one another for CO_2 sequestration modeling, specifically, the GEM, STOMP-EOR, and TOUGHREACT simulators. The TOUGHREACT simulator did not currently have the capacity to treat petroleum as a separate pore fluid. Thus, to the compare the performance of TOUGHREACT to the GEM and STOMP-EOR simulators, a scenario was chosen involving injection of CO_2 into the Morrow B where water was the only pore fluid present. This scenario has further value in that it provides a baseline for comparison. As noted above, the same parameter values (Tables 1–3), boundary conditions, and initial conditions were used with all three simulators to model this scenario.

3.3. Model Scenario 2: Injection of CO_2 into a Depleted Hydrocarbon Reservoir

This scenario investigates the effects of the coexistence of petroleum and water on CO_2 sequestration in the Morrow B Sandstone. The parameter values, boundary conditions, and initial conditions were the same as in Model scenario 1, except that the initial water saturation was reduced from 99% to 73% and oil saturation was raised from 0 to 16%. The initial gas saturation remained the same as in Model scenario 1 at 1%.

The composition of the petroleum used in the models is shown in Table 4 and is based on the FWU petroleum composition reported by Gunda et al. [23]. The presence of petroleum with water in a porous medium can significantly alter the sequestration behavior of CO_2 compared to the case when only water is initially present in the porous medium. A significant fraction of the injected CO_2 is expected to dissolve into the petroleum, leaving less CO_2 to exist as a separate immiscible gas phase and to dissolve into the formation water, affecting the physical flow behavior of the pore fluids through altered relative permeability values. Lower CO_2 concentration in the formation water will raise its pH, fundamentally affecting the concentrations of other aqueous species and the precipitation and dissolution of minerals. As noted above, only GEM and STOMP-EOR were used to investigate Model scenario 2, as TOUGHREACT did not currently have the capability to treat a separate petroleum fluid phase.

Table 4. Model petroleum component properties and initial mole fractions.

Component	Mole Fraction	Molar Weight (kg/kmol)	Critical Temperature (K)	Critical Pressure (bar)
CO_2	0.0	44.01	304.21	73.77
CH_4	0.385	16.04	188.85	46.00
C2	0.039	30.07	197.45	48.83
C3	0.025	44.10	247.19	42.44
C4+	0.028	58.12	289.89	37.76
C5+	0.020	72.15	328.13	33.76
C6	0.018	86.18	365.70	29.68
HC1	0.335	189.95	577.54	22.48
HC2	0.150	545.65	864.34	16.25

Component definitions, properties, and abundances derived from Sun et al. [8].

4. Results

4.1. Model Scenario 1

4.1.1. Temperature and Pressure Distributions

Figures 4 and 5 show plan views of the evolution of reservoir pressure and temperature predicted in the STOMP-EOR, TOUGHREACT, and GEM models in the middle layer of the model grid as a result of water and CO_2 introduced through the injection well at the lower left corner of each plot. All three models predicted an increase in fluid pressure from the initial value of 30 MPa during the 25 years of injection, reaching a maximum of ~33 MPa. After the injection period ended, all three models predicted fluid pressures to decline, but not at the same rate. The fastest fluid pressure decline was predicted in the TOUGHREACT model, where fluid pressures returned to initial reservoir values within 100 years. The STOMP-EOR model predicted a slower decline in pressure, requiring several centuries for fluid pressure to return to the initial reservoir value. In the GEM model, fluid pressure had not yet returned to the initial reservoir value by the end of the 1000 year simulation time, reaching a minimum value of ~31.5 MPa.

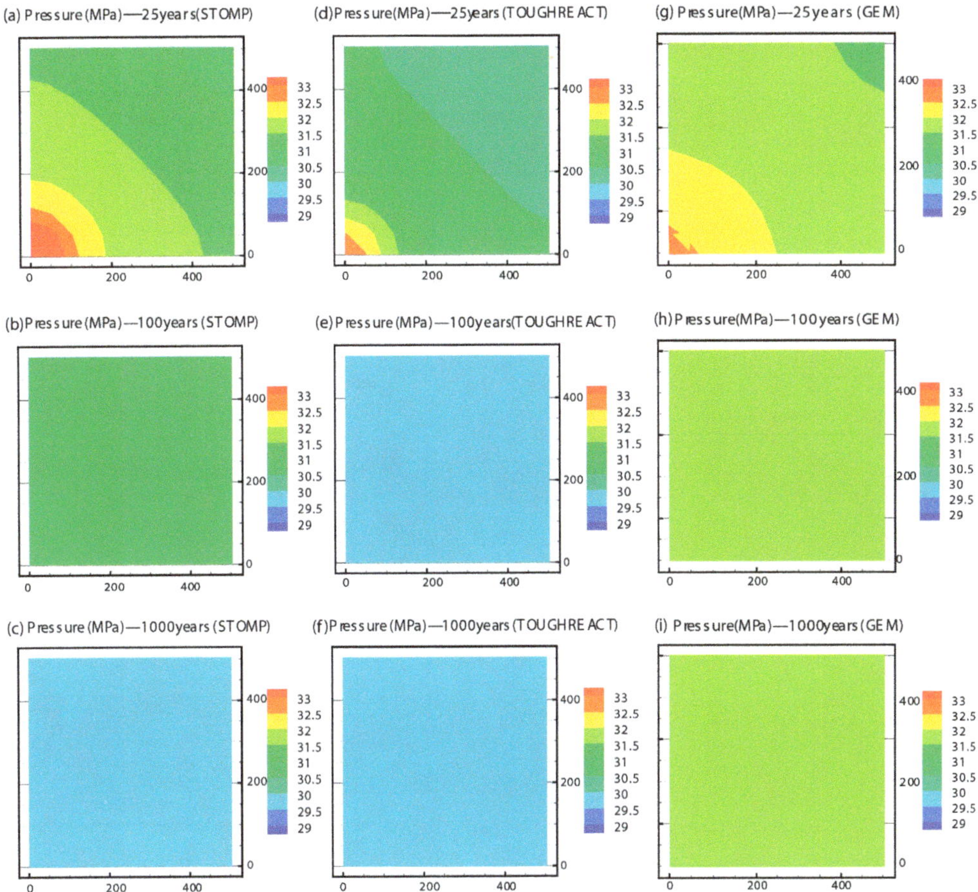

Figure 4. Plan views of the evolution of fluid pressure in the middle cell layer of the model grid predicted by the STOMP-EOR models (**a–c**), TOUGHREACT models (**d–f**), and GEM models (**g–i**) after 25, 100, and 1000 years.

The temperature evolutions predicted by all three models were similar. In each model, temperature near the injection well dropped from the initial reservoir temperature of 75 °C to 40 °C, the temperature of the injected fluids. Lower temperatures gradually propagated across the model domain toward the production well located at the upper left corner of the plots, continuing until the end of the simulations after 1000 years. However, after the injection well was shut in, temperature near the injection well gradually began to rise, reaching ~60 °C in the TOUGHREACT and GEM models and ~70 °C in the STOMP-EOR model after 1000 years.

4.1.2. Evolution of Pore Fluid and Mineral Composition

In addition to altering the pressure and temperature distribution in the reservoir, the injected water and CO_2 alter the pore fluid composition of the reservoir. Figure 6 shows CO_2 gas saturation after 25, 100, 600, and 1000 years along a vertical profile between the injection well and production well as predicted by the STOMP-EOR, TOUGHREACT, and GEM models. The results of the models were most dissimilar at early times (see Figure 6a–c at 25 years), but all show CO_2 to concentrate in the upper part of the profile, which is due to buoyancy. Similar maximum gas saturations around 0.38 are also predicted by all three models. Over time the model CO_2 gas saturations converged to a similar, vertically differentiated pattern with saturations decreasing from the top of the model domain to the bottom. The CO_2 gas saturations in each model also largely stabilized after 100 years, changing little until the end of the simulations at 1000 years. However, even at longer times some differences are visible in the results. The STOMP-EOR model predicted the highest overall gas saturations and the TOUGHREACT model predicted the CO_2 gas plume not to migrate all of the way to the production well on the right boundary of the plots. The differences in gas saturation may be a function of different CO_2 solubility relationships used in the three simulators [24–26].

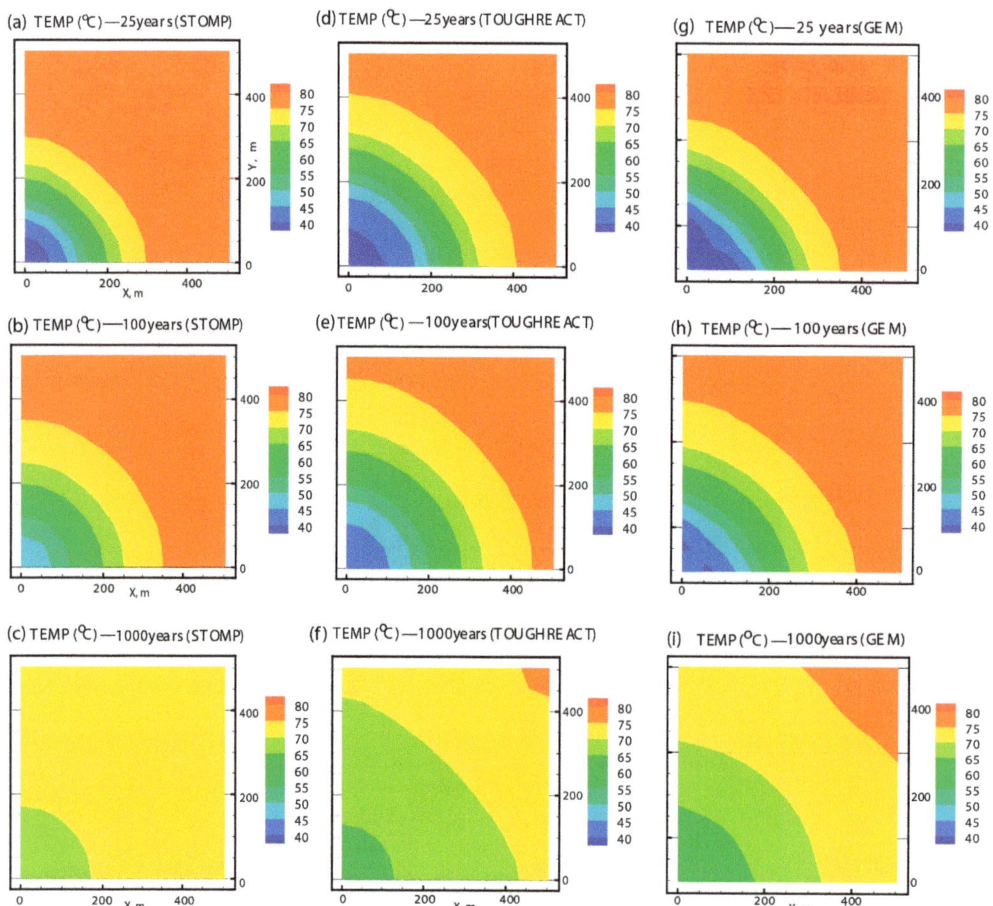

Figure 5. Plan views of the evolution of temperature in the middle cell layer of the model grid predicted by the STOMP-EOR models (**a–c**), TOUGHREACT models (**d–f**), and GEM models (**g–i**) after 25, 100, and 1000 years.

CO_2 gas saturation as a function of time is shown in Figure 7a. All three models show a sharp increase in gas saturation during the injection period. In addition, in all three models, gas saturation remains significantly elevated over the entire 1000-year simulation period, though in the TOUGHREACT model, gas saturation steadily declines after the injection well is shut in, whereas in the STOMP-EOR and GEM models, gas saturation remains relatively constant at its maximum level. The long-term persistence of this immiscible CO_2 gas phase is largely a product of the no-flow boundary conditions that encompass the model domain. Once the production well is turned off after 25 years, the CO_2 can no longer exit the model domain and can only diminish in abundance by dissolving into the formation water. Much of the injected CO_2 dissolves into the Morrow B formation water and this has a direct effect on its pH and HCO_3^- concentration, as indicated by Equation (5) and as shown in Figure 7b,c. The injection of CO_2 causes an immediate drop in the pH and an increase in the HCO_3^- concentration, though not by the same amounts. The GEM, TOUGHREACT and STOMP-EOR models predict minimum pH values of 4.6, 4.7, and 4.8, respectively, at the onset of CO_2 injection. All three models predicted a gradual increase in pH over time; this occurs within the first few years after the injection ceased until the end of the simulation. The eventual pH increase is probably caused by reactions with various minerals in the Morrow B that consume H^+.

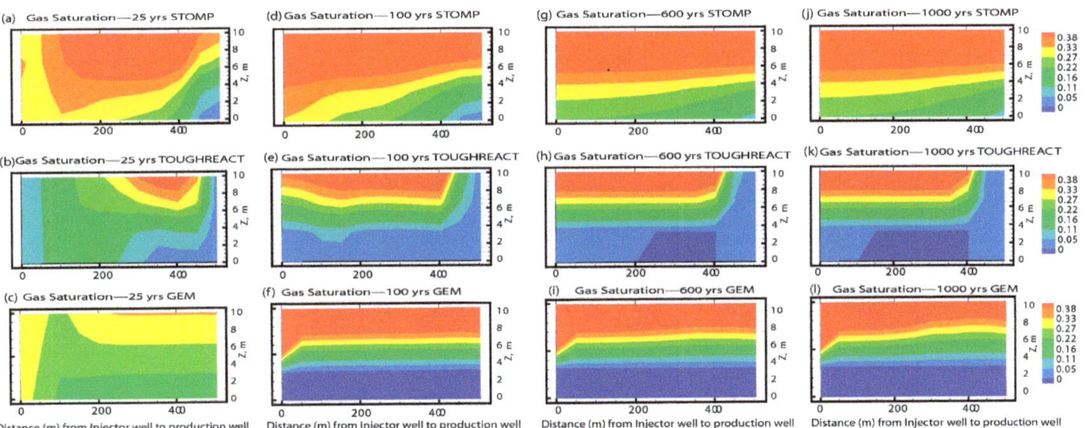

Figure 6. CO_2 gas saturation along a vertical cross section between the injection well and the producing well after 25 years predicted by the STOMP-EOR, TOUGHREACT, and GEM models after (**a–c**) 25 years, (**d–f**) 100 years, (**g–i**) 600 years, and (**j–l**) 1000 years.

All three models predicted an initial increase in HCO_3^- concentration over time, though these concentrations varied by ~1.5 orders of magnitude among the three simulators, with the lowest concentrations predicted by the GEM model, followed by the STOMP-EOR and TOUGHREACT models, respectively. In the GEM and TOUGHREACT models, HCO_3^- concentration continued to rise throughout the rest of the simulation, whereas in the STOMP-EOR simulation, HCO_3^- concentration began to decline after about 200 years. Several competing factors affect the concentration of HCO_3^-. Some HCO_3^- may be generated through the gradual dissolution of residual CO_2 gas into the formation water and by the dissolution of calcite, a native reservoir mineral, which also neutralizes pH (Equations (4) and (5); Figures 8 and 9). Some HCO_3^- is removed from the formation water by the precipitation of other carbonate minerals such as dolomite, magnesite, and ankerite (Equations (6)–(8)).

$$CO_{2(aq)} + H_2O \leftrightarrow H^+ + HCO_3^- \tag{4}$$

$$H^+ + CaCO_3(\text{Calcite}) \rightarrow Ca^{2+} + HCO_3^- \tag{5}$$

$$Ca^{2+} + Mg^{2+} + HCO_3^- \rightarrow CaMgCO_3(\text{Dolomite}) + H^+ \tag{6}$$

$$Mg^{2+} + HCO_3^- \rightarrow MgCO_3(\text{Magnesite}) + H^+ \tag{7}$$

$$Ca^{2+} + 0.3Mg^{2+} + 0.7Fe^{2+} + 2HCO_3^- \rightarrow CaMg_{0.3}Fe_{0.7}(CO_3)_2(\text{Ankerite}) + 2H^+ \tag{8}$$

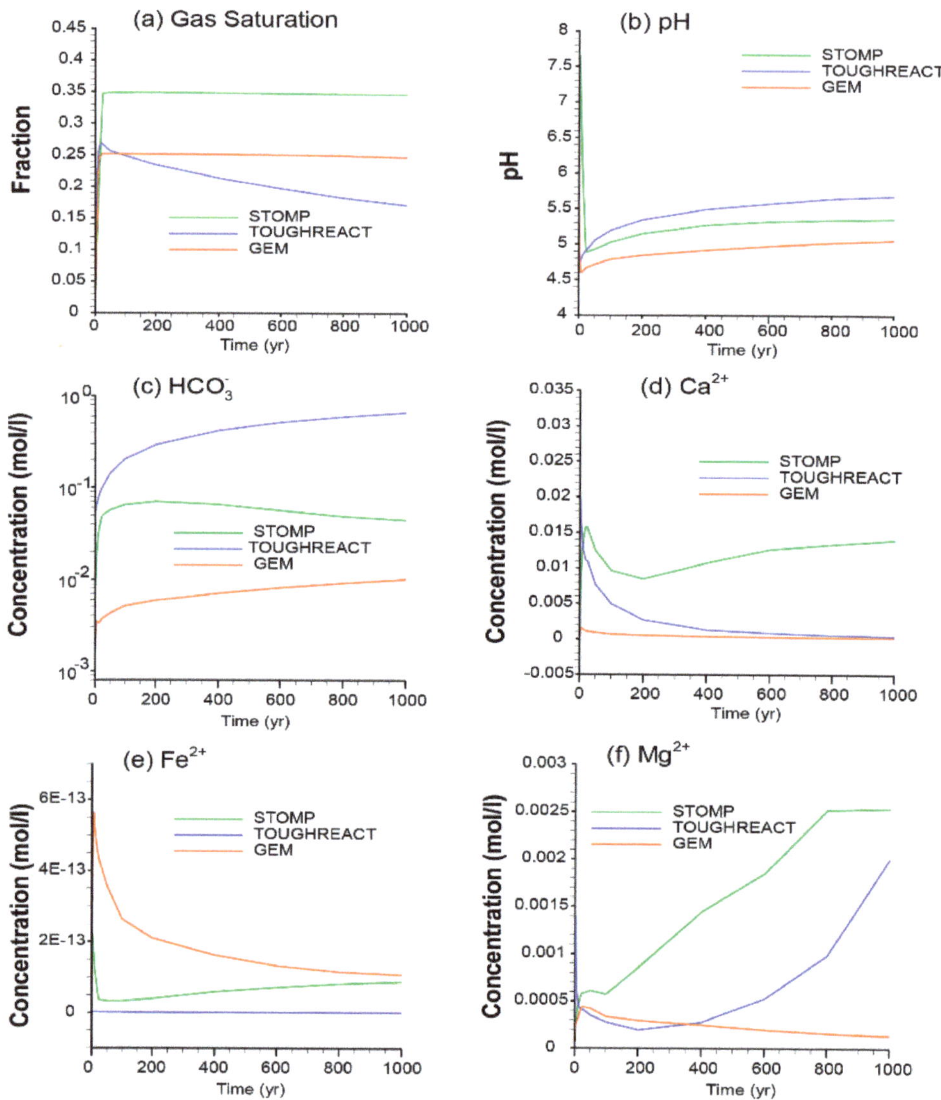

Figure 7. Simulated (**a**) gas saturation; (**b**) pH; and concentrations of (**c**) HCO_3^-, (**d**) Ca^{2+}, (**e**) Fe^{2+}, and (**f**) Mg^{2+} as a function of time integrated over the entire model domain.

The sharp initial drop in pH predicted by all three models during the first years of the injection period leads to the dissolution of the native reservoir minerals, calcite, albite, clinochlore, and illite in all three models. In addition, STOMP-EOR and GEM predict kaolinite to dissolve, whereas TOUGHREACT predicts kaolinite to precipitate. By the end of the 1000-year simulation period, all three models predict calcite, clinochlore, and illite to have continued to dissolve. STOMP-EOR and TOUGHREACT also predict albite to have continued to dissolve, whereas GEM predicts albite to have begun precipitating. STOMP-EOR and GEM predict kaolinite to continue to dissolve and TOUGHREACT predicts kaolinite to continue to precipitate until the end of the simulation at 1000 years.

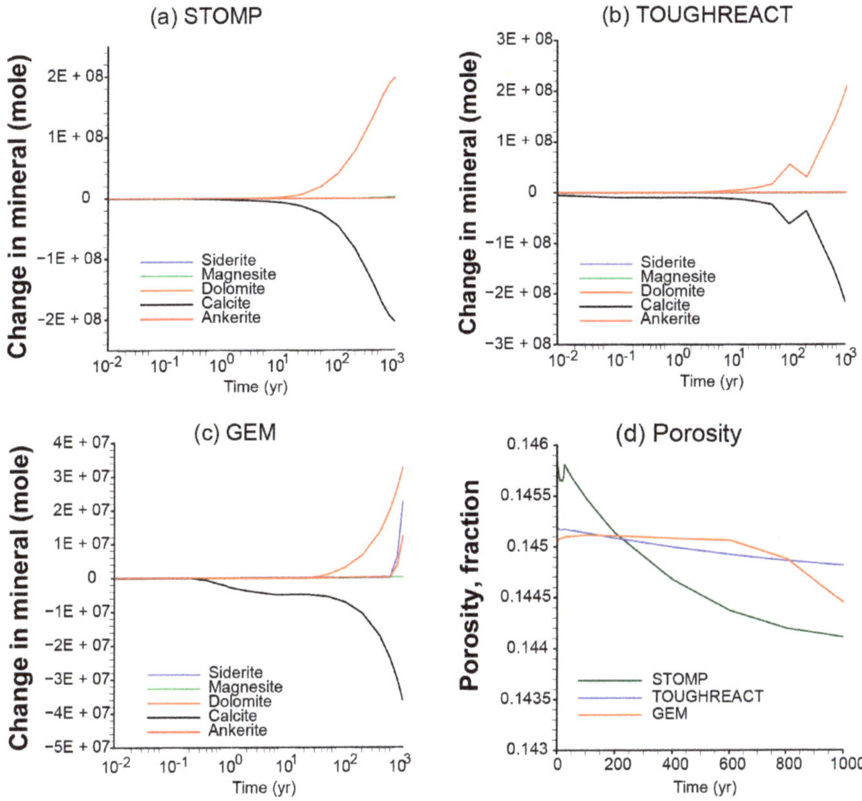

Figure 8. Simulated changes in carbonate minerals abundances (**a**) STOMP-EOR, (**b**) TOUGHREACT, (**c**) GEM, and changes in fraction. (**d**) Porosity as a function of time integrated over the entire model domain.

The predicted steady dissolution of calcite by all three models does not lead to a corresponding steady increase in Ca concentration in the formation water. Rather, Ca concentration gradually decreased over time in the TOUGHREACT and GEM models, and in the STOMP-EOR model, Ca concentration increased for the first 10 years, decreased from 10 to 200 years, and then gradually increased for the remainder of the simulation. Decreases in Ca concentration were driven by the precipitation in all three models of the Ca minerals, dolomite, and Ca-montmorillonite. In the STOMP-EOR and GEM models, further Ca was removed from the formation water by the precipitation of ankerite, siderite, and magnesite. No Fe minerals were predicted to precipitate in the TOUGHREACT model. Thus, the Fe concentration of the formation water in the TOUGHREACT model remained constant over time. However, ankerite and siderite were predicted to precipitate in the STOMP-EOR and GEM models, though much more so in the GEM model. This was enough to cause a steady decrease over time in the concentration of Fe in the formation water in the GEM model, but not enough to prevent a slight increase over time in the STOMP-EOR model. The GEM model predicted a brief initial period of increase in Mg concentration in the formation water over the first 25 years, followed by a gradual decline. The initial increase in Mg concentration is probably caused by the dissolution of clinochlore. However, with increasing time, the precipitation of dolomite, ankerite, magnesite, and Ca-montmorillonite led to a net decrease in Mg concentration. In the TOUGHREACT model, the absence of dissolution of any Mg minerals during the early years of the simulation and the precipitation of dolomite caused a decrease in Mg concentration in the formation water. At later times, high levels of clinochlore dissolution caused Mg concentration in the formation water to increase. In the STOMP-EOR model, Mg concentration increased almost continuously throughout the simulation, with the exception of a brief decline between approximately 50 and 100 years. The dissolution of clinochlore is likely the main source for the Mg concentration increase.

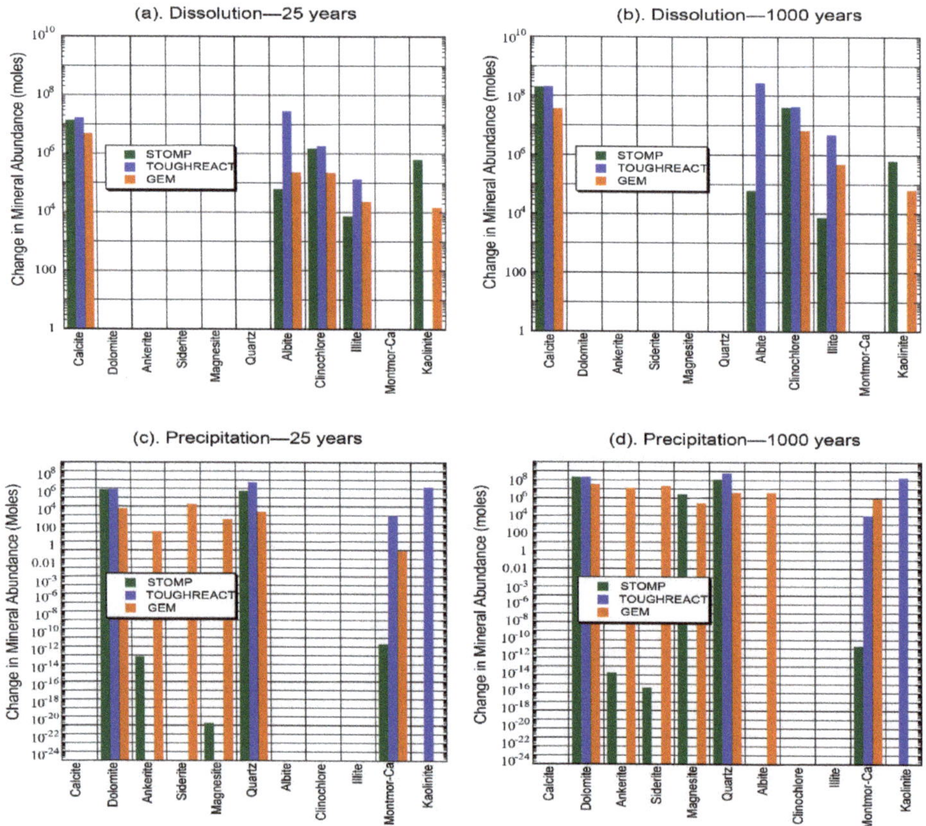

Figure 9. Simulated changes in mineral volume integrated over the entire model domain due to dissolution after (**a**) 25 years, (**b**) 1000 years, and due to precipitation after (**c**) 25 years and (**d**) 1000 years predicted by the STOMP-EOR, TOUGHREACT, and GEM models.

Quartz is the only native reservoir mineral that was predicted to precipitate in all three models (Figure 9c,d). In contrast, kaolinite, another native reservoir mineral, was predicted to dissolve in the GEM and STOMP-EOR models but was predicted to precipitate in the TOUGHREACT model.

All three of the models predicted very small, nearly continuous decreases in porosity over the course of the simulations (Figure 8d). The largest porosity decrease was predicted by the STOMP-EOR model at 0.0017. These porosity changes are likely to be too small to have a significant impact on the hydraulic properties of the Morrow B Sandstone and its capacity to sequester CO_2 in either the formation water or as an immiscible gas phase.

Figure 10 shows how the three models predict the injected CO_2 to be distributed among an immiscible gas phase, the formation water, and carbonate minerals. The STOMP-EOR and GEM models predict most of the injected CO_2 to be sequestered within an immiscible gas phase throughout the 1000 years of the simulation. All three models similarly predict the formation water to be the next most important sink for the injected CO_2. Carbonate minerals are a negligible sink for the injected CO_2 in the first years of the simulation. Indeed, in the GEM model, calcite dissolution outpaces the precipitation of other carbonate minerals for the first 200 years, after which net carbonate mineral sequestration of CO_2 begins to occur. However, in all three models, carbonate minerals become an increasingly important sink for injected CO_2 over time, and by the end of the simulation after 1000 years, they approach the formation water in importance as a mineral sink. The TOUGHREACT model predicts a relatively small amount of CO_2 to be partitioned into an immiscible gas phase compared to the other two models. In fact, by the end of the 1000-year simulation period, the TOUGHREACT model predicts the immiscible gas phase to be the smallest sink for the injected CO_2. As noted for Figure 6, the differences in the amounts of immiscible CO_2 gas predicted by the three simulators may be a function of the different CO_2 solubility functions that they employ [24–26].

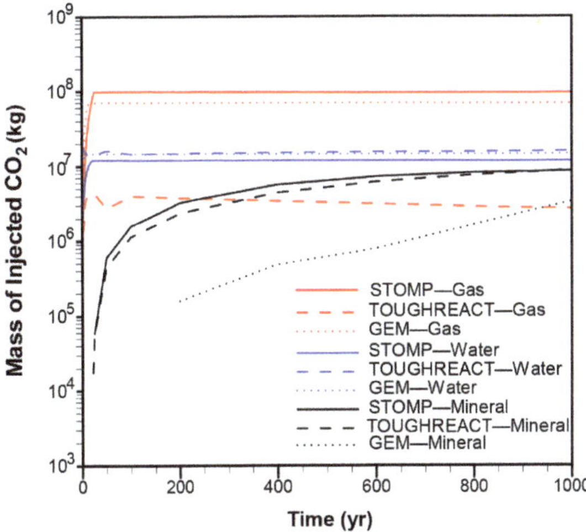

Figure 10. Mass of injected CO_2 that is sequestered in an immiscible gas phase, in the formation water, and in carbonate minerals as predicted by the STOMP-EOR, GEM, and TOUGHREACT models in Model Scenario 1.

4.2. Model Scenario 2

Figures 11 and 12 show plan views of the evolution of reservoir pressure and temperature predicted in the STOMP-EOR and GEM models in the middle layer of the model grid as a result of water and CO_2 introduced through the injection well at the lower left corner of each plot. The two models predicted an increase in fluid pressure from the initial value of 30 MPa during the 25 years of injection, reaching a maximum of ~34 MPa in STOMP-EOR and GEM. After injection ceased, fluid pressure was predicted to decline continuously in the STOMP-EOR model, reaching 32 MPa after 100 years and returning to the initial reservoir pressure of 30 MPa after 1000 years. A similar pressure evolution pattern was produced by the GEM model as long as siderite and ankerite were omitted from the model, as is the case for the results shown in Figure 11. Although the geochemical input parameters for siderite and ankerite were the same in the GEM and STOMP-EOR models, including siderite and ankerite in the GEM model caused pressure to continue to increase over time instead of returning to the initial value after injection ceased. This result seems to represent a limitation of the GEM model. A comparative analysis showed that the presence or absence of siderite and ankerite in the GEM model did not significantly impact the results for any model outputs except kaolinite, magnesite, and temperature. Compared to the case when siderite and ankerite were present, when siderite and ankerite were absent the GEM model predicted kaolinite abundance to be ~65% lower, magnesite precipitation was almost entirely prevented, and temperature was approximately 2.5 to 6 °C lower. The GEM results shown in the remaining plots are for the case when siderite and ankerite were included in the model.

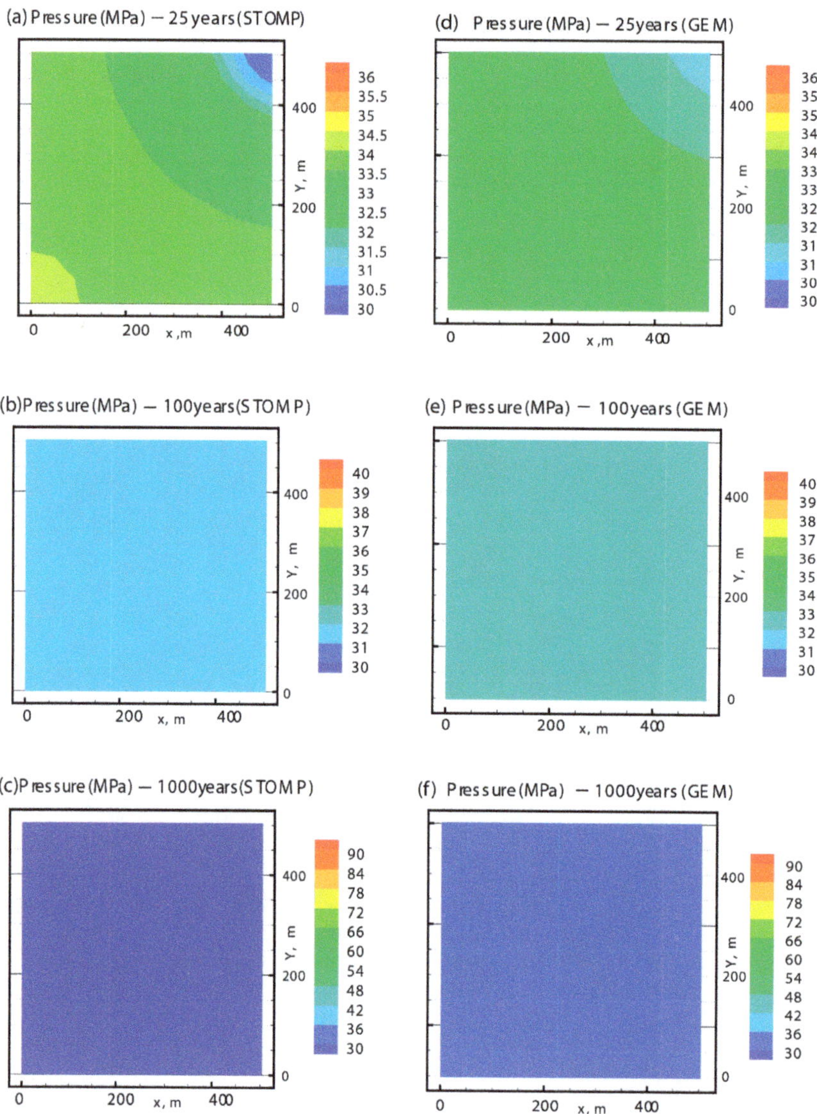

Figure 11. Plan views of the evolution of fluid pressure in the middle cell layer of the model grid predicted by the STOMP-EOR models (**a–c**) and GEM models (**d–f**) after 25, 100, and 1000 years, respectively. The GEM results are for ankerite and siderite omitted from the model, as the inclusion of these minerals in the GEM model produced unrealistically high pressures that steadily increased over time.

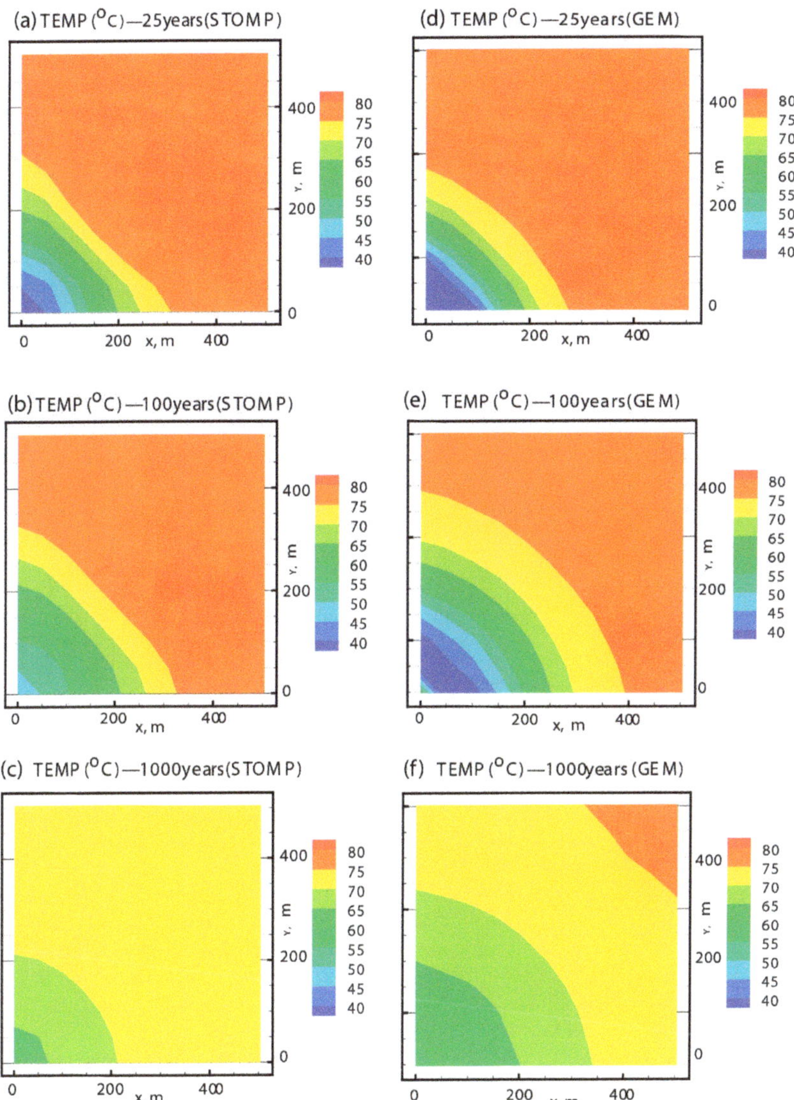

Figure 12. Plan views of the evolution of temperature in the middle cell layer of the model grid predicted by the STOMP-EOR models (**a–c**) and GEM models (**d–f**) after 25, 100, and 1000 years, respectively.

The temperature evolutions predicted by the STOMP-EOR and GEM models in Scenario 2 are similar to one another and to the results in Scenario 1. In each model, temperature near the injection well dropped from the initial reservoir temperature of 75 to 40 °C during the injection period. After injection ceased, temperatures gradually rose and became more homogeneous across the model domain, though they had not yet completely returned to the initial reservoir temperature of 75 °C after 1000 years.

Figure 13 shows the changes in CO_2 gas saturation along a cross section between the injection well and the production well as a function of time for the STOMP-EOR and GEM models. The results of the STOMP-EOR and GEM models differ significantly in detail but have some broad similarities in that they both show a plume of CO_2 gas to migrate about halfway across the cross section by about 25 years and to remain relatively stationary thereafter. The model results of Scenario 2 differ significantly from those of Scenario 1. In Scenario 2, CO_2 gas saturation never develops a strong vertical differentiation as in Scenario 1, and the injected CO_2 plume in Scenario 2 does not arrive at the production well by the end of the simulations at 1000 years. Maximum gas saturations in Scenario 2 are higher than in Scenario 1. However, comparing Figure 14a to Figure 7a, both of which show gas saturation averaged over the entire volume of the model domain, reveals that gas saturation in Scenario 2 is overall lower than in Scenario 1. This is probably because in Scenario 2, CO_2 can dissolve in both oil and water, whereas in Scenario 1, the only pore fluid into which CO_2 can dissolve is water.

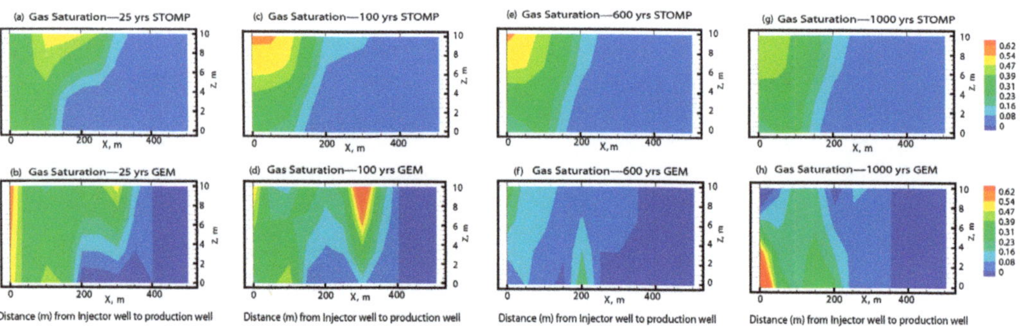

Figure 13. CO_2 gas saturation and porosity along a vertical cross section between the injection well and the producing well predicted by the STOMP-EOR and GEM models, respectively, after (**a,b**) 25 years, (**c,d**) 100 years, (**e,f**) 600 years, and (**g,h**) 1000 years.

Figure 14 also shows the pH and concentrations of HCO_3^-, Ca^{2+}, Fe^{2+}, and Mg^{2+} in the Morrow B formation water as a function of time, also averaged over the entire volume of the model domain. The trends in pH and concentrations of HCO_3^-, Ca^{2+}, and Mg^{2+} over time predicted by the STOMP-EOR and GEM models in Model Scenario 2 are qualitatively similar to those in Model Scenario 1 but differ significantly in their numerical values. In Model Scenario 2, the STOMP-EOR model consistently predicts higher pH and concentrations of HCO_3^-, Ca^{2+}, and Mg^{2+} compared to the GEM model, whereas the GEM model predicts higher concentrations of Fe^{2+} than in the STOMP-EOR model at early times, and the STOMP-EOR model predicts higher Fe^{2+} concentrations at later times. In Model Scenario 2, after an early increase in the GEM model, gas saturation decreases and largely parallels that predicted by the STOMP-EOR model.

The Scenario 2 STOMP-EOR and GEM models made some similar predictions about the evolution of carbonate mineral abundance (Figures 15 and 16). Both models predicted continuous dissolution of calcite. The two models differ further in that large amounts of siderite, magnesite, and ankerite precipitated in the GEM model but only tiny amounts of ankerite and no siderite precipitated in the STOMP-EOR model. Minimal magnesite precipitated at early times in the STOMP-EOR model but by the end of the simulation after 1000 years, considerable magnesite had precipitated. The carbonate mineral abundances predicted in Scenario 2 resemble those in Scenario 1 in some respects. Dolomite continued to be the main carbonate mineral predicted to be precipitated in Scenario 2 in the STOMP-EOR model. In contrast, in the GEM model, siderite was the most abundant mineral precipitated followed by ankerite, dolomite, and magnesite. The presence of oil did not greatly impact the patterns of carbonate mineral precipitation in the STOMP-EOR model but appears to have greatly increased the precipitation of siderite and ankerite in the GEM model. Carbonate mineral precipitation and dissolution trends were monotonic, reaching their highest levels at the end of the 1000-year simulation period.

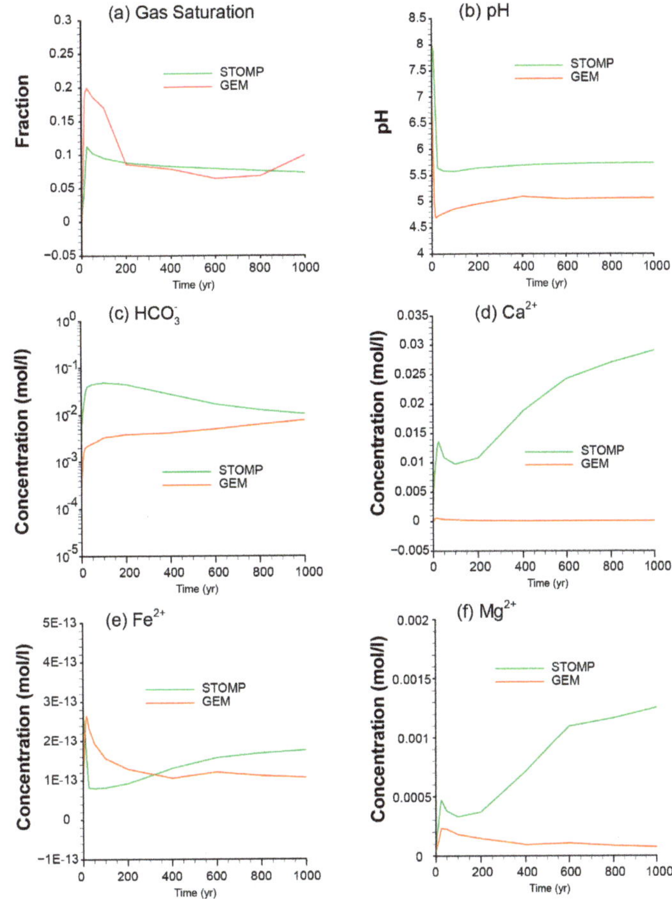

Figure 14. Simulated (**a**) gas saturation; (**b**) pH; and concentrations of (**c**) HCO_3^-, (**d**) Ca^{2+}, (**e**) Fe^{2+}, and (**f**) Mg^{2+} as a function of time averaged over the entire model domain.

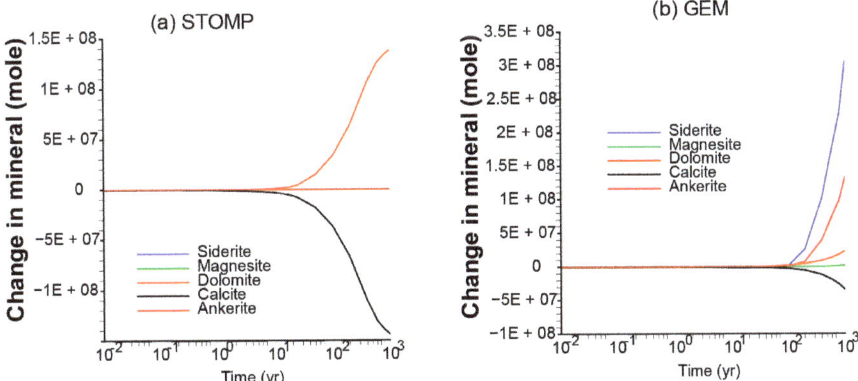

Figure 15. Simulated changes in carbonate mineral abundances for the (**a**) STOMP-EOR model and (**b**) GEM model as a function of time averaged over the entire model domain.

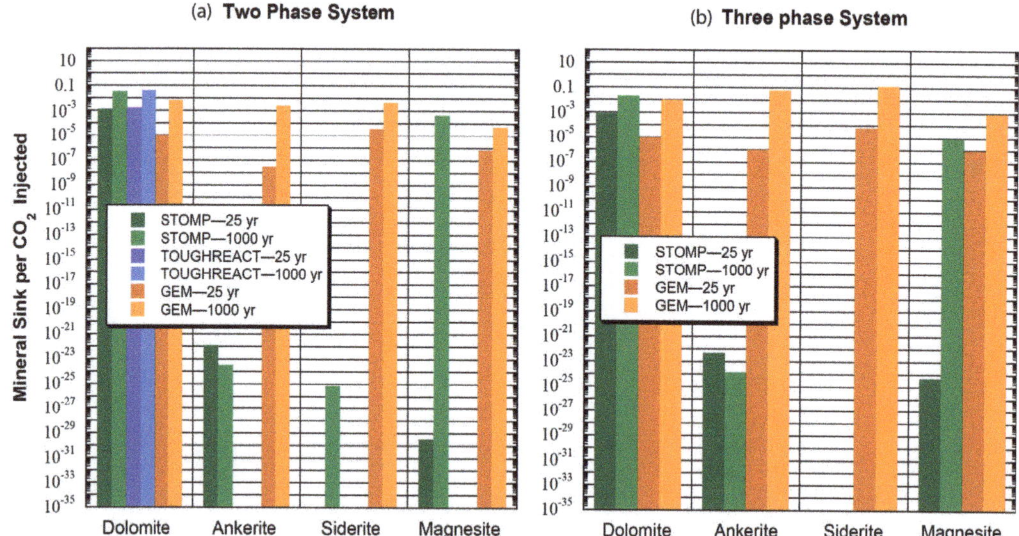

Figure 16. Predicted carbonate mineral abundances for (**a**) Scenario 1 and (**b**) Scenario 2 after 25 and 1000 years.

Changes in the abundances of non-carbonate minerals predicted by the GEM and STOMP-EOR models are shown in Figure 17. The two models predicted progressive clinochlore dissolution over time, though the GEM model predicted a much smaller amount. However, the STOMP-EOR model predicted relatively constant abundances of illite, Ca-montmorillonite, and albite, whereas the GEM model predicted a large decrease in illite abundance and a large increase in Ca-montmorillonite and albite abundance. For kaolinite, the STOMP-EOR model predicted an initial sharp dissolution event followed by relatively constant abundance, whereas the GEM model predicted precipitation throughout most of the simulation. Both models predicted continuous quartz precipitation, though the STOMP-EOR model predicted a much larger amount. The combined effects of mineral precipitation and dissolution were nearly in balance in the STOMP-EOR model, with a small porosity increase of ~0.001 predicted during the 25 year injection period, about half of which was then gradually eliminated over the remainder of the 1000 year simulation (Figure 18). The GEM model did not predict porosity to change during the injection period, but after about 100 years, porosity decreased steadily from an initial value of ~0.145 to ~0.138 after 1000 years.

Figure 19 shows how the injected CO_2 is distributed among an immiscible gas phase, oil, formation water, and carbonate minerals. Both the GEM and STOMP-EOR model predicted oil to be the largest sink for the injected CO_2, with the GEM model predicting higher amounts of CO_2 dissolution in oil than the STOMP-EOR model. Both models predicted less CO_2 to occur as immiscible gas than to dissolve in oil, and less CO_2 to dissolve in the formation water than to occur as immiscible gas. In addition, the temporal trends of CO_2 occurring in immiscible gas strongly resembled one another in the two models, as did the temporal trends of CO_2 dissolved in formation water. As for Model Scenario 1, in Model Scenario 2 carbonate minerals sequester only a small fraction of the injected CO_2, though the amount that they sequester continuously increases over time. By the end of the 1000-year simulation, the STOMP-EOR model predicts the amount of CO_2 to be sequestered in carbonate minerals to be close to the amounts occurring in immiscible gas and the formation water. In the GEM model, CO_2 sequestration in carbonate minerals has increased so much by the end of the simulation that this amount exceeds the amounts in all other CO_2 sinks except oil.

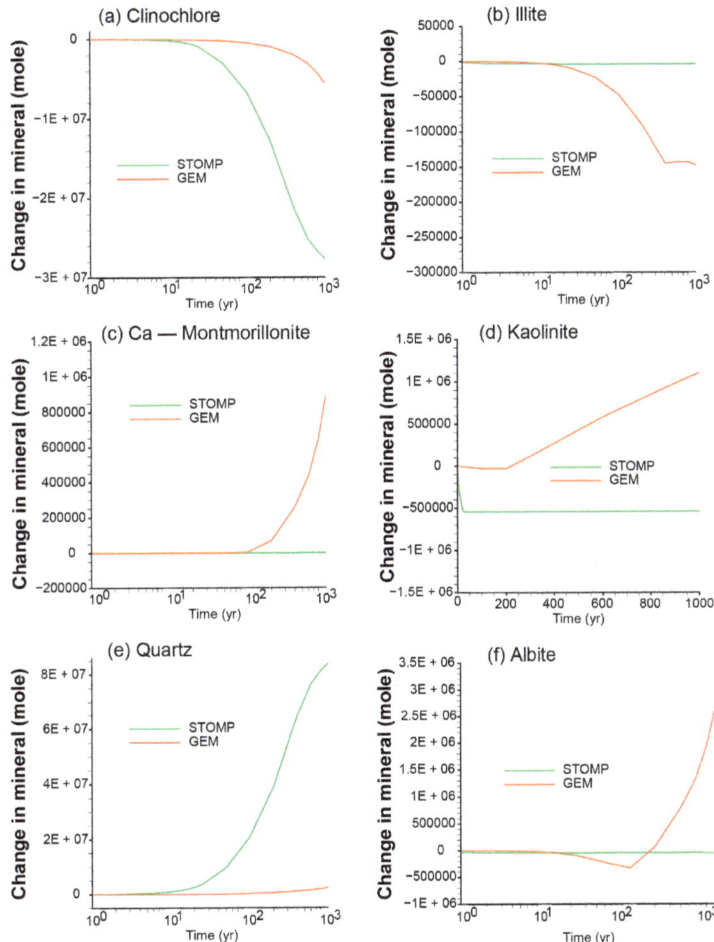

Figure 17. Changes in mineral abundance predicted by the STOMP-EOR and GEM models for (**a**) clinochlore, (**b**) illite, (**c**) Ca-montmorillonite, (**d**) kaolinite, (**e**) quartz, and (**f**) albite.

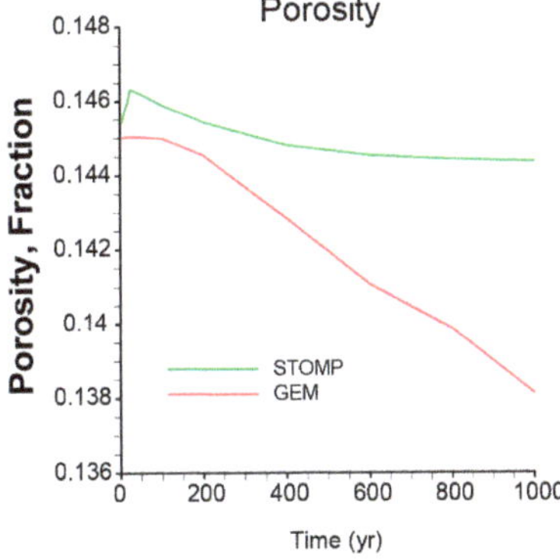

Figure 18. Simulated changes in porosity as a function of time averaged over the entire model domain as predicted by the STOMP-EOR and GEM models.

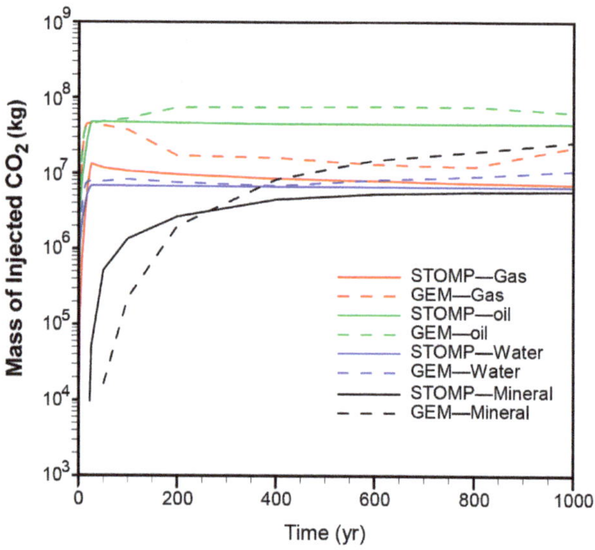

Figure 19. Mass of injected CO_2 that is sequestered in oil, in an immiscible gas phase, in the formation water, and in carbonate minerals as predicted by the STOMP-EOR and GEM models in Model Scenario 2.

5. Discussion

The present study showed that although the models for the different simulators in each scenario were set up in the same way, they produced some significantly different results. In Model Scenario 1, the STOMP-EOR, TOUGHREACT, and GEM models predicted similar evolutions in temperature and pressure, though the pressures in the GEM model were approximately 1–2 MPa higher after 1000 years than in the other two models. This may have driven slightly higher concentrations of CO_2 into solution in the formation water in the GEM model, which would then account for its generally slightly lower pH. All three models in Scenario 1 predicted calcite to dissolve, which is a consequence of the lowering of the pH due to the injection of CO_2. However, the lower pH predicted by the GEM model corresponds with its lower overall amount of carbonate mineral precipitation compared to the other two models.

In Scenario 1, the three models also predicted significant differences in the abundances of non-carbonate minerals. The overall net changes in mineral abundances, though, were similar enough to cause similar decreasing trends in porosity over time, amounting to only about a tenth of a percent over the 1000 years of the simulation for the reservoir as a whole. Such a small change in porosity would cause a similarly small change in permeability of only tenths of a percent, meaning that the hydraulic properties and behavior of the Morrow B reservoir as a whole would not be expected to change significantly as a result of the planned CO_2 injection.

In Scenario 2, despite the differences in predicted pressures, the amount of CO_2 predicted by GEM to dissolve into water and oil does not differ much from the amount predicted by STOMP-EOR. The pH is consistently lower in the GEM model than in the STOMP-EOR model, but this does not consistently suppress the precipitation of carbonate minerals in the GEM models. Instead, overall, more carbonate mineral precipitation is predicted to occur in the GEM model than in the STOMP-EOR model. This contributes to the greater decrease in porosity predicted by the GEM model than by the STOMP-EOR model, though in both models the porosity decrease is relatively small and not enough to cause significant changes in the hydraulic properties of the reservoir as a whole.

Some of the differences in predicted mineral abundances between the GEM model and the other two models may pertain to GEM's use of damping factors for mineral precipitation and dissolution reactions. The damping factors are multipliers applied to the reaction rates in order to aid convergence and are allowed to vary between 0 and 1000. According to the CMG-GEM user's guide [27], damping factors for mineral reactions are justified because published reaction rate parameters in the literature, such as the rate constant and reactive surface area, are often measured on a core scale in the lab. To use these parameter values appropriately in a field-scale model, the values must be upscaled corresponding to the larger grid block sizes and the scale of the model. When laboratory-derived reaction parameters are used without upscaling, then the resultant reaction rates become spuriously high, which causes numerical convergence difficulties during the simulation. Thus, specifying mineral reaction damping factors reduces the reaction rates to more realistic values, which also helps the model to converge better. In the present study, a damping factor of 0.001 was used in the GEM models, the maximum value that allowed the models to converge. In contrast, the STOMP-EOR and TOUGHREACT models converged without damping factors.

6. Summary and Conclusions

Two model scenarios for CO_2 injection into the Morrow B Sandstone in the Farnsworth Unit were investigated in the present study. In Model Scenario 1, water was the only pore fluid initially present. In Model Scenario 2, water and petroleum were both initially present as pore fluids. Model Scenario 1 allowed a comparison of the performance of the TOUGHREACT, STOMP-EOR, and GEM simulators to be made. Model Scenario 2 allowed a comparison of the STOMP-EOR and GEM simulators to be made. Both model scenarios also provided fundamental insights into the behavior and effects of the injected CO_2. In Model Scenario 1, the models from the three simulators predicted a similar rise in pressure up to ~33 MPa during the 25-year injection period but predicted different rates of pressure decline after injection ceased. The three models predicted similar patterns of reservoir cooling to a minimum temperature of 40 °C near the injection well, followed by similar patterns of temperature homogenization after injection ceased. All three models predicted the long-term persistence of an immiscible CO_2 gas phase but differed by up to approximately a factor of two in the amounts that persisted. All three models predicted sharp declines in pH from the initial value of 7 to between approximately 4.6 and 4.9, gradually rising with increasing time due to water–rock reactions. All three models predicted calcite to dissolve through the simulations and for dolomite to be the main carbonate mineral sink for the injected CO_2. However, the models differed in the amounts of other carbonate minerals (siderite, magnesite, and ankerite) that were predicted to precipitate. The three models differed more strongly in terms of their predictions about silicate minerals. All three models consistently predicted quartz and Ca-montmorillonite to precipitate and clinochlore and illite to dissolve, but in significantly different amounts. However, the predicted differences in neither silicate nor carbonate mineral abundance were sufficient to cause large changes in porosity, which showed a slight decreasing trend in all three models. The STOMP-EOR and GEM models predicted similar amounts of immiscible CO_2 gas to be the main sink for the injected CO_2 over the 1000 years of the simulations, while the amount predicted by the TOUGHREACT model was much lower and not the main injected CO_2 sink. All three models predicted similar amounts of injected CO_2 to be sequestered in aqueous solution. Carbonate minerals were predicted by all three models to be a smaller sink for injected CO_2 than the formation water, though carbonate minerals were the only CO_2 sink that grew in magnitude over time.

In Model Scenario 2, only the GEM and STOMP-EOR simulators were tested. Both the GEM and STOMP-EOR models made similar predictions of initial cooling around the injection well followed by thermal homogenization that were made in Model Scenario 1. The STOMP-EOR model made a qualitatively similar prediction of pressure evolution as in Model Scenario 1. However, the GEM model predicted an ongoing increase in pressure after injection ceased until the end of the simulation at 1000 years unless the minerals siderite and ankerite were removed from the model. The GEM model predicted gas saturations to be about twice as high as the STOMP-EOR model during the early years of the simulation, but after about 200 years the two models predicted similar gas saturations. Both models predicted lower overall gas saturations in Model Scenario 2 than in Model Scenario 1. Both the GEM and the STOMP-EOR models predicted a sharp decrease in pH during injection in Model Scenario 2. However, whereas the GEM model predicted a similar minimum pH of 4.6 in the two model scenarios, the STOMP-EOR model predicted a significantly higher minimum pH of 5.6 in Model Scenario 2 compared to the minimum pH of 4.9 it predicted in Model Scenario 1. Both models continued to predict calcite to dissolve continuously in Model Scenario 2. The STOMP-EOR model again predicted dolomite to be the main carbonate mineral sink for injected CO_2, whereas the GEM model predicted siderite and ankerite to be more important mineral sinks. The two models predicted significant differences in silicate mineral abundance. Together, the differences in carbonate and silicate mineral abundances led to significant differences in porosity, with the STOMP-EOR model predicting an overall porosity decrease to 0.1445 and the GEM model to about 0.1385. Overall, CO_2 injection was predicted to have a small impact on porosity over 1000 years. Both models in Scenario 2 predicted oil to be the main sink for injected CO_2. Both models predicted immiscible gas and the formation water, respectively, to be smaller sinks for injected CO_2. STOMP-EOR predicted carbonate minerals to be the smallest sink for injected CO_2. For GEM this was also true for about the first 300 years of the simulation, but the end of the 1000 years of the simulation, GEM predicted carbonate minerals to be the second most important sink for injected CO_2 after oil.

Although the models in each scenario were set up the same and although the model results have many qualitative similarities, the models differ in many of the details of their results. The results indicate that executing models on multiple simulators can more clearly identify areas of confidence as well as uncertainty in projected outcomes in the field.

Author Contributions: Conceptualization, E.J.K. and M.S.A.; Methodology, E.J.K. and M.S.A.; Software, M.D.W.; Validation, E.J.K. and M.S.A.; Formal Analysis, E.J.K. and M.S.A.; Investigation, E.J.K. and M.S.A.; Resources, M.S.A.; Data Curation, E.J.K.; Writing—Original Draft Preparation, E.J.K.; Writing—Review and Editing, M.S.A.; Review and Editing, M.D.W. and W.A.; Visualization, E.J.K. and M.S.A.; Supervision, M.S.A.; Project Administration, M.S.A.; Funding Acquisition, M.S.A. All authors have read and agreed to the published version of the manuscript.

Funding: This research was funded by the U.S. Department of Energy's (DOE) National Energy Technology Laboratory (NETL) through the Southwest Regional Partnership on Carbon Sequestration (SWP) under Award No. DE-FC26-05NT42591.

Data Availability Statement: Publicly available datasets were analyzed in this study. This data can be found here: https://mospace.umsystem.edu/ accessed on 20 August 2021.

Conflicts of Interest: The authors declare no conflict of interest. The funders had no role in the design of the study; in the collection, analyses, or interpretation of data; in the writing of the manuscript; or in the decision to publish the results.

References

1. Balch, R.; McPherson, B.; Grigg, R. Overview of a Large Scale Carbon Capture, Utilization, and Storage Demonstration Project in an Active Oil Field in Texas, USA. *Energy Procedia* **2016**, *114*, 5874–5887. [CrossRef]
2. Balch, R.; McPherson, B. Integrating Enhanced Oil Recovery and Carbon Capture and Storage Projects: A Case Study at Farnsworth Field, Texas. In *SPE Western Regional Meeting*; OnePetro: Anchorage, AK, USA, 2016.
3. Ahmmed, B. *Numerical Modeling of CO₂-Water-Rock Interactions in the Farnsworth, Texas Hydrocarbon Unit, USA*. 2015, Volume 3, pp. 54–67. Available online: https://mospace.umsystem.edu/xmlui/bitstream/handle/10355/46985/research.pdf?sequence=2 (accessed on 3 May 2021).
4. Xu, T.; Sonnenthal, E.; Spycher, N.; Pruess, K. *TOUGHREACT User's Guide: A Simulation Program for Non-Isothermal Multiphase Reactive Geochemical Transport in Variably Saturated Geologic Media*; V1.2.1-LBNL-55460-2008; Lawrence Berkeley National Lab. (LBNL): Berkeley, CA, USA, 2008; Volume 32, pp. 1–206.

5. Pan, F.; McPherson, B.J.; Esser, R.; Xiao, T.; Appold, M.S.; Jia, W.; Moodie, N. Forecasting evolution of formation water chemistry and long-term mineral alteration for GCS in a typical clastic reservoir of the Southwestern United States. *Int. J. Greenh. Gas Control* **2016**, *54*, 524–537. [CrossRef]

6. Gallagher, S.R. Depositional and Diagenetic Controls On Reservoir Heterogeneity: Upper Morrow Sandstone. Master's Thesis, Farnsworth Unit, Ochiltree County, TX, USA, 2014; pp. 1–233.

7. Khan, R.H. Evaluation of the geologic CO_2 sequestration potential of the Morrow B sandstone in the Farnsworth, Texas hydrocarbon field using reactive transport modeling. *Am. Geophys. Union* **2017**. [CrossRef]

8. Sun, Q.; Ampomah, W.; Kutsienyo, E.J.; Appold, M.; Adu-Gyamfi, B.; Dai, Z.; Soltanian, M.R. Assessment of CO_2 trapping mechanisms in partially depleted oil-bearing sands. *Fuel* **2020**, *278*, 118356. [CrossRef]

9. Nghiem, L.; Sammon, P.; Grabenstetter, J.; Ohkuma, H. Modeling CO_2 Storage in Aquifers with a Fully-Coupled Geochemical EOS Compositional Simulator. In *SPE/DOE Symposium on Improved Oil Recovery*; No. SPE 89474 Modeling; OnePetro: Tulsa, OK, USA, 2004; pp. 1–16.

10. CMG. GEM Compositional and Unconventional Simulator. 2021. Available online: https://www.cmgl.ca/gem (accessed on 10 August 2021).

11. White, M.; McPherson, B.; Grigg, R.; Ampomah, W.; Appold, M. Numerical Simulation of Carbon Dioxide Injection in the Western Section of the Farnsworth Unit. *Energy Procedia* **2014**, *63*, 7891–7912. [CrossRef]

12. Span, R.; Wagner, W. A New Equation of State for Carbon Dioxide Covering the Fluid Region from the Triple-Point Temperature to 1100 K at Pressures up to 800 MPa. *J. Phys. Chem. Ref. Data* **1996**, *25*, 1509–1596. [CrossRef]

13. Xu, T.; Apps, J.A.; Pruess, K. Reactive geochemical transport simulation to study mineral trapping for CO_2 disposal in deep arenaceous formations. *J. Geophys. Res. Solid Earth* **2003**, *108*. [CrossRef]

14. Munson, T.W. Depositional, Diagenetic, and Production History of the upper Morrowan Buckhaults Sandstone, Farnsworth Field Ochiltree County Texas. *OCGS-Shale Shak. Dig. XII* **1989**, *XXXX–XXXXI*, 2–20.

15. Heath, J.E.; Dewers, T.A.; Mozley, P.S. *Characteristics of the Farnsworth Unit, Ochiltree County*; Southwest Partnership CO_2 Storage-EOR Project: Ochiltree, TX, USA, 2015.

16. Ross-Coss, D.; Ampomah, W.; Cather, M.; Balch, R.S.; Mozley, P.; Rasmussen, L. An Improved Approach for Sandstone Reservoir Characterization. In Proceedings of the SPE Western Regional Meeting, Anchorage, AK, USA, 23–26 May 2016.

17. Trujillo, N.A. *Influence of Lithology and Diagenesis on Mechanical and Sealing Properties of the Thirteen Finger Limestone and Upper Morrow Shale, Farnsworth Unit, Ochiltree County, Texas*; ProQuest: Ann Abor, MI, USA, 2017.

18. White, M.D.; Oostrom, M. User Guide: Subsurface Transport Over Multiple Phases. June 2006 Contract: DE-AC05-76RL01830. Available online: https://www.pnnl.gov/main/publications/external/technical_reports/PNNL-15782.pdf (accessed on 20 August 2021).

19. Xu, T.; Apps, J.A.; Pruess, K. Numerical simulation of CO_2 disposal by mineral trapping in deep aquifers. *Appl. Geochem.* **2004**, *19*, 917–936. [CrossRef]

20. CMG. CMOST User's Guide: CO_2 Sequestration Using GEM. 2018. Available online: https://www.cmgl.ca/training/co2-sequestration-using-gem (accessed on 20 August 2021).

21. Ahmmed, B.; Appold, M.S.; Fan, T.; McPherson, B.J.O.L.; Grigg, R.B.; White, M.D. Chemical Effects of Carbon Dioxide Sequestration in the Upper Morrow Sandstone in the Farnsworth, Texas, hydrocarbon unit. *Environ. Geosci.* **2016**, *23*, 81–93. [CrossRef]

22. Palandri, J.L.; Kharaka, Y.K. A Compilation of Rate Parameters of Water-Mineral Interaction Kinectics for Application to Geochemical Modelling. *U.S. Geol. Surv. Open File Rep.* **2004**, *271*, 1–70.

23. Gunda, D.; Ampomah, W.; Grigg, R.; Balch, R. *Reservoir Fluid Characterization for Miscible Enhanced Oil Recovery. Carbon Management Technology Conference*; OnePetro: Sugarland, TX, USA, 2015.

24. Battistelli, A.; Calore, C.; Pruess, K. The simulator TOUGH2/EWASG for modelling geothermal reservoirs with brines and non-condensible gas. *Geothermics* **1997**, *26*, 437–464. [CrossRef]

25. Spycher, N.; Pruess, K. CO_2-H_2O Mixtures in the Geological Sequestration of CO_2. II. Partitioning in Chloride Brines at 12–100 °C and up to 600 bar. *Geochim. Cosmochim. Acta* **2005**, *69*, 3309–3320. [CrossRef]

26. Harvey, A. Semiempirical correlation for Henry's constants over large temperature ranges. *AIChE J.* **1996**, *42*, 1491–1494. [CrossRef]

27. CMG-GEM. GHG-GEM Users Guide: GHG Option—Damping Factor for Reactions Other than Chemical *MRDAMP-ALL, *MRDAMP. 2020. Available online: https://www.cmgl.ca/resources (accessed on 20 August 2021).

Article

Time-Lapse Integration at FWU: Fluids, Rock Physics, Numerical Model Integration, and Field Data Comparison

Robert Will [1], Tom Bratton [2], William Ampomah [1,*], Samuel Acheampong [1], Martha Cather [1] and Robert Balch [1]

[1] New Mexico Tech/PRRC, Socorro, NM 87801, USA; robert.will@nmt.edu (R.W.); samuel.acheampong@student.nmt.edu (S.A.); Martha.Cather@nmt.edu (M.C.); robert.balch@nmt.edu (R.B.)

[2] Tom Bratton LLC, Littleton, CO 80127, USA; tom@tombrattonllc.com

* Correspondence: william.ampomah@nmt.edu

Abstract: We present the current status of time-lapse seismic integration at the Farnsworth (FWU) CO_2 WAG (water-alternating-gas) EOR (Enhanced Oil Recovery) project at Ochiltree County, northwest Texas. As a potential carbon sequestration mechanism, CO_2 WAG projects will be subject to some degree of monitoring and verification, either as a regulatory requirement or to qualify for economic incentives. In order to evaluate the viability of time-lapse seismic as a monitoring method the Southwest Partnership (SWP) has conducted time-lapse seismic monitoring at FWU using the 3D Vertical Seismic Profiling (VSP) method. The efficacy of seismic time-lapse depends on a number of key factors, which vary widely from one application to another. Most important among these are the thermophysical properties of the original fluid in place and the displacing fluid, followed by the petrophysical properties of the rock matrix, which together determine the effective elastic properties of the rock fluid system. We present systematic analysis of fluid thermodynamics and resulting thermophysical properties, petrophysics and rock frame elastic properties, and elastic property modeling through fluid substitution using data collected at FWU. These analyses will be framed in realistic scenarios presented by the FWU CO_2 WAG development. The resulting fluid/rock physics models will be applied to output from the calibrated FWU compositional reservoir simulation model to forward model the time-lapse seismic response. Modeled results are compared with field time-lapse seismic measurements and strategies for numerical model feedback/update are discussed. While mechanical effects are neglected in the work presented here, complementary parallel studies are underway in which laboratory measurements are introduced to introduce stress dependence of matrix elastic moduli.

Keywords: 4D; time lapse; CO_2; EOR; WAG; sequestration; monitoring

Citation: Will, R.; Bratton, T.; Ampomah, W.; Acheampong, S.; Cather, M.; Balch, R. Time-Lapse Integration at FWU: Fluids, Rock Physics, Numerical Model Integration, and Field Data Comparison. *Energies* **2021**, *14*, 5476. https://doi.org/10.3390/en14175476

Academic Editor: Ricardo J. Bessa

Received: 1 May 2021
Accepted: 30 July 2021
Published: 2 September 2021

Publisher's Note: MDPI stays neutral with regard to jurisdictional claims in published maps and institutional affiliations.

1. Introduction

1.1. Farnsworth Site Background

The Southwest Regional Partnership on Carbon Sequestration (SWP) is one of seven large-scale CO2 sequestration projects sponsored by the U.S. Department of Energy [1]. The primary objective of the SWP effort is to exhibit and evaluate an active commercial-scale carbon capture, utilization, and storage (CCUS) operation, and demonstrate associated effective site characterization, monitoring, verification, accounting, and risk assessment. The SWP field site is located within the Farnsworth Unit CO_2 WAG (water-after-gas) EOR (Enhanced Oil Recovery) project at Ochiltree County, northwest Texas which is undergoing conversion to a CO_2 flood. All CO_2 utilized by the project is anthropogenic, sourced from a fertilizer and an ethanol plant, and this CO_2 would otherwise be vented to the atmosphere (Figure 1). The CO2 WAG field development scheme being applied at FWU is a is a popular form of tertiary hydrocarbon recovery which also holds promise as a large-scale CO_2 utilization and storage (CCUS) mechanism. The CO_2 WAG process [2] involves cyclic alternation between CO_2 and water injection phases for optimal mobilization and sweep of liquid hydrocarbons remaining in pattern drilling developments after primary recovery and waterflood (secondary recovery). As a potential carbon sequestration mechanism, CO_2 WAG projects will be subject to some degree of monitoring and verification, either as a regulatory requirement or to qualify for economic incentives. Time-lapse (or "4D") seismic provides a robust method for wide-scale fluid monitoring which has been widely applied in petroleum resource development for decades [3]. More recently time-lapse method has been used to monitor evolution of the CO_2 "plume" on most carbon capture and sequestration (CCS) projects [4]. The Southwest Partnership has conducted time-lapse seismic monitoring at Farnsworth using the 3D Vertical Seismic Profiling (VSP) method.

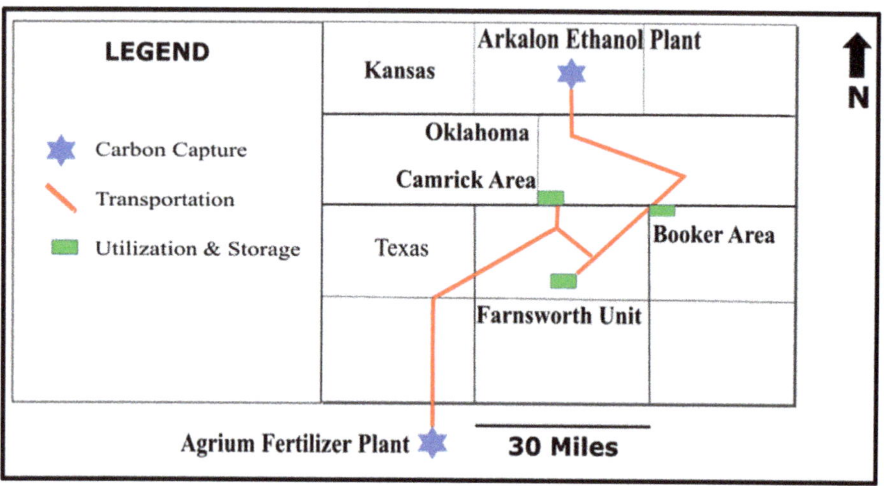

Figure 1. Location map of the SWP study area at Farnsworth, Texas and anthropogenic CO_2 sources.

The efficacy of time-lapse seismic depends on a number of key factors which vary widely from one application to another. Most important among these factors are the thermophysical properties of the original fluid in place and the displacing fluid, followed by the petrophysical properties of the rock matrix which together determine the effective elastic properties of the rock fluid system. Here, is where monitoring of CO_2 WAG systems varies greatly from other oilfield and brine aquifer (CCS) CO_2 storage owing to the thermodynamic conditions dictated by the properties of the original and displacing fluids, reservoir temperature, and pressure. Geologic sequestration of supercritical CO_2 into a brine aquifer, which is the typical case for CCS projects, results in a fluid system with effectively binary fluid properties and relatively simple interface between original and displacing fluids by comparison with miscible CO_2 WAG systems. By contrast, CO_2 WAG operations result in a thermodynamically complex fluid system with multiple fluid contacts and a high degree of ambiguity in thermophysical properties. Further, under miscible conditions an additional phase is introduced. We use the data from the SWP Farnsworth West project in an extensive fluid and rock physics modeling study to understand the unique monitoring challenges presented by miscible CO_2 WAG operations. The rock physics models are applied to compositional reservoir simulator output and the resulting elastic predictions compared to time-lapse 3D VSP surveys acquired at the FWU site.

1.2. Literature Review

There are numerous case histories in the literature documenting the application of time-lapse monitoring for petroleum resource management and carbon sequestration. Here, we focus on the distinct fluid systems presented by the various applications. CCS (brine aquifer storage) projects such as Sleipner [5–7], Aquistore [8–10], Illinois Basin-

Decatur Project (IBDP) [11–13], and Ketzin [14] are all non-miscible fluid systems which are thermodynamically stable, have relatively favorable fluid mobility ratios, and in which the single displacing (injected) CO_2 volume is monotonically increasing. The result is a continuous "plume" of CO_2 emanating from the injection well, the interface of which with the aquifer brine may be characterized as a "front". Similarly, most documented successful (non-EOR) hydrocarbon reservoir surveillance applications of time-lapse seismic have the objective of monitoring an encroaching brine aquifer during field depletion, progression of a water injection front in waterflood operations, or evolution of light hydrocarbon gas upon pressure drawdown, all of which present essentially binary fluid discrimination problems which may be characterized (and modeled) using "plume", "front", or "cap" concepts. While not being trivial by any means, these fluid property and displacement scenarios have the added advantage of well-established empirical correlations, analytical models, and reasonable conceptual approximations to simplify and/or constrain interpretation. The binary nature of this problem is further evidenced by the evolution of discrete analytical methods using classification and analytical integration approaches [15–17].

By comparison there are relatively few documented case histories of time-lapse seismic monitoring on CO_2 EOR projects. Extensive fluid analysis and rock physics modeling performed by Brown at Weyburn [18] show an estimated maximum of 6.3% variation in compressional velocities between endpoint saturations of 100% brine and 100% CO_2 in high porosity (24%) fractured marly dolostone. While this degree of sensitivity is favorable from a seismic detection perspective, the prediction for intermediate fluid mixtures in a 29% porosity rock with oil and 40% CO_2 are on the order of 1%, which is marginal for seismic detection. Quantitative time-lapse seismic data integration at Weyburn was performed using a novel methodology which optimized a penalty function formed of the distance between simulated CO_2 "front" from ensemble models and that interpreted from thresholded time lapse seismic anomaly maps [19]. At Cranfield Gosh [20] applied a pressure dependent effective media model (PDEM) to invert time lapse seismic data for CO_2 saturation. While success is reported in mapping the extent of CO_2 migration, the author acknowledges the need for additional constraints on gas distribution in order to accurately predict CO_2 movement. Alfi and Housenni [21] compared time-lapse seismic interpretations to simulator predictions at Cranfield. As Cranfield is not a WAG operation the CO_2 "plume" was continuous. Limited success in the comparison was attributed to model deficiencies and uncertainties. No attempt was reported to optimize the simulation model through time-lapse seismic constraints. At the Denbury Bell Creek [22] project researchers performed extensive rock physics and forward elastic modeling from reservoir simulator output. The time-lapse seismic interpretation provided insights into reservoir connectivity which was fed back to simulation modelers for consideration in the subsequent reservoir model update. Incorporated discrepancies were fed back to reservoir engineers for use in model updates. While these case histories show high levels of analytical rigor, all rely either explicitly or implicitly on the concept of a CO_2 "plume" or a distinct CO_2 saturation "front".

Our literature review also reveals the wide variety of geophysical analyses and resulting attributes which have been utilized for time-lapse seismic data interpretation and integration. These range from computationally simple but robust seismic data transforms which can be extracted without external conditioning, to highly sophisticated and computationally expensive specialized attributes which require conditioning to high quality geophysical logs and/or numerical models. At the Hall-Gurney field in Kansas [23] researchers experimented with use of low cost noninversion attributes for monitoring the effectiveness of EOR in thin, shallow carbonates and were able to identify "overall area effected by injected CO_2". Time or depth shift (sag, displacement) in events underlying a storage zone is a popular and robust attribute which requires no external constraint for use as a qualitative indicator [24,25]. With suitable constraint on reservoir properties and rock physics, time delay may be used to estimate saturation changes in the injection zone. Time delay has been successfully used at Sleipner [6] and Norne [26]. The next level of analytical rigor is seismic post stack (acoustic inversion), followed by pre-stack (elastic) inversion, both of which have been applied at Sleipner [5,7]. Yet, more sophistication is introduced by implementing wave equation constraints [27]. In rocks displaying anisotropy such as the naturally fractured Weyburn dolostone, multi-component seismic attributes may be used to produce semi-quantitative maps of elastic anisotropy for more representative rock physics modeling [19].

2. Materials and Methods

2.1. General Methodology

We use the extensive body of site characterization data and reservoir modeling performed at FWU as the basis for a comprehensive fluid property and rock physics modeling study, followed by comparison with time-lapse VSP surveys. The calibrated FWU compositional simulation model is used to characterize expected fluid distributions in the injection formation at times corresponding to seismic survey times. These results are used to perform systematic fluid equation of state modeling for representative ranges of anticipated reservoir fluid mixtures. These fluid EOS models are then used with a site-specific petrophysical model for systematic fluid substitution modeling over ranges of representative saturated reservoir rock conditions. These systematic modeling studies reveal important characteristics of the rock-fluid system which are critical for the following comparison of reservoir scale simulations with field time-lapse measurements.

The petrophysical and elastic properties used in our compositonal models and subsequent rock physics computations are correlated through a common 3D porosity distribution which has been developed through analysis of log and core data and interpolated using spatial trends extracted from the available 3D seismic data. First, the 3D permeability distribution required for reservoir simulation was created through poro-perm relationships extracted from the logging/core dataset and interpolated through geostatistical integration with the 3D porosity distribution using the same seismic data for spatial trends. Next, the 3D elastic property distributions required for rock physics computations were developed through correlations with porosity and interpolated through geostatistical integration with the same common 3D porosity distribution, using the same seismic data for spatial trends. In this way we feel that we have maintained correlation across petrophysical and elastic properties throughout our process.

2.2. FWU Geological Model

Site characterization efforts at FWU have produced a rich collection rock and fluid samples, geophysical logs, and multiple time-lapse seismic datasets [28,29]. Core from several wells, including characterization wells drilled specifically for this project, underwent comprehensive petrographic analysis, flowthrough, and mechanical testing. Three wells were drilled as characterization wells (or "science" wells) by the partnership. Approximately 250 ft of core was obtained from each of the new characterization wells. Cored intervals include the entire Morrow B reservoir interval, as well as Morrow shale that underlies and overlies the Morrow B, the B1 sandstone interval, and the Thirteen Finger limestone which forms the remainder of the primary seal. The geophysical logging program for the science wells was designed to support the anticipated geophysical, petrophysical, geomechanical, and geochemical studies as well as coupled process modeling.

Although geotechnical data acquisition activities at the FWU site have transitioned from site-characterization activities to monitoring data collection, ongoing improvements have been made to the site characterization through application of improved processing and analytical methods. Figure 2 shows the type well log for the Farnsworth Unit, the unit boundary, and location of SWP characterization wells. Specialized integration of geophysical logs and mechanical core tests have resulted in creation of detailed wellbore Mechanical Earth Models (MEMs). Pre-stack depth imaging of the 3D seismic dataset has resulted in both higher fidelity structural and stratigraphic imaging and improved image gathers for elastic inversion. Ongoing core flooding experiments and petrographic studies have resulted in definition of distinct hydraulic flow units (HFU's) [29] within the Morrow B formation as well as corresponding porosity dependencies and porosity-permeability relations. The results of these data analyses have been integrated through a combination of geostatistical and machine learning methods to provide enhanced geologic description, hydrodynamic property estimation (Figure 3), and development of a 3D mechanical property model for support of ongoing numerical elastic and mechanical modeling studies. Although numerical simulations were performed in the Morrow B production interval, porosity and permeability have been interpolated in from the Thirteen Finger to base of the Morrow formation to support ongoing related studies. The property interpolation workflow applied to each formation depended on the data available and the formation characteristics. Integration methods included artificial neural network facies identification from well logs and core, spatial variogram analysis, discrete and continuous distributions, and co-simulation with elastic inversion properties. Due to the limited well log data in all formations except the Morrow B, spatial variograms from seismic impedance were used as proxies for well log data variograms in property interpolation. Such use of variogram proxies, and the use of secondary variables in co-simulation, were justified by observed correlations in available well log data.

The method applied within the Morrow B was distinct from other formations due to availability of legacy well logs and research into HFUs conducted by SWP. A "Winland R35" transform was derived from analysis of core porosity and permeability for 51 wells. Eight different sub populations were identified in poro-perm space and used to create R35 cut-offs defining HUF. Poro-perm relationships were derived for each HFU sub population from core data. The R35 transformation was used to compute R35 logs for the 51 wells with data used in core analysis. For poro-perm interpolation, porosity logs were upscaled into the grid and interpolated through Gaussian co-simulation with seismic acoustic impedance

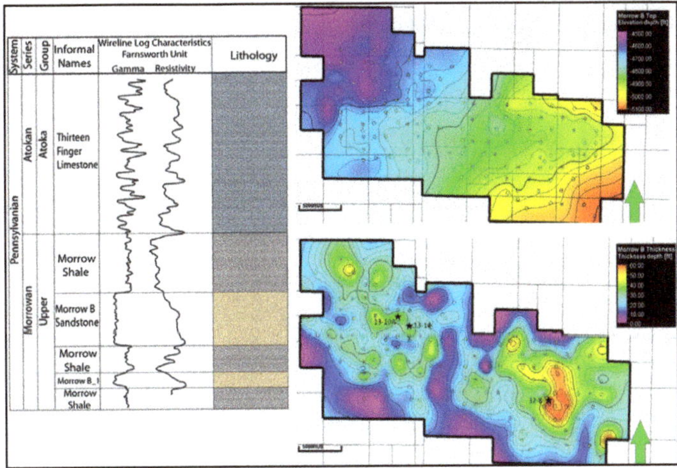

Figure 2. (**Left**) type log of FWU caprock and reservoir. (**Upper right**) surface contour of Morrow B top with 200 ft depth contour intervals. (**Lower right**) thickness map of Morrow B sands and location of characterization wells with 2 ft thickness contour intervals.

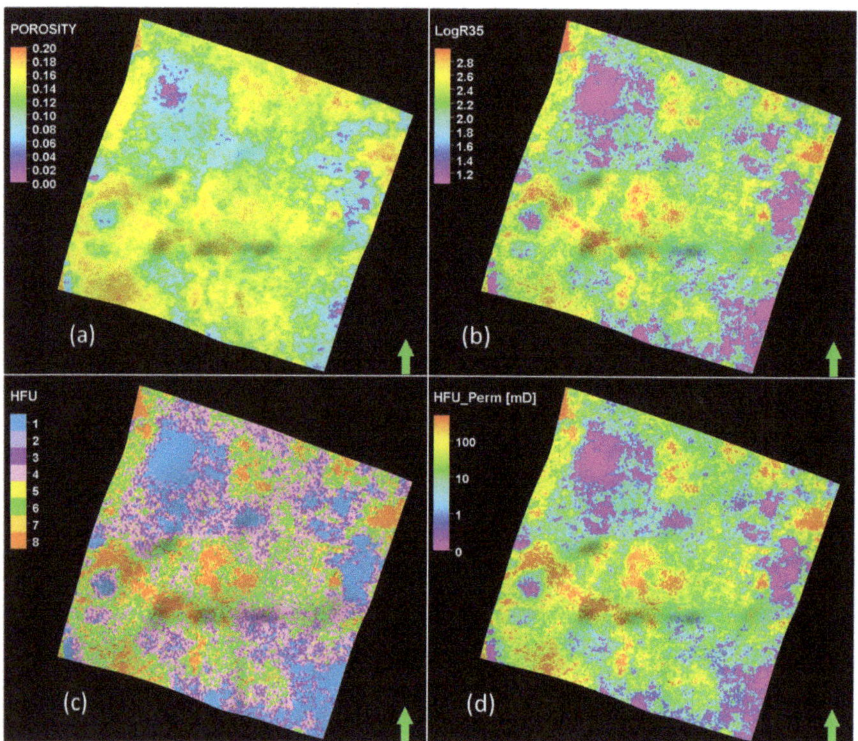

Figure 3. (**a**) Interpolated porosity logs, (**b**) R35 created by co-simulation of R35 logs with interpolated porosity, (**c**) HFU property created by applying R35 cut-offs, (**d**) Permeability created by application of HFU-specific poro-perm relationships.

2.3. FWU Numerical Simulation Model

A field-scale Eclipse 300 compositional reservoir flow model has been developed by SWP researchers for assessing the performance history of the CO_2 flood, and optimizing oil production and CO_2 storage at FWU [30–32].

2.3.1. Compositional Fluid Model

The compositional fluid model was constructed from laboratory fluid experiments tuned to an equation of state (EOS) [33,34]. The mixing rules of Pedersen [35] were followed to split C7+ fractions into two pseudo-components using the average molecular weight, average specific gravity and the total mole percent. The 3- parameter Peng Robinson equation of state [36] with Peneloux volume correction [37] was used to perform all the calculations with the resulting composition shown in Table 1 and phase envelope shown in Figure 4. The viscosity was modeled using the Lohrenz-Bray-Clark correlation [38]. After calibrating the fluid model to equation of state, a slim tube simulation experiment was conducted to obtain the minimum miscible pressure (MMP) for FWU. A one-dimensional 200 cell model was used for the experiment with a CO_2 injection volume of 1.2 pore volume. The MMP of 4009 psia realized from the simulation as compared to an MMP value 4200 psia derived from laboratory experiments provided by the operator represents a less than 5% error.

Table 1. Fluid composition sampled from the FWU.

Components	Molecular fraction %	Molecular weight gm/mol	Critical Temperature °F	Critical Pressure Psi
CO_2	0	44.01	87.89	1069.8
C1	38.49	16.04	−116.59	667.17
C2	3.86	30.07	90.05	708.36
C3	2.46	44.1	205.97	615.83
C4's	1.95	58.12	453.65	430.62
C5's	1.79	72.15	301.12	547.81
C6's	2.83	86.18	380.71	489.79
HC1 (7–38)	33.48	189.95	802.94	326.19
HC2 (38–70)	15.13	545.65	1077.75	235.69

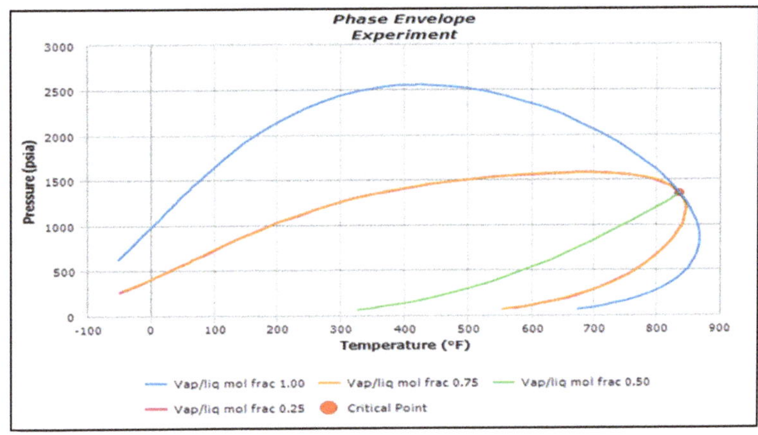

Figure 4. Regressed phase envelope for the 9 component FWU compositional fluid model.

2.3.2. Numerical Simulation Model

The simulation model for the Morrow B formation was calibrated through primary (depletion), secondary (waterflood), and tertiary (CO_2 WAG) recovery periods using a machine learning assisted methodology and available pressures and injection/production rates and pressures [32]. The original geological model uses a mesh-grid with a dimension of 100 ft by 100 ft. Top and base of the Morrow B formation were modeled as no-flow boundaries. Running simulation using such a fine grid system takes relatively expensive computational cost. Our preliminary investigations indicate one run using the fine-meshed model could take hours, which results in difficulties to the history matching process considering the demands of running hundreds of simulation cases to find the history matching solution. Therefore, the first step of the history matching work is to upscale the model using a coarsened grid of 200 ft by 200 ft mesh without sacrificing the accuracy. Rasmussen et al. [39] presented relative permeability curves based on laboratory experiment which corresponds to each hydraulic flow units from the FWU which is used in the numerical modeling. During the history matching process, the uncertain parameters considered included lateral permeability, vertical permeability anisotropy, and relative permeability curve inputs. A total of 100 simulations were run by randomly combine the uncertainty parameters and neural network based proxies were developed to improve robustness of history matching workflow and save computational time. Particle swarm optimization is employed and coupled with the expert proxy models to minimize the history matching error considering the oil and water rate agreements between the simulated and field observation data. Sun et al. [40] has presented the detailed history matching model and results utilized for this study. The various periods timesteps that coincided with the VSP time-lapse acquisition data were extracted for the rock physics analysis.

2.4. Rock Physics

2.4.1. Properties of Hycrocarbon-CO_2-Brine Mixtures—Analytical Study

Thermophysical properties hydrocarbon-CO_2 fluid mixtures for systematic rock physics sensitivity investigations were computed using NIST REFPROP [41] and SUPERTRAPP [42] databases and FORTRAN subroutines, integrated with python scripts. We use REFPROP for calculating the bulk modulus and density of CO_2 as a function of pressure (P) and temperature (T) because of the established REFPROP's accuracy for PVT modeling of CO_2 [43]. SUPERTRAPP is used for calculating the bulk moduli and bulk density of hydrocarbon mixture with and without dissolved CO_2. Fluid phase velocities computed with REFPROP and SUPERTRAPP were used to calculate bulk moduli for each phase. CO_2 was in supercritical state at all reservoir temperature and pressure ranges investigated. Depending on modeling pressure and temperature the fluid mixture has as many as 4 phases; original oil, water, CO_2-oil miscible mixture, and free (supercritical) phase CO_2.

The calibrated reservoir simulation model was used to establish representative hydrocarbon-CO_2-Brine mixtures for investigation by extracting fluid composition along injector-producer profiles from WAG simulations. It was determined that, as a result of many decades of primary depletion and waterflood leaving the remaining oil is "dead", no significant quantities of hydrocarbon gas evolve during WAG production. The native hydrocarbon fluid reaches saturation with CO_2 and additional CO_2 exists as a separate (supercritical) phase. Further, hydrocarbon components do not exhibit selective stripping as a result of the miscible extraction process as has been reported by some experimentalists [44]. Based on this observation we constructed our hydrocarbon-CO_2-brine mixture EOS computations such that as the feed mole fraction of CO_2 and brine in the mixture increase, the feed mole fractions of hydrocarbon components are reduced in proportion with the original composition.

Given the original oil composition (without CO_2);

$$\Sigma_1^{ncomp} xHC_i = 1 \tag{1}$$

CO_2 is introduced to the feed incrementally, such that,

$$\Sigma_1^{ncomp} x\prime HC_i = 1 - xCO_2, \tag{2}$$

and,

$$x\prime HC_i = xHC_i * (1 - xCO_2). \tag{3}$$

where x' are the adjusted hydrocarbon mole fractions.

For this analysis a python wrapper script was constructed to systematically sample the relevant EOS and fluid substitution parameter space, execute compiled SUPERTRAPP FORTRAN flash subroutines, calculate thermophysical properties of the hydrocarbon-CO_2-water mixture with additional input from the REFPROP database, and perform fluid substitution. The python script generates a database of fluid properties and fluid substitution results for graphical and statistical analysis. The output from python script was used as input to the rock physics coupled with the numerical simulation model through series of user-defined workflows developed within Schlumberger Petrel platform. All EOS results are converted to volume fractions for conversion to liquid and vapor saturations. Oil and miscible CO_2 Oil mixtures are liquids because the reservoir is above MMP. Free phase CO_2 is in supercritical state. There is no hydrocarbon gas because the reservoir pressure is above the bubble point at the current temperature.

2.4.2. Properties of Hydrocarbon-CO_2-Brine Mixtures—Numerical Model Integration

For integration with the numerical compositional simulation model the fluid properties were obtained from Eclipse 300 EOS calculations. Bulk reservoir fluid density was computed using predicted oil, water, and gas saturations and densities (keywords SOIL, SWAT, SGAS, DENO, DENW, DENG). Bulk fluid modulus was computed as the inverse of the Eclipse 300 solution for total fluid compressibility (keyword TOTCOMP).

2.4.3. Fluid Substitution

The elastic properties of saturated rock were calculated using Gassmann's relation [45].

$$K_{sat} = K_{dry} + \frac{\left(1 - \frac{K_{dry}}{K_s}\right)^2}{\frac{\varnothing}{K_f} + \frac{1-\varnothing}{K_s} - \frac{K_{dry}}{K_s^2}} \tag{4}$$

where;

Φ = Porosity;
$Ksat$ = Bulk modulus of the saturated rock;
$Kdry$ = Bulk modulus of the dry rock;
Ks = Bulk modulus of the mineral constituents;
Kf = Bulk modulus of the pore fluid.

Gassmann's equation provides a fundamental relationship to relate a fluid saturated formation with an idealized dry frame formation. However, to apply Gassmann's equation to solve fluid substitution problems the geophysicist must supply dry rock properties. Biot [46] identified systematic relationship between solid matrix moduli (K_s, G_s) and moduli of the material with porosity.

$$K_{dry} = K_s(1 - \alpha). \tag{5}$$

Introducing the "Biot" coefficient

$$\alpha = 1 - \frac{K_{dry}}{K_s}, \tag{6}$$

Gassmann's relation may be rewritten as

$$K_{sat} = K_{dry} + \frac{\alpha^2}{\frac{\varnothing}{K_f} + \frac{(\alpha-\varnothing)}{K_s}} \tag{7}$$

Applying Wood's mixing law [47];

$$\frac{1}{K_{fl}} = \sum_{i=1}^{n} \frac{S_i}{K_i} \tag{8}$$

where i = fluid phase.

Matrix elastic properties were determined through analysis of the available geophysical logging suite which included magnetic resonance, dipole sonic, spectral gamma ray, pulsed neutron, array induction, formation image, and spontaneous potential. Petrophysical analysis yielded formation intrinsic properties enabling evaluation of elastic moduli for the Morrow B formation. Figure 5 shows the calibration of formation properties to available core data. Figure 6 shows the analysis of shear and bulk moduli used to determine K_{dry}, G, and α for application of Gassmann's Equation (3) and elastic modeling (10). The K_{dry}, G, and K_s obtained from the geophysical log analysis, served as input parameters for the Gassmann fluid substitution model (4). The saturated bulk modulus, expressed as a function of the dry rock frame and pore fluid properties, was computed using the Gassmann Equation (4). A linear regression model was fitted to the data to identify the relationship between the changes in the elastic properties to the rock physical properties. The solid modulus, K_s, was determined by extrapolating porosity to zero on the K_{dry} vs. porosity plot.

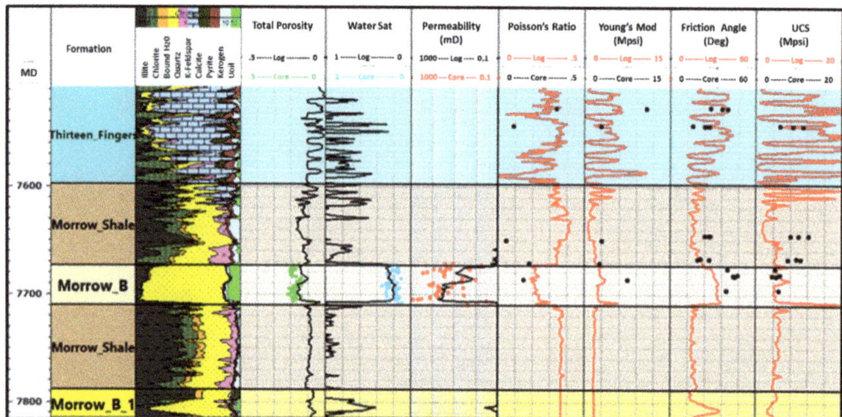

Figure 5. Well 13-10A petrophysics and mechanical properties. Track 1: measured depths; track 2: formation names; track 3: mineralogy; tracks 4, 5, and 6; total porosity, water saturation and permeability; tracks 7, 8, 9, and 10: Poisson's ratio, Young's modulus, friction angle, and unconfined compressive strength. Dots show core measurements used to calibrate correlations.

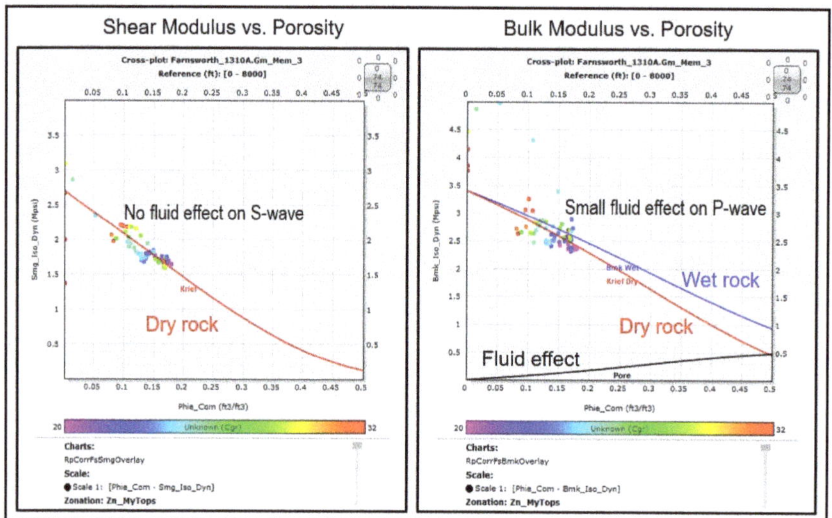

Figure 6. Morrow B geophysical log based evaluation of shear (**left**) and bulk (**right**) moduli showing Krief model fit for determination of K_s, G_s (y-intercept) and K_d, G_d vs. porosity relationships.

Shear and compressional seismic velocities were computed as;

$$V_s = \sqrt{\frac{\mu}{\rho_b}} \tag{9}$$

$$V_p = \sqrt{\frac{\mu + \frac{4}{3}K_{sat}}{\rho_b}} \tag{10}$$

where;

$$\rho_b = \phi * \rho_{fl} + (1 - \phi) * \rho_{matrix} \tag{11}$$

$$\rho_{fl} = \sum_{i=1}^{n}(S_i * \rho_i) \tag{12}$$

For analytical studies, these relationships were implemented by python script. For numerical model integration, fluid substitution was performed on the 3D reservoir model and simulator output grids using the PETREL process manager.

2.5. Time-Lapse Seismic Surveys

The project has acquired multiple seismic data sets for site characterization and monitoring. In addition to a full-field surface 3D dataset acquired for site characterization purposes, 3D-VSP surveys were acquired in all three characterization wells [47]. A total of four 3D VSP surveys (pre-CO_2 baseline and 3 monitor surveys), were acquired in well 13-10A, which is the injector in the 5-spot pattern used for this study. The baseline survey was acquired in February 2014 during the well shut-in for conversion to WAG injection. Monitor surveys were acquired January 2015, December 2016, and December 2017. All monitor surveys were acquired after the CO_2 leg of the WAG cycle. Baseline and monitor surveys were processed through an identical three-component (3C) processing workflow. Pre-processing included source/receiver geometry quality control, receiver selection, 3 component orientation, noise attenuation, Surface Consistent Amplitude Compensation (SCAC), 3C wavefield separation, deterministic trace-by-trace wave-shaping deconvolution, and static correction. Images were created from upgoing wavefields using a Generalized Radon Transform (GRT) imaging algorithm. Figure 7 shows maps of depth shift below the injection interval computed for the three p-wave baseline-monitor survey pairs using the method proposed by Nickel and Sonneland [48]. In consideration of acquisition geometry, and image point coverage, images were cropped outside 1000 ft radial distance from the survey well (13–10A). It should be noted here that this is an extremely challenging seismic detection problem due to the relative thinness of the reservoir (~40–45 ft), the estimated wavelength of p waves (~150 ft) in the data at the reservoir interval, and what we will later see to be very subtle fluid effects.

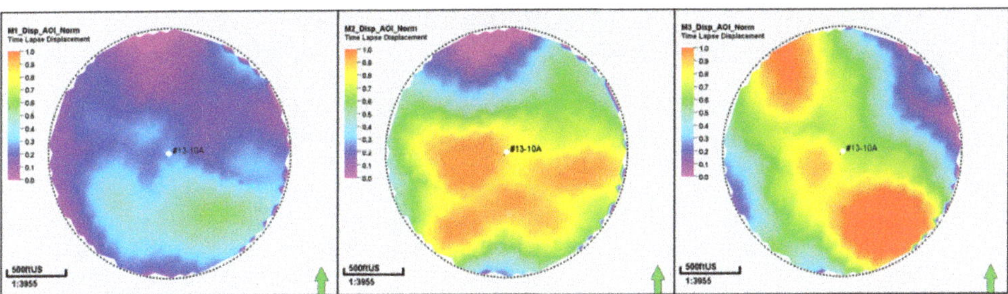

Figure 7. Normalized measured VSP depth shift below the injection interval in injector 13–10A for; (**left**) monitor 1, (**center**) monitor 2, and (**right**) monitor 3.

2.6. Numerical Simulaton Model Integration

In addition to aiding in selection of parameter ranges for the analytical fluid EOS and rock physics study, the numerical simulation model was used to model the time-lapse change on elastic properties for comparison with the recorded time-lapse VSP data. Reservoir simulations were configured to output liquid and vapor phase mole fractions for each fluid component, oil, gas, and water densities, and total fluid compressibility as computed by the E300 compositional fluid model. In order to minimize computational effort a sector model was constructed encompassing the 13-10A injector study pattern. Figure 8 shows the full model and sector model domains, the 3D VSP image area, and the well 13-10A/13-12 profile which is used for diagnostic purposes. The reservoir property model was supplemented with elastic moduli based on log correlations as shown in Figure 8. Fluid substitution and elastic modeling was performed on the four-layer Morrow B formation. A shift attribute map was computed for each baseline-monitor pair by vertically integration of the fluid substitution effects within the reservoir. The statistical and spatial characteristics of modeled and measured shifts were compared on absolute value and normalized bases.

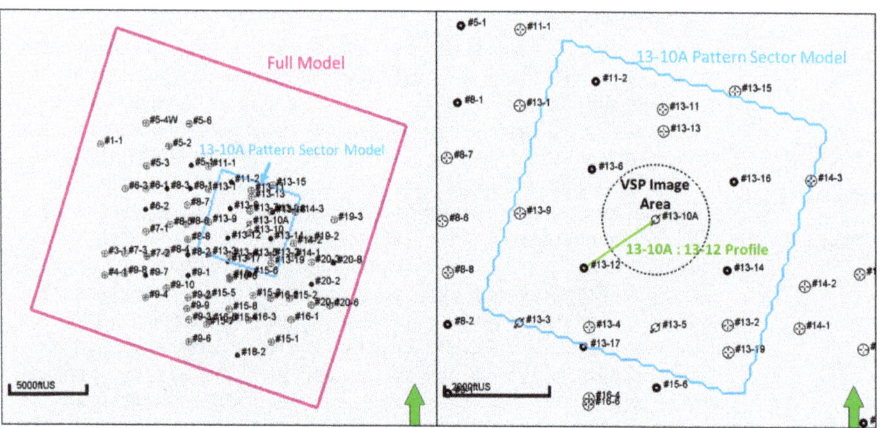

Figure 8. (**left**) Full simulation model (magenta) and 13-10A pattern sector model (blue) boundaries. (**right**) 13-10A pattern sector model boundary (blue), well 13-10A: 13-12 study profile (green), VSP image area (black stippled).

3. Results

3.1. WAG Operational Factors

Figure 9 shows the CO_2 and Water injection rates and cumulative volumes injected by well 13-10A during the time-lapse monitoring period. We have adopted the labeling convention that WAG cycles (labeled 1–8) commence with the water leg of the cycle. The baseline survey was acquired just prior to commencement of the cycle 1 water leg, the monitor 1 survey was acquired at the end of cycle 1, just before the cycle 2 water leg, Monitor 2 was acquired at the end of cycle 5, just prior to the cycle 6 water leg, and Monitor 3 was acquired at the end of cycle 8. It can be seen that the durations and injection rates are variable, often due to operational or economic factors. Note that in Figure 9 volumes are reported in STB for liquid (water) and mscf (thousand standard cubic ft) for gas (CO_2). While these are standard conventions for reservoir engineering and production management purposes these units to not adequately represent the relative volumetric proportions within the reservoir which may impact seismic imaging. To normalize the volumes to reservoir conditions we use the gas formation volume factor (FVF) B_g.

$$B_g = \frac{V_{g,r}}{V_{g,sc}}$$

(13)

where;

$V_{g,r}$ = Volume of gas at reservoir conditions

$V_{g,sc}$ = Volume of gas at standard conditions

At reservoir conditions of 168 °F and 4500 PSI, B_{CO_2} equals 0.001455 while $B_{H_2O} \sim 1$. Volumes will also be adjusted to bulk reservoir volume using the relationship;

$$bulk\ reservoir\ volume = \frac{fluid\ Volume}{\phi * (1 - S_{wirr})}$$

(14)

where *Swirr* = Irreducible water saturation.

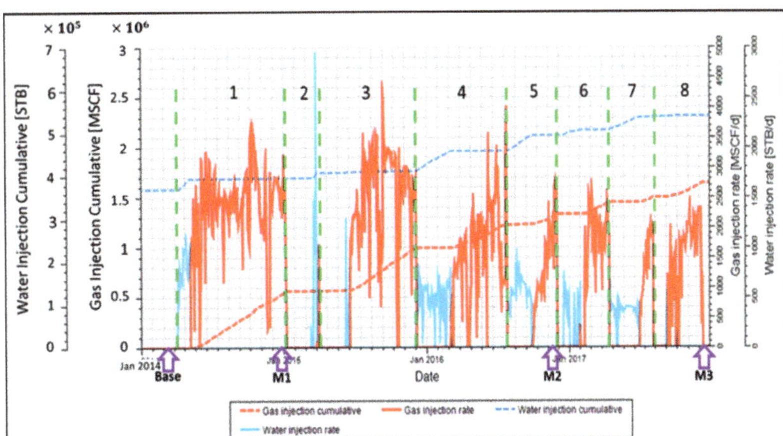

Figure 9. CO_2 and water WAG injection cycles in well 13-10A during time-lapse monitoring. Purple arrows indicate VSP survey dates. Red lines are CO_2 injection rate (solid) and cumulative volume (dashed). Blue lines are water injection rate (solid) and cumulative volume (dashed). Vertical green dashed lines delineate WAG cycles for reference.

Table 2 lists cumulative volumes of water and gas injected from the time of the baseline survey at surface and reservoir conditions, and in equivalent bulk reservoir volumes using a porosity of 0.15 and S_{wirr} = 0.3. After the short initial WAG cycle the Water: CO_2 bulk reservoir volume injection ratio is approximately 1:3 (0.29–0.33). There are several intervening WAG cycles within the M1–M2, and M2–M3 survey times.

Table 2. Cumulative volumes of water and gas injected from the time of the baseline survey at surface and reservoir conditions, and in equivalent bulk reservoir volumes using a porosity of 0.15 and S_{wirr} = 0.3.

		Standard Conditions		Reservoir Volume		
Monitor	Date	Water (stb)	CO_2 (mscf)	Water (cf)	CO_2 (cf)	W/G Ratio
1	1/17/2015	16,550	553,100	885,100	7,665,000	0.12
2	12/3/2016	99,100	1,314,000	5,298,000	18,020,000	0.29
3	1/1/2018	141,000	1,623,000	789,100	22,490,000	0.33

In order to achieve a perspective of fluid volume with respect to time-lapse survey timing we invoke a simplistic conceptual model of non-mixing (piston displacement), annular rings representing the sequential fluid phase injection cycles to compute an "equivalent cylindrical radius" for each hypothetical phase front. Figure 10 shows the hypothetical equivalent annular fluid fronts using a reservoir thickness of 45 ft at each time lapse monitor survey time.

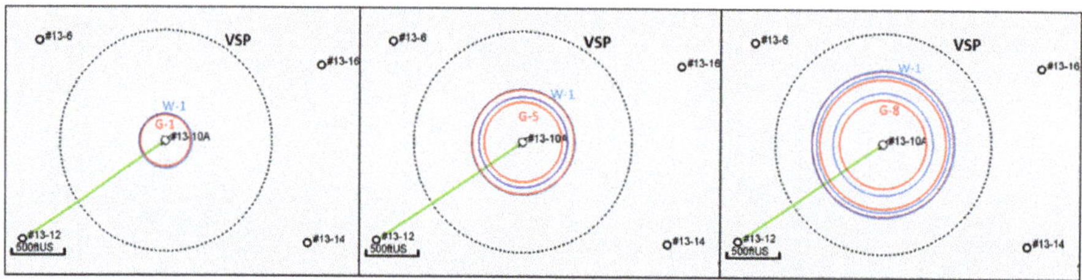

Figure 10. Illustration of hypothetical annular phase fronts at time lapse monitor surveys 1 (**left**), 2 (**center**), and 3 (**right**). The VSP image area is shown as the black dashed line.

While the non-mixing piston front assumption allows us to gain spatial and relative volumetric perspective with regard to injected volumes and pattern pore volume, a more accurate representation of the thermodynamics and hydrodynamics of the system are reflected in numerical simulations. Figure 11 shows streamline (top) and compositional (bottom) simulation results for the monitor survey times. Streamlines clearly show preferential drainage in the pore space between the injector and producing wells. Compositional simulations of gas saturation are consistent with streamlines showing preferential flow of CO_2 toward injectors, albeit exhibiting asymmetry which suggests either differences in well control parameters and/or performance characteristics, or heterogeneity in reservoir properties. Inspection of production data (Figure 12) shows breakthrough in all four producers shortly after acquisition of M2. Figure 12 shows the simulated mole fraction of CO_2 (top) and oil saturation (bottom) along the study profile between wells 13-10A and 13-12. Simulations show no CO_2 breakthrough at the time of M1. Interestingly, oil saturation increases in front of the CO_2 front at the time of the M1 survey (Figure 13). It is hypothesized that this is the result of banking of oil by the leading cycle 1 waterfront before cycle 1 CO_2 moves in and the miscibility process develops.

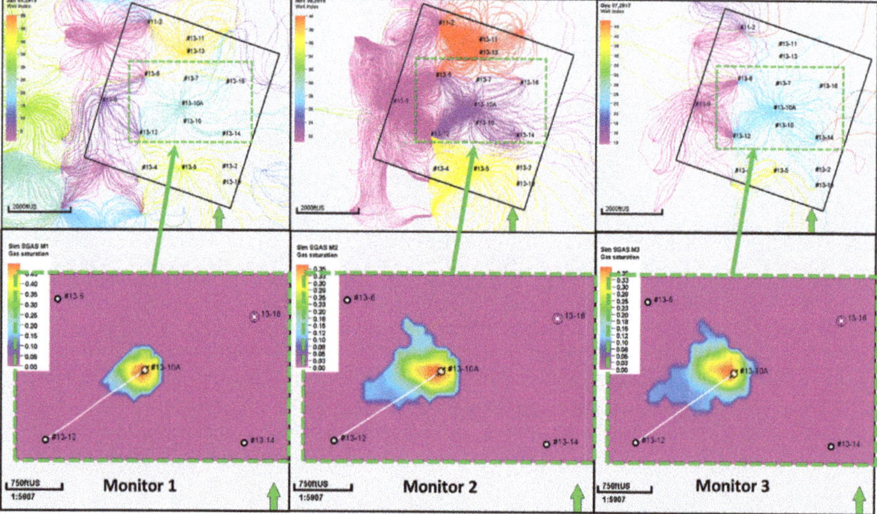

Figure 11. Streamlines (**top**) and compositional simulations of CO_2 saturation (**bottom**) in the 13-10A pattern at monitor survey times.

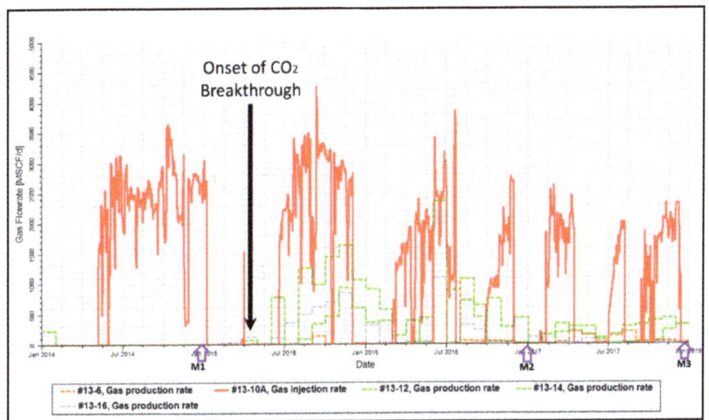

Figure 12. Gas injection and production in 13-10A pattern. CO_2 breakthrough occurs in early 2015 just after M1.

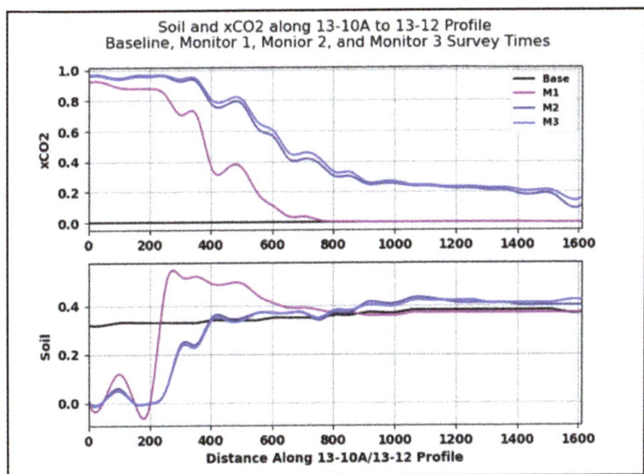

Figure 13. Simulated mole fraction of CO_2 (**top**) and oil saturation (**bottom**) along the study profile between wells 13-10A and 13-12 shown in Figures 10 and 12. Oil banking is suggested by the increase in Soil in front of the cycle 1 waterfront at the M1 survey time. The distance is in feet.

As a final step in our review of operational factors we inspect the variation of predicted hydrocarbon compositions in time and space as the oil is mobilized and extracted by the miscible process. We normalized hydrocarbon composition by component mole fractions along the study profile (Figure 14) at each survey time and observe that the composition of the remaining oil stays nearly constant, suggesting that there is no significant selective stripping of components by the miscible process.

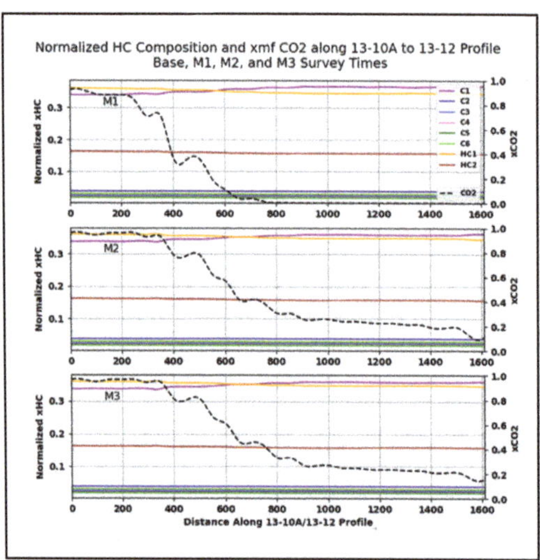

Figure 14. Simulated CO_2 mole fraction and normalized hydrocarbon mole fractions along the 13-10A-13-12 study profile at M1 (**top**), M2 (**center**), and M3 (**bottom**) survey times.

3.2. Analytical Model Rock Physics Investigations

Based on our data and numerical model review we designed a systematic analytical fluid EOS and rock physics study to investigate the expected elastic response of the reservoir within representative ranges of fluid composition, rock properties, and reservoir thermodynamic state. Parameter ranges for the study are listed in Table 3.

Table 3. Fluid EOS and rock physics study parameter ranges.

Parameter	Min	Max
Porosity (fraction)	0.075	0.175
Water Saturation (fraction)	0.3	0.75
Oil Saturation (fraction)	0.27	0.7
CO2 Feed Fraction	0	1
Hydrocarbon Fractions	Proportional	
Temperature (°F)	163	173
Pressure (psi)	4000	6000

Used together, our petrophysical data analysis, the python implementation of NIST subroutines, and databases facilitated a comprehensive investigation of a broad range of realistic fluid substitution scenarios providing valuable insights into the time-lapse integration problem. Here, we show a selection of graphical results which capture many of the important behaviors of the rock-fluid system resulting from WAG implementation at FWU.

Figure 15 shows contour plots of reservoir fluid compositions which represent stiffness endpoints. The top row with the maximum possible water saturation and no CO_2 represents the stiffest fluid in the system. The bottom row with the minimum possible water saturation and maximum CO_2 represents the softest fluid in the system. Contours are relative to the attribute value at user selected reference temperature which is annotated on each plot. For this comparison we used nominal reservoir temperature and pressure (168 °F, 4500 psi) as the reference point. These plots show that the saturated rock properties have little sensitivity to temperature for both fluids. Further, the stiff fluid shows negligible variation with pressure from a seismic detection perspective, but can vary as much as 20% from nominal conditions for highest anticipated pressures. However, this interpretation must be put in the perspective that such high pressures are not optimal for the production scheme and should be considered anomalous.

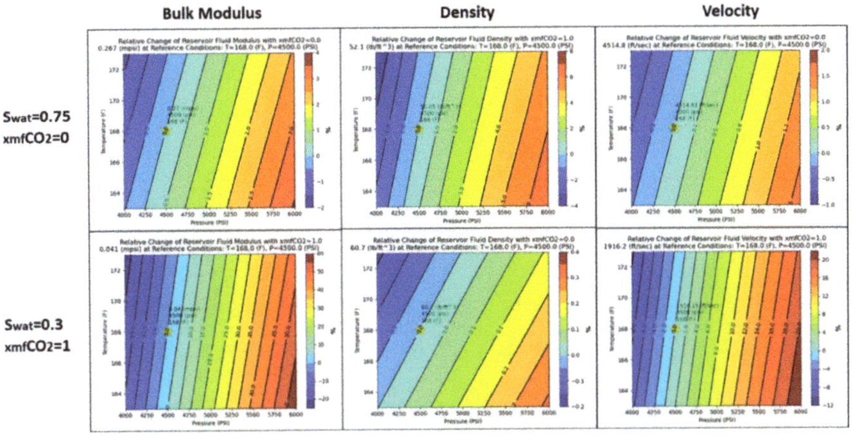

Figure 15. Contoured variations of fluid bulk modulus (**left**), density (**center**), and velocity (**right**) vs. temperature and pressure for stiffest fluid mix (**top**) and softest fluid mix (**bottom**). Contours are percent change relative to reference points shown.

Figure 16 illustrates another informative visualization tool, again comparing (endpoint) stiff versus soft fluids at a selected nominal reference temperature and pressure. The four plots in Figure 16 show the variation of normalized elastic properties for hydrocarbon-CO_2-water mixtures (left column), saturated reservoir rock (right column), stiff fluid (top row), soft fluid (bottom row) as a function of CO_2 molar (feed) fraction. All curves are normalized to their value at $xCO_2 = 0$. Additionally, shown are the fluid mixture vapor (red) and liquid (blue) fraction curves which clearly show the saturation point of 0.75 mole fraction of CO_2. Inspection of the top and bottom plots on the left shows the approximately 7% drop in velocity of the reservoir fluid at zero mole fraction CO_2 (owing to the difference in water saturation), and a much-amplified response to the addition of CO_2 for the softer fluid, again owing to replacement of water with potentially miscible hydrocarbon. Comparing left columns to right illustrated the extreme dampening effect of the rock frame on the elastic response. The difference in velocity between stiff and soft fluids (maxim vs. minimum water saturation) at zero CO_2 mole fraction is only ~0.5%. Velocity of the rock saturated with the stiff fluid drops by only ~0.6% at CO_2 fractions up to saturation before the existence of free phase CO_2 begins to take effect on the bulk fluid properties. For the soft fluid the response to additional CO_2 is somewhat stronger but still at just under 1% at CO_2 saturation.

Finally, we looked at elastic property variations in saturated rock at nominal temperature and temperature (168 °F, 4500 psi). Figure 17 illustrates contoured variations of p wave velocity with CO_2 mole fraction and porosity (left), and water saturation (right). The left plot shows a high degree of sensitivity with respect to porosity, which is a source of significant statistical uncertainty in 3D models, while showing little sensitivity to the quantity of CO_2 except for the inflection at saturation ($xCO_2 = 0.75$). Variation with water saturation and CO_2 mole fraction (right) shows a complex response surface with most significant variations related to the phase transition and at lower water saturations.

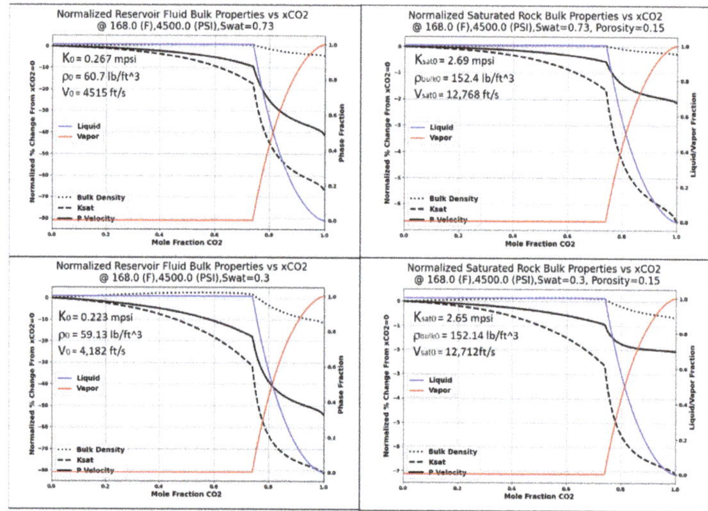

Figure 16. Normalized reservoir fluid (**left**) and saturated rock (**right**) elastic properties for stiff fluid scenario where Swat = 1-Sor and xCO2 = 0 (**top**), and soft fluid scenario where Swat = Swirr (**bottom**) at nominal conditions (T = 168 °F, p = 4500 psi, and φ = 0.15).

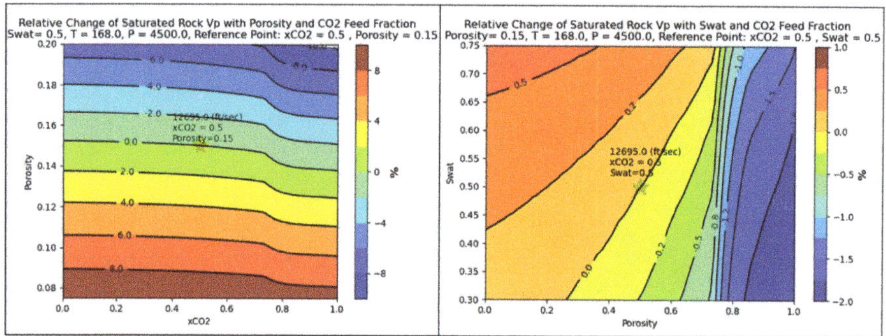

Figure 17. Contoured variations of saturated rock P velocity vs. CO_2 mole fraction and porosity (**left**), and water saturation (**right**) at reservoir conditions T = 168 °F, and p = 4500 psi, xCO2 = 0.5. Contours are percent change relative to reference points shown.

3.3. Model Rock Physics from Compositional Simulator

Elastic properties were computed from 3 dimensional arrays of reservoir porosity and reservoir fluid properties at each time-lapse survey time. Computed velocities from the baseline survey and each monitor survey were used to compute synthetic time-lapse depth shifts. Figure 18 shows the computed time-shifts (color maps) and the contour for mole fraction of CO_2 at the saturation level of 0.75. Figure 19 shows distributions of computed monitor survey depth shifts. Monitor 1 survey has the fewest depth shifted samples (simulator blocks) with depth shifts. The number of shifted cells increases over time as expected injection of additional CO_2. Depth shift magnitude for M1 is distributed uniformly by comparison to those for M2 and M3 which show apparent shifting of magnitudes downward. It is possible that this is due to a greater amount of free phase CO_2 in the early stages of miscibility development, with more complete mixing over time yielding lower magnitude depth shifts. In all cases the computed depth shifts are extremely small from a seismic detection perspective.

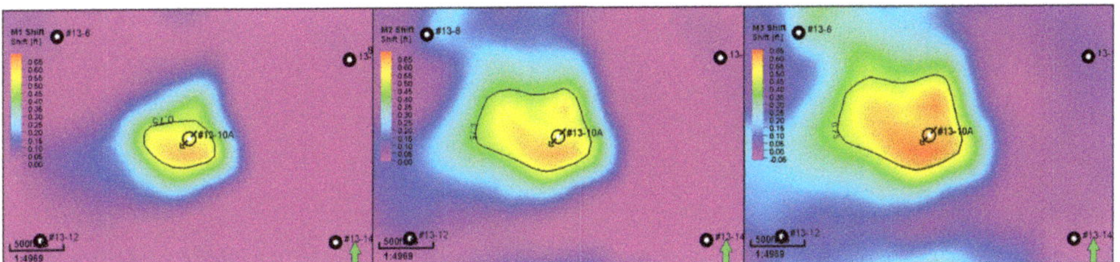

Figure 18. Simulated time-lapse depth shifts mole fraction at monitor times M1 (**left**), M1 (**center**), and M3 (**right**). The $xCO_2 = 0.75$ contour is shown (black).

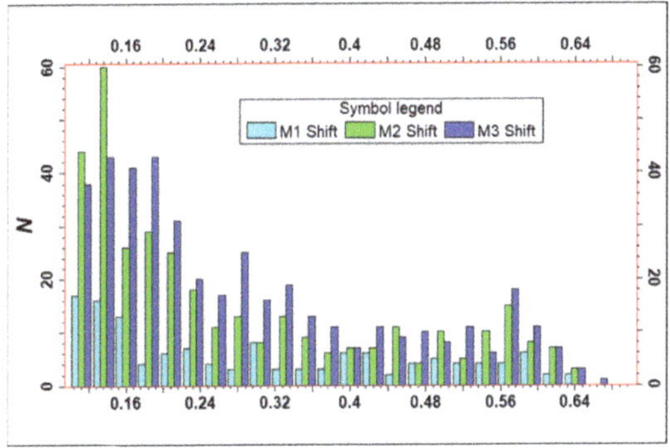

Figure 19. Magnitude distributions of simulated time-lapse depth shifts for monitor survey times M1 (light blue), M1 (green), and M3 (dark blue).

3.4. Comparison with Time-Lapse Measurements

We compare simulated depth shifts with Schlumberger displacement attribute computations for the three monitor survey times. Figure 20 shows maps of simulated depth shifts (top) and Schlumberger displacement attribute (bottom) in the zone beneath the Morrow B at M1 (left), M2 (center), and M3 (right) survey times. Color scales have been adjusted to normalize the visual comparison between simulated and measured values. Although the visual comparison is not encouraging, we note that the spatial trends in the measured displacement maps are suggestive of preferential fluid saturation changes toward producing wells, consistent with the known hydrodynamics of the 5-spot pattern. Differences from the simulated distribution may be due to local heterogeneities not captured in the reservoir model and not corrected by the calibration which was achieved through optimization at a global parameter scale.

Figure 21 compares the frequency distributions of absolute and normalized, measured and simulated time-lapse shifts for the three monitor surveys. Simulated and measured datasets were normalized independently to their maximum values. Maximum shifts were 0.68 ft for simulations and 3.8 ft for the measured displacement attribute for a ratio of 5.6. Ratios of average distribution values were 7.5, 10.4, and 3.7, respectively for M1, M2, and M3.

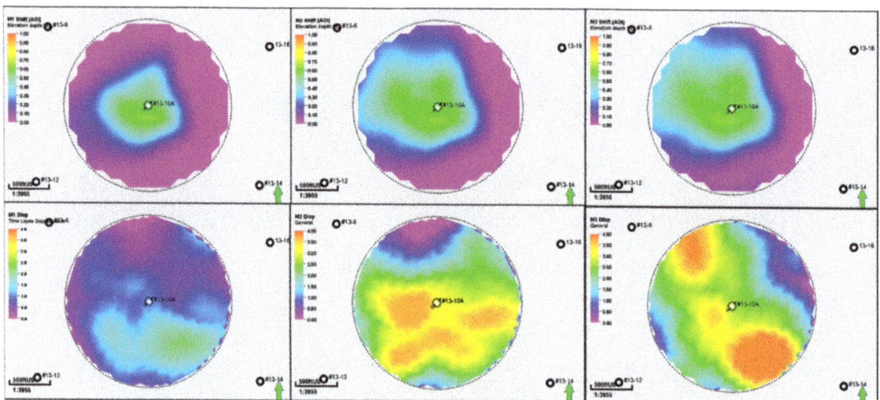

Figure 20. Simulated depth shifts (**top**) and Schlumberger displacement attribute (**bottom**) in the zone beneath the Morrow B at monitor M1 (**left**), M2 (**center**), and M3 (**right**) survey times. Color scales are adjusted to normalize the visual comparison between simulated and measured values.

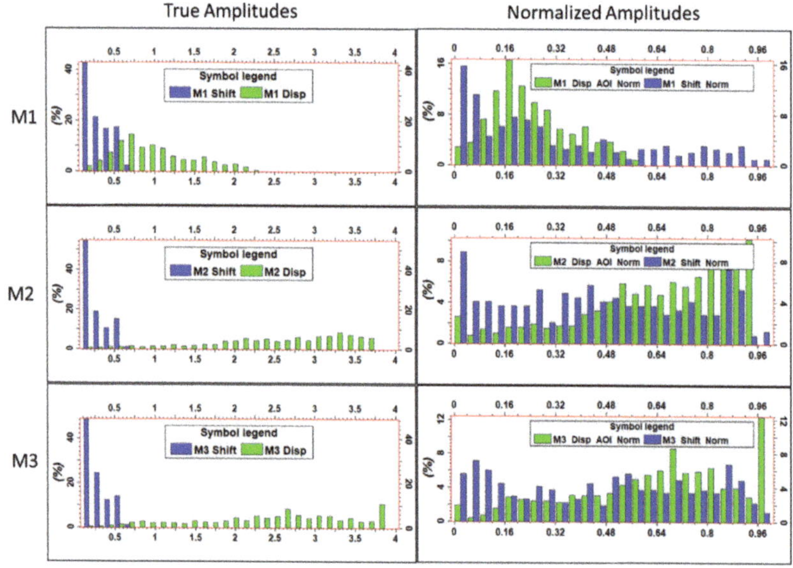

Figure 21. Magnitude distributions of simulated time-lapse depth shifts (blue) and Schlumberger Displacement attribute computed from VSP data (green) for monitor surveys M1 (**top**), M2 (**Mid**), and M2 (**bottom**). and M3. Data in the left column is true amplitude. Data in the right column has been normalized to the maximum observed value for each data type.

4. Discussion and Conclusions

We are tasked with interpretation and integration of time-lapse VSP data optimizing the FWU reservoir simulation model. From previous experience and literature review we know that WAG operations result in fluid properties and seismic detection scenarios that are much more challenging than immiscible systems. Because these processes are difficult to isolate for investigation in a complex compositional reservoir simulation, we first performed an analytical fluid study which enabled us to isolate and investigate complex thermodynamic processes. We reviewed the field operations at FWU in terms of the fluids in place, and the volumes and timing of injected fluids. We discovered that, contrary to the concept of a discrete CO_2 plume, as a conformance management technique the WAG operation is specifically designed to create an optimal CO_2-hydrocarbon mixture with water as an intermediate phase. We used available data and models to perform a systematic EOS analysis of WAG fluid thermodynamics. We found that the FWU WAG operational scenario results in a thermodynamically complex transient distribution of fluid mixtures with largely similar thermophysical properties while the hydrocarbon undersaturated with respect to CO_2. We then used Gassmann's model to perform a systematic rock physics study which showed that already subtle differences in fluid properties are heavily muted by the rock framework. From these analytical studies we conclude that the most likely seismic detection scenario at FWU is for the existence of free phase CO_2 which exists where the available oil is saturated, or the miscibility process has not yet fully developed.

We applied our rock physics model to compositional simulations using the production data calibrated FWU reservoir model. These simulations verify a continuum of saturations rather than a distinct distribution of high concentration CO_2 which might be characterized as a plume, or as forming a discrete fluid front. These gradational compositional variations are reflected in similarly gradational spatial distributions of simulated depth shifts. Poor spatial correlations between simulated and measured depth shifts may be attributed to either data resolution or model error. Clearly, strict application of seismic imaging fundamentals tells us that these shifts are all well below the seismic resolution. However, it is not yet clear whether or not we are able to achieve the less stringent goal of merely detecting changes in bulk reservoir properties with the given data. As regards to seismic data fidelity, interpretating estimates of effective media properties at seismic wavelengths is a well-studied problem. Worthington [49] reports ratios of up to 10:1 for estimates of bulk reservoir properties from multi-scale measurements of fractured media compliance using cross-well seismic and laboratory ultra-sonic measurements. On the side of the model accuracy, the model used was calibrated at the global (full field) scale. As such, the calibration at the pattern scale may be non-unique with respect to production data only. Recognition of such ambiguities and non-uniqueness in geostatistical model property distributions is one of the main motivators for introduction of time-lapse seismic data as a calibration constraint. If the fidelity of our measured data can be verified, then the spatial discrepancies between simulated and measured shifts can be used to drive model updates. This subject is currently under study.

Other potential sources of simulated versus measured discrepancies owing to numerical simulations but not investigated here include: the accuracy of the Eclipse 300 assumption of first contact miscibility, the effects of simulation grid scale on the modeling of miscible process development, and the correspondence between the Eclipse calculation of total fluid system compressibility and the assumptions used by the Wood's law mixing equation and Gassmann's equation. We also assume that the time-lapse response of the reservoir is independent of pore pressure and effective stress. While we are aware of ultra-sonic measurements under stress on Morrow B reservoir samples which indicate potentially significant velocity versus stress sensitivity, inclusion of such effects was not within the scope of funding for this project.

5. Avenues for Future Work

It is clear that successful integration of the time-lapse seismic data at FWU depends on establishing the confidence level for the seismic data. This is not a black and white issue and should not be considered a binary decision. All quantitative integrations provide a mechanism for scaling the contribution of each observation in computation of model updates. Unfortunately, a rigorous estimate of confidence in seismic data is difficult to establish. It is possible that methods based on data signal to noise characteristics may be implemented. However, these are based on seismic amplitude data so a strategy for application to displacement data would be required.

Quantitative integration of the time-lapse seismic data for reservoir model updating presents a number of spatial and temporal sampling challenges. One of these is the discrepancies between spatio-temporal sampling characteristics of the different observation domains. Seismic data are finely and spatially sampled (~50' × 50') but coarsely sampled temporally (1–2 years). Production data are coarsely sampled spatially (~1000' × 1000') but finely sampled temporally (hourly, daily). Further, the dynamic processes dictating fluid compositional changes and movement in 3D space over time are very complex. This is why that it is essential to use a transient compositional simulation model as the basis for integration. Although integration of simulation models for 3D interpretation is not new, recent advances in distributed computing and machine learning techniques (such as proxy model based optimization) are fuelling progress. Another challenge is the superposition of effects from sub-pattern scale fluvial system heterogeneity and the thermodynamically active fluid system effects. First, we feel that further lateral refinement of the simulation grid scale is needed. An update and optimization scheme such as the one implemented by Ampomah et al. [32] or Sun et al. [40] for production data calibration of the reservoir model and optimization of development schemes at FWU may be used for time-lapse data integration. In these studies, machine-learning technologies such as response surface models (RSM), multi-layer neural networks (MLNN), and support vector machines (SVM) are used to develop proxy models for numerical simulation results against production data. These proxy models are coupled with suitable optimization strategy (such as the evolutionary strategy) to achieve optimization of a penalty function which includes historical data. For time-lapse integration, the proxy model would include both production data from pattern wells and time-lapse seismic measurements. The required proxy for EOS and rock physics response could be developed through integration of NIST databases in a similar manner as was used in our analytical fluid and rock physics study. Although the full field reservoir model property distribution was developed through a pixel based geostatistical method, it may be necessary to implement an object-based property population method using a fluvial system model in order to impose and systematically update representative anisotropy in model porosity and permeability updates. These methods are currently being applied for calibration of coupled hydro-chemical simulation models by SWP researchers [unpublished]. The proxy model process for time-lapse integration may be preconditioned with a geologically realistic and equiprobable realization ensemble strategy such as implemented by Souza et al. [17].

Author Contributions: Conceptualization, R.W., W.A.; methodology, R.W., T.B., W.A., S.A.; validation, T.B., S.A., R.B.; formal analysis, R.W., W.A. and S.A.; resources, R.B., M.C. and W.A.; writing—original draft preparation, R.W., T.B., W.A. and S.A.; writing—review and editing, M.C. and W.A.; visualization, R.W., S.A., T.B. and W.A.; supervision, W.A. and R.B.; project administration, M.C. and R.B.; funding acquisition, R.B. All authors have read and agreed to the published version of the manuscript.

Funding: Funding for this project is provided by the U.S. Department of Energy's (DOE) National Energy Technology Laboratory (NETL) through the Southwest Regional Partnership on Carbon Sequestration (SWP) under Award No. DE-FC26-05NT42591.

Acknowledgments: The authors would like to acknowledge the many current and past SWP researchers who have contributed to the foundational work for this study. Funding for this project is provided by the US Department of Energy's (DOE) National Energy Technology Laboratory (NETL) through the Southwest Regional Partnership on Carbon Sequestration (SWP) under Award No. DE-FC26-05NT42591. Additional support has been provided by site operator and Schlumberger.

Conflicts of Interest: The authors declare no conflict of interest.

Disclaimer: This report was prepared as an account of work sponsored by an agency of the United States Government. Neither the United States Government nor any agency thereof, nor any of their employees, makes any warranty, express or implied, or assumes any legal liability or responsibility for the accuracy, completeness, or usefulness of any information, apparatus, product, or process disclosed, or represents that its use would not infringe privately owned rights. Reference herein to any specific commercial product, process, or service by trade name, trademark, manufacturer, or otherwise does not necessarily constitute or imply its endorsement, recommendation, or favoring by the United States Government or any agency thereof. The views and opinions of authors expressed herein do not necessarily state or reflect those of the United States Government or any agency thereof.

References

1. Balch, R.S.; Mcpherson, B.; Grigg, R. Overview of a large scale carbon capture, utilization, and storage demonstration project in an active oil field in Texas, USA. In Proceedings of the 13th International Conference on Greenhouse Gas Control Technologies, Lausanne, Switzerland, 14–18 November 2016.
2. Christensen, J.R.; Stenby, E.H.; Skauge, A. Review of WAG field experience. *SPE Reserv. Eval. Eng.* **2001**, *4*, 97–106. [CrossRef]
3. Johnston, D.H. *Practical Applications of Time-Lapse Seismic Data*; Society of Exploration Geophysicists, 2013; p. 289. Available online: https://library.seg.org/doi/epdf/10.1190/1.9781560803126 (accessed on 27 February 2021).
4. Lumley, D.; Sherlock, D.; Daley, T.; Huang, L.; Lawton, D.; Masters, R.; Verliac, M.; White, D. *Highlights of the 2009 SEG Summer Research Workshop on CO_2 Sequestration*; The Leading Edge, 2010; Volume 29, pp. 138–145. Available online: https://library.seg.org/doi/10.1190/tle29020138.1 (accessed on 27 February 2021).
5. Ghosh, R.; Sen, M.K.; Vedanti, N. Quantitative interpretation of CO_2 plume from Sleipner (North Sea) using post-stack inversion and rock physics modeling. *Int. J. Greenh. Gas Control* **2015**, *32*, 147–158. [CrossRef]
6. Chadwick, R.A.; Williams, G.A.; Falcon-Suarez, I. Forensic mapping of seismic velocity heterogeneity in a CO_2 layer at the Sleipner CO_2 storage operation, North Sea, using time-lapse seismics. *Int. J. Greenh. Gas Control* **2019**, *90*, 102793. [CrossRef]
7. Raknes, E.B.; Weibull, W.; Arntsen, B. Seismic imaging of the carbon dioxide gas cloud at Sleipner using 3D elastic time-lapse full waveform inversion. *Int. J. Greenh. Gas Control* **2015**, *42*, 26–45. [CrossRef]
8. Roacha, L.; White, D.J. Evolution of a deep CO_2 plume from time-lapse seismic imaging at the Aquistore storage site, Saskatchewan, Canada. *Int. J. Greenh. Gas Control* **2018**, *74*, 79–86. [CrossRef]
9. White, D.J.; Roach, L.A.N.; Roberts, B. Time-lapse seismic performance of a sparse permanent array: Experience from the Aquistore CO_2 storage site. *Geophysics* **2015**, *80*, WA35–WA48. [CrossRef]
10. Gorecki, C. *Plains CO_2 Reduction Partnership Phase III Final Report (No. DOE-EERC-42592)*; Energy & Environmental Research Center University of North Dakota, 2019. Available online: https://www.osti.gov/biblio/1580755-plains-co2-reduction-partnership-phase-iii-final-report (accessed on 27 February 2021).
11. Gollakota, S.; McDonald, S. Commercial-Scale CCS Project in Decatur, Illinois—Construction Status and Operational Plans for Demonstration. *Energy Procedia* **2014**, *63*, 5986–5993. [CrossRef]
12. Finley, R.J.; Frailey, S.M.; Leetaru, H.E.; Senel, O.; Couëslan, M.L.; Scott, M. Early operational experience at a one-million tonne CCS demonstration project, decatur, Illinois, USA. *Energy Procedia* **2013**, *37*, 6149–6155. [CrossRef]
13. Senel, O.; Will, R.; Butsch, R.J. Integrated reservoir modeling at the Illinois Basin—Decatur Project. *Greenh. Gases Sci. Technol.* **2014**, *4*, 662–684. [CrossRef]
14. Ivanova, A.; Kashubin, A.; Juhojuntti, N.; Kummerow, J.; Henninges, J.; Juhlin, C.; Lüth, S.; Ivandic, M. Monitoring and volumetric estimation of injected CO_2 using 4D seismic, petrophysical data, core measurements and well logging: A case study at Ketzin, Germany. *Geophys. Prospect.* **2012**, *60*, 957–973. [CrossRef]
15. Obidegwu, D.; Chassagne, R.; MacBeth, C. Seismic assisted history matching using binary maps. *J. Nat. Gas Sci. Eng.* **2017**, *42*, 69–84. [CrossRef]
16. Grude, S.; Landrø, M.; White, J.; Torsæter, O.; Torsaeter, O. CO_2 saturation and thickness predictions in the Tubåen Fm., Snøhvit field, from analytical solution and time-lapse seismic data. *Int. J. Greenh. Gas Control* **2014**, *29*, 248–255. [CrossRef]
17. Souza1, R.; Lumley, D.; Shragge, J.; Davolio, A.; Schiozer, D.J. Analysis of time-lapse seismic and production data for reservoir model classification and assessment. *J. Geophys. Eng.* **2018**, *15*, 1561–1587. [CrossRef]
18. Brown, L.T. Integration of Rock Physics and Reservoir Simulation for the Interpretation of Time-Lapse Seismic Data at Weyburn Field, Saskatchewan. Master's Thesis, Colorado School of Mines, Golden, CO, USA, 2002.
19. Leeuwenburgh, O.; Meekes, S.; Vandeweijer, V.; Brouwer, J. Stochastic history matching to time-lapse seismic of a CO_2 -EOR project sector model. *Int. J. Greenh. Gas Control.* **2016**, *54*, 441–453. [CrossRef]
20. Ghosh, R. Monitoring field scale CO_2 injection from time-lapse seismic and well log, integrating with advanced rock physics model at Cranfield EOR site. *Acta Geophys.* **2017**, *65*, 1207–1218. [CrossRef]
21. Alfi, M.; Hosseini, S.A. Integration of reservoir simulation, history matching, and 4D seismic for CO2-EOR and storage at Cranfield, Mississippi, USA. *Fuel* **2016**, *175*, 116–128. [CrossRef]
22. Mur, A.; Barajas-Olalde, C.; Adams, D.C.; Jin, L.; He, J.; Hamling, J.A.; Gorecki, C.D. Integrated simulation to seismic and seismic reservoir characterization in a CO_2 EOR monitoring application. *Lead. Edge* **2020**, *39*, 668–678. [CrossRef]
23. Raef, A.E.; Miller, R.D.; Byrnes, A.P.; Harrison, W.E. 4D seismic monitoring of the miscible CO_2 flood of Hall-Gurney Field, Kansas, U.S. *Geophysics* **2004**, *23*, 1171–1176. [CrossRef]
24. Ji, L.; MacBeth, C.; Mangriotis, M.-D. A Critical Comparison of Three Methods for Time-Lapse Time-Shift Calculation. *Math. Geol.* **2021**, *53*, 55–80. [CrossRef]
25. MacBeth, C.; Mangriotis, M.; Amini, H. An interpretation and evaluation of post-stack 4D seismic time-shifts. In Proceedings of the 80th EAGE Conference and Exhibition, Copenhagen, Denmark, 11–14 June 2018.
26. Aarre, V. Estimating 4D velocity changes and contact movement on the norne field, OTC 19049. In Proceedings of the 2007 Offshore Technology Conference, Houston, TX, USA, 30 April–3 May 2007.

27. Barajas-Olalde, C.; Haffinger, P.; Gisolf, D.; Zhang, M.; Droujinina, A.; Doulgeris, P.; Khatibi, S.; Jin, L.; Burnison, S.A.; Hamling, J.A.; et al. Simultaneous time-lapse WEB-AVO inversion for seismic reservoir monitoring: Application to CO_2 enhanced oil recovery at the Bell Creek oil field. In *2019 SEG Technical Program Expanded Abstracts*; 2019; pp. 564–568. Available online: https://library.seg.org/doi/abs/10.1190/segam2019-3216895.1 (accessed on 27 February 2021). [CrossRef]
28. Cather, M.; Rose-Coss, D.; Galagher, S.; Trujillo, N.; Cather, S.; Hollingworth, R.S.; Leary, J. Deposition, diagenesis, and sequence stratigraphy of the pennsylvanian morrowan and atokan intervals at farnsworth unit. *Energies* **2021**, *14*, 1057. [CrossRef]
29. Gallagher, S.R. Depositional and Diagenetic Controls on Reservoir Heterogeneity: Upper Morrow Sandstone, Farnsworth Unit, Ochiltree County, Texas. Master's Thesis, New Mexico Tech, Socorro, NM, USA, 2014.
30. Moodie, N.; Ampomah, W.; Jia, W.; Heath, J.; McPherson, B. Assignment and calibration of relative permeability by hydrostratigraphic units for multiphase flow analysis, case study: CO2-EOR operations at the Farnsworth Unit, Texas. *Int. J. Greenh. Gas Control* **2019**, *81*, 103–114. [CrossRef]
31. Balch, R.S.; McPherson, B.; Will, R.A.; Ampomah, W. Recent Developments in Modeling: Farnsworth Texas, CO_2 EOR Carbon Sequestration Project. In Proceedings of the15th International Conference on Greenhouse Gas Control Technologies, Abu Dhabi, United Arab Emirates, 5–8 October 2020.
32. Ampomah, W.; Balch, R.; Grigg, R.B.; Cather, M.; Gragg, E.; Will, R.A.; White, M.; Moodie, N.; Dai, Z. Performance assessment of CO2-enhanced oil recovery and storage in the Morrow reservoir. *Géoméch. Geophys. Geo-Energy Geo-Resour.* **2017**, *3*, 245–263. [CrossRef]
33. Gunda, D.; Ampomah, W.; Grigg, R.; Balch, R. Reservoir fluid characterization for miscible enhanced oil recovery. In Proceedings of the Carbon Management Technology Conference, Sugar Land, TX, USA, 17–18 November 2015.
34. Grigg, R.B.; Ampomah, W.; Gundah, D. Integrating CO_2 EOR and CO_2 Storage in Farnsworth Field. In Proceedings of the 2015 DOE Carbon Storage Meeting, Pittsburgh, PA, USA, 18–20 August 2015.
35. Pedersen, K.; Milter, J.; Sørensen, H. Cubic Equations of State Applied to HT/HP and Highly Aromatic Fluids. *SPE J.* **2004**, *9*, 186–192. [CrossRef]
36. Peng, D.-Y.; Robinson, D.B. A New Two-Constant Equation of State. *Ind. Eng. Chem. Fundam.* **1976**, *15*, 59–64. [CrossRef]
37. Péneloux, A.; Rauzy, E.; Fréze, R. A consistent correction for Redlich-Kwong-Soave volumes. *Fluid Phase Equilibria* **1982**, *8*, 7–23. [CrossRef]
38. Lohrenz, J.; Bray, B.G.; Clark, C.R. *Calculating Viscosities of Reservoir Fluids From Their Compositions*; Society of Petroleum Engineers, 1964. Available online: https://www.semanticscholar.org/paper/Calculating-Viscosities-of-Reservoir-Fluids-From-Lohrenz-Bray/ed2d581ab5899d5a7f16fae0d455c68c2f777798 (accessed on 27 February 2021). [CrossRef]
39. Rasmussen, L.; Fan, T.; Rinehart, A.; Luhmann, A.; Ampomah, W.; Dewers, T.; Heath, J.; Cather, M.; Grigg, R. Carbon storage and enhanced oil recovery in pennsylvanian morrow formation clastic reservoirs: Controls on oil–brine and oil–CO_2 relative permeability from diagenetic heterogeneity and evolving wettability. *Energies* **2019**, *12*, 3663. [CrossRef]
40. Sun, Q.; Ampomah, W.; You, J.; Cather, M.; Balch, R. Practical CO_2—WAG field operational designs using hybrid numerical-machine-learning approaches. *Energies* **2021**, *14*, 1055. [CrossRef]
41. Lemmon, E.W.; Huber, M.L.; McLinden, M.O. *NIST Standard Reference Database 23: Reference Fluid Thermodynamic and Transport Properties—REFPROP Version 8.0*; National Institute of Standards and Technology, Standard Reference Data Program: Gaithersburg, MD, USA, 2007.
42. Huber, M.L. *NIST Thermophysical Properties of Hydrocarbon Mixtures, NISTt4 (Supertrapp), v3.1*; Standard Reference Data, National Institute of Standards and Technology: Gaithersburg, MD, USA, 2003.
43. Altundas, B.; Chugunov, N.; Ramakrishnan, T.S.; Will, R. Quantifying the effect of CO_2 dissolution on seismic monitoring of CO_2 in CO2-EOR. In *SEG Technical Program Expanded Abstracts*; Society of Exploration Geophysicists, 2017; pp. 3771–3775. Available online: https://library.seg.org/doi/abs/10.1190/segam2017-17792996.1 (accessed on 27 February 2021).
44. Holme, L.; Josendal, V.A. Effect of oil displacement on miscible-type displacement by carbon dioxide. *SPE J.* **1982**, *22*, 87–98.
45. Mavko, G.; Mukerji, T.; Dvorkin, J. *The Rock Physics Handbook*; Cambridge University Press: Cambridge, UK, 2009.
46. Biot, M.A. Theory of propagation of elastic waves in a fluid saturated porous solid. II. Higher frequency range. *J. Acoust. Soc. Am.* **1956**, *28*, 179–191. [CrossRef]
47. El-Kaseeh, G.; Czoski, P.; Will, R.; Balch, R.; Ampomah, W.; Li, X. Time-lapse vertical seismic profile for CO_2 monitoring in carbon capture, utilization, and sequestration/EOR, Farnsworth project. In *SEG Technical Program Expanded Abstracts 2018*; 2018; pp. 5377–5381. Available online: https://library.seg.org/doi/10.1190/segam2018-2995747.1 (accessed on 27 February 2021). [CrossRef]
48. Nickel, M.; Sonneland, L. Non-rigid matching of migrated time-lapse seismic. In *69th Ann. Internat. Mtg., SEG, Expanded Abstracts*; 1999; p. 872. Available online: https://library.seg.org/doi/abs/10.1190/1.1821191 (accessed on 27 February 2021).
49. Worthington, M. Interpreting seismic anisotropy in fractured reservoirs. *First Break.* **2008**, *26*, 57–63. [CrossRef]

Article

A Gate-to-Gate Life Cycle Assessment for the CO_2-EOR Operations at Farnsworth Unit (FWU)

Anthony Morgan *, Reid Grigg * and William Ampomah

Petroleum Recovery Research Center, New Mexico Tech, Socorro, NM 87801, USA; William.Ampomah@nmt.edu
* Correspondence: anthony.morgan@student.nmt.edu (A.M.); reid.grigg@nmt.edu (R.G.)

Abstract: Greenhouse gas (GHG) emissions related to the Farnsworth Unit's (FWU) carbon dioxide enhanced oil recovery (CO_2-EOR) operations were accounted for through a gate-to-gate life cycle assessment (LCA) for a period of about 10 years, since start of injection to 2020, and predictions of 18 additional years of the CO_2-EOR operation were made. The CO_2 source for the FWU project has been 100% anthropogenically derived from the exhaust of an ethanol plant and a fertilizer plant. A cumulative amount of 5.25×10^6 tonnes of oil has been recovered through the injection of 1.64×10^6 tonnes of purchased CO_2, of which 92% was stored during the 10-year period. An LCA analysis conducted on the various unit emissions of the FWU process yielded a net negative (positive storage) of 1.31×10^6 tonnes of CO_2 equivalent, representing 79% of purchased CO_2. An optimized 18-year forecasted analysis estimated 86% storage of the forecasted 3.21×10^6 tonnes of purchased CO_2 with an equivalent 2.90×10^6 tonnes of crude oil produced by 2038. Major contributors to emissions were flaring/venting and energy usage for equipment. Improvements on the energy efficiency of equipment would reduce emissions further but this could be challenging. Improvement of injection capacity and elimination of venting/flaring or fugitive gas are methods more likely to be utilized for reducing net emissions and are the cases used for the optimized scenario in this work. This LCA illustrated the potential for the CO_2-EOR operations in the FWU to store more CO_2 with minimal emissions.

Keywords: life cycle analysis; CO_2-enhanced oil recovery; anthropogenic CO_2; global warming potential; greenhouse gas (GHG); carbon storage

Citation: Morgan, A.; Grigg, R.; Ampomah, W. A Gate-to-Gate Life Cycle Assessment for the CO_2-EOR Operations at Farnsworth Unit (FWU). *Energies* **2021**, *14*, 2499. https://doi.org/10.3390/en14092499

Academic Editor: Rouhi Farajzadeh

Received: 25 February 2021
Accepted: 21 April 2021
Published: 27 April 2021

Publisher's Note: MDPI stays neutral with regard to jurisdictional claims in published maps and institutional affiliations.

1. Introduction

Carbon dioxide (CO_2) atmospheric concentrations are high compared to the last 400 centuries and are still rising [1]. About 50% of the increase has been in the last forty years and is mainly attributed to human activities [1]. This has led to rising temperatures and climate change globally [2]. One (1) megawatt (MW) electrical coal fire plant releases up to eight (8) megatons of CO_2 yearly. About 75% and 50% of this amount is released by oil fired and natural gas combined-cycle electrical plants, respectively [3]. In the US, about 86% of anthropogenic greenhouse gas (GHG) emissions are from energy production, which includes principally the generation of power and transportation [4]. The oil and gas industry globally accounts for about 8% of anthropogenic carbon dioxide (CO_2) and 15% of the methane gas (CH_4) with 3% coming from upstream operations [5]. This increase in GHG emissions is believed to have an adverse impact on the environment. Improved energy efficiency of production equipment, use of renewable energy and low carbon fuels, and storage/sequestration of captured CO_2 are all potential emission reduction approaches with each having their inherent pros and cons [5]. From the works of Farajzadeh et al., CO_2-EOR incorporating carbon capture consumes a high amount of energy compared to the amount of crude produced [2]. This also significantly leads to an increase in emissions.

Geologic formations are estimated to have storage capacities of about 9×10^{11} tonnes of CO_2 globally with oil and gas fields alone offering capacities of about 1.3×10^{11} tonnes of CO_2 [6]. Carbon dioxide can be sequestrated as part of CO_2-enhanced oil recovery (EOR), a process used to increase oil production in use since the early 1970's [7]. Injection of pressurized CO_2 into oil reservoirs causes crude oil to swell, decreases viscosity thus increasing crude mobility, and develops miscibility as interfacial tension is reduced [8]. CO_2-EOR is common in the US in the Permian Basin, Rocky Mountains, Northern Plains, and Louisiana-Mississippi, all regions that have access to natural CO_2 and/or large natural gas processing plants that may produce high volumes of CO_2 as a by-product [9]. International CO_2-EOR projects include the Weyburn-Midale CO_2 project in Saskatchewan, Canada and the Lula field in the Santos Basin, Brazil (offshore) [10]. Notable international CCS include the Sleipner in Norway, In Salah in Algeria, Ketzin in Germany, k12B in Netherland [11] and the Gorgon project in Western Australia by Chevron, ExxonMobil and Shell [12].

Assuming a reservoir has the requisite caprock integrity, using CO_2 for EOR in low-performing or depleted oil reservoirs presents a number of advantages including incremental oil recovery, stabilization of the storage formation by repressurizing, and sequestration of CO_2 that can reduce the net CO_2 emission of the EOR project [13]. Carbon capture and storage (CCS) is a highly capital-intensive operation that becomes more economically viable when incorporated as part of an EOR project by producing crude oil which may not be extracted by primary and secondary recovery processes [14]. EOR can use natural or anthropogenic CO_2, but for GHG reduction anthropogenic sources from industrial processing plants (e.g., gas processing, fertilizer, and ethanol plants) and power-generating plants (e.g., coal, oil and natural gas power plants) would be used.

This paper does not consider the detailed technologies and processes involved in CO_2-EOR operations, but instead focuses on a GHG emission Life Cycle Assessment (LCA), of CO_2-EOR operations. This is a necessary analysis, because CO_2-EOR involves operations that may contribute to GHG emissions. These operations include processes from the upstream sector (capture and separation of carbon dioxide, facility construction and pipeline transportation), gate to gate (carbon dioxide dehydration, drilling of wells, oil production and processing, constructions of facilities, land usage, gas separation as well as venting and flaring) and down-stream sector (crude transport, refining and fuel combustion) [15].

Calculating and evaluating all inputs and outputs of environmental stressors and products potential impact on the environment describes LCA. By ISO 14000 environmental management standards, LCA is performed in phases: the scope/goal definition, the Life Cycle Inventory (LCI), Impact Assessment (Classification and Characterization), and interpretation/application [16]. The extent of these phases depends on the goal/scope defined. The issues addressed are the energy balance of the integrated system, substances that are emitted at a higher rate, and the part of the system linked to these emissions [17].

Thus, to undertake LCA, specific boundaries or areas of interest are to be determined for the analysis. There have been a number of studies that have looked into GHG emissions and its relation to the operations of CO_2-EOR [1,4,14–17]. The focus of this LCA is to estimate the total GHG emissions from the Farnsworth Unit (FWU) CO_2-EOR operation and further forecast emissions for another period of 18 years, a period proposed by the field operators to continue CO_2 injection. Within this projected operational period, the operators seek to incorporate a number of improved conditions to reduce emissions as will be discussed following sections. In this paper the focus is on emissions that contribute to Global Warming potential (GWP, kg CO_2eq/bbl. of crude produced), and as such the greenhouse emissions that will be considered are CO_2, CH_4, and Nitrous Oxide (N_2O). It is good to emphasize that 100% of the CO_2 used for EOR at the FWU came from anthropogenic sources (ethanol and fertilizer plants) or CO_2 from field production that is essentially 100% anthropogenic CO_2 previously injected into the field from the same anthropogenic sources. There was no CO_2 detected in the original reservoir oil [18].

1.1. Geological and Reservoir Description of FWU

The Farnsworth Unit is located in the northern part of Texas in Ochiltree County, situated on the northwestern shelf of the Anadarko Basin (Figure 1). The producing reservoir is Morrow B sandstone, an incised valley fluvial sandstone deposited during the Regional Stage of the Lower -Pennsylvanian Morrowan period [18,19]. Overlying and underlying the Morrow B formation are late Pennsylvanian through to mid-Permian shales and limestones. The primary seals are the overlying Morrow shale and Thirteen Finger limestone with secondary seals overlying these units that include thousands of feet of limestones and interbedded dolomite, and evaporites. Morrow B sandstone is about 18 m thick and occurs at a depth of about 2301 m to 2423 m, with about 55 m to 61 m of the primary seal overlying the reservoir [19]. The Morrow B at FWU is a relatively coarse-grained sandstone interpreted as a fluvial deposit within an incised valley [20]. Rocks are subarkosic in composition and typically exhibit one or more sequences of basal conglomerates overlain by coarse-to-fine grained sandstone and capped by very fine sands that grade into the mudstones of the Morrow shale and then transition abruptly into the alternating limestones and shales that characterize the Thirteen Finger limestone. Porosity and permeability for the Morrow B averages 15% and 35 mD respectively [20]. Caprock integrity testing using mercury porosimetry and geo-mechanical (Brazil tension, unconfined compression, triaxial compression, and multi-stress compression) tests indicate the integrity of the caprock and its ability to ensure safe storage of injected CO_2 [21].

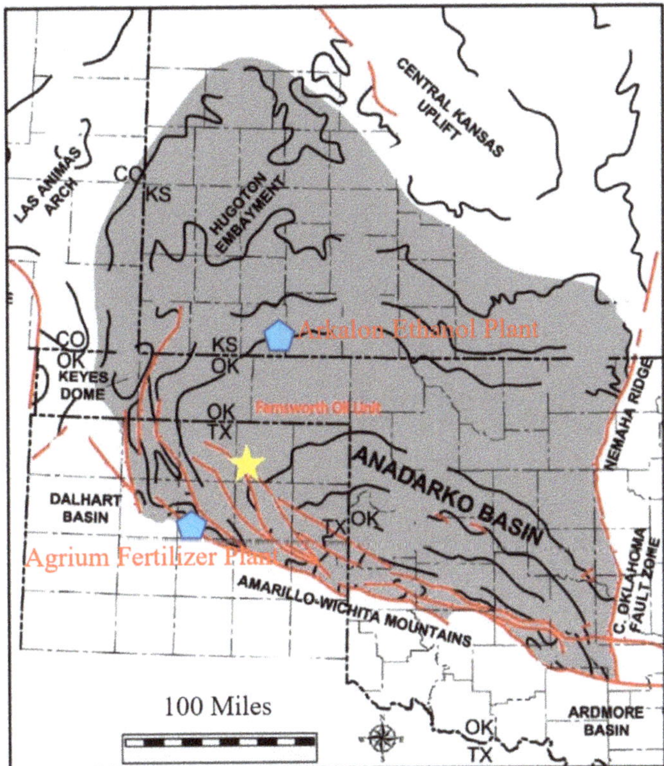

Figure 1. Location of the Farnsworth Unit (FWU) on the Northwest Shelf of the Anadarko Basin in West Texas. Red lines are approximate locations of faults that have been documented in the region. Blue pentagons indicate locations of Anthropogenic CO2 sources.

Reservoir discovery was listed as 26 October 1955 and the FWU was unitized on 6 December 1963 by the operator with the initiation of water-flooding with fresh-water from the Ogallala Formation shortly thereafter. Table 1 gives a summary of initial reservoir conditions. The field has two distinct sections that seemed to behave differently over its lifetime. Primary production was initially better in the eastern side in comparison to the western side, but the western performed much better during waterflooding [22]. This work focuses on the western side of the FWU.

Table 1. Initial reservoir conditions.

Reservoir Properties	Values
Oil initially in place	120 MM STB
Gas initially in place	41.48 BSCF
Reservoir pressure	2217.7 psia
Bubble point	2073.7 psia
Formation volume factor	1.192 RB/STB
Reservoir temperature	168 °F
Reservoir drive	Solution gas drive

1.2. Overview of CO_2-EOR Operations on the FWU

CO_2-EOR was initiated in the western portion of FWU in December 2010, and the present operator intends to continue CO_2-EOR until the economic limit of the field is reached. Initially, CO_2 came via pipeline from both the Arkalon ethanol plant in Liberal Kansas and the Agrium fertilizer plant in Borger, Texas. Currently the only source of CO_2 is from the Arkalon ethanol plant.

Figure 2 illustrates a simplified flow chart of the facilities, equipment and the CO_2-EOR processes at FWU. Three major processes are involved once CO_2 is delivered to the unit: CO_2 distribution, produced liquid handling, and produced gas handling.

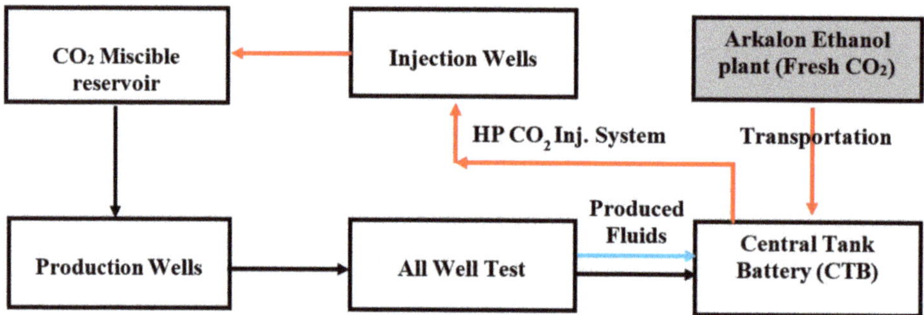

Figure 2. Simplified flow chart representing an overview of the CO_2-EOR Operations of FWU (Red lines indicate injection fluids, black: produced fluids and blue: gas from AWT).

Delivered CO_2 and recycled CO_2 are transported via pipeline from the Central Bank Battery (CTB) CO_2 distribution units to water alternating gas (WAG) injectors. Produced fluids go to several central gathering system (All Well Test (AWT)) locations where each site consists of a central gathering line, vessels for fluid separation, and individual well test separators, after which fluids are transferred to the CTB. At the CTB the separation of gas, crude, and brine continues using a series of vessels and storage tanks based on density differences and resident time to separate the fluids. Flow meters are used to record both daily purchased and produced volumes of CO_2.

The FWU currently operates thirty two producing wells and seventeen injection wells with three injection manifolds which have valves to switch between water and CO_2. Separated crude oil has 2930 ppm CO_2 (0.293%) and is sold out of tanks. The gas (89–93% CO_2) mixture is produced with less than 690 ppm i.e., 0.069% water and is reinjected using reciprocal compression and high-pressure horizontal pumps. Thus, no dehydration is required. The minimum miscible pressure (MMP) of the reservoir crude oil and CO_2 was determined to be 4200 psia [23].

Figure 3 shows CO_2 production and injection volumes from the FWU between December 2010 and 1 September 2020 as well as oil and water production and disposition As of this period the purchased CO_2 is 1.64×10^6 tonnes, with 1.51×10^6 tonnes (92% of purchased) being stored and a crude oil production of 4.76×10^6 barrels.

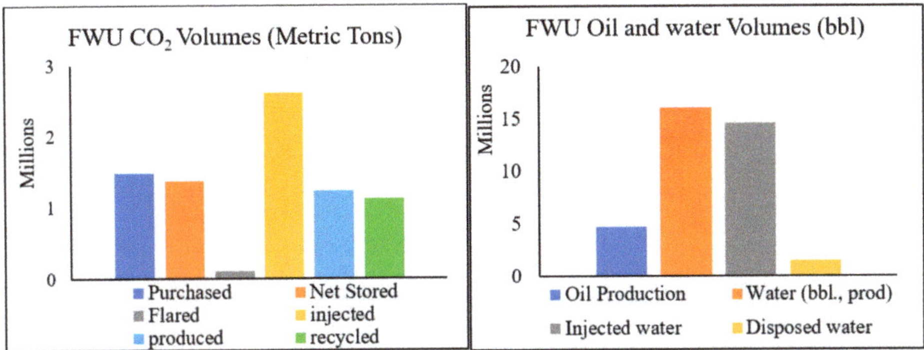

Figure 3. Production and injection data gathered from the FWU from December 2010 to August 2020.

2. Materials and Methods

The LCA approach employed in this project followed in part the framework established by ISO [16] as mentioned in previous sections, and the Plains CO_2 Reduction (PCOR) LCA approach by DOE-NETL [24]. The PCOR approach comes with a spreadsheet model allowing users to modify specific fields and CO_2-EOR operations to suit their needs. The model includes a cycle analysis from a coal fired power plant retrofitted with CO_2 capture through to a CO_2-EOR operation to the transportation of crude to refineries and finally to end users, usually for combustion. This LCA is a complete cradle-to-grave cycle, which has the upstream, gate to gate and downstream representing the end nodes of the cycle. Our work focuses on the operations within the FWU or a gate to gate analysis of the CO2-EOR operations. Where the required FWU data were unavailable they were estimated from the literature [25–27] and the National Energy Technology laboratory (NETL) databases [28,29] (Figure 3). In addition, because this model was quite generic and lacked certain variations with respect to FWU operations, a number of modifications were made based on a couple of scenarios:

Scenario 1: Perform an LCA specific to the FWU and compare with a more generalized CO_2-EOR operations from 2010 to 2020 (period for which CO_2-EOR has been in operation at the FWU).

○　For general CO_2-EOR, gas separation would be considered but not in the case for FWU because all recycled gases are reinjected.

○　The percentage of water content in gas is insignificant hence a dehydration component is not included in study.

○　There is an insignificant percentage of CO_2 and lighter hydrocarbons in separated crude oil and water hence estimates of gases or volatile oil components (VOC) vented on storage are omitted.

○　Based on the geological description and study, it is very unlikely for formation leakages to occur.

Scenario 2: Perform a FWU LCA for a forecasted CO_2-EOR for a period of 18-year (2020 to 2038) model run with bottomhole pressure and oil rate target constraints as proposed by the field operator. This would also look at two scenarios; an optimized operational condition to reduce emissions or reach a net-zero carbon operational condition and to ensure higher percent of injected CO_2 storage as encouraged by the US government 45Q incentives versus current operational practices as used in task one.

- A flexible compressor capacity—expanded to meet large volumes of recycled CO_2, thus flaring or venting of excess recycled gases would occur only during maintenance periods (due to high cost of backup compressors).
- Conversion of existing water injectors to WAG wells to add to existing WAG wells.
- All purchased and produced gases would be reinjected within the 12-year period.
- All produced water is reinjected in the WAG process, hence treatment of produced water is omitted.
- Surveillance is put in place (pipelines, wellheads, wells and other surface equipment) to meet requirements in the Texas Administrative Code (TAC) rules for the Texas Railroad Commission (TRRC) Oil and Gas Division to report and quantify leaks, and to minimize leakage of GHG from surface equipment.
- The option for gas powered/energy efficient compressors other than electric power is also considered.

Based on a maximum monitoring area (MMA) defined by the operator as the boundary of the FWU with $\frac{1}{2}$ mile buffer zone (minimum required by Subpart RR). Figure 4 shows a simulated tertiary CO_2 flood for 22 years (2010 to 2020 plus the additional 12 years projected operational period) shown in Figure 4A in addition to 5 years post-injection monitoring shown in Figure 4B. These demonstrate the stored CO_2 remains within the boundary of the FWU with little change during the 5 years post-injection. The geologic seals are expected to contain the injected CO_2 within the Morrow B formation. Abandoned wells are properly plugged and very unlikely to have any leaks. Mechanical integrity testing (MIT) as per the Underground Injection Control (UIC) program is also conducted and provides evidence of mechanical integrity, therefore no leakage is expected through injection/production wells [30,31]. Further, regular analysis of fluids from Ogallala aquifer wells around FWU as well as soil gas and atmospheric monitoring by the Southwest Regional Partnership on Carbon Sequestration (SW) shows no indications or unusual occurrences of CO_2, brine or hydrocarbons since 2013.

2.1. Life Cycle Inventory

Site-specific data inputs from the FWU within a set system boundary were used in this analysis. A flow chart indicating major processes within the scope of the analysis is presented in Figure 5. Data collection and treatment, allocation (impact of products or processes operations on the environment) and calculation, and data quality checks would be done at this stage. Rates of fluid production were major inputs to the model, as well as other process key to GHG relation to the CO_2-EOR operation, such as fluid injection (CO_2 and brine), gas and liquid separation, gas compression, crude and brine storage, gas venting and flaring, produced gas, gas combustion for heat, and gas separation. Initially gas separation was taken into consideration analyzing three common processing techniques; Ryan-Homes, refrigeration/fractionation, and membrane [15]. They were each taken into account in this gate to gate LCA. Specific FWU fluid volumes, incremental oil recovery, produced brine, gas injection, and production were used for Task 1. Other comparable data to DOE-NETL (2013) [15,28], and NETL Unit Process Library [32] were also utilized in the study.

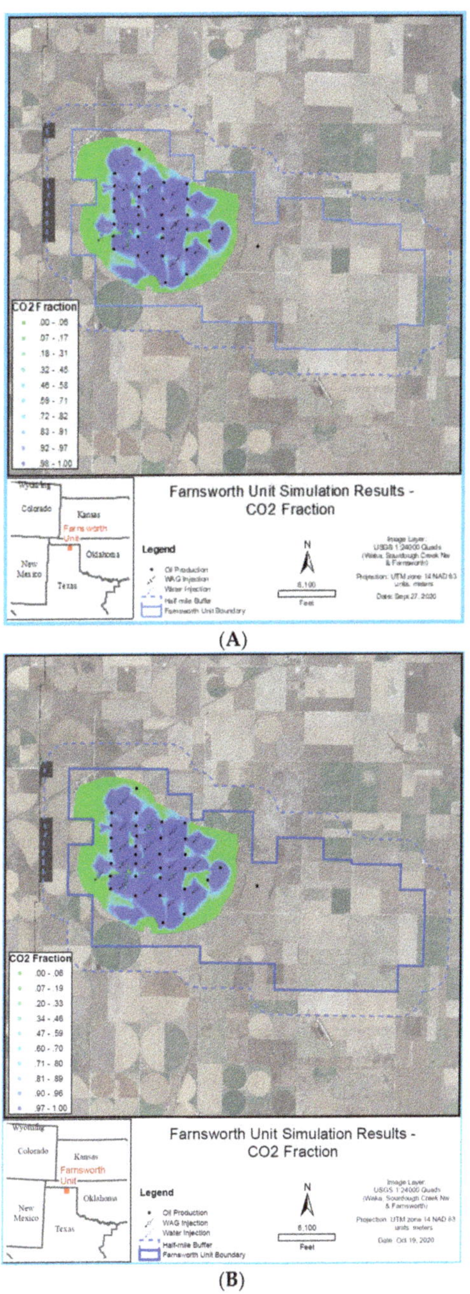

Figure 4. CO_2 Plume Extent (**A**) Before; (**B**) 5 years post-injection.

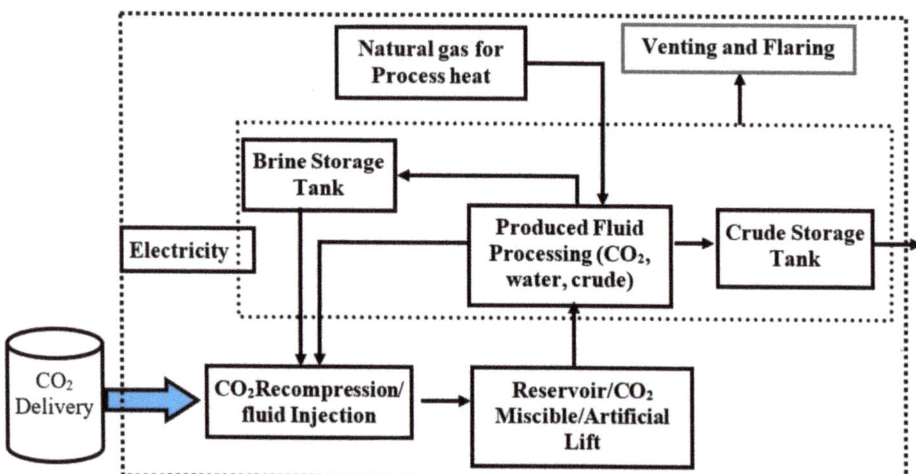

Figure 5. A simplified boundary/scope of Life Cycle Inventory for CO_2- EOR at the FWU.

Fractionation/refrigeration operates by chilling a gas stream, which separates CO_2 from hydrocarbon (HC) gases. Distillation columns are then used to separate the HC gases. This technique can be reconfigured to bypass distillation columns, thus reducing the energy consumption. The Ryan-Holmes separation technique also involves the separation of CO_2, lighter HC, and NGL by a refrigeration vessel, a de-methanizer, and HC separation columns. For the membrane process, the rate of permeation through a porous medium between two different gases is the principle utilized for its separation technique. This comes with a pre-treatment; compression of gas to about 3.45 MPa, dehydration, and chilling. Energy requirements and material usage vary widely among these techniques [15].

For Scenario 2, volumes or fluid injection and production rates used were from forecasted simulation data. The Electric Reliability Council of Texas (ERCOT) grid mix emission factor of 411 kg CO_{2e}/MWh [33] was specified as the electricity delivered to the FWU [28]. Other parameters such as brine and hydrocarbon gas production, which might not readily be available for the forecasting aspect of this analysis (Scenario 2) were estimated using various correlations [28].

Summary of Forecasting Model Description

Fluid transport dynamics were investigated through a compositional numerical reservoir simulation model. The model was used to perform a history matching simulation for primary, secondary and tertiary recovery processes for the FWU. The duration for the primary and secondary processes was 55 years, and for tertiary (CO_2 flood) the duration was between December 2010 and August 2019. Hydraulic flow units (HFU) as delineated by Rose-Coss et al. [20] were utilized to characterize heterogeneity of the reservoir. Porosity and permeability relationships were also established based on depositional and diagenetic facies described from cores and thin sections from 51 wells. The HFUs and parameters of Corey's correlations corresponding to each permeability porosity relation were used as parameters to history match the primary and secondary production data through a machine learning based methodology [34]. X, Y, and Z directional permeability multipliers and Corey coefficients of three phase relative permeabilities were parameters considered for tuning. Comparing simulation results with field data, it was observed that the history-matched case was consistent with oil production, gas injection, and production data. The Table 2 summarizes estimated volumes from the model.

Table 2. Summary of predicted volumes.

Parameter	Unit	Value
Max cumulative oil production	MM bbl	19.3
Max cumulative CO_2 storage	MM tonnes	2.98
Max % Storage of purchased CO_2	percentage	92.9%

2.2. Emissions/Emission Factor Estimations

The contributions of each unit process to GHG emissions within the boundary of the CO_2-EOR operations were estimated using a functional unit of kilograms -CO_2—equivalent (kgCO_2eq) to signify the quantity of CO_2 per barrel of crude oil produced. The 100-year GWP coefficients of 298 for N_2O and 34 for CH_4, to convert amounts of N_2O and CH_4 to equivalent CO_2, were applied [16].

Separation of gas and liquids could lead to the release of volatile organic compounds (VOC) due to the changes in temperature and pressure conditions, which are either flared or vented. Storage of crude and brine may result in VOC emissions as a result of flashing, working, and standing losses. These are recovered by vapor recovery equipment and directed to flaring and venting units. A 99% conversion efficiency of flared gases to CO_2 was used in this study (for each kg of CH_4 flared, 0.99 kg is converted to CO_2eq) [25]. The required amount of fuel and electricity for each unit process is estimated based on the amount of product to be processed. The equivalent mass of carbon dioxide emitted as a result of generating these amounts of energy is estimated. Table 3 gives a summary of key parameters, units, and their associated values used in modelling the LCA for CO_2-EOR. Aside from site-specific data, all other data sets (mostly emission factors to specific unit operations) were gathered from literature as indicated in the table representing tempo-geographical and associated technical characteristics of CO_2-EOR. Venting and flaring volumes for Scenario 2 were assumed to be the difference between the purchased and stored CO_2, arising mainly as results of compressor maintenance/break down.

Table 3. Summary of parameter/factors and Input values.

Parameters	Unit	Values	Reference
Crude oil produced	bbl crude		Operator/forecast
Crude oil density	kg crude/bbl	135	Operator/forecast
Net CO_2 utilization rate	Mscf CO_2/bbl		Operator/forecast
Purchased CO_2 requirement	kg CO_2		Operator/forecast
Fugitive loss rate of purchased CO_2	%	2.0%	[13]
CO_2 produced (recycled)	kg CO_2		Operator/forecast
CO_2 injected	kg CO_2		Operator/forecast
CO_2 stored	kg CO_2		Operator/forecast
CO_2 leakage rate from storage over 100-year time period	% CO_2 stored	0.5%	[25]
Hydrocarbon gas production rate	kg gas/kg crude		Operator/forecast
Brine production rate	kg brine/kg crude		Operator/forecast
Well footprint	Acre	0.25	[25]
Number of wells	Count	49	Operator/forecast
Emissions per m^2 of repurposed land	kg CO_2eq/m2	7.5	[32]
Water disposal well construction	kg CO_2eq/bbl	1.0	[32]
Injection well construction	kg CO_2eq/bbl	1.2	[32]
artificial lift pump electricity rate	kWh/kg crude	1.18×10^{-1}	[32]
Compressor power factor	MW/[tonne recycled CO_2/day]	2.70×10^{-3}	[25]
CO_2 pump power factor	MW/[tonne injected CO_2/day]	1.91×10^{-4}	[25]
Compressor CO_2 emissions rate (direct to atmosphere)	kg CO_2eq/MW-day	63.6	[25]
Brine injection pump electricity rate	kWh/kg brine injected	7.87×10^{-4}	[34]
VOC uncontrolled emissions rate to venting and flaring	kg VOC/kg crude	8.70×10^{-3}	[25]

Table 3. *Cont.*

Parameters	Unit	Values	Reference
Flare rate (% of vented VOC that is flared)	%	95%	[25]
Combustion efficiency	%	99%	[15]
Natural gas usage rate	kg natural gas/kg crude	3.09×10^{-3}	[32]
Natural gas delivered CO_2 emissions factor	kg CO_2/kg natural gas	1.68×10^{-1}	[32]
Natural gas delivered CH_4 emissions factor	kg CH_4/kg natural gas	1.81×10^{-2}	[32]
Natural gas delivered N_2O emissions factor	kg N_2O/kg natural gas	4.60×10^{-6}	[34]
Natural gas combustion CO_2 emissions factor	kg CO_2/kg natural gas combusted	2.75	[32]
Natural gas combustion CH_4 emissions factor	kg CH_4/kg natural gas combusted	5.26×10^{-5}	[32]
Natural gas combustion N_2O emissions factor	kg N_2O/kg natural gas combusted	5.03×10^{-5}	[32]
Produced water methane content	kg CH_4/bbl water	1.50×10^{-3}	[32]
Brine disposal pump electricity rate	kWh/kg brine disposal injected	3.30×10^{-3}	[32]
ERCOT mix, electricity delivered carbon emission factor	kg CO_2eq/MWh	4.11×10^2	[32]

3. Results and Discussion

This study focuses on the estimations of the GHG emissions for the CO_2-EOR operations at the FWU for a period of about ten years from December 2010 to September 2020 for which CO_2 injection has already occurred, and for another projected 18-year period (2020–2038) with optimized operational conditions. These estimates account for emissions that are direct functions of the mass of CO2 and oil production volume, hence the functional unit of kgCO_2eq/bbl of crude oil produced. The first scenario accounted for the emissions of the FWU and compared these to a base case, which included gas separation options. The second task also focused on a projected optimized operation and a comparison of GHG emissions to a base case of current existing conditions on the FWU. Emissions were estimated and presented on the basis of various units within the CO_2-EOR field that are key contributors to emissions within the system. The goal here was to identify specific units to optimize to aid in the reduction of GHG emissions. Cumulated purchased CO_2 amounted to 1.64×10^6 tonnes, with 1.51×10^6 tonnes (92% of purchased) stored and a corresponding crude oil production of 4.76×10^6 barrels represent estimates from the operator which were utilized in Scenario 1 (Figure 3). Scenario 2 utilized the forecast (Table 2).

3.1. Scenario 1

(a) Gas Separation

Economics dictate whether hydrocarbon gas and CO_2 are separated from the produced gas. There are at least two reasons to separate hydrocarbon gas before reinjecting the produced gas. The first reason is if the impurities in the produced CO_2 increase the MMP in the reservoir above the fracture pressure or high enough to significantly increase cost, and the second is if the value of the recovered gases is more than the cost of separation, or more likely a combination of the two. In the base case in Scenario 1 gas separation techniques considered are the fractionation/refrigeration, Ryan–Holmes, and membrane. The energy requirements and material usage vary widely among these techniques. Table 4 represents emission factors and Table 5 represents total mass emissions for refrigeration/fractionation, Ryan–Holmes and membrane gas processing techniques, respectively. For both Ryan–Holmes and membrane separation, natural gas usage accounts for the majority of emissions in their operations with electricity being the key source of emission for refrigeration/fractionation. The natural gas upstream represents the emissions from the recovery of natural gas delivered to the plant; this in many situations is a small quantity since the plant utilizes part of the gases separated in the combustion processes.

These are estimates generalized for the processes of these techniques with actual production and injection data from FWU, and the results are similar to published studies [9,15,24].

Table 4. Emission Factors of major components of Gas Separation units.

	Factors (kgCO$_2$eq/bbl)		
	Fract/Ref	**Ryan-Holmes**	**Membrane**
Electricity	1.4988	-	2.3641
Natural gas upstream	0.0004	1.3608	33.3343
Natural gas combustion	0.0014	12.0493	15.6464
Diesel Usage	-	0.1933	-
Fugitive emissions	-	-	0.1815
SUM	1.5006	13.6035	51.5262

Table 5. Mass Emissions of major components of gas Separation units.

	Emission (kgCO$_2$eq)		
	Frac/Refr	**Ryan-Holmes**	**Membrane**
Electricity	7,137,323	-	11,257,801
Natural gas upstream	1892	64,80,307	158,738,878
Natural gas combustion	6661	57,379,327	74,508,567
Diesel Usage	-	920,532	-
Fugitive emissions	-	-	864,146
SUM	7,145,875	64,780,167	245,369,392

(b) FWU CO$_2$-EOR Processes

Table 6 highlights the emission factors and mass emissions of key unit processes as defined in the boundary of the LCA. Gas compression and injection electricity accounted for 47% (7.41 kgCO$_2$eq/bbl. and 35.31 × 10^6 kgCO$_2$eq) of GHG emissions from equipment in the CO$_2$-EOR system. Thus, improving the energy efficiency of compression would significantly reduce the life-cycle GHG emissions. Unfortunately, increasing the efficiency of compressors is technically challenging [9]. Differences in the life cycle emissions of compressors, however, may differ depending on the energy source since each source has its emission factor (660 kgCO$_2$/MWh for coal powered plant, 423 kgCO$_2$/MWh for natural gas powered, etc.). The ERCOT power factor of 411 kgCO$_2$/MWh is lower due to the inclusion of renewable (wind, hydro) energy source components as part of its power generation mix and probably represents the source of power to the FWU. Artificial lifting of crude oil and associated produced water and gases comes next with estimates of about 4.44 kgCO$_2$eq/bbl and 21.12 × 10^6 kgCO$_{2eq}$. These estimates were made based on the volumes of fluids produced. A factor of 0.118 kWh/kg crude lifted [25] was used. From this the amount of energy required to lift the volumes of fluids produced and the associated quantity of potential emissions was estimated. These values are not exact representations of emissions from artificial lift from the FWU, since there are different kinds of lifting mechanisms employed in the field (sucker rods and submersible pumps) and in many of the wells artificial lift was not initially required. Thus, this value is expected to be an overestimate, but the idea presented here is to show how much CO$_2$ would have been emitted if all of the produced fluids were acquired through artificial lift. Artificial lift is quite energy intensive and is used throughout the production period once a well is put on artificial lift. Construction and land use also account for significant GHG emissions directly through energy use, construction of facilities, well drilling, and other processes and indirectly in land use effects on existing vegetation, repurposing land and so on. Using a factor for an emission per square meter of repurposed land of 7.46 kgCO$_2$/m^2 [29], emissions were estimated at 2.98 kgCO$_2$eq/bbl corresponding to a mass of 14.18 × 10^6 kgCO$_2$eq. These processes and their GHG emissions could be classified as indirect as associated GHG emissions are due to the processes or energy usage. Flaring and venting which in this study is classified as a direct emission accounted for an estimated 60% of GHG emissions through the 10-year period of CO_2-EOR. This was highest in the early stages of CO_2 injection, but reduced as compressors for reinjection came online. On a longer-term scale, most venting and flaring occurs during times when compressors are offline for maintenance or repairs. The cumulative quantity of equivalent CO_2 flared within the period from December 2010 to August 2020 was 117.52×10^6 $KgCO_2eq$ with an emission factor of 24.68 $kgCO_2eq$/bbl. This is the highest source of emissions amongst the CO_2-EOR processes. Though there could be challenges with respect to improving on the efficiency of equipment to reduce GHG emissions, analysts have suggested that reduction of fugitive GHG emissions of vented and flared CO_2 and methane would be more easily achieved. This effect is reflected in Scenario 2 of this study. Based on the geological description and the mechanical integrity tests performed on the field's reservoir cap rocks, emissions due to leakages from the geologic storage formation were estimated to be zero (0%) of the stored CO_2 [21].

Table 6. Mass Emissions and Factors for EOR units (Task 1).

Unit Processes	Emission 10^6 kg CO_2eq	Factors kg CO_2eq/bbl
Construction/Land use	14.18	2.98
Artificial lift	21.12	4.44
CO_2 compression, and injection (Electricity)	35.31	7.41
Brine injection (Electricity)	3.43	0.72
Brine disposal (Electricity)	1.11	0.23
Flared/Vented	117.52	24.68

(c) Comparative Analysis of Gate to Gate Results

Table 7 (mass emission) and Table 8 (emissions factors), sums up the overall total emissions and emission factors for the CO_2-EOR processes both with (Base cases) and without (FWU) gas separation techniques. Total net storage factors and net CO_2 storage, were estimated as; total emission factors minus initial storage factor, and total emissions minus initial CO_2 storage. For the base cases, the total emissions from CO_2-EOR operations in order of increasing total emissions and emission factors are Fractionation/ Refrigeration, Ryan-Holmes, and Membrane. In comparison, FWU without gas separation recorded the lowest. The net storage and net storage factors resulted in negative net values for all scenarios. This is an indication of a pay-off to global warming reduction and/or a positive outcome to environmental intervention, that is, much more CO_2 is stored in the formations than is emitted to the atmosphere. The major difference is the use of gas separation. Refrigeration/fractionation has a greater advantage with regards to emissions due to its low energy consumption. This is because the fractionation process could be configured to bypass distillation columns, thus reducing energy that would have been consumed by such columns. However, when it comes to efficiency in separation of gases, the Ryan-Holmes and membrane processes both are highly effective in recovering natural gas liquids but come with a higher energy penalty, as can be seen from the results.

Table 7. Total Emission and Net Storage for cases considered.

	Total Emissions 10^6 kgCO_2eq	Net Storage 10^6 kgCO_2eq	Purchased Stored %
Refrigeration/fractionation	217.53	−1161.86	78.80
Ryan-Holmes	275.16	−1104.22	74.95
Membrane	454.93	−924.46	62.94
FWU (No Gas Separation)	210.38	−1169.00	79.28

Table 8. Summary of emission factors for Task 1 and Task 2.

Process	Emission Factor kgCO$_2$eq/bbl	Net Storage Factor kgCO$_2$eq/bbl
Refrigeration/fractionation	42	−247.70
Ryan-Holmes	54	−235.60
Membrane	92	−197.85
FWU (No Gas Separation)	40	−249.20
FWU (Forecast-opt)	10	−130.01
FWU (Forecast-Base)	28	−112.00

3.2. Scenario 2

(a) Forecasted FWU CO$_2$-EOR with Optimized Conditions

Scenario 2 considers an 18-year forecast of the CO$_2$-EOR operations, a time range chosen because that concludes with the probable expiration date of a 45Q tax credit. The scenario assumes all purchased CO$_2$ (2.91 × 10^6 Metric tons) will be injected and stored within the 18-year period. Assumptions also consider that adequate compressor capacity precludes venting or flaring and (during compressor optimum performance except failure or maintenance), injection of all produced gas and water. Our emission estimates were made factoring all these conditions. A base case of the current condition as applied in Scenario 1 for FWU was also applied to this forecasted CO$_2$ purchased volume. For the base case, venting and flaring accounted for the majority of emissions with 17.9 kgCO$_2$eq/bbl (345.28 × 10^6 kgCO$_2$eq), and compression and artificial lift energy being the next major source of GHG contributors in both the optimized and base case. For the same volume of crude oil produced, the energy requirement for artificial lift is likely to be the same in both cases for the same period of time. Differences between the two cases arise as a result of fugitive emissions from equipment as well as from venting and flaring. Total estimated emission factors for both the forecasted base and optimized scenarios are shown in Table 8 with detailed estimates on individual operations in Figure 6. A net negative storage factor of −130 kgCO$_2$eq/bbl and −112 kgCO$_2$eq/bbl corresponding to 86% of the purchased CO$_2$ for the optimized case and 74.3% for base case was found. As this LCA excludes all fugitive emissions, this is an indication that energy consumption by process equipment is a key contributor to GHG emissions. Flaring and venting, being a direct emission of GHG, is a major component in the CO$_2$-EOR process which increased emissions in all cases. A reduction in this one key source could significantly reduce total emissions. This could be achieved through a reduction in time needed for repair and maintenance or through other operational methods.

Table 8 summarizes the emission factors of the various case studies, using volume estimates in Figure 3 for Scenario 1 and Table 2 for Scenario 2 (projected). The factors in both cases are not directly comparable due to different volumes of CO$_2$ and crude oil used in their estimates. These estimates (emission factors) can be compared to other gate to gate CO$_2$-EOR GHG LCA even with the inclusion of a gas processing facility. Figure 7 gives a number of CO$_2$-EOR operations and their estimated emission factors that range 60 kgCO$_{2ee}$ to 175 kgCO$_2$eq compared to Scenario 1 for FWU of about 40 kgCO$_2$eq.

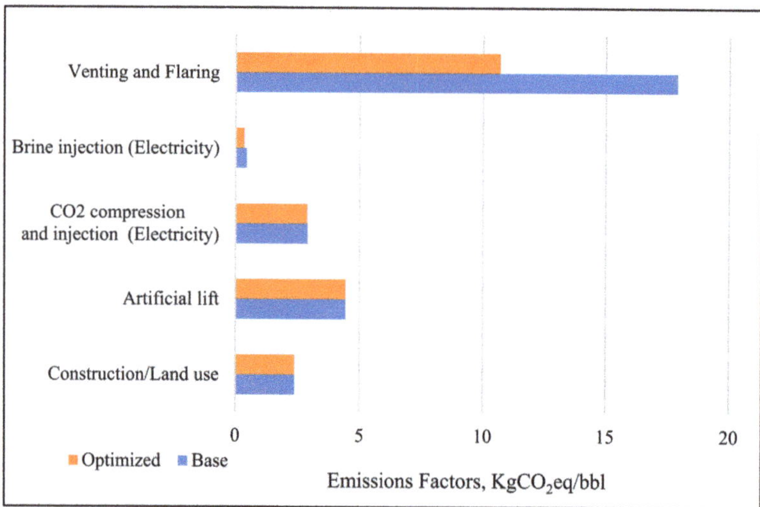

Figure 6. FWU emission factors for Base and forecasted optimized cases.

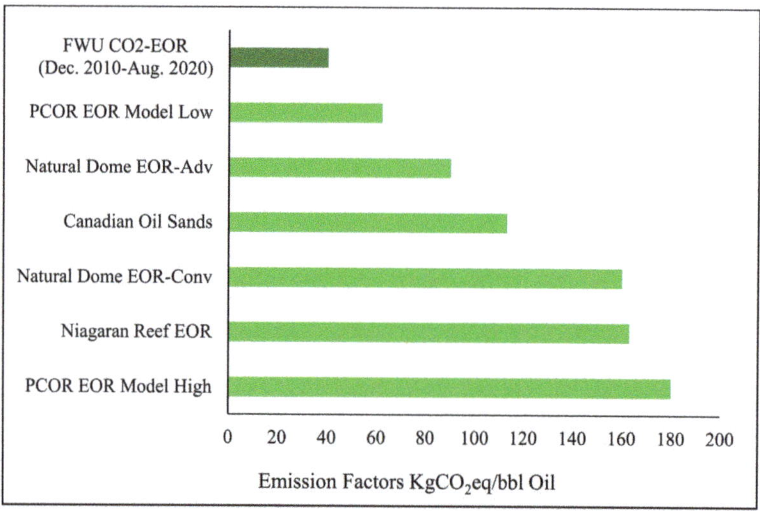

Figure 7. Reproduced gate-to-gate emission factors of other CO_2-EOR fields [9], in comparison to FWU CO_2-EOR.

4. Summary and Conclusions

Emissions of GHGs such as CO_2, CH_4, and N_2O are considered as major pollutions and have become a source of concern in efforts to slow the pace of global warming. This study presents a GHG LCA for the FWU for a period during which CO_2-EOR has been in operation (about 10 years) and a projected future operation of 18 years. For these cases, GHG emissions were estimated for a number of CO_2-EOR processes, fugitive emissions, and from flaring/venting. Data from the operator as well as simulated and forecasted fluid volumes were utilized in estimating emissions. Data gathered by the operator through monitoring and metering recorded 1.64×10^6 tonnes (1.49×10^9 kgCO$_2$eq) of CO_2 purchased, with 92% being stored during the 10 years CO_2-EOR has been in operation. A gate-to-gate LCA of the GHG emissions estimated a net negative 2.25×10^6 tonnes (1.19×10^6 kgCO$_2$eq) of CO$_2$ representing storage of 79% of purchased CO$_2$. For this period, venting and flaring accounted for the highest source of emissions with compressor energy consumption being the next-highest. Improving conditions through an optimized process (Scenario 2) for a forecasted period of operations lowering or eliminating fugitive emissions and including flaring/venting only during maintenance of compressors yielded 86% (130 kgCO$_2$eq/bbl) of projected purchased volumes of 3.21×10^6 tonnes of CO$_2$ compared to a forecasted base case estimated at 74% of the purchased CO$_2$.

A very significant consideration for the implementation of the forecasted optimized scenario would be economics and technology that dictate the amount of CO$_2$ to be purchased as well as to be stored. These factors can also dictate how a CO$_2$-EOR operation might be designed. Higher crude oil prices might create a favorable condition for the use of CO$_2$ due to the ratio between the cost of oil price and cost of CO$_2$. A drop in CO$_2$ cost might encourage CO$_2$ use for EOR, and a change in tax policy might change either side of the equation. The financial incentive of the 45Q tax credit from the current U.S. tax policy for both capture and storage of CO$_2$ is currently providing another motivation for CO$_2$-EOR operators to retain more CO$_2$.

This GHG life-cycle assessment is an indication that the integration of CO$_2$-EOR and carbon storage, such as seen at FWU, is one approach to minimize net GHG emissions to the atmosphere. This study presents specific emission estimates for the FWU and could give useful information to field operators with regards to GHG emissions of their operations. The basic processes and methodologies could easily be followed in other fields.

Key unit processes have been demonstrated to be major contributors to emissions that operators need to take notice of and seek to improve for reduced GHG emissions. Energy consumption for process equipment is a significant input and a major cause of GHG emissions. Improving on the energy efficiency of equipment and the use of alternative clean energy sources are sure ways of reducing emissions from this source. These changes may be technologically challenging, or in some cases beyond the control of field operators. In our study, flaring and venting accounted for the largest source of emissions in all scenarios examined. Reducing emissions from this source is believed to be easier compared to the challenging issues of improving the energy efficiency of equipment. Proper monitoring, smart and quick sealing of fugitive sources and the avoidance of flaring/venting as much as possible could reduce emissions. Expanding compressor capacities and/or backup compressors are ways of ensuring sufficient gas recycle capacity and hence emission reductions, though these could be capital intensive and project economics would play a major role in this decision. The FWU has a simpler operational boundary, hence its low gate-to-gate emission factors as compared to other studies mentioned earlier. Added complexity of operational processes as seen in some fields could lead to an increase in sources of emissions; however, reducing the direct emission of GHGs via venting and flaring should provide beneficial in almost all cases.

Author Contributions: Conceptualization, W.A. and R.G.; methodology, A.M.; software, A.M.; validation, A.M., R.G. and W.A.; formal analysis, A.M.; resources, W.A. and R.G.; writing—original draft preparation, A.M.; writing—review and editing, R.G. and W.A.; visualization, A.M.; supervision, W.A. and R.G.; project administration, W.A.; funding acquisition, W.A. All authors have read and agreed to the published version of the manuscript.

Funding: Funding for this project is provided by the U.S. Department of Energy's (DOE) National Energy Technology Laboratory (NETL) through the Southwest Regional Partnership on Carbon Sequestration (SWP) under Award No. DE-FC26-05NT42591.

Conflicts of Interest: The authors declare no conflict of interest.

References

1. Khoo, H.H.; Tan, R.B.H. Environmental Impact Evaluation of Conventional Fossil Fuel Production (Oil and Natural Gas) and Enhanced Resource Recovery with Potential CO2 Sequestration. *Energy Fuels* **2006**, *20*, 1914–1924. [CrossRef]
2. Farajzadeh, R.; Eftekhari, A.A.; Dafnomilis, G.; Lake, L.W.; Bruining, J. On the sustainability of CO2 storage through CO_2—Enhanced oil recovery. *Appl. Energy* **2020**, *261*, 114467.
3. Herzog, H.; Golomb, D. Carbon Capture and Storage from Fossil Fuel Use. *Encycl. Energy* **2004**, *1*, 277–287. [CrossRef]
4. United States Environmental Protection Agency. *Inventory of U.S. Greenhouse Gas Emissions and Sinks*; United States Environmental Protection Agency: Washington, DC, USA, 1990.
5. Aycaguer, A.-C.; Lev-On, M.; Winer, A.M. Reducing Carbon Dioxide Emissions with Enhanced Oil Recovery Projects: A Life Cycle Assessment Approach. *Energy Fuels* **2001**, *15*, 303–308. [CrossRef]
6. Gale, J.; Metz, B.; Davidson, O.; de Coninck, H.; Loos, M.; Meyer, L. *Carbon Dioxide Capture and Storage. Intergovernmental Panel on Climate Change*; Cambridge University: Cambridge, UK, 2005.
7. National Petroleum Council. *Enhanced Oil Recovery, An Analysis of the Potential for Enhanced Oil Recovery from Known Fields in the United State*; Library of Congress Catalog No. 76-62538; National Petroleum Council: Washington, DC, USA, 1976.
8. Kuuskraa, V.; Wallace, M. CO2-EOR set for growth as new CO2 supplies emerge. *Oil Gas J.* **2014**, *112*, 92.
9. Sminchak, J.R.; Mawalkar, S.; Gupta, N. Large CO2 Storage Volumes Result in Net Negative Emissions for Greenhouse Gas Life Cycle Analysis Based on Records from 22 Years of CO2-Enhanced Oil Recovery Operations. *Energy Fuels* **2020**, *34*, 3566–3577. [CrossRef]
10. Imanovs, E.; Krevor, S.; Mojaddam Zadeh, A. CO2-EOR and Storage Potentials in Depleted Reservoirs in the Norwegian Continental Shelf NCS. In Proceedings of the 82nd EAGE Annual Conference & Exhibition, Amsterdam, The Netherlands, 18–21 October 2021.
11. Chamwudhiprecha, N.; Blunt, M.J. CO2 Storage Potential in the North Sea. In Proceedings of the International Petroleum Technology Conference, Bangkok, Thailand, 15–17 November 2011.
12. Wickramathilaka, S.; Morrow, N.; Howard, J. Effect of Salinity on Oil Recovery by Spontaneous Imbibition. In Proceedings of the 24th International Symposium of Core Analysts, Halifax, NS, Canada, 4–7 October 2010.
13. Melzer, L.S. *Carbon Dioxide Enhanced Oil Recovery (CO2 EOR): Factors Involved in Adding Carbon Capture, Utilization and Storage (CCUS) to Enhanced Oil Recovery*; Center for Climate and Energy Solutions: Arlington, VA, USA, 2012.
14. Gaspar, A.T.F.S.; Suslick, S.B.; Ferreira, D.F.; Lima, G.A.C. Economic Evaluation of Oil-Production Project with EOR: CO2 Sequestration in Depleted Oil Field. In *SPE Latin American and Caribbean Petroleum Engineering Conference*; Society of Petroleum Engineers: Rio de Janeiro, Brazil, 2005; p. 13.
15. Skone, T.J.; James, R.E., III; Cooney, G.; Jamieson, M.; Littlefield, J.; Marriott, J. *Gate-to-Gate Life Cycle Inventory and Model of CO_2-Enhanced Oil Recovery*; National Energy Technology Laboratory: Pittsburgh, PA, USA, 2013.
16. International Standards Organization. *ISO 14041: Environmental Management—Life Cycle Assessment—Goal and Scope Definition and Inventory Analysis*; International Standards Organization: Geneva, Switzerland, 1998.
17. Baumann, H.; Tillman, A.-M. *The Hitch Hicker's Guide to LCA. An Orientation in Lifecycle Assessment Methodology and Application*; Studentlitteratur: Lund, Sweden, 2004.
18. CORE LABORATORIES, INC. *Petroleum Reservoir Engineering, Core Summary and Calculated Recoverable Oil*; Lawton State No. 2 well; CORE LABORATORIES, INC.: Dallas, TX, USA, 1956; Available online: https://geoinfo.nmt.edu/libraries/subsurface/search/scans/API/30-025-01613/CA_30-025-01613.pdf (accessed on 23 April 2021).
19. Ampomah, W.; Balch, R.S.; Grigg, R.B.; Will, R.; Dai, Z.; White, M.D. Farnsworth Field CO2-EOR Project: Performance Case History. In *SPE Improved Oil Recovery Conference*; Society of Petroleum Engineers: Tulsa, OK, USA, 2016; p. 18.
20. Ross-Coss, D.; Ampomah, W.; Cather, M.; Balch, R.S.; Mozley, P.; Rasmussen, L. An Improved Approach for Sandstone Reservoir Characterization. In *SPE Western Regional Meeting*; Society of Petroleum Engineers: Anchorage, AK, USA, 2016; p. 15.
21. Sun, Q.; Ampomah, W.; Kutsienyo, E.; Appold, M.; Adu-Gyamfi, B.; Dai, Z.; Soltanian, M. Assessment of CO2 trapping mechanisms in partially depleted oil-bearing sands. *Fuel* **2020**, *278*, 118356. [CrossRef]
22. Ampomah, W.; Balch, R.S.; Grigg, R.B.; McPherson, B.; Will, R.A.; Lee, S.-Y.; Dai, Z.; Pan, F. Co-optimization of CO2-EOR and storage processes in mature oil reservoirs. *Greenh. Gases Sci. Technol.* **2017**, *7*, 128–142. [CrossRef]
23. Gunda, D.; Ampomah, W.; Grigg, R.; Balch, R. Reservoir Fluid Characterization for Miscible Enhanced Oil Recovery. Presented at the Carbon Management Technology Conference, Sugar Land, TX, USA, 17–19 November 2015; p. 17.
24. Azzolina, N.A.; Hamling, J.A.; Peck, W.D.; Gorecki, C.D.; Nakles, D.V.; Melzer, L.S. A Life Cycle Analysis of Incremental Oil Produced via CO2 EOR. *Energy Procedia* **2017**, *114*, 6588–6596. [CrossRef]
25. Cooney, G.; Littlefield, J.; Marriott, J.; Skone, T.J. Evaluating the Climate Benefits of CO2-Enhanced Oil Recovery Using Life Cycle Analysis. *Environ. Sci. Technol.* **2015**, *49*, 7491–7500. [CrossRef] [PubMed]
26. Koornneef, J.; Van Keulen, T.; Faaij, A.; Turkenburg, W. Life cycle assessment of a pulverized coal power plant with post-combustion capture, transport and storage of CO2. *Int. J. Greenh. Gas Control.* **2008**, *2*, 448–467. [CrossRef]
27. Suebsiri, J.; Wilson, M.; Tontiwachwuthikul, P. Life-Cycle Analysis of CO2EOR on EOR and Geological Storage through Economic Optimization and Sensitivity Analysis Using the Weyburn Unit as a Case Study. *Ind. Eng. Chem. Res.* **2006**, *45*, 2483–2488. [CrossRef]

28. Azzolina, N.A.; Peck, W.D.; Hamling, J.A.; Gorecki, C.D.; Ayash, S.C.; Doll, T.E.; Nakles, D.V.; Melzer, L.S. How green is my oil? A detailed look at greenhouse gas accounting for CO2-enhanced oil recovery (CO2-EOR) sites. *Int. J. Greenh. Gas Control.* **2016**, *51*, 369–379. [CrossRef]
29. DOE/NETL. Advanced Carbon Dioxide Capture R&D Program: Technology Update, May 2011 edition (p. 118). National Energy Technology Laboratory. Available online: http://www.netl.doe.gov/technologies/coalpower/ewr/pubs/CO2Handbook (accessed on 23 April 2021).
30. Xiao, T.; McPherson, B.; Pan, F.; Esser, R.; Jia, W.; Bordelon, A.; Bacon, D. Potential chemical impacts of CO2 leakage on underground source of drinking water assessed by quantitative risk analysis. *Int. J. Greenh. Gas Control.* **2016**, *50*, 305–316. [CrossRef]
31. Xiao, T.; McPherson, B.; Bordelon, A.; Viswanathan, H.; Dai, Z.; Tian, H.; Esser, R.; Jia, W.; Carey, W. Quantification of CO2-cement-rock interactions at the well-caprock-reservoir interface and implications for geological CO2 storage. *Int. J. Greenh. Gas Control.* **2017**, *63*, 126–140. [CrossRef]
32. DOE/NETL. *Complete Unit Process Library Listing.* Available online: https://netl.doe.gov/node/2573 (accessed on 23 April 2021).
33. Electricity Reliability Council of Texas. *ERCOT Analysis of the Impacts of the Clean Power Plan*; Electricity Reliability Council of Texas: Austin, TX, USA, 2014.
34. You, J. Multi-Objective Optimization of Carbon Dioxide Enhanced Oil Recovery Projects Using Machine Learning Algorithms. Ph.D. Thesis, New Mexico Tech, Socorro, NM, USA, 2020. Unpublished doctoral dissertation.

Article

Risk Assessment and Management Workflow—An Example of the Southwest Regional Partnership

Si-Yong Lee [1,*], Ken Hnottavange-Telleen [2], Wei Jia [3,4], Ting Xiao [3,4], Hari Viswanathan [5], Shaoping Chu [5], Zhenxue Dai [6], Feng Pan [7], Brian McPherson [3,4] and Robert Balch [8]

[1] Schlumberger, Denver, CO 80202, USA
[2] GHG Underground, Arrowsic, ME 04530, USA; kenht@ghgu.net
[3] Energy & Geoscience Institute, The University of Utah, Salt Lake City, UT 84108, USA; wei.jia@utah.edu (W.J.); ting.xiao@utah.edu (T.X.); b.j.mcpherson@utah.edu (B.M.)
[4] Department of Civil & Environmental Engineering, The University of Utah, Salt Lake City, UT 84112, USA
[5] Los Alamos National Laboratory, Earth and Environmental Sciences Division, Los Alamos, NM 87545, USA; viswana@lanl.gov (H.V.); spchu@lanl.gov (S.C.)
[6] College of Construction Engineering, Jilin University, Changchun 130026, China; dzx@jlu.edu.cn
[7] Utah Division of Water Resources, Salt Lake City, UT 84116, USA; Fpan@utah.gov
[8] New Mexico Tech, Socorro, NM 87801, USA; robert.balch@nmt.edu
* Correspondence: slee55@slb.com

Abstract: This paper summarizes the risk assessment and management workflow developed and applied to the Southwest Regional Partnership on Carbon Sequestration (SWP) Phase III Demonstration Project. The risk assessment and management workflow consists of six primary tasks, including management planning, identification, qualitative analysis, quantitative analysis, response planning, and monitoring. Within the workflow, the SWP assembled and iteratively updated a risk registry that identifies risks for all major activities of the project. Risk elements were ranked with respect to the potential impact to the project and the likelihood of occurrence. Both qualitative and quantitative risk analyses were performed. To graphically depict the interactions among risk elements and help building risk scenarios, process influence diagrams were used to represent the interactions. The SWP employed quantitative methods of risk analysis including Response Surface Method (RSM), Polynomial Chaos Expansion (PCE), and the National Risk Assessment Partnership (NRAP) toolset. The SWP also developed risk response planning and performed risk control and monitoring to prevent the risks from affecting the project and ensure the effectiveness of risk management. As part of risk control and monitoring, existing and new risks have been tracked and the response plan was subsequently evaluated. Findings and lessons learned from the SWP's risk assessment and management efforts will provide valuable information for other commercial geological CO_2 storage projects.

Keywords: risk assessment; workflow; Farnsworth; workshop; process influence diagram; response surface model; polynomial chaos expansion; NRAP

Citation: Lee, S.-Y.; Hnottavange-Telleen, K.; Jia, W.; Xiao, T.; Viswanathan, H.; Chu, S.; Dai, Z.; Pan, F.; McPherson, B.; Balch, R. Risk Assessment and Management Workflow—An Example of the Southwest Regional Partnership. *Energies* **2021**, *14*, 1908. https://doi.org/10.3390/en14071908

Academic Editor: Alexei Milkov

Received: 9 February 2021
Accepted: 22 March 2021
Published: 30 March 2021

Publisher's Note: MDPI stays neutral with regard to jurisdictional claims in published maps and institutional affiliations.

1. Introduction

Storage of CO_2 in geologic formations is one of the most applicable options for mitigating anthropogenic CO_2 emissions contributing to climate change [1–3]. Particularly, CO_2-enhanced oil recovery (CO_2-EOR, an oil production method in which oil recovery is enhanced by CO_2 injection) and storage have gained specific interest for its potential economic benefits of increasing hydrocarbon recovery and reducing risks of overpressure [4–8]. However, due to the nature of the deep subsurface environment and its uncertainties, geologic CO_2 storage (GCS) projects require appropriate assessment and management of risks for safe operation.

The Southwest Regional Partnership (SWP) Phase III project is an industrial research collaboration focused on an active CO_2-EOR and storage field, the Farnsworth Unit (FWU), located within the Anadarko Basin of northernmost Texas along the Oklahoma border [9]. The FWU is both a CO_2 sequestration demonstration and a research project managed by the SWP, one of seven regional carbon sequestration partnerships (RCSP's) instigated and supported by the U.S. Department of Energy (DOE) and its National Energy Technology Laboratory (NETL) [10].

The FWU project seeks to predict and to monitor the effects of injecting CO_2 into an actively producing oilfield. The outcomes of the FWU project research are of value to oilfield operators, who seek CO_2-EOR and storage, as well as other stakeholders interested in long-term GCS.

For the SWP's FWU project, risk information was routinely applied toward both strategic and tactical design and adjustment of project activities so that risk of failure (in any dimension) can be minimized. Risks associated with GCS include CO_2 leakage (to shallow subsurface and/or atmosphere), geomechanical risks (e.g., fault reactivation and induced seismicity), storage/injectivity loss, production decrease in CO_2-EOR, etc. This paper summarizes the risk assessment workflow developed and applied to the FWU project. The risk assessment workflow is an iterative process where potential risks are identified and monitored. The likelihood and severity of the risks are quantified, and a response plan is subsequently established and updated.

2. Risk Assessment Workflow

To accomplish the effective risk assessment and risk management, SWP formed a risk assessment working group (RAWG) from the initial stage of the project and established and applied the continuous and iterative risk assessment and management workflow shown in Figure 1. Other major working groups within the SWP include monitoring, verification, and accounting (MVA), characterization, and simulation. Each working group comprises of active project personnel of each discipline.

Figure 1. Risk assessment and management workflow diagram depicting six primary tasks and how they relate.

The SWP's risk assessment approach consists of six primary tasks: risk management planning, risk identification, qualitative risk analysis, quantitative risk analysis, risk response planning, and risk control and monitoring. Tools essential to risk communication are applied within several of these tasks.

2.1. Risk Management Planning

The risk management plan comprises roles and responsibilities of personnel, budget assignment, and timing and frequency of risk assessment tasks. In this task the RAWG defined the best methodologies, tools, and data sources for both technical and programmatic risks. This task also sets how RAWG coordinated with other working groups. For example, RAWG coordinated with the simulation group to ensure models were developed on identified risk pathways (e.g., wellbore leakage, shallow groundwater impacts, caprock integrity, induced seismicity, and performance goals). Coordination among working groups in regard to risk management planning is a key aspect of internal risk communication.

2.2. Risk Identification

Risk identification (also known as risk source assessment) is the process of determining project risks and their characteristics. The RAWG first developed an initial (draft) risk registry that included programmatic/operational risks as well as technical/sequestration risks by using a Features, Events, and Processes (FEPs)-based approach. A FEP is a broad entity: an individual FEP could create negative impact on the project values through various chains of events (scenarios) and could function in combination with various other FEPs. Based on the initial FEPs registry, the SWP performed its risk survey from 2014 to 2017 annually. The purpose of this risk survey was to identify and evaluate current risks to project objectives. Participants of the risk survey evaluated FEPs, each of which was associated with one or more example scenarios.

On 13 January 2014, a live web-based risk workshop was held to evaluate and rank the identified risks and to newly identify additional risks associated with the SWP Phase III project. Then, the evaluation of 24 additional FEPs collected from the online workshop was completed by a follow-up spreadsheet. Initially, a total of 405 FEPs were identified for the SWP Phase III project. Twenty-three (23) project professionals participated in the workshop and all the personally attributed data were collected and evaluated. Project professionals provided self-assessments of areas of expertise. After FEPs were screened for redundancy and relevance to the SWP, 103 FEPs were ranked by risk with the expertise-weighted evaluation method.

In August 2015, the second risk survey was conducted. The 2015 survey was executed entirely via email and telephone communication, with spreadsheets as the primary information tool, whereas the 2014 assessment included a "live" online workshop with discussion and real-time visibility of charted data. The risk elements evaluated in the 2015 risk assessment exercise were the 50 highest risks as determined in 2014, of the 103 total risks evaluated at that time. Respondents in 2015 (to a large degree, the same individuals who participated in 2014) were invited to add "new" risks or re-nominate other "old" risks for evaluation, but in practice no risks beyond the 2014 top-50 set were identified.

In September 2016, substantially the same group of project professionals again reevaluated the project risks. The process focused on 69 risk entities comprising the 50 FEPs evaluated in 2015 plus additional FEPs nominated or agreed by risk-workshop participants. The additional FEPs grew out of information gathering that was used to identify the important new or potentially higher-risk elements.

On 14 December 2017, the fourth risk workshop was held as a one-day face-to-face session during the SWP annual meeting at New Mexico Tech, Socorro, New Mexico. The 23 workshop participants included nearly all persons actively involved in the project. The risk workshop followed a day of plenary meetings in which the staff shared information on project technical and managerial topics. This shared information formed a solid basis for exposing and exploring sources of risk to project objectives.

2.2.1. Risk Calculation

Risk was calculated as the product of Severity (S) and Likelihood (L), each of which factors was judged on a categorical 5-point scale (Table 1) by the risk workshop participants. S is defined as "severity of potential negative impact to defined project values," and L is defined as "likelihood that the specified severity level will occur during the project lifespan." In the 2014 and 2016 risk surveys, participants provided two values of S and one of L for each FEP. The S values are upper-bound Severity (S_{ub}) and best-guess Severity

(S_{bg}), and the L value is Likelihood of the best-guess Severity (L of S_{bg}, or simply L). The S_{ub} or worst-case severity value is of specific interest. In addition, the process of its elicitation provides a measure of self-calibration for participants. For each risk element, the S_{ub} value is elicited first, followed by the S_{bg} value. In this way, each participant's best-guess severity value is appropriately related to the just-provided "worst case" aspect of the same risk element.

Table 1. Severity and likelihood scales.

Ranking Factor		Severity of Negative Impact (S)
5	Catastrophic	Multiple fatalities. Damages exceeding $100M. Project shut down.
4	Serious	One fatality. Damages $10M–$100M. Project lost time greater than 1 year.
3	Significant	Injury causing permanent disability, Damages exceeding $1M to $10M. Project lost time greater than 1 month. Permit suspension. Area evacuation.
2	Moderate	Injury causing temporary disability. Damages $100k to $1M. Project lost time greater than 1 week. Regulatory notice.
1	Light	Minor injury or illness. Damages less than $100k. Project lost time less than 1 week.
Ranking Factor		Likelihood of Impact or Failure Occurring (L)
5	Very Likely	Happens every year, or more often. Nearly sure to happen during Farnsworth Project.
4	Likely	Happens every few years. Probably will happen during the project.
3	Unlikely	Happens every few decades. Might not happen during the project even if nothing is done.
2	Very Unlikely	Would happen less often than every century, in projects similar to this one.
1	Incredible or Impossible	If these projects like this went on forever, would not happen in a thousand years.

Whereas, in the 2015 and 2017 workshop, a single S_{bg} and the L of S_{bg} values were collected for each FEP, rather than separate values for S_{ub} versus S_{bg} in contrast to 2014 and 2016 evaluations. This saved workshop time and recognized the fact that worst-case severity data collected during previous workshops was likely to be sufficiently representative. A "worst-case severity" ranking was computed from the 2017 workshop data based on the provided severity values plus one standard deviation.

Using the data gathered from all the participants including experts and non-experts, various risk values and rankings can be constructed by using different weightings of expert and non-expert views. As noted earlier, participants self-rated their areas of technical expertise. In 2014, the selected expertise-weighted ranking used gradational weighting based on gradations of expertise. In 2015, the selected ranking used "triple-weighted" values from experts and single-weighted values from non-experts, based on a binary ("yes/no") designation of subject-matter expertise. Given that experts' presumably greater accuracy in estimating risk cannot be confirmed until the project is well advanced, there are no clear criteria for "optimal" weighting. However, as for 2015 and similar to 2014, a "triple-weighted experts" risk calculation and ranking based on binary expertise have been constructed from the 2016 data.

2.2.2. Risk Rankings

Table 2 compares FEP rankings in 2017 ("all participants" ranking) to rankings in the prior years. Of the 69 FEPs evaluated in 2017, 57 were from the previous evaluations and 13 were newly added. Among the 69 FEPs evaluated in 2016, 46 were evaluated in 2015, and all but two (new in 2016) were evaluated in 2014. Table 2 shows risk rankings for those four successive evaluations. In 2015, only the highest-ranking 50 FEPs from 2014 were evaluated. Only a single ranking method from each year is shown for comparison.

Table 2. Features, Events, and Processes (FEPs) rankings of four annual evaluations since 2014. Thirteen FEPs evaluated in 2016, but not in 2017, are listed at bottom. FEPs whose titles were worded slightly differently in prior years are indicated by an asterisk. FEPs not evaluated during a particular year are shown ranked as "N/A." For most FEPs the related specific risks are evident to CCS practitioners; certain FEPs followed by a letter in parentheses are further explained below the table.

2017 FEP *	Rank 2017	Rank 2016	Rank 2015	Rank 2014
Price of oil (or other related commodities)	1	1	1	6
DOE financial support (a)	2	N/A	N/A	N/A
On-road driving	3	16	28	35
Change of field owner and/or operator	4	N/A	N/A	N/A
CO_2 supply adequacy	5	4	7	2
EOR oil recovery (b)	6	7	2	37
Operating and maintenance costs	7	5	3	7
Legislation affecting CO_2 injection or CO_2-EOR *	8	2	18	29
Simulation and modeling—parameters * (c)	9	23	36	1
Well component failure (tubing, seals, wellhead, etc.)	10	N/A	N/A	N/A
Reservoir heterogeneity (d)	11	29	15	16
Accidents and unplanned events	12	3	8	18
Workovers: Damage to instrumentation	13	N/A	N/A	N/A
Defective hardware *	14	24	16	48
Simulation of geomechanics	15	25	6	9
Seismic method effectiveness *	16	39	25	12
Severe weather	17	10	N/A	84
Undetected features	18	51	N/A	52
Project execution strategy (DOE project, not EOR or production) *	19	31	9	21
Over pressuring	20	41	10	10
Workovers: Costs, hazards, interruptions	21	N/A	N/A	N/A
Release of compressed gases or liquids	22	19	13	3
Economic competition (for hardware, staff, etc.)	23	N/A	N/A	N/A
Relative-permeability and capillary-pressure curves	24	45	N/A	N/A
EOR early CO_2 breakthrough	25	18	5	25
Ignition of flammable gases or liquids	26	12	N/A	72
Blowouts	27	9	26	8
Simulation of coupled processes	28	22	19	5
Simulation and modeling—Numerical model resolution	29	N/A	N/A	N/A
Simulation of fluid dynamics	30	6	17	15
Drilling *	31	42	44	14
Fault valving and reactivation	32	30	N/A	57
Operator training	33	37	N/A	62
Injection and production well pattern and spacing *	34	8	4	45
Fractures and faults (CO_2 leakage via new or existing)	35	34	N/A	90
Contracting	36	50	27	42
Seismicity (natural earthquakes)	37	35	N/A	101
Caprock lateral extent and continuity	38	13	N/A	80
Contractors: Unavailability of major contractor	39	N/A	N/A	N/A
Well lining and completion	40	32	31	38
Caprock fracture pressure	41	43	N/A	82
Conflicts in monitoring methods (instrument space, power, interference, etc.) *	42	14	N/A	51
Co-migration of other gases	43	61	35	27
Moving equipment	44	40	30	39
CO_2 leakage through existing wells	45	N/A	N/A	N/A
Simulation and modeling—software	46	N/A	N/A	N/A
Geomechanical characterization	47	36	32	4
Operator error in pipeline operation	48	20	49	31
Caprock heterogeneity	49	65	40	11

Table 2. *Cont.*

2017 FEP *	Rank 2017	Rank 2016	Rank 2015	Rank 2014
Fluid chemistry	50	62	47	20
Leaks and spills (related to oil and chemicals other than CO_2)	51	28	23	44
Hydrogen sulfide, H_2S * (e)	52	27	21	13
Mineral deposition (porosity or perm loss)	53	N/A	N/A	N/A
Permit compliance	54	58	N/A	92
Seismic survey execution *	55	59	33	50
Fluid samples and sampling *	56	53	38	24
Integration of technical learnings	57	N/A	N/A	N/A
Relations among major project proponents and parties *	58	63	37	43
Safety coordination and integration	59	54	N/A	77
EOR viscosity relations (f)	60	46	22	47
Management team	61	64	34	41
Competing project objectives	62	17	42	46
Working in confined areas or spaces	63	60	N/A	68
Permit modifications	64	38	24	40
CO_2 release to the atmosphere	65	47	N/A	74
Propagation of project learnings beyond SWP	66	N/A	N/A	N/A
Exploitation of caprock or reservoir by non-project activities *	67	48	46	28
Health and safety inspections	68	44	43	23
Mineral reactivity *	69	68	N/A	N/A
EOR oil reservoir heterogeneity		11	11	19
Reservoir exploitation		15	N/A	97
Seal failure		21	14	22
Injection well components		26	41	33
Competition		33	12	49
On-site facilities for EOR		49	N/A	86
Pipeline supervisory control and data system		52	N/A	98
Modeling and simulation—software		55	20	17
Storage Complex definition		56	N/A	83
Workover		57	29	30
Mineral dissolution		66	N/A	94
Desiccation of clay		67	N/A	99
CO_2 exsolution from formation fluids		69	N/A	102

(* Different Wording in Prior Year/s); (a) Risk of loss of financial support from the principal funder, U.S. Dept. of Energy. (b) Risks related to changes in (mainly declines in) EOR-related oil recovery. (c) Risks related to incorrect or misleading modeling results, due to inaccurate, imprecise, or overly precise inputs of parameter values and parameter ranges. (d) Risks related to reduced ability to predict and/or control plume migration, due to reservoir heterogeneity and/or incorrect model specification of heterogeneity. (e) Risks related mainly to H_2S toxicity; also to potential metal embrittlement. (f) Risks related to inability to predict oil movement changes caused by viscosity reduction from CO_2 injection.

The dominance of red to yellow colors in the upper part of the list shows that many high-risk issues (especially programmatic issues) have remained high-risk throughout the project. In some cases, progressively cooler colors suggest technical learnings that reduced the perceived risk; examples include geomechanical characterization, H_2S, and Health and Safety Inspections.

Rankings from 2017 show that programmatic issues remain top concerns: oil price (as in previous years) and DOE financial support (new FEP in 2017) were rated as the highest project risks. After the steep drop of oil price from the second half of 2014, price of oil has been placed at the first in the risk ranking. Other operational risks related to CO_2-EOR were also relatively ranked high due to the concerns about oil price. Except for those newly included in 2017, most FEPs evaluated in multiple years have maintained roughly consistent rank positions.

Among the scientific issues, relatively high risk is ascribed to the parameters used for simulation and modeling and to reservoir heterogeneity. This may imply that for this project, given geologic heterogeneity, modeling indicates that the extensive available field data have not constrained model outputs to the degree anticipated. Some FEPs re-included in 2016 were evaluated as relatively high risk in 2016. The re-included FEPs had not been evaluated in 2015 based on their low risk in 2014, but were included for 2016 because of high 2014 severity values. Their emergence as high risk in 2016 (e.g., ranks #10, 12–15) is surprising and may suggest recent changes in project status (including funding and economics) or information that had not been articulated during the information-gathering process that preceded the workshop.

Based on the four risk workshops and analyses, we have learned that the results of FEPs-based risk evaluation can be applied toward risk management as follows:

- Using risk rankings and other statistics, select a set of FEPs for which action (treatment) will be undertaken. It is often useful to select roughly 15–35% of the evaluated FEPs.
- Parse the selected FEPs by FEP group and assign a risk-treatment coordinator for each group.
- For each FEP, clarify the specific scenarios (chains of events) by which impact would occur. Develop risk treatments to lower the likelihood of their occurrence and/or the severity of impact in case of occurrence. Assign and track treatment execution, and periodically evaluate the effectiveness in treating the target risks (residual risk level).
- Evaluate the effectiveness of treatments in also treating/controlling the *non-target* (lower) risks; confirm that all identified risks are adequately controlled.
- Re-evaluate risk whenever there is a substantial change to project information or objectives.

2.3. Qualitative Risk Analysis

The RAWG applied both qualitative and quantitative analysis processes. The qualitative risk analysis is mainly for prioritizing the identified risks according to their potential effects on the project objectives and for identifying interactions between FEPs, which includes:

- Continue (update) relative ranking or prioritization of project risks,
- Risk categorization by root cause and potential impacts,
- Define interactions between FEPs,
- Identify risks that require responses in the near-term,
- Identify risks that require more analysis or investigation, and
- Develop watchlists for lower risks for monitoring.

The significant risk matrix components necessary for the risk assessment were first identified to define the FEPs interaction. For example, for the risks to oil recovery, reservoir temperature, reservoir pressure, oil composition, and oil viscosity were defined as four independent (or uncertain) parameters and associated dependent variables (or risk factors) include oil production, water cut, and methane production. Similarly, we identified independent and dependents variables for three additional categories such as CO_2 storage, geomechanics, and CO_2 leakage which are risk areas in the quantitative risk assessment. Table 3 summarizes the uncertain parameters and dependent variables identified in the four different risk areas; CO_2 storage, oil recovery, geomechanics, and CO_2 leakage.

Based on the FEPs identification, relevant classification and ranking, and risk matrix development, the SWP utilized process influence diagrams (PIDs). PIDs graphically depict the interactions between FEPs and help in building risk scenarios, which form an instrument for effective risk communication. The initial site-specific PID for the SWP project was based on the PIDs developed for a typical CCUS-EOR project through a previous DOE-funded project (DE-FE0001112). Appropriate scenarios identified throughout the PIDs can subsequently be used for the quantitative risk analysis.

Figure 2 shows the PIDs for CO_2 storage and CO_2-EOR risks/FEPs. Similarly, the PID associated with the geomechanics and CO_2 leakage is illustrated in Figure 3. The risks (dependent variables) that can be quantified in terms of probability density function (PDF) or cumulative distribution function (CDF) are highlighted in the PIDs. In the PIDs, an arrow represents the influence path showing cause and effect. The circle indicates the interaction between FEPs. No loop or chain start with Events in the PIDs. In other words, there should be an appropriate cause for an effect. PID only considers direct impact.

Table 3. Risk matrix components including independent and dependent variables.

Risk Area	Independent Variables (Uncertain Parameters)	Dependent Variables
CO$_2$ Storage	Reservoir properties (porosity and permeability, Kv/Kh ratio) Relative permeability (e.g., irreducible water saturation) WAG (including well pattern and spacing, and injection rate) CO$_2$ miscibility (e.g., minimum miscibility pressure) Boundary conditions Model uncertainty (e.g., simulation of coupled processes, simulation of fluid dynamics) CO$_2$ impurity Initial water, oil, and gas saturations Mineralogical composition	Amount of CO$_2$ stored (or CO$_2$ recovered or Net CO$_2$ stored) Early CO$_2$ Breakthrough time CO$_2$ Retention (or residence) CO$_2$ Injectivity reduction (Net CO$_2$ injection amount) CO$_2$ storage capacity loss -Amount of CO$_2$ mineral trapping -Mineral alteration and porosity evolution AOR (CO$_2$ plume size and pressure buildup)
Oil Recovery	Reservoir temperature Reservoir pressure Oil composition, gravity Oil viscosity	Oil production Water cut (or net water injection) Gas (CH$_4$) production
Geomechanics	Fault density and distributions Stress and mechanical properties Coefficient of friction (fault properties) Caprock geomechanical properties Mechanical processes and conditions	Pressure Buildup Induced seismicity (seismic magnitude) Injection-induced faults reactivation
CO$_2$ Leakage	Caprock geometry (discontinuity) and heterogeneity Caprock capillary entry pressure Initial water chemistry CO$_2$ migration (point and non-point source) Distributions of leaky wells	pH change in the overlying aquifer CO$_2$ concentration or total carbon concentration Heavy metal concentration TDS change in the overlying aquifer Trace metal mobilization CO$_2$ migration through caprock Caprock sealing quality evolution (porosity change)

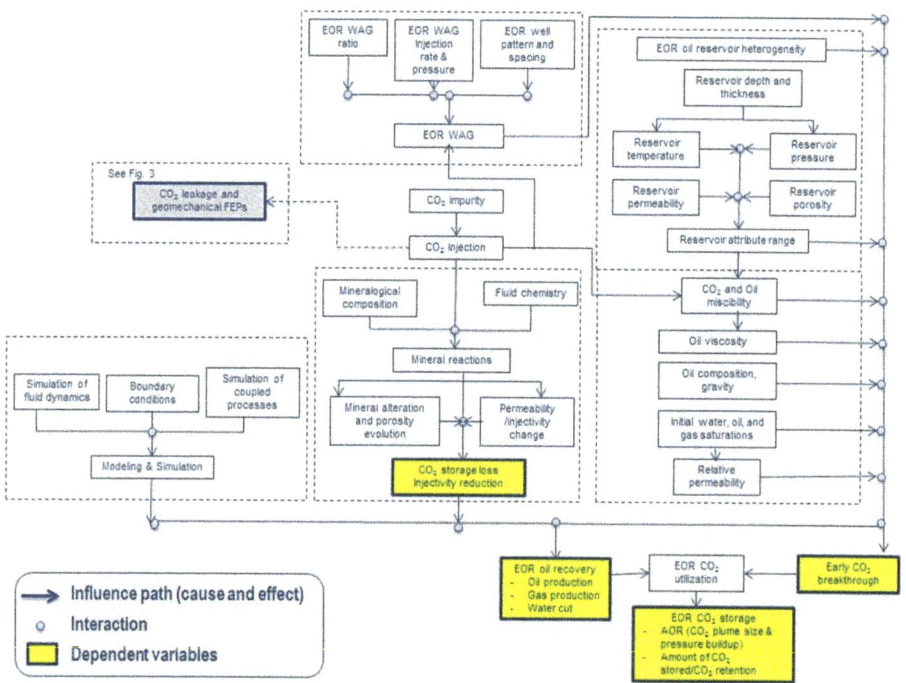

Figure 2. Process influence diagram (PID) for CO$_2$ storage and CO$_2$-EOR risks/FEP in Southwest Regional Partnership (SWP) project. An arrow shows the influence path and each connection point is represented by a filled circle.

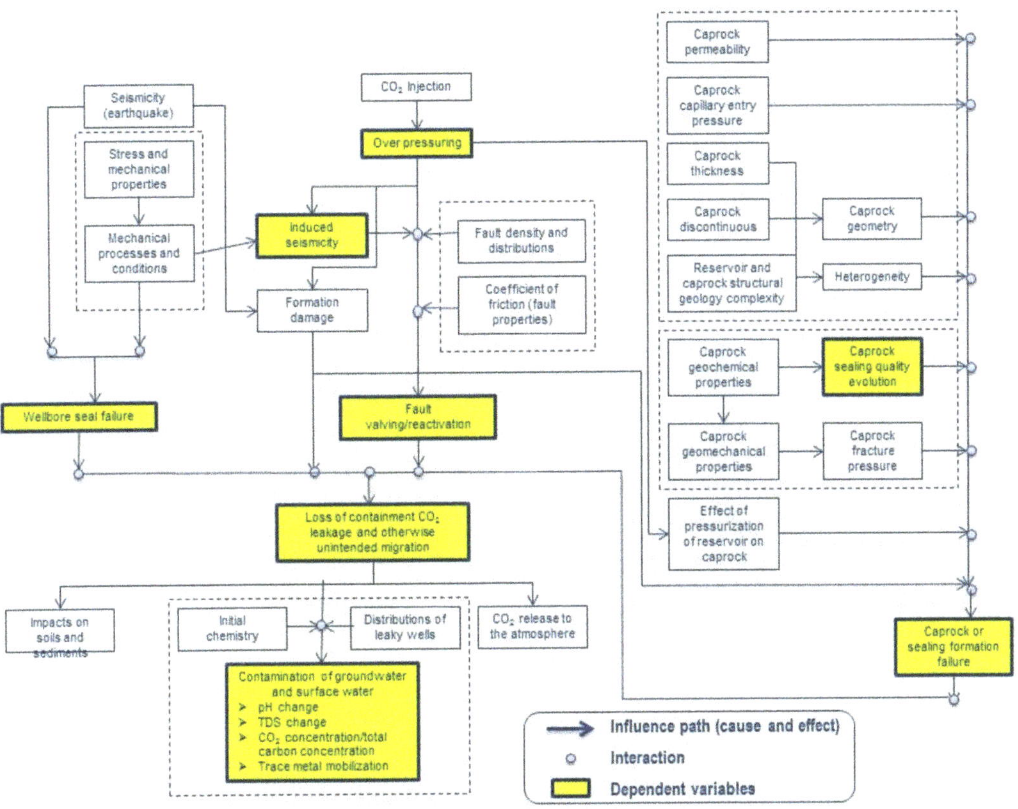

Figure 3. Process influence diagram for geomechanical and CO$_2$ leakage risks/FEP in SWP Farnsworth project. An arrow shows the influence path and each connection point is represented by a filled circle.

For example, if we inject CO$_2$, combined with mineralogical composition and fluid chemistry, it would affect the mineral reactions which might lead to mineral alteration and subsequently porosity/permeability change. Therefore, as a result there could be a storage loss or injectivity reduction. This risk scenario identified with PID was used in the quantitative risk analysis.

2.4. Quantitative Risk Analysis

Quantitative risk analysis for the FWU project has been conducted to numerically quantify the effect of risk scenarios on the project objectives. In general, integral aspects of risk assessment involve:

- Formalism and comprehensiveness of identified risks, which add confidence to the risk assessment;
- Development of common framework and approaches, which allow inter-comparison of probabilities for different elements or sites;
- Explicit treatment of uncertainties, which arise from factors such as incomplete parameters and process constraints, heterogeneities in natural systems, incomplete knowledge of the natural systems at the site, etc.

The SWP Farnsworth project employed formal quantitative methods of risk analysis specified in the following sub-tasks (based on evaluation of uncertainty):

- Quantify critical elements or variables that may affect the risk in question;

- Define the scenarios or conceptual model for each risk;
- Conduct probabilistic risk assessment with an appropriate tool for each potential risk;
- Synthesize the overall risk assessment using National Risk Assessment Partnership (NRAP) tools (formerly CO2-PENS and other newly developed tools), to evaluate CO_2 and brine fate and associated impact.

Response Surface Method (RSM), Polynomial Chaos Expansion (PCE), and NRAP toolset are the main computational tools used for quantitative risk assessment in the Farnsworth project.

2.4.1. Response Surface Method

RSM (also known as proxy model) was used for developing PDFs or CDFs for each critical risk factor of interest. The RSM with appropriate experimental design has been applied to reservoir engineering applications such as performance prediction, sensitivity analyses, upscaling, history matching, and optimization studies [11–13]. Comprehensive simulations with a conventional Monte Carlo approach may be computationally expensive given the uncertainties in model parameters, whereas RSM with a statistically linear model uses only a small number of runs at specified sampling points. We applied RSM combined with Monte Carlo sampling to efficiently provide probabilistic assessment.

Figure 4 summarizes the RSM workflow, which first determines independent variables/factors to construct the design of experiment (DoE), followed by the numerical experiments. Then, the response surface (regression equation or proxy model) is delineated with a stepwise regression technique applied to eliminate insignificant factors from the regression equation. Then, several goodness-of-fit measurements examine the performance of the regression model. Lastly, Monte Carlo samplings of mutually independent input parameters were used in the obtained response surface models in order to generate the CDFs of output responses from the given input distributions without running numerical simulations

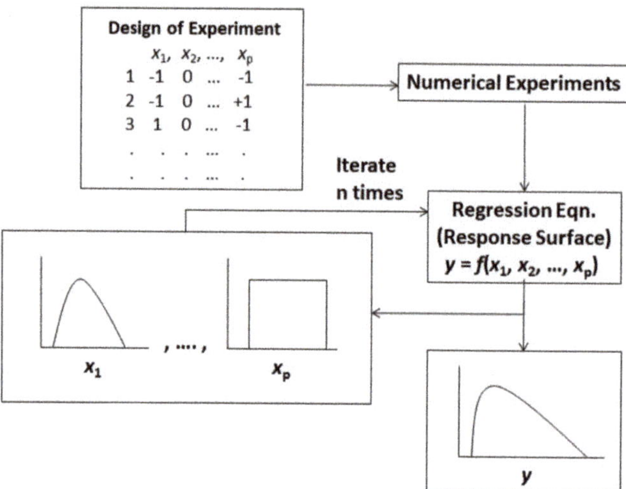

Figure 4. Workflow for the response surface methodology combined with Monte Carlo simulation. For example, reservoir permeability, anisotropy ratio of permeability (kv/kh), water-alternating-gas (WAG) time ratio are x_1, x_2, x_3, and x_4. In addition, net CO_2 storage and oil production are y_1 and y_2 in Pan et al. [14].

The RSM consists of mathematical and statistical techniques to develop a functional relationship between a response or dependent variable (y) of interest and associated independent variables or factors $(x_1, x_2, \ldots, x_k)$. The response surface method is typically a polynomial approximation to the responses (y) obtained with a linear regression given the input/design variables (x_i) in a chosen DoE. For example, the Box-Behnken design (BBD), which is a particular subset of the factorial combinations from the 3^k factorial designs, consists of three levels $(-1, 0, 1)$ corresponding to (lower, middle, upper endpoint) for each factor [15]. Each factor is placed at one of the three equally spaced values. The BBD has been widely used because of its economical design (smaller number of runs) compared to the full factorial designs. Full factorial design with 2-level $(-1, 1)$ or 3-level $(-1, 0, 1)$ is fully crossed design requiring 2^k and 3^k runs, respectively. In addition, the BBD contains not only the interaction terms of factors but also the higher-order quadratic effects. We utilized the stepwise regression technique to eliminate the insignificant factors from the regression equation. The RSM was efficiently implemented within a Monte Carlo framework to assess uncertainty.

Pan et al. [14] applied RSM with Monte Carlo sampling to quantify the uncertainties in the key reservoir parameters of Farnsworth project. Forecasted net CO_2 storage and oil production were predicted by the CDFs given the uncertainty in key reservoir parameters such as reservoir permeability, anisotropy ratio of permeability (kv/kh), water-alternating-gas (WAG) time ratio, and initial oil saturation. Similarly, with three independent parameters (CO_2 saturation, reservoir pressure, wellbore fracture proportion), Xiao et al. [16] quantified potential risks of CO_2 and brine leakage into the overlying USDW (the Ogallala aquifer) with RSM and identified water chemistry parameters as early detection indicators based on up-to-date site monitoring data.

Dai et al. [17] developed a multiscale statistical framework for CO_2 accounting and risk analysis at the FWU. A set of geostatistical-based Monte Carlo simulations were conducted for risk and global sensitivity assessment of CO_2-hydrocarbon-water flow in the Morrow B formation. The major risk metrics include CO_2/water injection/production rates, cumulative net CO_2 storage, cumulative oil/gas productions, and CO_2 breakthrough time. A response-surface-based economic framework was also derived to calculate the CO_2-EOR profitability for the FWU with an oil price of \$38/bbl, suggesting that approximately 31% of the 1000 realizations can be profitable.

Our RSM-based work [14,16,17] demonstrated useful tools which can be used to numerically and probabilistically quantify the effect of risk scenarios on the project objectives. In terms of computational time, the RSM was efficient compared to the conventional Monte Carlo simulation. However, simulation processes inherently contain uncertainty. Thus, it would be critical to correctly define the ranges and distribution of uncertain parameters to significantly reduce the uncertainty. In addition, the RSM applied for quantifying risks are tested and verified with numerical outputs rather than actual data, thus they unavoidably contain epistemic uncertainty.

2.4.2. Polynomial Chaos Expansion

In addition to the RSM described in the previous section, non-intrusive polynomial chaos expansion (PCE) was also used in the Farnsworth project, as it only requires a small number of sampling and does not modify the governing equations. RSM uses a polynomial regression to model the response, $y = (y_1, y_2, \cdots, y_N)^T$ where input parameters, $x = (x_1, x_2, \cdots, x_M)^T$. Whereas, if the input parameters vector x is uncertain [18], an element y_i in the vector y can be represented by a PCE as follows:

$$y_i = M(x) = \alpha_0 B_0 + \sum_{j=1}^{M} \alpha_j B_1(x_j) + \sum_{j=1}^{M}\sum_{k=1}^{j} \alpha_{jk} B_2(x_j, x_k) + \sum_{j=1}^{M}\sum_{k=1}^{j}\sum_{h=1}^{k} \alpha_{jkh} B_3(x_j, x_k, x_h) \quad (1)$$

where the α's are coefficients and the B's are multivariate polynomial basis functions. Hermite polynomials basis functions are generally used for normally distributed parameters [19]. Once y_i is simulated from geo-cellular models, the coefficients can be solved with linear inversion, e.g., $\alpha = (B^T B)^{-1} B^T y_i$. Then reduced order models (ROMs) can be developed by substituting the calculated coefficients into Equation (1). With the flexibility regarding the basis functions and not requiring the DoE, PCE is capable of constructing ROMs for a variety of properties of interest, such as the pressure and CO_2 saturation in each cell of the reservoir model. For the reviews and details of PCE techniques, see the previous studies such as [20–22].

Jia et al. [23] evaluated primary CO_2 trapping mechanisms of Morrow B sandstones at the FWU. In particular, the heterogeneity of petrophysical properties (porosity and permeability) was considered as the source of parameter uncertainty. Their impacts were analyzed using PCE-derived ROMs combined with Monte Carlo simulations. Model outputs of interest in CDFs include amounts of CO_2 trapped by three different trapping mechanisms: hydrodynamic trapping, oil dissolution trapping, and aqueous dissolution trapping. The wide ranges of CDFs (as shown in Figure 5 in Jia et al. [23]) demonstrate significant variations in CO_2 storage at FWU due to the uncertain reservoir porosity and permeability. However, the results of the uncertainty analysis suggest that the hydrodynamic trapping is the dominant trapping mechanism at FWU.

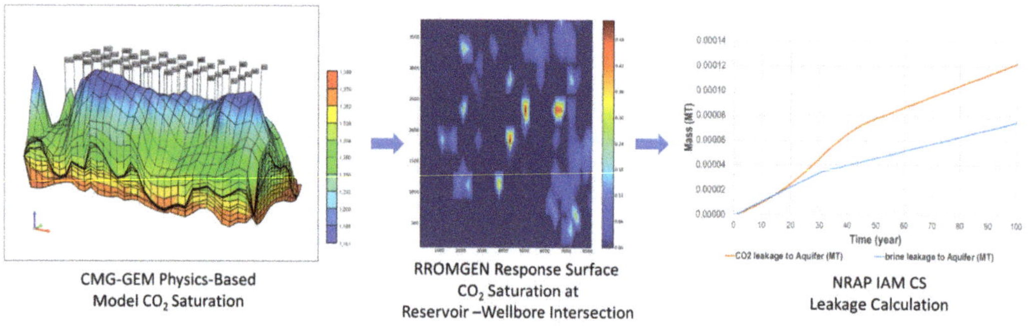

Figure 5. Workflow from physics-based simulators to leakage calculations using National Risk Assessment Partnership (NRAP) tools.

2.4.3. National Risk Assessment Partnership Toolset

As part of the quantitative risk assessment for the SWP, the NRAP toolset is being applied to evaluate CO_2 and brine leakage risks at the FWU. The NRAP toolset is a computational toolkit that includes ten science-based computational tools that predict environmental risk performance of geologic CO_2 storage sites [24,25]. In order to conduct a quantitative risk assessment of wellbore leakage at the FWU multiple realizations must be run to span to parameter space of key parameters such as wellbore permeability and the time evolution of CO_2 and water saturation in the reservoir. The physics-based simulators of the reservoir and detailed simulations of the wellbore are computationally intensive and cannot be practically coupled and run 1000s of times to bound the uncertainty in the system. The NRAP toolset is being used to establish a comprehensive workflow between physics-based simulators of the reservoir and physics-based simulators of the wellbore using the concept of ROMs [26].

NRAP tool RROM-Gen is used for generating response surfaces for the time evolution of CO_2 and water saturation at the depth where wellbores intersect the reservoir. The response surfaces are generated from output data of a reservoir simulator that simulate CO_2 injection into the reservoir. Response surfaces of CO_2 saturation and pressure are generated as a function of time using RROM-Gen [27]. These response surfaces were then used to estimate the leakage risk from wellbores using NRAP-IAM-CS (formerly CO2-PENS, [28]) from the NRAP tool kit. The workflow is shown in Figure 5 for a generic CO_2 sequestration site.

Chu et al. [29] summarized the leakage risk assessment work for FWU. NRAP tools were used for risk assessment and uncertainty quantification of wellbore leakage that covers the full parameter range of ECLIPSE reservoir simulations at FWU representing various reservoir conditions with different assignments for relative permeability and capillary pressure to control the CO_2 injection amount and fluid mobility for leakage potential. Various wellbore integrity distribution scenarios were also examined including several different wellbore permeability probability distribution models such as Alberta, Gulf of Mexico, and FutureGen low/high flow rate scenarios. The results show that the highest possible leakage scenario (open well) could result in ~0.1% cumulative CO_2 leakage after 25-year CO_2 injection and 50-year post-injection.

2.5. Risk Response Planning

To avoid delays, underperformance, or failure of a project, risk factors need to be identified and promptly addressed. Therefore, the development of a risk response plan is crucial for the success of a project. Risk responses, or treatments, focus on reducing the occurrence probability (prevention) and/or the consequences (mitigation) of a risk to the project objectives and values. Risk transfer is a type of mitigation in which occurrence probability is unchanged, but negative consequences would be borne by a third party (such as an insurer), by contract.

Risk identification and analysis tasks discussed in the previous sections provided the basis for developing a comprehensive risk response (risk treatment) program which consequently led to the update in project management plan. We updated risk prevention and mitigation treatments according to the top 40 FEPs identified from the annual risk survey. For example, Table 4 shows the 2017 top 10 risks for the Farnsworth site and corresponding risk prevention and mitigation treatments. Treatments are developed by and shared among all project staff, forming another key element of internal risk communication.

Table 4. Prevention and mitigation treatments for the top 10 FEPs of 2017 risk survey.

FEPs *	Rank 2017	Rank 2016	Rank 2015	Rank 2014	Risk Prevention	Risk Mitigation
Price of oil (or other related commodities)	1	1	1	6	Analyze trends in commodity prices Plan for worst case scenarios Hedge oil prices Establish a CO_2-EOR economical model to predict the possible profit and lost and to evaluate the economical risk	Control costs Shut in wells until prices recover Shift to backup CO_2 supplier
DOE financial support	2	N/A	N/A	N/A	Use conservative estimates Maintain good communications with DOE program manager	Prioritize expenses and exclude low priority costs Renegotiate the scope of work Try to obtain additional funding
On-road driving	3	16	28	35	Maintain vehicles in safe operating condition Implement safety training and standard procedures for operators Conduct regular safety audits during construction and operation Implement emergency response plan and risk management plan	Maintain safety training and standard procedures Document response to safety incidents Maintain emergency response planning Maintain risk management plan Maintain liability insurance
Change of field owner and/ or operator	4	N/A	N/A	N/A	Communicate with the operator continuously Download/backup the data regularly	Establish the relationship with the new owner/operator immediately Maintain consistent workflow with the new operator

Table 4. *Cont.*

FEPs *	Rank 2017	Rank 2016	Rank 2015	Rank 2014	Risk Prevention	Risk Mitigation
CO_2 supply adequacy	5	4	7	2	Maintain multiple sources of CO_2	Monitor CO_2 quality Cut back CO_2 injection on some patterns or compensate with increased water injection
EOR oil recovery	6	7	2	37	Fully characterize the reservoir for EOR attributes. Select EOR reservoirs that fall within the acceptable range of EOR attributes Model EOR operation and try to optimize oil recovery through reservoir engineering. Operate above the minimum miscibility pressure	Monitor EOR actual versus projected performance. Identify the cause of any variation. Adjust CO_2 EOR strategy to improve oil recovery if necessary Optimize WAG, injected water curtains, selective perforation, use of polymer gels or sealants, and CO_2 recycling to control CO_2 migration and utilization and increase oil recovery Optimize CO_2-EOR processes to maximize both net CO_2 storage and oil production simultaneously.
Operating and maintenance costs	7	5	3	7	Use historical O&M data and experienced cost estimators to prepare budgets Prepare budget for the unexpected/emergency costs or insurance	Implement a total productive maintenance (TPM) program
Legislation affecting CO_2 injection or CO_2-EOR *	8	2	18	29	Tie investment in GCS projects to passage of appropriate CO_2 legislation Implement public outreach program to educate stakeholders on the legislative needs of the project Shift from DSA to EOR or ECBM if CO_2 legislation does not get passed, is insufficient or too onerous for DSA	Monitor CO_2 legislation and analyze the impact of CO_2 legislation on the project Continue public outreach program Comply with CO_2 legislation
Simulation and modeling—parameters *	9	23	36	1	Understand the statistics (range, mean, variance, etc.,) of parameters Review simulation model results for accuracy and completeness using a cross-functional team of experts	Periodic review of available data and simulation results Parameter calibration based on monitoring data Parameter uncertainty quantification Global sensitivity analysis of independent parameters
Well component failure (tubing, seals, wellhead, etc.)	10	N/A	N/A	N/A	Use the proper materials/equipment compatible with CO_2 (corrosion) Maintain tight H_2S and H_2O specification on CO_2 stream Monitor CO_2 leakage Develop and adhere to schedule for inspections and maintenance	Stop injection and fix the leakage Monitor corrosion and scale buildup in injection wells Take corrective actions if necessary

(* Different Wording in Prior Year/s)

Risk treatments were developed by project staff with areas of technical expertise relevant to the treatment activities. Accordingly, these individuals are also likely to be tasked to carry out the treatments. Because many of the highest-ranked risks are programmatic in nature, many treatments were developed by management staff and working-group leaders. To help further inform the management on resource allocation for treatment activities, working group leaders were then requested to describe each treatment in terms of its degree of completion, its expected effectiveness in reducing risk, and its cost. Each of these attributes was rated on a categorical one-through-five scale. As of this writing, the treatment attributes work is not yet complete; however, generally it shows that most treatments suggested midway through the project have been largely completed, and most are deemed moderately to fully effective in reducing risks.

The SWP also participated in the RCSP Interpartnership Simulation and Risk Assessment Working Group to focus collaborative efforts on mitigation planning, as well as integration of monitoring with risk assessment, with an ultimate objective of updating the RCSP BPM for Simulation and Risk Assessment [30]. The SWP participation in this working group serves to support the external risk communication efforts.

2.6. Risk Control and Monitoring

Risk control and monitoring are needed in order to ensure the appropriate operation of the risk response plan previously developed in the Risk Response Planning task and to evaluate their effectiveness during the project execution. We iteratively continued risk identification, analysis, planning, and tracking of new and existing risks, including the watch list. In addition to characterization and MVA results, outcomes from Quantitative Risk Analysis Task, and Risk Response Planning Task, provided a basis for monitoring and controlling risks.

There is a strong need for risk communication which includes formalizing the links between the various qualitative and quantitative risk assessments performed at FWU and then conveying those risks to internal project staff, professionals working in other CCUS projects, and external stakeholders. High-risk elements identified during risk workshops need to be subjected to scenario modeling to define the pathways by which risk targets would be impacted, thereby specifying the quantities that could be usefully constrained through modeling. Evaluation of risk status before and after modeling work and its communication with other efforts (characterization, monitoring, and simulation) is important.

Internal risk communications for the FWU project were organized largely around technical working groups among which all staff members were assigned. In simplest form, each working group's weekly meeting provided opportunity to communicate risk status arising within the technical areas covered by other working groups. The SWP internal project report by Hnottavange-Telleen [31] identified a set of internal stakeholders among whom regular risk communications should be pursued; internal stakeholders consist of the existing working groups, plus a hypothetical "Operations" group that would involve a field operator, plus project management.

Communications (including about risk) with and among the external stakeholders were relatively well established at the start of the SWP Phase III demonstration project, given that FWU was a long established producing oilfield. Consequently an elaborate new scheme for external risk communication was not needed. Hnottavange-Telleen [31] tabulates external stakeholders with whom—in a greenfield or otherwise new CCUS project—risk communications would be needed. Table 5 lists these external stakeholders.

Table 5. List of external stakeholders.

External Stakeholders Involved in Risk Communication
CO_2 Sources (e.g., large emitters that capture CO_2)
CO_2 Transporters (e.g., pipeline company)
Project operator company
Principal subcontractors
Smaller subcontractors
Town governments
Landowners
Public funding agencies
Private funders; Investors
Insurers
CO_2 Taxing or Crediting Authority
State & Federal govt. (legisl., exec.)
Regulatory agencies
Interest groups/NGOs (Non-Governmental Organizations)

Internal and external stakeholders both *need* certain risk information, and *possess* or can generate key risk information needed by other stakeholders. These different types of information needed or offered are tabulated in the above referenced report, for a generic commercial-scale CCUS project. An effective network of risk communication is needed among all stakeholders, in order that each can best judge its own risks and knowledgeably fulfill its role in the project.

3. Findings and Lessons Learned

The risk identification task indicates that programmatic risks (oil price, legislation, CO_2 supply, operator) and issues specific to EOR were high in the rankings. Among GCS technical risks, concerns about simulation efforts remain high. With regard to geological aspects of site suitability, concerns about reservoir rock have remained moderate, which makes sense considering the pre-project knowledge of the Farnsworth Unit oilfield. Concern about caprock petrology and mineralogy (heterogeneity) has decreased in response to project-generated information, but concern about caprock stratigraphy (lateral extent and continuity) has increased. This observation is noteworthy as a demonstration that increased information (ostensibly decreasing uncertainty) does not necessarily lead to a decrease in judged risk; rather, increased information can reveal that hidden assumptions had been in play, leading to under-estimated risk.

Several FEPs ranked #10, 12–15 in 2016 and #2, 4, 10, 13 in 2017 were not evaluated in prior years because they ranked below #50 in 2014. However, they were re-instituted in 2016 or 2017 because of their "Black Swan" nature; their high upper-bound severity values. The high ranking of certain of these FEPs may reflect actual change in risk or probably changed the appreciation of risk. In designing risk responses, it may be useful to distinguish between those two potential sources of change.

Our quantitative risk analysis demonstrated useful tools to numerically and probabilistically quantify the effect of risk scenarios on the project objectives. However, simulation processes (especially geological ones) inherently contain aleatory uncertainty. Thus, it would be most helpful to correctly define the ranges and distribution of uncertain parameters to significantly reduce the uncertainty. In addition, reduced order models and tools applied for quantifying risks are tested and verified with the numerical outputs rather than real world data, thus they unavoidably contain epistemic uncertainty. Without a lot of real-world data, it is difficult to test whether a proxy model or ROM adequately represents the physics of a process. Thus, the validation of a model would require history matching, which cannot happen within a short time period.

To support risk management efforts effectively, risks should be re-assessed approximately annually, or more frequently when major changes occur in project circumstances or information. Common examples of possible substantial changes include passage into a new project phase, acquisition of new data that significantly alters the understandings of site attributes or potential project-induced effects, or changes in the external funding or regulatory environment that could affect the ability to reach project objectives.

For a path forward with the risked FEPs, the following five steps are recommended:

1. Identify a discrete set of high-ranking FEPs to be managed.
2. As necessary, further develop or clarify the scenarios under which each higher-risk FEP will plausibly create negative impacts within this specific project.
3. Among the higher-risk scenarios, distinguish those with especially high severity from those with especially high likelihood.
4. Develop at least one actionable prevention and one actionable mitigation treatment for each scenario. To the extent practical, prefer reducing high likelihoods (i.e., develop preventive actions); and next prefer low-cost efforts to reduce high severities (e.g., ensure that personal protective equipment is worn).
5. Assign responsibility for completing risk treatments and for tracking their effects on inferred risk levels.

In addition to the above findings on risk evaluation processes and results, some observations on risk communication can be drawn from the work at FWU:

1. Much internal communication about factors that influence risk has occurred informally and semi-formally, among the researchers and managers involved in the project. Capturing this information in a structured way requires additional effort from the researchers themselves as well as from at least one individual whose role is so tasked and resourced. Some level of additional resourcing to support formal internal risk communication is probably justified, but the optimal level is difficult to assess.
2. External communications about FWU work (on risk and other topics) have focused on extensive technical publications and presentations within the specialized CCS/CCUS community. Because the project has taken place within an operating oilfield whose activity, geographic footprint, and risk have not materially changed, the previously established relationships with neighboring landowners have been largely sufficient for external risk communications.

Author Contributions: Conceptualization, S.-Y.L. and K.H.-T.; methodology, S.-Y.L. and K.H.-T.; investigation, F.P., W.J., S.C., T.X. and Z.D.; writing—original draft preparation, S.-Y.L. and K.H.-T.; writing—review and editing, S.-Y.L., T.X. and K.H.-T.; supervision, B.M. and H.V.; project administration, R.B. All authors have read and agreed to the published version of the manuscript.

Funding: This research was funded by the U.S. Department of Energy's (DOE) National Energy Technology Laboratory (NETL) through the Southwest Regional Partnership on Carbon Sequestration (SWP) under Award No. DE-FC26-05NT42591.

References

1. International Energy Agency. *20 Years of Carbon Capture and Storage*; International Energy Agency: Paris, France, 2016.
2. Metz, B.; Davidson, O.; De Coninck, H.; Loos, M.; Meyer, L. *IPCC Special Report on Carbon Dioxide Capture and Storage*; Campridge University Press: New York, NY, USA, 2005.
3. Vitillo, J.G.; Smit, B.; Gagliardi, L. Introduction: Carbon capture and separation. *Chem. Rev.* **2017**, *117*, 9521–9523. [CrossRef]
4. Ampomah, W.; Balch, R.S.; Grigg, R.B.; McPherson, B.; Will, R.A.; Lee, S.Y.; Dai, Z.; Pan, F. Co-optimization of CO_2-EOR and storage processes in mature oil reservoirs. *Greenh. Gases Sci. Technol.* **2017**, *7*, 128–142. [CrossRef]
5. Dai, Z.; Middleton, R.; Viswanathan, H.; Fessenden-Rahn, J.; Bauman, J.; Pawar, R.; Lee, S.-Y.; McPherson, B. An integrated framework for optimizing CO_2 sequestration and enhanced oil recovery. *Environ. Sci. Technol. Lett.* **2014**, *1*, 49–54. [CrossRef]
6. Godec, M.L.; Kuuskraa, V.A.; Dipietro, P. Opportunities for using anthropogenic CO_2 for enhanced oil recovery and CO_2 storage. *Energy Fuels* **2013**, *27*, 4183–4189. [CrossRef]

7. Gozalpour, F.; Ren, S.R.; Tohidi, B. CO_2 EOR and storage in oil reservoir. *Oil Gas Sci. Technol.* **2005**, *60*, 537–546. [CrossRef]
8. Mac Dowell, N.; Fennell, P.S.; Shah, N.; Maitland, G.C. The role of CO_2 capture and utilization in mitigating climate change. *Nat. Clim. Change* **2017**, *7*, 243–249. [CrossRef]
9. Balch, R.; McPherson, B. Integrating Enhanced Oil Recovery and Carbon Capture and Storage Projects: A Case Study at Farnsworth Field, TEXAS. In Proceedings of the SPE Western Regional Meeting, Anchorage, AK, USA, 23–26 May 2016; Society of Petroleum Engineers: Houston, TX, USA, 2016.
10. Rodosta, T.; Bromhal, G.; Damiani, D. US DOE/NETL Carbon Storage Program: Advancing Science and Technology to Support Commercial Deployment. *Energy Procedia* **2017**, *114*, 5933–5947. [CrossRef]
11. Bu, T.; Damsleth, E. Errors and uncertainties in reservoir performance predictions. *SPE Form. Eval.* **1996**, *11*, 194–200. [CrossRef]
12. Chu, C. Prediction of steamflood performance in heavy oil reservoirs using correlations developed by factorial design method. In Proceedings of the SPE California Regional Meeting, Ventura, CA, USA, 4–6 April 1990; Society of Petroleum Engineers: Houston, TX, USA, 1990.
13. Willis, B.J.; White, C.D. Quantitative outcrop data for flow simulation. *J. Sediment. Res.* **2000**, *70*, 788–802. [CrossRef]
14. Pan, F.; McPherson, B.J.; Dai, Z.; Jia, W.; Lee, S.-Y.; Ampomah, W.; Viswanathan, H.; Esser, R. Uncertainty analysis of carbon sequestration in an active CO_2-EOR field. *Int. J. Greenh. Gas Control* **2016**, *51*, 18–28. [CrossRef]
15. Montgomery, D.C. *Design and Analysis of Experiments*; John Wiley & Sons. Inc.: New York, NY, USA, 2001; Volume 1997, pp. 200–201.
16. Xiao, T.; McPherson, B.; Pan, F.; Esser, R.; Jia, W.; Bordelon, A.; Bacon, D. Potential chemical impacts of CO_2 leakage on underground source of drinking water assessed by quantitative risk analysis. *Int. J. Greenh. Gas Control* **2016**, *50*, 305–316. [CrossRef]
17. Dai, Z.; Viswanathan, H.; Middleton, R.; Pan, F.; Ampomah, W.; Yang, C.; Jia, W.; Xiao, T.; Lee, S.-Y.; McPherson, B. CO_2 accounting and risk analysis for CO_2 sequestration at enhanced oil recovery sites. *Environ. Sci. Technol.* **2016**, *50*, 7546–7554. [CrossRef]
18. Wiener, N. The homogeneous chaos. *Am. J. Math.* **1938**, *60*, 897–936. [CrossRef]
19. Xiu, D.; Karniadakis, G.E. Modeling uncertainty in flow simulations via generalized polynomial chaos. *J. Comput. Phys.* **2003**, *187*, 137–167. [CrossRef]
20. Jia, W.; McPherson, B.J.; Pan, F.; Xiao, T.; Bromhal, G. Probabilistic analysis of CO_2 storage mechanisms in a CO_2-EOR field using polynomial chaos expansion. *Int. J. Greenh. Gas Control* **2016**, *51*, 218–229. [CrossRef]
21. Oladyshkin, S.; Class, H.; Nowak, W. Bayesian updating via bootstrap filtering combined with data-driven polynomial chaos expansions: Methodology and application to history matching for carbon dioxide storage in geological formations. *Comput. Geosci.* **2013**, *17*, 671–687. [CrossRef]
22. Zhang, Y.; Sahinidis, N.V. Uncertainty quantification in CO_2 sequestration using surrogate models from polynomial chaos expansion. *Ind. Eng. Chem. Res.* **2012**, *52*, 3121–3132. [CrossRef]
23. Jia, W.; Pan, F.; Dai, Z.; Xiao, T.; McPherson, B. Probabilistic risk assessment of CO_2 trapping mechanisms in a sandstone CO_2-EOR field in northern texas, USA. *Energy Procedia* **2017**, *114*, 4321–4329. [CrossRef]
24. Dilmore, R. *NRAP Phase I Tool Development and Quality Assurance Process*; National Energy Technology Laboratory (NETL): Pittsburgh, PA, USA; Morgantown, WV, USA, 2016.
25. Pawar, R.J.; Bromhal, G.S.; Chu, S.; Dilmore, R.M.; Oldenburg, C.M.; Stauffer, P.H.; Zhang, Y.; Guthrie, G.D. The National Risk Assessment Partnership's integrated assessment model for carbon storage: A tool to support decision making amidst uncertainty. *Int. J. Greenh. Gas Control* **2016**, *52*, 175–189. [CrossRef]
26. Viswanathan, H.S.; Pawar, R.J.; Stauffer, P.H.; Kaszuba, J.P.; Carey, J.W.; Olsen, S.C.; Keating, G.N.; Kavetski, D.; Guthrie, G.D. Development of a hybrid process and system model for the assessment of wellbore leakage at a geologic CO_2 sequestration site. *Environ. Sci. Technol.* **2008**, *42*, 7280–7286. [CrossRef]
27. King, S. *Reservoir Reduced-Order Model–Generator (RROM-Gen) Tool User's Manual Version 2016.11-1.2*; National Energy Technology Laboratory-Energy Data eXchange NETL: Morgantown, WV, USA, 2016. [CrossRef]
28. Stauffer, P.H.; Viswanathan, H.S.; Pawar, R.J.; Guthrie, G.D. A system model for geologic sequestration of carbon dioxide. *Environ. Sci. Technol.* **2009**, *43*, 565–570. [CrossRef] [PubMed]
29. Chu, S.; Viswanathan, H.; Moodie, N.; Jia, W. *SWP Milestone Report for MS8: Extended Quantitative Brine and CO2 Leakage Calculations*; LA-UR-20-24728; LANL: Los Alamos, NM, USA, 2020.
30. NETL. *Best Practices: Risk Management and Simulation for Geologic Storage Projects, DOE/NETL-2017/1846*; NETL: Pittsburgh, PA, USA, 2017; p. 114.
31. Hnottavange-Telleen, K. *Risk Communications among Project Stakeholders*; Southwest Regional Partnership (Internal Report); NETL: Pittsburgh, PA, USA, 2019; p. 20.

Article

Impact of Mineral Reactive Surface Area on Forecasting Geological Carbon Sequestration in a CO$_2$-EOR Field

Wei Jia [1,2], Ting Xiao [1,2], Zhidi Wu [1,2], Zhenxue Dai [3] and Brian McPherson [1,2,*]

[1]   Energy & Geoscience Institute, University of Utah, Salt Lake City, UT 84108, USA; wei.jia@utah.edu (W.J.); ting.xiao@utah.edu (T.X.); zhidi.wu@utah.edu (Z.W.)

[2]   Department of Civil & Environmental Engineering, University of Utah, Salt Lake City, UT 84112, USA

[3]   College of Construction Engineering, Jilin University, Changchun 130026, China; dzx@jlu.edu.cn

*   Correspondence: b.j.mcpherson@utah.edu

Abstract: Mineral reactive surface area (RSA) is one of the key factors that control mineral reactions, as it describes how much mineral is accessible and can participate in reactions. This work aims to evaluate the impact of mineral RSA on numerical simulations for CO$_2$ storage at depleted oil fields. The Farnsworth Unit (FWU) in northern Texas was chosen as a case study. A simplified model was used to screen representative cases from 87 RSA combinations to reduce the computational cost. Three selected cases with low, mid, and high RSA values were used for the FWU model. Results suggest that the impact of RSA values on CO$_2$ mineral trapping is more complex than it is on individual reactions. While the low RSA case predicted negligible porosity change and an insignificant amount of CO$_2$ mineral trapping for the FWU model, the mid and high RSA cases forecasted up to 1.19% and 5.04% of porosity reduction due to mineral reactions, and 2.46% and 9.44% of total CO$_2$ trapped in minerals by the end of the 600-year simulation, respectively. The presence of hydrocarbons affects geochemical reactions and can lead to net CO$_2$ mineral trapping, whereas mineral dissolution is forecasted when hydrocarbons are removed from the system.

Keywords: geological carbon sequestration; reactive surface area; mineral trapping; enhanced oil recovery with CO$_2$ (CO$_2$-EOR); geochemical reactions; risk assessment

Citation: Jia, W.; Xiao, T.; Wu, Z.; Dai, Z.; McPherson, B. Impact of Mineral Reactive Surface Area on Forecasting Geological Carbon Sequestration in a CO$_2$-EOR Field. *Energies* **2021**, *14*, 1608. https://doi.org/10.3390/en14061608

Academic Editors: Nikolaos Koukouzas and João Fernando Pereira Gomes

Received: 1 February 2021
Accepted: 11 March 2021
Published: 14 March 2021

Publisher's Note: MDPI stays neutral with regard to jurisdictional claims in published maps and institutional affiliations.

1. Introduction

Geological carbon sequestration (GCS) is a critical component in accomplishing the goal of net-zero or carbon neutrality set by governments and industries [1–3], as it provides an enormous estimated storage capacity, and its efficacy has been successfully demonstrated many times by pilot-scale and field-scale projects worldwide [4]. Among the storage options, GCS at depleted oil fields, especially enhanced oil recovery with CO$_2$ (CO$_2$-EOR), has drawn a lot of attention, because of the higher economic incentives (in addition to government subsidies, e.g., the 45Q tax credit in the U.S.) and the lower costs of characterization, construction, and deployment. While most CO$_2$ trapping mechanism analyses were performed for GCS at deep saline aquifers (e.g., [5]), the essence of the trapping mechanisms is similar for CO$_2$-EOR projects [6–8]. However, one particular trapping mechanism, the mineral trapping of CO$_2$, is often ignored in storage forecasts for CO$_2$-EOR projects, such as Jia et al., 2016 [6], even though mineral trapping is the most secure mechanism to sequester CO$_2$ in the long term. Due to the lack of geochemical modeling in these forecasts, the processes of CO$_2$ dissolution and its presence in aqueous ions are also ignored. However, understanding the geochemical interactions between CO$_2$, in situ fluid, and formation, is critical to ensure long term CO$_2$ conformance and to mitigate the risks of groundwater contamination due to CO$_2$ leakage.

Mineral reactions are numerically described via two key parameters: chemical equilibrium constant and mineral dissolution and precipitation reaction rate. For a given reaction, the chemical equilibrium constant is a function of temperature, whereas the mineral reaction rate is controlled by further factors. As defined in the Transition State Theory (TST), the mineral reaction rate is a function of reactive surface area (RSA), activation energy, temperature, and equilibrium constant [9]. Particularly, the mineral RSA is crucial as it determines how much mineral is accessible and can participate in the reactions, yet it has been a challenge for it to be fully characterized. There are several approaches to measuring the mineral RSA (or specific surface area) in the laboratory, such as the widely used Brunauer–Emmett–Teller (BET) method [10], and newer image-based methods [11,12]. However, neither of these methods is ideal for characterizing mineral RSA. For example, the geometry-based method is constrained to the assumption of uniform-sized mineral grains, and thus its performance is compromised for the heterogeneously shaped mineral grains; on the other hand, the image-based methods need a drastically large number of images to capture the evolving mineral RSA over time. Therefore, it is common to observe order of magnitude differences in RSA measurements for the same mineral [13]. Such wide ranges of uncertainty pose great challenges in selecting parameters for numerical simulations. Even worse, most GCS simulation tools that include geochemical modules (e.g., CMG-GEM, TOUGHReact) ignore the spatial and geometry heterogeneity of mineral RSA values and simplify the temporal variation of mineral RSA values.

Existing uncertainty studies for GCS primarily focus on uncertainties stemming from porosity, permeability, and operational factors (e.g., injection scheme) [6,14–20]. Only a few studies have examined the impact of uncertain reactive surface areas on GCS performances. Qin and Beckingham (2021) compared simulation results of a core-scale model using mineral RSA obtained from different methods [21]. Luo et al., 2012 investigated the effect of the RSA of calcite and anorthite using a two-dimensional (2D) generic deep saline aquifer model and confirmed that these parameters have a significant impact on CO_2 mineral trapping [22]. Bolourinejad et al., 2014 followed a similar approach to evaluate the impact of the RSA of seven minerals on CO_2 trapping for a depleted gas field in the Netherlands [23]. It was found that the RSA of quartz has the greatest impact on CO_2 mineral trapping; however, neither hydrocarbon components nor realistic field operations were taken into consideration in their numerical model. Moreover, both Luo et al., 2012 and Bolourinejad et al., 2014 employed only one well (the injector) in their simulations. Recently, Jia et al., 2019 studied CO_2 trapping mechanisms, including mineral trapping, for an ongoing CO_2-EOR project with multiple five-spot well patterns, but only briefly discussed the potential impacts of using different mineral RSA values in their simulations [22–24].

This work aims to evaluate the impact of mineral reactive surface area on GCS numerical simulations where hydrocarbons are present. The novelty of this work includes: (1) using a three-dimensional (3D) field-scale reservoir model to eliminate the boundary effects imposed by 2D models; (2) using site-specific formation and fluid properties as well as realistic operational activities, i.e., CO_2-EOR with multiple wells; and (3) taking hydrocarbon components into consideration, thus simulating reactive transport for the three-phase system (water, gas, and oil). The Farnsworth Unit (FWU) site in northern Texas is selected as a case study. As the study site of the Southwest Regional Partnership on Carbon Sequestration (SWP) Phase III, the FWU site has been investigated with many characterization and monitoring techniques (e.g., [25–28]) and numerical analyses (e.g., [7,8,14,15,24,29,30]).

The rest of this paper is structured as follows: Section 2 describes the FWU reservoir model and the methods to evaluate the impact of mineral reactive surface areas; Sections 3 and 4 present simulation results and discuss how the presence of hydrocarbon components and uncertain RSA values affect reactive transport and CO_2 mineral trapping; and Section 5 summarizes the findings of this work.

2. Materials and Methods

2.1. FWU Description

The FWU site is located in the Anadarko basin in northern Texas (Figure 1a). Following the primary and secondary production that started in the 1950s and 1960s, the FWU has been currently undergoing CO_2-EOR since December 2010. As of 2020, more than one million metric tons of CO_2 has been sequestered at the FWU [31].

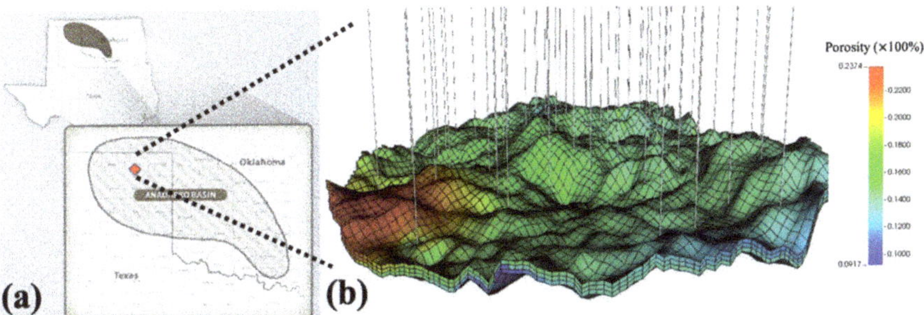

Figure 1. (**a**) Location of the Farnsworth Unit (FWU), and (**b**) a three-dimensional (3D) reservoir model of the Farnsworth Unit.

The CO_2 storage system at the FWU consists of late Pennsylvanian Morrow B sandstone as the storage formation and overlying Atokan Thirteen Finger limestone as the sealing layer. With the support of the U.S. Department of Energy and the National Energy Technology Laboratory, the SWP team has been conducting a variety of characterization; monitoring, verification, and accounting (MVA); simulation; and risk assessment research activities over the past decade. For the sake of brevity, please refer to previous publications for the details of CO_2 sequestration at the FWU. For example, Ross-Coss et al., 2015 and Ampomah et al., 2016 focused on geological characterization performed at the FWU [26,32]; Kumar et al., 2018 and Balch et al., 2017 presented MVA findings at the FWU [28,33]; Ampomah et al., 2016, Moodie et al., 2017, and Moodie et al., 2019 presented numerical simulation results for GCS forecast at the FWU [34–36]; and Dai et al., 2016, Pan et al., 2016, Xiao et al., 2016, Jia et al., 2017, and Xiao et al., 2020 addressed uncertainty analysis due to both geological and operational factors and quantitative risk assessment of CO_2 leakage and its impact on overlying underground sources of drinking water [16,17,37–39]. This study is built upon this previous FWU work and focuses on risk assessments associated with reactive transport, in particular mineral reactive surface areas.

2.2. Mineral Reactive Surface Area

The reactive surface area of minerals is a dynamic property that varies from mineral to mineral, changes over time as geochemical reactions develop, and depends on the heterogeneous distributions, various shapes, and complex contact interfaces of all minerals in the subsurface. A variety of methods are available to characterize mineral reactive surface areas, such as the well-known BET (Brunauer-Emmett-Teller) approach, analytical methods based on kinetic experiments, and the image-based approaches that rely on scanning electron microscopy (SEM) and computed tomography (CT) techniques [11,40]. These methods have been used for measuring mineral reactive surface areas for GCS purposes, e.g., [40,41]. As expected, the measured reactive surface area of the same mineral drastically varies from method to method and from site to site. For example, Bolourinejad et al., 2014 measured the specific surface area of kaolinite in the Rotliegend reservoir cores in the range of 1.2×10^6 m^2/m^3~3.4×10^7 m^2/m^3; Beckingham et al., 2016 reported different specific surface areas of calcite, ranging from 8.1×10^4 m^2/m^3, measured by the BET approach, to 7.6×10^5 m^2/m^3, measured by image-based methods [11,23]. The term specific surface area (SSA) is generally used to describe the reactivity of the pure mineral, while the term reactive surface area (RSA) is usually referred to as the average reactivity of each mineral in the porous media. Therefore, the SSA values need to be converted to RSA values with site-specific mineral volume fractions. At the FWU, there are seven key minerals identified from core analyses. A survey of the reactive surface area of these minerals was performed and is summarized in Table 1. Please note that mineral volume fractions (listed in Table 2) were used to convert the SSA values reported in other studies to FWU site-specific RSA values. The several orders of magnitude difference between the lower and upper values in Table 1 reiterate the primary research question of this study of whether the choice of a reactive surface area value impacts the results of reactive transport simulation and to what extent.

Table 1. Ranges of reactive surface area (RSA) of seven minerals from a literature survey [11,22–24,40,41].

RSA (m^2/m^3)	Calcite	Kaolinite	Dolomite	Quartz	Ankerite	Siderite	Illite
Low	88	17,600	560	607	521	2008	2528
High	6446	2,298,400	56,146	42,313	74,030	918,585	1,238,400

Table 2. Major minerals at the FWU and volume fractions [24].

Mineral	Chemical Formula	Volume Fraction
Quartz	SiO_2	80.75%
Kaolinite	$Al_2Si_2O_5(OH)_4$	6.76%
Siderite	$FeCO_3$	4.41%
Calcite	$CaCO_3$	3.86%
Illite	$K_{0.6}Mg_{0.25}Al_{1.8}(Al_{0.5}Si_{3.5}O_{10})(OH)_2$	2.58%
Ankerite	$CaMg_{0.3}Fe_{0.7}(CO_3)_2$	0.37%
Dolomite	$CaMg(CO_3)_2$	0.01%

While choosing a value from the wide range for a certain mineral is already difficult, describing the changing characteristics of the mineral reactive surface area is no easier. The common practice in reactive transport simulations is to calculate the reactive surface area with the following equation:

$$A_i = A_i^0 \times \frac{N_i}{N_i^0},$$

(1)

where A_i is the reactive surface area of mineral i at current time step, A_i^0 is the reactive surface area of mineral i at time 0 (i.e., initial value), N_i is the mole amount of mineral i per unit grid block volume at the current time step, and N_i^0 is the mole amount of mineral i per unit grid block volume at time 0. There are two main restrictions on using this approach. Firstly, it assumes a uniform distribution of reactive surface area for the same mineral across the entire model domain (which could be kilometers for GCS sites); in other words, a homogenous reactive surface area is assigned to each mineral. Secondly, the controlling factor of reactive surface area at the current time step, N_i, is determined by the current mineral dissolution and precipitation rate that is affected by the mineral reactive surface area at previous time steps, as shown in Equation (2):

$$r_i = A_i k_i \left(1 - \frac{Q_i}{K_{eq,i}}\right),$$

(2)

where r_i is the reaction rate for mineral i, k_i is the rate constant of mineral reaction i, Q_i is the activity product of mineral reaction i, and $K_{eq,i}$ is the chemical equilibrium constant for mineral reaction i. Therefore, deviations between the estimated and real reactive surface areas accumulate over time and may lead to significantly different mineral reaction predictions after hundreds or thousands of years.

Given that an exhaustive dataset for the mineral reactive surface area is not available, and neither is an advanced reactive transport numerical model framework, an uncertainty analysis is probably the best approach to investigate the impact of mineral reactive surface area on GCS predictions. Based on the ranges listed in Table 1, a seven-factor Box–Behnken Design was developed to cover the entire uncertain space with 85 cases. The low (−1), mid (0), and high (+1) values for each of the seven minerals are presented in Figure 2. Please note that the mid values are not the average of the low and high values, but rather the median taken from the RSA values of the literature survey. Two more cases, (−1, −1, −1) and (+1, +1, +1), i.e., all low values and all high values, were added to the simulation design. A full list of the cases is presented in Table A1. It is anticipated that the design of these permutations will discover the impact of mineral reactive surface areas, with the mutual dependencies among the mineral reactions taken into account.

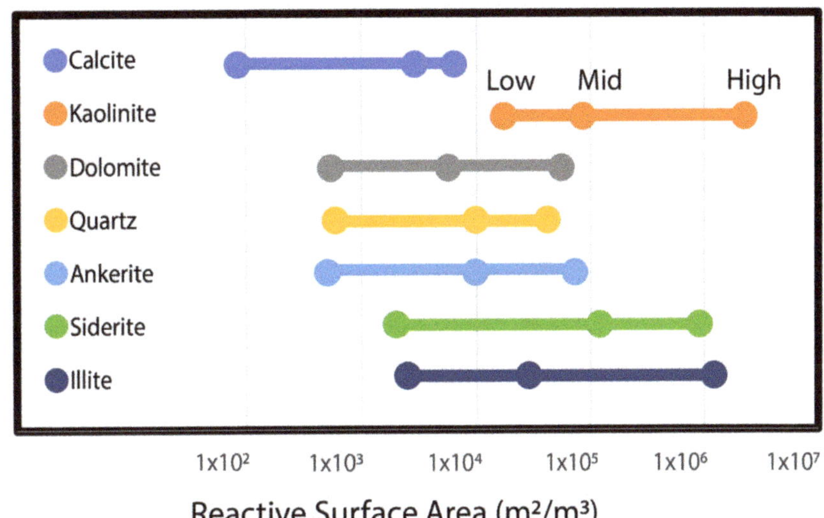

Figure 2. Low, mid, and high values of the reactive surface area of seven minerals.

2.3. Numerical Models

The 3D FWU reservoir model consists of 82 cells in the x-direction, 78 cells in the y-direction, and four cells in the z-direction, with a total of 25,584 grid blocks. Each grid block is 200 ft (or 60.96 m) by 200 ft (or 60.96 m) in x and y directions and about 8.67 ft (or 2.64 m) in the z-direction. This model only includes the Morrow B sandstone, as the integrity of the overlying Thirteen Finger limestone will not be compromised for at least 5000 years [31]. A total of 55 wells (23 injectors and 32 producers) were distributed across the model domain in 5-spot patterns (Figure 1b). This reservoir model has been calibrated with the FWU field history by the SWP [34,42]. Well schedules include a 10-year CO_2-EOR period, a subsequent 10-year post-EOR CO_2 injection period, and another 580-year monitoring period. In the CO_2-EOR period, CO_2 and water were alternatively injected (water alternating gas (WAG)), reflecting the site history. In the post-EOR CO_2 injection period, all producers were shut-in, and CO_2 was continuously injected at 5×10^6 ft^3/day (or 1.42×10^5 m^3/day) through all injectors. In the monitoring period, all wells were shut-in, and no CO_2 was injected into the model. The heterogeneous porosity and permeability distributions of the reservoir model were derived from the site characterization and geostatistical analysis of previous SWP work. Figure 1b presents the heterogeneous porosity distribution of the 3D FWU model and the locations of the wells.

Seven major minerals were identified from FWU cores, as listed in Table 2. The measured volume fractions for each mineral are used as the initial conditions for the reservoir simulations. The SWP team periodically monitored the water quality of samples from the shallow aquifer and produced water collected at the FWU site. In this work, the averaged measurements of the produced water are used as the initial conditions of the aqueous species, as shown in Table 3. The compositions of hydrocarbon components were analyzed from FWU oil samples and are also listed in Table 3.

Table 3. Initial conditions of aqueous species and hydrocarbon components.

Aqueous Species	Molality (mol/kg H_2O)	Hydrocarbon Components	Initial Global Composition
H^+	2.494×10^{-7}	C1	0.405
K^+	3.288×10^{-6}	C2	0.038
Na^+	2.585×10^{-2}	C3	0.024
Ca^{2+}	2.269×10^{-5}	IC4 + NC4	0.016
Mg^{2+}	2.930×10^{-5}	IC5 + NC5	0.027
SiO_2 (aq)	5.640×10^{-4}	C6	0.017
Fe^{2+}	1.598×10^{-6}	C7–C38	0.325
Cl^-	1.379×10^{-2}	C38–C70	0.147
OH^-	6.354×10^{-7}	CO_2	0.001
HCO_3^-	3.720×10^{-2}		
CO_3^{2-}	1.190×10^{-5}		

Seven mineral reactions and three aqueous reactions were modeled in this work, as shown in Table 4. The chemical equilibrium constants ($\log K_{eq}$) were calculated for reservoir temperature at 70 °C with the EQ3/6 database [43], and the activation energy and the standard TST rate constant ($\log k_{25°C}$) values were taken from Palandri and Kharaka (2004) [44]. In particular, fourth-order polynomials were fitted to the EQ3/6 database of chemical equilibrium constants of the reactions versus temperature (please see Figure A1).

Table 4. Mineral and aqueous reactions and parameters for reactive transport modeling [43,44].

Reaction	Activation Energy (kJ/mol)	$\log K_{eq}$ at 70 °C	$\log k_{25°C}$ (mol/m^{2}s)
Calcite + H^+ = Ca^{2+} + HCO_3^-	41.87	1.192	−5.810
Kaolinite + $6H^+$ = $5H_2O$ + $2SiO_2$(aq) + $2Al^{3+}$	62.76	3.103	−13.180
Dolomite + $2H^+$ = Ca^{2+} + Mg^{2+} + $2HCO_3^-$	73.75	2.278	−8.065
Quartz = SiO_2(aq)	89.09	−3.345	−13.680
Ankerite = Ca^{2+} + $0.95Fe^{2+}$ + $0.05Mg^{2+}$ + $2CO_3^{2-}$	62.76	0.125	−8.900
Siderite + H^+ = Fe^{2+} + HCO_3^-	62.76	−1.001	−8.900
Illite + $8.0H^+$ = $0.25Mg^{2+}$ + $0.6K^+$ + $2.3Al^{3+}$ + $3.5SiO_2$(aq) + $5.0H_2O$	35.00	4.677	−12.780
H_2O = H^+ + OH^-	-	−13.263	-
CO_2 + H_2O = H^+ + HCO_3^-	-	−6.322	-
CO_2 + H_2O = $2H^+$ + CO_3^{2-}	-	−16.556	-

Reactive transport models require more computational resources than multiphase flow models because of the additional governing equations for aqueous and mineral reactions. Therefore, simplified modes are often used for reactive transport simulations. A similar approach was followed in this work. Instead of using the abovementioned reservoir model for all 87 simulations, a screening study was conducted first, which used a simplified 3D model to evaluate the performances of all 87 combinations of mineral reactive surface areas. Then the selected combinations, which are either representative or extreme scenarios, were repeated with the full FWU reservoir model. The simplified 3D model effectively represents a batch simulation. In particular, there are three cells in each direction (x, y, and z), with infinite acting aquifers around all boundaries to avoid the boundary effect. A CO_2 injection well was placed in the center of the model to introduce supercritical CO_2 to the model at a bottom-hole pressure constrained injection rate. Except for the grid structure and well configurations, all other model settings are identical to the full-size reservoir model. Simulation results at the center of this simplified model can be used to screen the combinations of mineral reactive surface areas. In addition to the much lower computational cost, the other advantage of this approach is that it can isolate control factors to simulation results, e.g., the domain size and the interference of multiple wells, and therefore focus on discovering the impact of mineral reactive surface area.

All simulations were performed with the 2019.10 version of the CMG-GEM simulation package, the flow equations of which are listed in Appendix B.

3. Results

3.1. Screening the RSA Combinations

The 87 combinations were simulated to study the ranges of uncertainty in mineral precipitation/dissolution and CO_2 mineral trapping caused by using different mineral RSA values. Figures 3 and 4 present the simulation results of the simplified model and emphasize three RSA combinations, Case #85, Case #86, and Case #87, where all mineral RSA values were set at mid, low, and high, respectively (please refer to Table A1 for the detailed RSA combinations). The predicted pH values of all cases exhibit a similar pattern (Figure 3a). A drastic drop from the initial pH (~6.7) to the range of 4.0~4.4 during the first day is followed by a relatively slower recovery through the rest of the simulation period. While most of the curves overlap each other, the dashed dark green line, which indicates the results of Case #86, presents a deeper pH decrease in the beginning and a slower pH recovery rate. Comparing the much faster pH recovery rate of the other cases, it is obvious that the prolonged recovery time of Case #86 results from the lack of buffers, i.e., fewer accessible minerals to react with the H^+ introduced by CO_2.

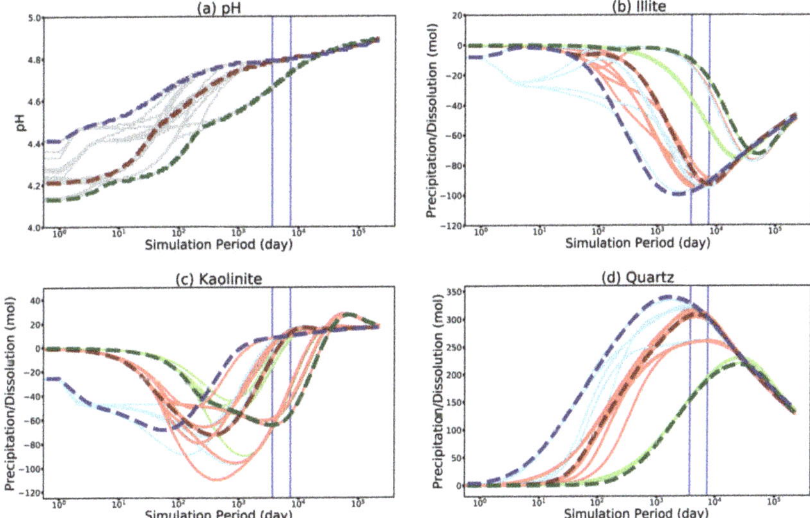

Figure 3. (**a**) Predicted pH and mineral precipitation or dissolution of (**b**) illite, (**c**) kaolinite, and (**d**) quartz at the center of the model. Light green, light red, and light blue indicate low, mid, and high RSA of the particular mineral in each subplot; dashed lines in dark green, dark red, and dark blue indicate cases with all low RSA (Case#86), all mid RSA (Case #85), and all high RSA (Case #87), respectively; two vertical lines denotes the end of 10 years and the end of 20 years.

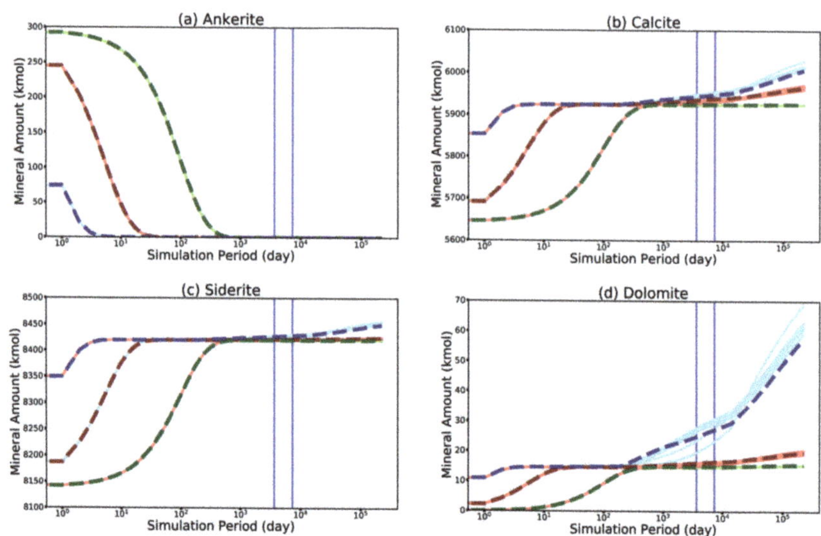

Figure 4. Predicted total mineral amount in the simplified model for (**a**) ankerite, (**b**) calcite, (**c**) siderite, and (**d**) dolomite. Light green, light red, and light blue indicate low, mid, and high RSA of the particular mineral in each subplot; dashed lines in dark green, dark red, and dark blue indicate Case#86, Case #85, and Case #87, respectively; two vertical lines denote the end of 10 years and the end of 20 years.

Because the changes of the silicate minerals are much smaller than the total amount of silicate minerals in the model, the mineral reaction rate is a better parameter to illustrate the silicate mineral reactions. The predicted precipitation or dissolution of silicates at the center of the model are shown in Figure 3b–d. For illite and quartz, Case #87 (dashed dark blue line) and Case #86 (dashed dark green line) is the most and the least reactive scenario for these two minerals, whereas Case #85 (dashed dark red line) overlaps with most of the medium cases. On the other hand, changes in kaolinite are more complicated. There are many crossovers and overlaps among results using different kaolinite RSA values, and the deviations between lines in the same color are much greater than those in Figure 3b or Figure 3d. This suggests that the reaction of kaolinite is controlled by not only the RSA of kaolinite but also the RSA of other minerals. Nevertheless, the high RSA cases (blue) predicted prompt and more intense reactions, while the low RSA cases (green) predicted delayed and more restricted reactions.

The simulation results of four carbonate minerals, ankerite, calcite, siderite, and dolomite, are shown in Figure 4a–d, respectively. Please note that all 87 cases started with the same initial mineral amounts, and Figure 4 presents mineral amounts since Day 1; therefore, the differences between the starting points of the lines in each subplot indicate the difference in the reactions that occurred during the first day of simulation. For example, the total amount of ankerite in the model was initially about 295 kmol, and dropped to 74 kmol in Case #87 (dashed dark blue line), whereas it dropped to about 245 kmol in Case #85 (dashed dark red line) and to 292 kmol in Case #86 (dashed dark green line), after the first day of simulation.

For all 87 cases, a monotonic dissolution of ankerite is observed, while all three other minerals show monotonic precipitation within the 600 years of simulation. The low RSA cases (green lines) once again exhibit the least reactivity, i.e., the slowest to deviate from the initial values. Interestingly, the prediction results fall into three clusters for each of these four minerals in the first year. In other words, the RSA value of the particular mineral, being either low, mid, or high, determines the reaction of this mineral in the first year, regardless of what RSA values are used for the other six minerals. Moreover, there is zero or small difference among predicted mineral amounts using the three levels of RSA values for ankerite (Figure 4a) and siderite (Figure 4c) in the later stage of the simulation period. The differences among the three clusters of lines are more obvious for calcite (Figure 4b) and dolomite (Figure 4d) after 300 days. At the end of the simulations, the cases with the high dolomite RSA value forecast much greater dolomite precipitation than the other cases.

Compared to the silicate minerals (Figure 3b–d), a stronger clustering effect is observed in the carbonate minerals, especially in the first two years of simulation, suggesting that the reaction of the carbonate mineral (Figure 4a–d) is more controlled by its own RSA value. The interference (or mutual dependence) between mineral reactions is weaker for the carbonate minerals than for the silicate minerals. As a result, Cases #85, #86, and #87 seem to be able to represent the median, minimum, and maximum reaction scenarios of all tested cases for all minerals.

The mineral trapping of CO_2 in the simplified model was evaluated with all 87 cases, and the results are presented in Figure 5. There is a much greater deviation among total mineral trapping amount predictions than the predictions for each mineral, as shown in Figures 3 and 4. No clear clustering is observed in Figure 5. While the impact of RSA on each individual mineral is apparent and easier to interpret, its impact on the CO_2 trapping mechanism is rather complicated. This is likely due to the different abilities of each mineral to sequester CO_2. For example, one mole of precipitated calcite effectively secures one mole of CO_2, while for dolomite the ratio becomes 1:2. At the end of the simulation, the amount of CO_2 trapped in the minerals ranges from 2 kmol to 200 kmol. The results of three representative cases (#86, #86, and #87) are very close to the median, minimum, and maximum values of all cases, as shown in the dashed lines in Figure 5. Therefore, these cases were selected for reactive transport simulations with the FWU reservoir model.

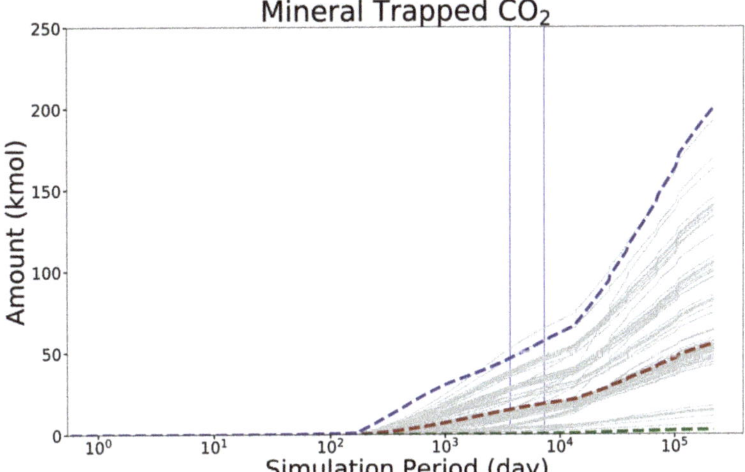

Figure 5. The predicted amount of mineral trapped CO_2 in the simplified 3D model. Dashed lines in dark green, dark red, and dark blue indicate Case#86, Case #85, and Case #87, respectively; gray lines denote all other cases, and two vertical lines denote the end of 10 years and the end of 20 years.

3.2. Reactive Transport Prediction with the Reservoir Model

As minerals precipitate or dissolve, it can change the porosity of the storage formation and thus affect fluid flow patterns and subsequent mineral reactions at new fluid-rock contacts. Therefore, we analyzed the porosity change due to mineral reactions in the FWU model. Figures 6–8 present the changes in porosity at three critical time steps: the end of the CO_2-EOR period, the end of the post-EOR CO_2 injection period, and the end of the simulation period, respectively.

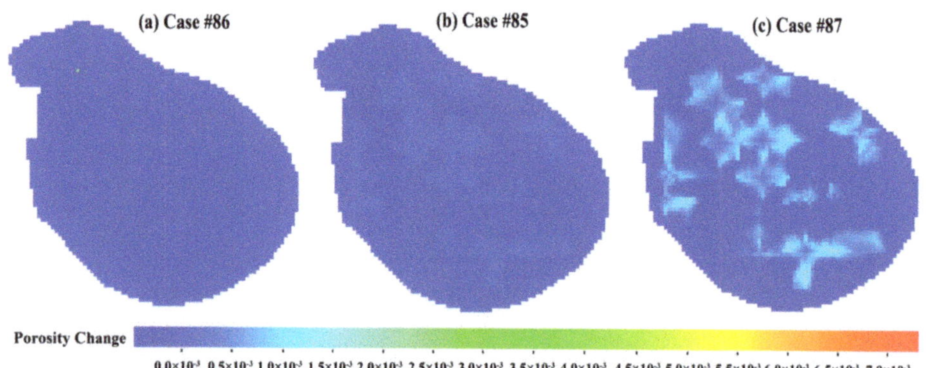

Figure 6. Estimated porosity loss at the end of the CO_2-EOR period compared to the initial values for (**a**) Case #86, (**b**) Case #85, and (**c**) Case #87.

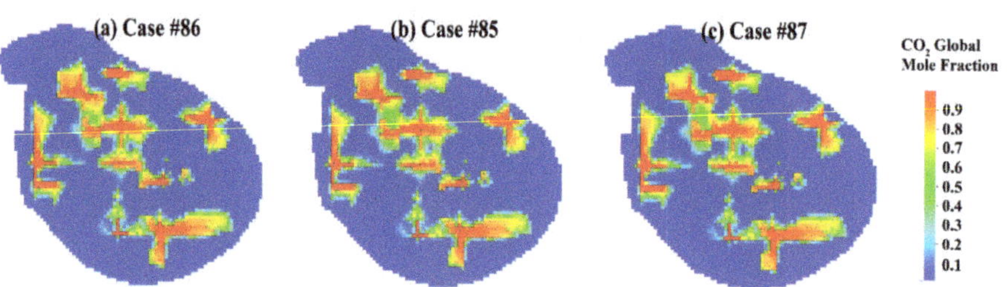

Figure 7. Simulated CO_2 global mole fraction at the end of the CO_2-EOR period for (**a**) Case #86, (**b**) Case #85, and (**c**) Case #87.

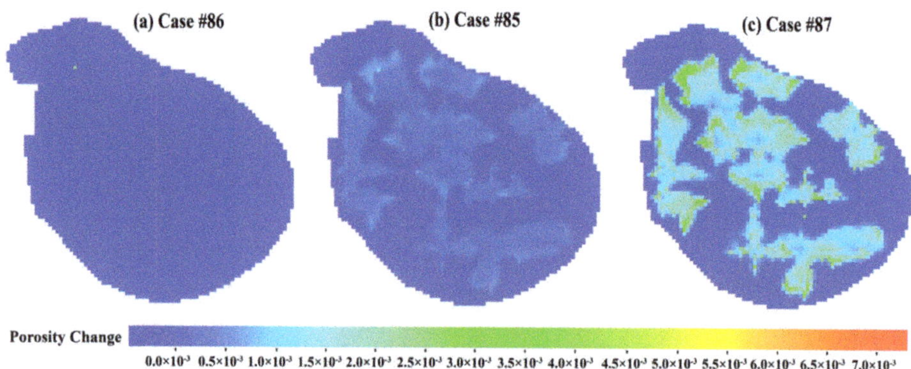

Figure 8. Estimated porosity loss at the end of the post-EOR CO_2 injection period compared to the initial values for (**a**) Case #86, (**b**) Case #85, and (**c**) Case #87.

After ten years of CO_2-EOR operation, porosity reduction is observed in all three cases. Figure 6 shows changes of porosity due to mineral reactions at the top layer of the Morrow B sandstone using the RSA values of Case #86 (all low RSA values of seven minerals), Case #85 (all mid RSA values), and Case #87 (all high RSA values), respectively. In general, a dominant porosity reduction was observed for all cases. Specifically, Case #86 (Figure 6a) exhibits the most restricted variation, while Case #87 (Figure 6c) presents the greatest porosity change across the layer, showing a maximum porosity reduction of 1.86×10^{-3} (or 0.9% of the initial porosity). However, very similar CO_2 global mole fraction distributions were predicted by three models (Figure 7). This suggests that the mineral reactions have an insignificant impact on the CO_2 migration. Given that the CO_2 plume shapes are mostly identical among the three cases, the difference in porosity change is attributed to the mineral RSA values.

The CO_2-EOR period was followed by a post-EOR CO_2 injection period, during which all production wells were shut-in, and CO_2 was continuously injected via all injection wells for ten years. Figure 7 presents the simulation results of porosity changes due to mineral reactions. There are visible changes in porosity reduction comparing Figures 8b and 6b, and Figures 8c and 6c. A clear expansion of areas with porosity loss is observed. The greater porosity reductions occur along the edges of the CO_2 plumes (Figure 8c). Specifically, after ten years of CO_2 injection, the maximum porosity loss has been increased to 0.7% in Case #85 and 2.5% in Case #87, from 0.3% and 0.9%, respectively. However, the CO_2 flow patterns are still almost identical for all three cases (figure not shown). Therefore, the impact of mineral reactions on CO_2 flow remains insignificant during the post-EOR CO_2 injection period.

Figure 9 presents the estimated porosity loss at the end of the 600-year simulation period. The simulation results of using all low RSA values present almost no change in porosity due to mineral reactions (Figure 9a). On the other hand, using all mid and high RSA values leads to greater porosity reduction, as shown in Figure 9b,c. While there is only a slight expansion of porosity loss areas during the no-injection period, the maximum porosity loss due to mineral reactions increased to 1.19% and 5.04% for Case #85 and Case #87, respectively. It is worth noting that the predicted porosity changes in Figures 6c and 8c are more profound than those in Figure 9a,b, suggesting that using high RSA values leads to dramatically different porosity change predictions, hence mineral trapping of CO_2, over even a short time period. Comparing the distributions of CO_2 global mole fraction (Figure 10) and porosity change (Figure 9), it is clear that the mineral reactions had a nominal impact on the forecast of CO_2 migration in 600 years. It is interesting to note that there are lower porosity changes in the centers of the CO_2 plumes, where they exhibit very high (greater than 0.8) CO_2 global mole fractions (see Figures 7 and 10). The areas with greater porosity reduction are located on the edges (or fronts) of the CO_2 plumes, suggesting that there are more intense geochemical reactions and thus more CO_2 trapped in minerals or aqueous ions. This is because mineral reactions require sufficient contacts between the minerals and the formation fluids, which is less likely to be present in areas with very high CO_2 saturation.

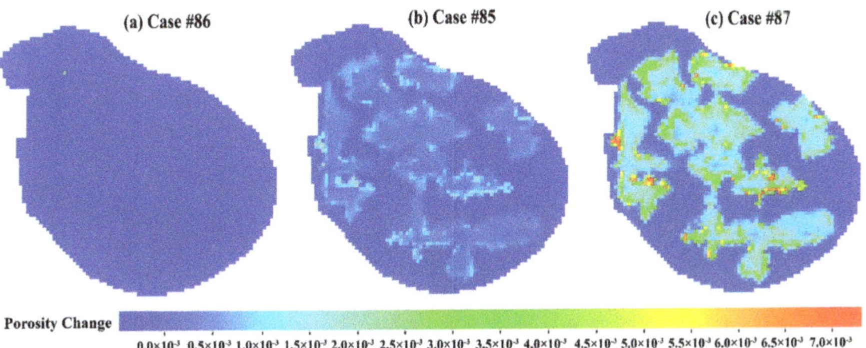

Figure 9. Estimated porosity loss at the end of the simulation period compared to the initial values for (**a**) Case #86, (**b**) Case #85, and (**c**) Case #87.

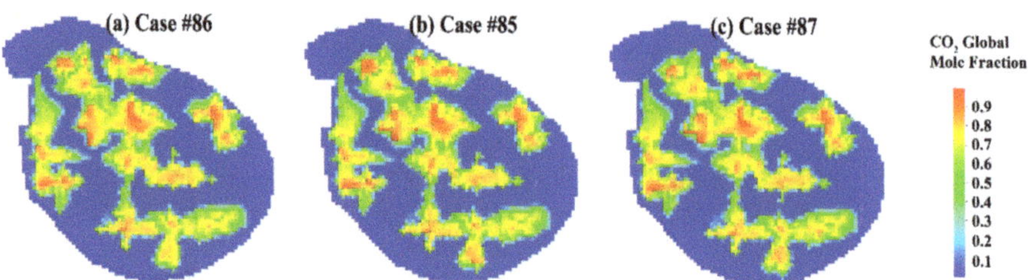

Figure 10. Simulated CO_2 global mole fraction at the end of the simulation period for (**a**) Case #86, (**b**) Case #85, and (**c**) Case #87.

The performance of trapping mechanisms that are directly related to the geochemical reactions was evaluated, as shown in Figure 11. As expected, and similar to the simplified model result (Figure 5), there is a significant difference between the amount of CO_2 trapped in minerals in the FWU field-scale reservoir model (Figure 11a). The mineral trapping first appeared as early as about 200 days (for Case #87), accelerated during the post-EOR CO_2 injection period (between the two vertical blue lines), and kept growing at a slower rate after CO_2 injection stopped. At the end of 600 years, the estimated amounts of CO_2 trapped in minerals are 3.25×10^6 kmol (1.43×10^5 metric tons), 0.8×10^6 kmol (3.52×10^4 metric tons), and 0.05×10^6 kmol (2.2×10^3 metric tons) for Case #87, Case #85, and Case #86, respectively. In other words, mineral trapping with all high RSA values is about four times more effective than with all mid RSA values, and 65 times more effective than with all low RSA values. At the end of 600 years, mineral trapping would contribute to 0.15%, 2.46%, and 9.44% of the total sequestered CO_2 at the FWU when using the low, mid, and high mineral RSA values, respectively. However, the mineral RSA values have much less impact on the CO_2 trapped in aqueous ions (Figure 11b). The maximum difference between predictions is only about 0.02×10^6 kmol (880 metric tons), and only 0.12×10^6 kmol (5.28×10^3 metric tons) of CO_2 was sequestered in aqueous ions by the end of the simulation. Nevertheless, mineral reactions and aqueous ions are able to sequester at least around 3000 metric tons of CO_2 (in Case #86), which would be otherwise presented in other forms (i.e., supercritical phase or dissolved) if reactive transport was not taken into consideration.

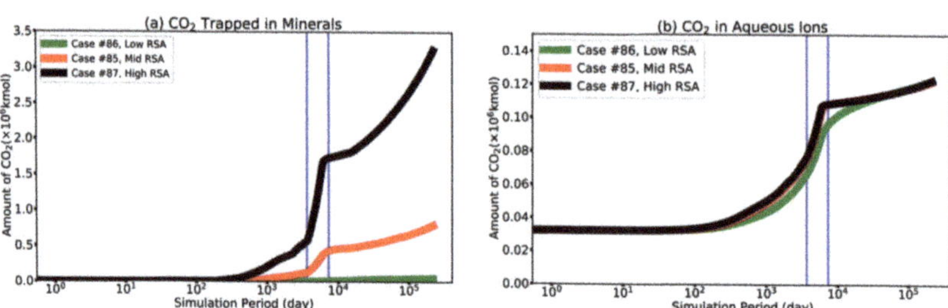

Figure 11. Performance comparison of (**a**) CO_2 mineral trapping and (**b**) CO_2 trapped in aqueous.

4. Discussion

Geochemical reactions, especially mineral reactions, are critical in predicting the fate and behavior of injected CO_2 into the subsurface and are involved in many aspects of a GCS project, such as the risk assessment of CO_2 leakage and caprock integrity. While most reactive transport studies were focused on CO_2 storage in deep saline aquifers, only a few incorporated the geochemical module into their simulations for CO_2 storage in depleted oil fields. Possible reasons for this could include: (1) mineral reactions in a depleted oil field are assumed to be similar to those in a deep saline aquifer, but in a much-restricted approach because of the presence of hydrocarbons; (2) the mineral trapping at depleted oil fields is considered to be less effective than other trapping mechanisms in a short period of time, thus it can be ignored for the purpose of storage estimation; and (3) most CO_2 storage numerical simulation packages are incapable of simulating reactive transport in three-phase systems (oil/water/CO_2). The results of this work demonstrate that neither the porosity reduction due to mineral reactions nor the mineral trapping of CO_2 at a CO_2-EOR field is negligible. Using the mid RSA values of all seven minerals, up to 0.7% of porosity loss was forecasted after 20 years of CO_2 injection, and up to 1.19% of porosity loss was forecasted after 600 years. This is similar to the porosity change estimated by Pan et al., 2016, which reported up to 0.7% of porosity reduction and up to 2.7% of porosity increase after 1000 years due to mineral reactions [45]. In that work, a modified FWU model was used, where there was only one five-spot well pattern and a deep saline storage scenario had been adopted (i.e., no hydrocarbon in the model). Nevertheless, similar results in estimated porosity change suggest that mineral reactions in three-phase systems probably are not as limited as it is assumed.

We have performed a preliminary study comparing the differences of mineral reactions in two-phase (Figures A2 and A3) and three-phase systems (Figures 3 and 4) using the simplified 3D model. When hydrocarbon components were replaced with water, different predictions were found in most minerals. While the changes of illite and quartz slowed down in the monitoring period and tended to recover to the initial values in the three-phase models, the dissolution of illite and precipitation of quartz in the two-phase model was accelerated. Moderate differences were observed in the carbonate minerals. In the two-phase models, an equilibrium state was reached during the first year of CO_2 injection and maintained throughout the entire simulation period (Figure A3), leading to almost no CO_2 being trapped in minerals (Figure 12); however, the three-phase models predicted that more carbonate minerals would precipitate in the monitoring period (Figure 4), which would then result in a significant amount of CO_2 being present in the minerals (Figure 5). There is a clear impact of the presence of hydrocarbons on mineral reactions with this simplified model. Therefore, the assumption that the mineral reactions in three-phase systems are only a scaled-down version of those in two-phase systems is not necessarily correct. When considering the interactions between hydrocarbon components and minerals and aqueous species (which are not incorporated in this work), it would be inevitable to develop a new paradigm for the reactive transport modeling of CO_2 storage in depleted oil fields.

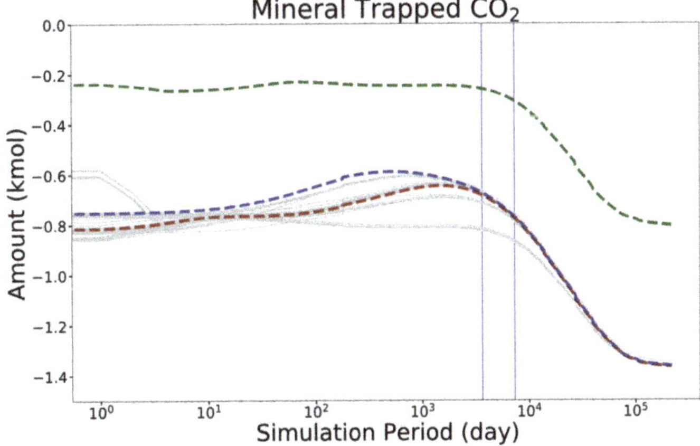

Figure 12. The predicted amount of mineral trapped CO_2 in the simplified 3D model with hydrocarbons removed. Dashed lines in dark green, dark red, and dark blue indicate Case#86, Case #85, and Case #87, respectively; two vertical lines denote the end of 10 years and the end of 20 years.

In order to identify and quantify the roles of mineral RSA, many other important factors were fixed in this work. The simulation results will be affected by varying these factors, which include geological and hydrogeological properties (e.g., porosity and permeability values), multiphase flow models (e.g., relative permeability, hysteresis, and capillary pressure models), and operational factors (e.g., well spacing and depths, well operational limits, and CO_2-EOR and WAG configurations). The impacts of mineral RSA will also be subject to other site-specific conditions when transferring the specific findings of this work to other case studies. However, it is believed that the importance of mineral RSA and the necessity of adding reactive transport modeling to CO_2-EOR simulations are applicable to other GCS projects.

Additionally, reactive transport modeling is an indispensable component in understanding and mastering the coupled thermal-hydro-mechanical-chemical (THMC) processes in the subsurface porous media [46–48]. A thorough investigation of mineral RSA is vital for coupling mineral reactions with other critical subsurface phenomena, for example, mechanical and thermal deformation. On the other hand, the thermal and mechanical in situ conditions will affect mineral reactions on the whole. Given the capability of existing simulation tools, improvements in geochemical modeling and a better coupling of THMC processes are in urgent need. The geochemical properties should be treated the same as the flow or reservoir properties (e.g., porosity, permeability, pressure, saturation, etc.). The heterogeneity in mineral distributions and the temporal evolution of mineral properties should be incorporated into the reactive transport modeling. This work employed 87 combinations of mineral RSA values as a workaround to take both spatial and temporal variations in mineral RSA into account. While this approach has been effective in showing the impact of mineral RSA values on mineral reactions, it does not demonstrate the geochemical-hydrological coupling effect primarily due to the fact that the uniform distribution of mineral compositions across the model inhibits the ability of mineral reactions to affect the fluid flow. With the recent advances in RSA measurement techniques, it is vital to make better use of lab observations in numerical simulations for more accurate predictions. The absence of the geochemical-hydrological coupling effect could also be attributed to the spatial resolution (200 ft, or 60.96 m) of the numerical model employed in this work, which might be too coarse to capture reactions and fluid flow at small-scales (from centimeter to meter). Even with paralleled high-performance computing nodes, each of the 600-year simulations on the model with 200-ft resolution took more than eight hours. Refining the spatial resolution will no doubt increase the computational cost significantly, and very likely cause numerical convergence issues as the reservoir heterogeneity will be more complicated after adding more small-scale local geological features. Moreover, obtaining a reasonable down-scaled heterogeneity renders its own challenges and will be tackled in our future research.

5. Conclusions

In this work, the impact of mineral reactive surface areas on the CO_2 storage forecast was evaluated for the GCS project in the Farnsworth Unit in northern Texas. In order to reduce the computational cost, a simplified 3D model was used to screen representative RSA combinations from 87 cases. Three cases were chosen for the FWU reactive transport simulation. The main conclusions are drawn as follows:

(1) The inter-dependency effects of mineral RSA values are stronger in the silicate mineral reactions and almost not observed in the carbonate mineral reactions;

(2) The impact of mineral RSA values on CO_2 mineral trapping, on the whole, is more complex than it is on individual geochemical reactions. However, the three selected cases (with all low, all mid, and all high mineral RSA values) are representative for predicting CO_2 trapped in minerals;

(3)   While the low RSA case predicted negligible porosity change and an insignificant amount of CO_2 mineral trapping for the FWU model, the mid and high RSA cases forecasted up to 1.19% and 5.04% of porosity reduction due to mineral reactions, and 2.46% and 9.44% of total CO_2 trapped in minerals by the end of the 600-year simulation, respectively;

(4)   The presence of hydrocarbons affects geochemical reactions and can lead to net CO_2 mineral trapping, whereas negative CO_2 mineral trapping is forecasted when hydrocarbons are removed from the system.

Author Contributions: Methodology, W.J. and T.X.; investigation, W.J. and T.X.; resources, Z.W. and Z.D.; supervision, B.M.; writing—original draft preparation, W.J.; writing—review and editing, all authors. All authors have read and agreed to the published version of the manuscript.

Funding: Funding for this project was provided by the U.S. Department of Energy's (DOE) National Energy Technology Laboratory (NETL) through the Southwest Regional Partnership on Carbon Sequestration (SWP) under Award No. DE-FC26-05NT42591.

Data Availability Statement: Publicly available datasets were analyzed in this study. This data can be found here: https://edx.netl.doe.gov/group/rcsp-swp (accessed on 7 February 2021).

Conflicts of Interest: The authors declare no conflict of interest.

Nomenclature

D	Depth
n_c	Number of components
N_i	Moles of Component i per unit of gridblock volume
N_{n_c+1}	Moles of water per unit of gridblock volume
p	Pressure
P_{cog}	Oil-gas capillary pressure
P_{cwo}	Water-oil capillary pressure
q	Injection/production rate
t	Time
T_j	Transmissibility of phase j
V	Gridblock volume
y_{ij}	Mole fraction of Component i in phase j
γ	Specific gravity
Δt	Timestep
ρ_m	Molar density of phase m
ϕ	Porosity
ψ	Function
Superscripts	
(k)	Iteration level
n	Old time level
$n+1$	New time level
Subscripts	
i	component
j	phase
o	oil
g	gas
w	water

Appendix A

Table A1. List of cases consists of a seven-factor Box–Behnken design (low: -1; mid: 0; high: 1) and two extreme permutations (all low values and all high values).

Case No.	Calcite	Kaolinite	Dolomite	Quartz	Ankerite	Siderite	Illite
1	−1	−1	0	0	0	0	0
2	1	−1	0	0	0	0	0
3	−1	1	0	0	0	0	0
4	1	1	0	0	0	0	0
5	−1	0	−1	0	0	0	0
6	1	0	−1	0	0	0	0
7	−1	0	1	0	0	0	0
8	1	0	1	0	0	0	0
9	−1	0	0	−1	0	0	0
10	1	0	0	−1	0	0	0
11	−1	0	0	1	0	0	0
12	1	0	0	1	0	0	0
13	−1	0	0	0	−1	0	0
14	1	0	0	0	−1	0	0
15	−1	0	0	0	1	0	0
16	1	0	0	0	1	0	0
17	−1	0	0	0	0	−1	0
18	1	0	0	0	0	−1	0
19	−1	0	0	0	0	1	0
20	1	0	0	0	0	1	0
21	−1	0	0	0	0	0	−1
22	1	0	0	0	0	0	−1
23	−1	0	0	0	0	0	1
24	1	0	0	0	0	0	1
25	0	−1	−1	0	0	0	0
26	0	1	−1	0	0	0	0
27	0	−1	1	0	0	0	0
28	0	1	1	0	0	0	0
29	0	−1	0	−1	0	0	0
30	0	1	0	−1	0	0	0
31	0	−1	0	1	0	0	0
32	0	1	0	1	0	0	0
33	0	−1	0	0	−1	0	0
34	0	1	0	0	−1	0	0
35	0	−1	0	0	1	0	0
36	0	1	0	0	1	0	0
37	0	−1	0	0	0	−1	0
38	0	1	0	0	0	−1	0
39	0	−1	0	0	0	1	0
40	0	1	0	0	0	1	0
41	0	−1	0	0	0	0	−1
42	0	1	0	0	0	0	−1
43	0	−1	0	0	0	0	1

Table A1. *Cont.*

44	0	1	0	0	0	0	1
45	0	0	−1	−1	0	0	0
46	0	0	1	−1	0	0	0
47	0	0	−1	1	0	0	0
48	0	0	1	1	0	0	0
49	0	0	−1	0	−1	0	0
50	0	0	1	0	−1	0	0
51	0	0	−1	0	1	0	0
52	0	0	1	0	1	0	0
53	0	0	−1	0	0	−1	0
54	0	0	1	0	0	−1	0
55	0	0	−1	0	0	1	0
56	0	0	1	0	0	1	0
57	0	0	−1	0	0	0	−1
58	0	0	1	0	0	0	−1
59	0	0	−1	0	0	0	1
60	0	0	1	0	0	0	1
61	0	0	0	−1	−1	0	0
62	0	0	0	1	−1	0	0
63	0	0	0	−1	1	0	0
64	0	0	0	1	1	0	0
65	0	0	0	−1	0	−1	0
66	0	0	0	1	0	−1	0
67	0	0	0	−1	0	1	0
68	0	0	0	1	0	1	0
69	0	0	0	−1	0	0	−1
70	0	0	0	1	0	0	−1
71	0	0	0	−1	0	0	1
72	0	0	0	1	0	0	1
73	0	0	0	0	−1	−1	0
74	0	0	0	0	1	−1	0
75	0	0	0	0	−1	1	0
76	0	0	0	0	1	1	0
77	0	0	0	0	−1	0	−1
78	0	0	0	0	1	0	−1
79	0	0	0	0	−1	0	1
80	0	0	0	0	1	0	1
81	0	0	0	0	0	−1	−1
82	0	0	0	0	0	1	−1
83	0	0	0	0	0	−1	1
84	0	0	0	0	0	1	1
85	0	0	0	0	0	0	0
86	−1	−1	−1	−1	−1	−1	−1
87	1	1	1	1	1	1	1

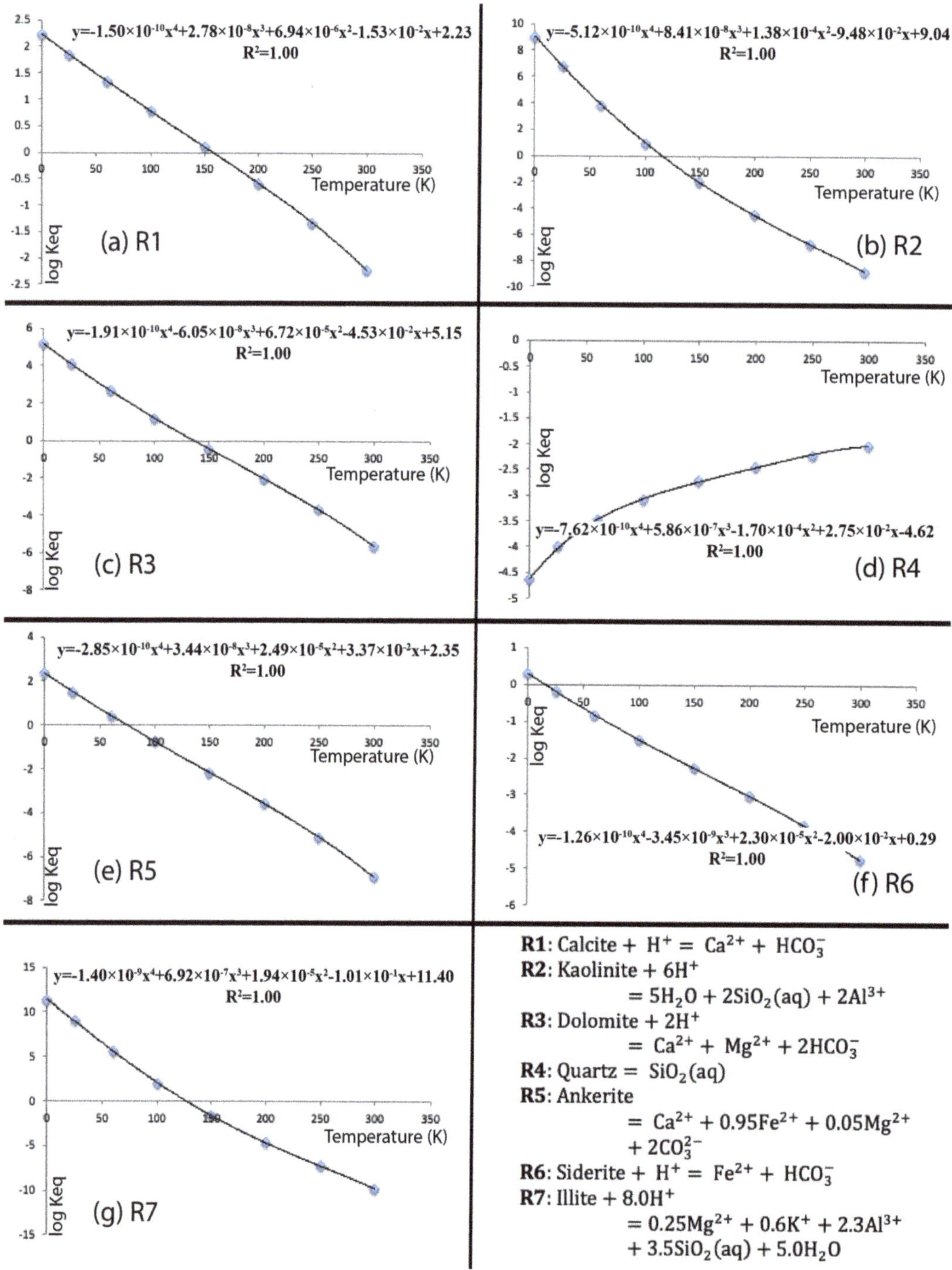

The reactions shown in the figure are:

R1: Calcite $+ H^+ = Ca^{2+} + HCO_3^-$

R2: Kaolinite $+ 6H^+ = 5H_2O + 2SiO_2(aq) + 2Al^{3+}$

R3: Dolomite $+ 2H^+ = Ca^{2+} + Mg^{2+} + 2HCO_3^-$

R4: Quartz $= SiO_2(aq)$

R5: Ankerite $= Ca^{2+} + 0.95Fe^{2+} + 0.05Mg^{2+} + 2CO_3^{2-}$

R6: Siderite $+ H^+ = Fe^{2+} + HCO_3^-$

R7: Illite $+ 8.0H^+ = 0.25Mg^{2+} + 0.6K^+ + 2.3Al^{3+} + 3.5SiO_2(aq) + 5.0H_2O$

Figure A1. Relationship between chemical equilibrium constants and temperature based on the EQ3/6 database (blue diamonds) for determining the chemical equilibrium constants of seven mineral reactions [43].

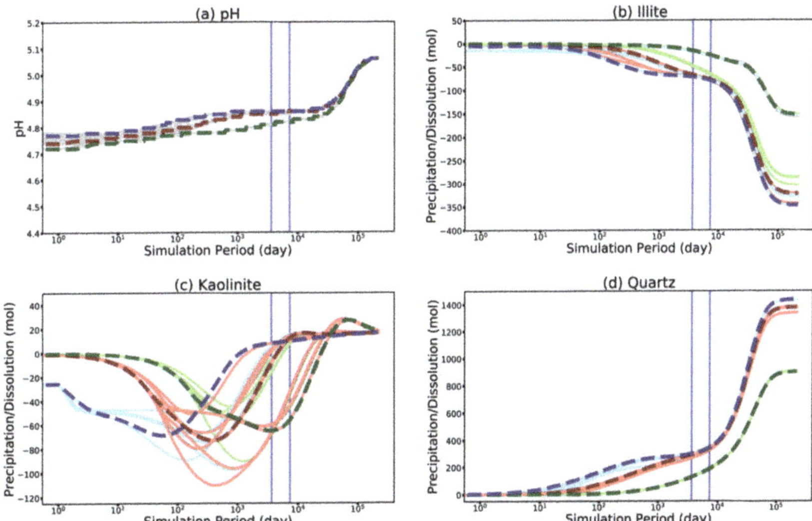

Figure A2. Predicted values from the 2-phase models: (**a**) pH, mineral precipitation or dissolution of (**b**) illite, (**c**) kaolinite, and (**d**) quartz at the center of the model. Light green, light red, and light blue indicate low, mid, and high RSA of the particular mineral in each subplot; dashed lines in dark green, dark red, and dark blue indicate Case#86, Case #85, and Case #87, respectively; two vertical lines denote the end of 10 years and the end of 20 years.

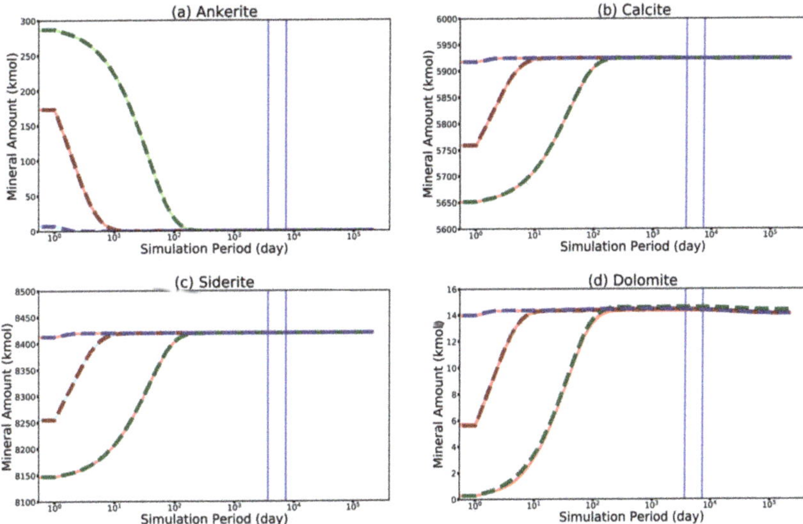

Figure A3. Predicted total mineral amount in the simplified model with only water and CO_2: (**a**) ankerite, (**b**) calcite, (**c**) siderite, and (**d**) dolomite. Light green, light red, and light blue indicate low, mid, and high RSA of the particular mineral in each subplot; dashed lines in dark green, dark red, and dark blue indicate Case#86, Case #85, and Case #87, respectively; two vertical lines denote the end of 10 years and the end of 20 years.

Appendix B

The flow equations of the CMG-GEM simulation package include:

$$\psi_i \equiv \Delta T_o^m y_{io}^m \left(\Delta p^{n+1} - \gamma_o^m \Delta D\right) + \Delta T_g^m y_{ig}^m \left(\Delta p^{n+1} + \Delta Pcogm - \gamma_g^m \Delta D\right) + q_i^m - \frac{V}{\Delta t}\left[N_i^{n+1} - N_i^n\right] = 0 \tag{A1}$$
$$i = 1, \cdots, n_c$$

$$\psi_{n_c+1} \equiv T_w^m \left(\Delta p^{n+1} + \Delta P_{cwo}^m - \gamma_g^m \Delta D\right) + q_{n_c+1}^m - \frac{V}{\Delta t}\left[N_{n_c+1}^{n+1} - N_{n_c+1}^n\right] = 0 \tag{A2}$$

$$N_i = \phi\left(\rho_o S_o y_{io} + \rho_g S_g y_{ig}\right) \tag{A3}$$

$$N_{n_c+1} = \phi\rho_w S_w \tag{A4}$$

References

1. Lackner, K.S. A guide to CO_2 sequestration. *Science* **2003**, *300*, 1677–1678. [CrossRef] [PubMed]
2. Metz, B.; Davidson, O.; De Coninck, H.C.; Loos, M.; Meyer, L.A. IPCC special report on carbon dioxide capture and storage. In *Working Group III of the Intergovernmental Panel on Climate Change*; IPCC: Geneva, Switzerland; Cambridge University Press: Cambridge, UK, 2005; Volume 4.
3. Pacala, S.; Socolow, R. Stabilization wedges: solving the climate problem for the next 50 years with current technologies. *Science* **2004**, *305*, 968–972. [CrossRef]
4. Jia, W.; McPherson, B. Chapter 6—Multiphase Flow Associated with Geological CO_2 Storage. In *Science of Carbon Storage in Deep Saline Formations*; Newell, P., Ilgen, A.G., Eds.; Elsevier: Amsterdam, The Netherlands, 2019; pp. 117–143. ISBN 978-0-12-812752-0.
5. Han, W.S.; McPherson, B.J.; Lichtner, P.C.; Wang, F.P. Evaluation of trapping mechanisms in geologic CO_2 sequestration: Case study of SACROC northern platform, a 35-year CO_2 injection site. *Am. J. Sci.* **2010**, *310*, 282–324. [CrossRef]
6. Jia, W.; McPherson, B.J.; Pan, F.; Xiao, T.; Bromhal, G. Probabilistic analysis of CO_2 storage mechanisms in a CO_2-EOR field using polynomial chaos expansion. *Int. J. Greenh. Gas Control* **2016**, *51*, 218–229. [CrossRef]
7. Kutsienyo, E.J.; Ampomah, W.; Sun, Q.; Balch, R.S.; You, J.; Aggrey, W.N.; Cather, M. *Evaluation of CO-EOR Performance and Storage Mechanisms in an Active Partially Depleted Oil Reservoir*; Society of Petroleum Engineers (SPE): London, UK, 2019.
8. Sun, Q.; Ampomah, W.; Kutsienyo, E.J.; Appold, M.; Adu-Gyamfi, B.; Dai, Z.; Soltanian, M.R. Assessment of CO_2 trapping mechanisms in partially depleted oil-bearing sands. *Fuel* **2020**, *278*, 118356. [CrossRef]
9. Bethke, C.M. *Geochemical Reaction Modeling: Concepts and Applications*; Oxford University Press: Oxford, UK, 1996; ISBN 0195094751.
10. Brunauer, S.; Emmett, P.H.; Teller, E. Adsorption of Gases in Multimolecular Layers. *J. Am. Chem. Soc.* **1938**, *60*, 309–319. [CrossRef]
11. Beckingham, L.E.; Mitnick, E.H.; Steefel, C.I.; Zhang, S.; Voltolini, M.; Swift, A.M.; Yang, L.; Cole, D.R.; Sheets, J.M.; Ajo-Franklin, J.B.; et al. Evaluation of mineral reactive surface area estimates for prediction of reactivity of a multi-mineral sediment. *Geochim. Cosmochim. Acta* **2016**, *188*, 310–329. [CrossRef]
12. Beckingham, L.E. Evaluation of Macroscopic Porosity-Permeability Relationships in Heterogeneous Mineral Dissolution and Precipitation Scenarios. *Water Resour. Res.* **2017**, *53*, 10217–10230. [CrossRef]
13. Bourg, I.C.; Beckingham, L.E.; DePaolo, D.J. The Nanoscale Basis of CO_2 Trapping for Geologic Storage. *Environ. Sci. Technol.* **2015**, *49*, 10265–10284. [CrossRef]
14. Jia, W.; McPherson, B.; Pan, F.; Dai, Z.; Xiao, T. Uncertainty quantification of CO_2 storage using Bayesian model averaging and polynomial chaos expansion. *Int. J. Greenh. Gas Control* **2018**, *71*. [CrossRef]
15. Dai, Z.; Viswanathan, H.; Xiao, T.; Middleton, R.; Pan, F.; Ampomah, W.; Yang, C.; Zhou, Y.; Jia, W.; Lee, S.-Y.; et al. CO_2 Sequestration and Enhanced Oil Recovery at Depleted Oil/Gas Reservoirs. *Energy Procedia* **2017**, *114*, 6957–6967. [CrossRef]
16. Xiao, T.; McPherson, B.; Pan, F.; Esser, R.; Jia, W.; Bordelon, A.; Bacon, D. Potential chemical impacts of CO_2 leakage on underground source of drinking water assessed by quantitative risk analysis. *Int. J. Greenh. Gas Control* **2016**, *50*. [CrossRef]
17. Xiao, T.; McPherson, B.; Esser, R.; Jia, W.; Dai, Z.; Chu, S.; Pan, F.; Viswanathan, H. Chemical Impacts of Potential CO_2 and Brine Leakage on Groundwater Quality with Quantitative Risk Assessment: A Case Study of the Farnsworth Unit. *Energies* **2020**, *13*, 6574. [CrossRef]
18. You, J.; Ampomah, W.; Sun, Q. Co-optimizing water-alternating-carbon dioxide injection projects using a machine learning assisted computational framework. *Appl. Energy* **2020**, *279*, 115695. [CrossRef]
19. Jia, W.; Mcpherson, B.; Pan, F.; Dai, Z.; Moodie, N.; Xiao, T. Impact of Three-Phase Relative Permeability and Hysteresis Models on Forecasts of Storage Associated with CO_2-EOR. *Water Resour. Res.* **2018**, *54*, 1109–1126. [CrossRef]
20. Dai, Z.; Xu, L.; Xiao, T.; McPherson, B.; Zhang, X.; Zheng, L.; Dong, S.; Yang, Z.; Soltanian, M.R.; Yang, C.; et al. Reactive chemical transport simulations of geologic carbon sequestration: Methods and applications. *Earth-Sci. Rev.* **2020**, *208*, 103265. [CrossRef]
21. Qin, F.; Beckingham, L.E. The impact of mineral reactive surface area variation on simulated mineral reactions and reaction rates. *Appl. Geochem.* **2021**, *124*, 104852. [CrossRef]

22. Luo, S.; Xu, R.; Jiang, P. Effect of reactive surface area of minerals on mineralization trapping of CO_2 in saline aquifers. *Pet. Sci.* **2012**, *9*, 400–407. [CrossRef]

23. Bolourinejad, P.; Shoeibi Omrani, P.; Herber, R. Effect of reactive surface area of minerals on mineralization and carbon dioxide trapping in a depleted gas reservoir. *Int. J. Greenh. Gas Control* **2014**, *21*, 11–22. [CrossRef]

24. Jia, W.; Xiao, T.; Moodie, N.; McPherson, B. Uncertainty Analysis of Impact of Geochemical Reactions on Forecasting CO_2 Storage at a Depleted Oil Field. In Proceedings of the 2019 AIChE Annual Meeting, Orlando, FL, USA, 13 November 2019.

25. Wu, Z.; Luhmann, A.J.; Rinehart, A.J.; Mozley, P.S.; Dewers, T.A.; Heath, J.E.; Majumdar, B.S. Chemo-mechanical Alterations Induced from CO_2 Injection in Carbonate-Cemented Sandstone: An Experimental Study at 71 °C and 29 MPa. *J. Geophys. Res. Solid Earth* **2020**, *125*, e2019JB019096. [CrossRef]

26. Ampomah, W.; Balch, R.S.; Ross-Coss, D.; Hutton, A.; Cather, M.; Will, R.A. An Integrated Approach for Characterizing a Sandstone Reservoir in the Anadarko Basin. In Proceedings of the Offshore Technology Conference, Houston, TX, USA, 2–5 May 2016.

27. Ross-Coss, D.; Ampomah, W.; Cather, M.; Balch, R.S.; Mozley, P.; Rasmussen, L. An improved approach for sandstone reservoir characterization. In Proceedings of the SPE Western Regional Meeting, Anchorage, AK, USA, 23–26 May 2016.

28. Kumar, A.; Chao, K.; Hammack, R.; Harbert, W.; Ampomah, W.; Balch, R.; Garcia, L. Surface-seismic monitoring of an active CO_2-EOR operation in the Texas Panhandle using broadband seismometers. In *SEG Technical Program Expanded Abstracts 2018*; Society of Exploration Geophysicists: Huston, TX, USA, 2018; pp. 3027–3031. ISBN 1949-4645.

29. Dai, Z.; Middleton, R.; Viswanathan, H.; Fessenden-Rahn, J.; Bauman, J.; Pawar, R.; Lee, S.-Y.; McPherson, B. An integrated framework for optimizing CO_2 sequestration and enhanced oil recovery. *Environ. Sci. Technol. Lett.* **2013**, *1*, 49–54. [CrossRef]

30. Xiao, T.; McPherson, B.; Bordelon, A.; Viswanathan, H.; Dai, Z.; Tian, H.; Esser, R.; Jia, W.; Carey, W. Quantification of CO_2-cement-rock interactions at the well-caprock-reservoir interface and implications for geological CO_2 storage. *Int. J. Greenh. Gas Control* **2017**, *63*, 126–140. [CrossRef]

31. Xiao, T.; Xu, H.; Moodie, N.; Esser, R.; Jia, W.; Zheng, L.; Rutqvist, J.; McPherson, B. Chemical-Mechanical Impacts of CO_2 Intrusion into Heterogeneous Caprock. *Water Resour. Res.* **2020**, *56*, e2020WR027193. [CrossRef]

32. Rose-Coss, D.; Wampomah, W.; Hutton, A.C.; Gragg, E.; Mozley, P.; Balch, R.S.; Grigg, R. Geologic Characterization for CO_2-EOR Simulation: A Case Study of the Farnsworth Unit, Anadarko Basin, Texas. In Proceedings of the AAPG Annual Convention and Exhibition, Socorro, NM, USA, 2 June 2015.

33. Balch, R.; Grigg, R.; McPherson, B. Overview of a Large Scale Carbon Capture, Utilization, and Storage Demonstration Project in an Active Oil Field in Texas, USA. *Energy Procedia* **2017**, *114*, 5874–5887. [CrossRef]

34. Ampomah, W.; Balch, R.; Cather, M.; Rose-Coss, D.; Dai, Z.; Heath, J.E.; Dewers, T.; Mozley, P. Evaluation of CO_2 Storage Mechanisms in CO_2 Enhanced Oil Recovery Sites: Application to Morrow Sandstone Reservoir. *Energy Fuels* **2016**, *30*, 8545–8555. [CrossRef]

35. Moodie, N.; Pan, F.; Jia, W.; McPherson, B. Impacts of relative permeability formulation on forecasts of CO_2 phase behavior, phase distribution, and trapping mechanisms in a geologic carbon storage reservoir. *Greenh. Gases Sci. Technol.* **2017**, *7*, 958–962. [CrossRef]

36. Moodie, N.; Ampomah, W.; Jia, W.; Heath, J.; McPherson, B. Assignment and calibration of relative permeability by hydrostratigraphic units for multiphase flow analysis, case study: CO_2-EOR operations at the Farnsworth Unit, Texas. *Int. J. Greenh. Gas Control* **2019**, *81*. [CrossRef]

37. Dai, Z.; Viswanathan, H.; Middleton, R.; Pan, F.; Ampomah, W.; Yang, C.; Jia, W.; Xiao, T.; Lee, S.-Y.; McPherson, B. CO_2 Accounting and Risk Analysis for CO_2 Sequestration at Enhanced Oil Recovery Sites. *Environ. Sci. Technol.* **2016**, *50*, 7546–7554. [CrossRef]

38. Pan, F.; McPherson, B.J.; Dai, Z.; Jia, W.; Lee, S.-Y.; Ampomah, W.; Viswanathan, H.; Esser, R. Uncertainty analysis of carbon sequestration in an active CO_2-EOR field. *Int. J. Greenh. Gas Control* **2016**, *51*. [CrossRef]

39. Jia, W.; Pan, F.; Dai, Z.; Xiao, T.; McPherson, B. Probabilistic Risk Assessment of CO_2 Trapping Mechanisms in a Sandstone CO_2-EOR Field in Northern Texas, USA. *Energy Procedia* **2017**, *114*, 4321–4329. [CrossRef]

40. Landrot, G.; Ajo-Franklin, J.B.; Yang, L.; Cabrini, S.; Steefel, C.I. Measurement of accessible reactive surface area in a sandstone, with application to CO_2 mineralization. *Chem. Geol.* **2012**, *318–319*, 113–125. [CrossRef]

41. Waldmann, S. Geological and Mineralogical Investigation of Rotliegend Gas Reservoirs in the Netherlands and Their Potential for CO_2 Storage. Ph.D. Thesis, Friedrich-Schiller-Universität Jena, Jena, Germany, 2011.

42. Ampomah, W.; Balch, R.S.; Grigg, R.B.; McPherson, B.; Will, R.A.; Lee, S.; Dai, Z.; Pan, F. Co-optimization of CO_2-EOR and storage processes in mature oil reservoirs. *Greenh. Gases Sci. Technol.* **2016**, *7*, 128–142. [CrossRef]

43. Wolery, T.J. *EQ3/6, a Software Package for Geochemical Modeling of Aqueous Systems: Package Overview and Installation Guide (Version 7.0)*; Lawrence Livermore National Lab.: Livermore, CA, USA, 1992.

44. Palandri, J.L.; Kharaka, Y.K. *A Compilation of Rate Parameters of Water-Mineral Interaction Kinetics for Application to Geochemical Modeling*; Geological Survey: Menlo Park, CA, USA, 2004.

45. Pan, F.; McPherson, B.J.; Esser, R.; Xiao, T.; Appold, M.S.; Jia, W.; Moodie, N. Forecasting evolution of formation water chemistry and long-term mineral alteration for GCS in a typical clastic reservoir of the Southwestern United States. *Int. J. Greenh. Gas Control* **2016**, *54*. [CrossRef]

46. Tao, J.; Wu, Y.; Elsworth, D.; Li, P.; Hao, Y. Coupled Thermo-Hydro-Mechanical-Chemical Modeling of Permeability Evolution in a CO_2-Circulated Geothermal Reservoir. *Geofluids* **2019**, *2019*, 5210730. [CrossRef]
47. Fan, C.; Elsworth, D.; Li, S.; Zhou, L.; Yang, Z.; Song, Y. Thermo-hydro-mechanical-chemical couplings controlling CH_4 production and CO_2 sequestration in enhanced coalbed methane recovery. *Energy* **2019**, *173*, 1054–1077. [CrossRef]
48. Tahmasebi, P.; Kamrava, S. A pore-scale mathematical modeling of fluid-particle interactions: Thermo-hydro-mechanical coupling. *Int. J. Greenh. Gas Control* **2019**, *83*, 245–255. [CrossRef]

Article

Chemical Impacts of Potential CO_2 and Brine Leakage on Groundwater Quality with Quantitative Risk Assessment: A Case Study of the Farnsworth Unit

Ting Xiao [1,2], Brian McPherson [1,2,*], Richard Esser [1,2], Wei Jia [1,2], Zhenxue Dai [3], Shaoping Chu [4], Feng Pan [5] and Hari Viswanathan [4]

[1] Department of Civil and Environmental Engineering, The University of Utah, Salt Lake City, UT 84112, USA; ting.xiao@utah.edu (T.X.); rich.esser@utah.edu (R.E.); wei.jia@utah.edu (W.J.)
[2] Energy and Geoscience Institute, The University of Utah, Salt Lake City, UT 84108, USA
[3] College of Construction Engineering, Jilin University, Changchun 130026, China; dzx@jlu.edu.cn
[4] Earth and Environmental Sciences Division, Los Alamos National Laboratory, Los Alamos, NM 87545, USA; spchu@lanl.gov (S.C.); viswana@lanl.gov (H.V.)
[5] Utah Division of Water Resources, Salt Lake City, UT 84116, USA; panpanfeng797@hotmail.com
* Correspondence: b.j.mcpherson@utah.edu; Tel.: +1-801-585-7961

Received: 23 October 2020; Accepted: 9 December 2020; Published: 14 December 2020

Abstract: Potential leakage of reservoir fluids is considered a key risk factor for geologic CO_2 sequestration (GCS), with concerns of their chemical impacts on the quality of overlying underground sources of drinking water (USDWs). Effective risk assessment provides useful information to guide GCS activities for protecting USDWs. In this study, we present a quantified risk assessment case study of an active commercial-scale CO_2-enhanced oil recovery (CO_2-EOR) and sequestration field, the Farnsworth Unit (FWU). Specific objectives of this study include: (1) to quantify potential risks of CO_2 and brine leakage to the overlying USDW quality with response surface methodology (RSM); and (2) to identify water chemistry indicators for early detection criteria. Results suggest that trace metals (e.g., arsenic and selenium) are less likely to become a risk due to their adsorption onto clay minerals; no-impact thresholds based on site monitoring data could be a preferable reference for early groundwater quality evaluation; and pH is suggested as an indicator for early detection of a leakage. This study may provide quantitative insight for monitoring strategies on GCS sites to enhance the safety of long-term CO_2 sequestration.

Keywords: geologic CO_2 sequestration; CO_2 and brine leakage; underground source of drinking water; risk assessment; response surface methodology; early detection criteria

---

1. Introduction

Carbon dioxide capture and sequestration (CCS) in geologic formations is considered a promising approach for mitigating CO_2 emissions, by injecting CO_2 from stationary sources into deep geologic formations [1,2]. After the policy of United States emphasized "utilization" of carbon capture, utilization, and storage (CCUS), CO_2-enhanced oil recovery (CO_2-EOR) and storage has gained specific interest for its potential benefits of increasing oil production and reducing CO_2 storage costs [3–5].

It is believed that the risks of geologic CO_2 sequestration (GCS) to the environment and human health are minimized with monitoring and managements of the sites, especially for operational reservoirs with pressure managements [6–8]. However, the concern of reservoir fluids, especially CO_2 leakage to overlying underground sources of drinking water (USDWs) cannot be completely ruled out [9–12]. Carbon dioxide itself is not hazardous to groundwater quality, but it triggers pH reduction, water-sediment interactions, and potential toxic trace metal release from sediments [13–15]. Reservoir brine leakage may significantly increase shallow groundwater salinity and introduce hazardous reservoir substances into overlying USDWs [16,17].

Wells are usually identified as a greater risk of potential leakage pathways than geological features of faults and/or fractures [18,19]. Specifically, reservoir fluids may leak through wellbore cement and well casing. Wellbore cement degradation caused by CO_2 intrusion is a complex function of cement properties, fluid dynamics, reaction kinetics, and stress state of the wellbore environment [20–23]. It is believed that moderate exposure to CO_2 could provide a less permeable front of the cement with calcite precipitation to avoid further acid intrusion for centuries [19,21,24,25]. Even with microcracks occurring in wellbores, most leaked CO_2 could be trapped by the cement, and leakage flux of reservoir fluids (CO_2 and brine) is neglectable to be considered as a potential risk [26]. In most risk assessment studies, abandoned legacy wells with hypothetical open boreholes and/or wellbore failures are assumed as the most likely leakage pathways [27,28]. It represents the worst-case scenario for area of review (AoR) evaluations stipulated by the U.S. Environmental Protection Agency (EPA) Underground Injection Control (UIC) guidance, where wells with stability problems, casing failure, and/or abandoned wells not identified with site characterization and monitoring approaches [27]. Maximum CO_2 leakage rates between 10^{-7} and 10^{-1} kg/s are used for most risk assessment approaches, which is usually up to 0.4% of the cumulative CO_2 injected in the reservoir [27,29,30].

Regulatory policy emphasizes the protections against reservoir leakage to USDWs with assessments of risks to water quality for the USDWs and groundwater monitoring prior to, during, and after injection phases [8]. To date, quantitative risk assessments of modeling approaches combined with field observations play an essential role for site-specific studies of potential leakage and its impact on USDW quality, to forecast the long-term response of groundwater, and help the operators to make effective and efficient plans of monitoring strategies [26,27,31–33]. Monte Carlo method is a straightforward simulation approach for risk assessment [34], but its high computational cost motivates the applications of reduced order models (ROMs) to replace original simulations with surrogate models [35]. Response surface methodology (RSM) is a widely-used statistical and mathematical technique to generate ROMs, and has been applied for many risk assessment approaches for GCS research [26,34,36]. Thus, RSM was selected to quantify risks on shallow groundwater quality in this study.

In this manuscript, we present quantitative assessment of potential risks to overlying groundwater quality due to CO_2 and brine leakage at an on-going CO_2-EOR and storage site, the Farnsworth Unit (FWU). Considerable operational, geological, and geochemical data available of the reservoir and the overlying USDW aquifer (the Ogallala aquifer) largely improved the reliability with reduced uncertainties. Specific objectives include: (1) to quantify potential risks to the overlying USDW quality due to CO_2 and brine leakage from the operational reservoir; and (2) to identify water chemistry indicators for early detection criteria. Results of this study may provide a useful perspective of combining numerical simulations, field observations, and ROMs for site-specific risk assessments to enhance the safety of GCS projects.

2. Materials and Methods

2.1. Site Description

The FWU site (Figure 1) located in northern Texas is a mature hydrocarbon reservoir undergoing active CO_2-EOR and sequestration since December 2010 [37,38]. It is the study site of the Southwest Regional Partnership on Carbon Sequestration (SWP) Phase III, sponsored by the U.S. Department of Energy (DOE) and the National Energy Technology Laboratory (NETL) [39]. The primary goal of this project is to exhibit and evaluate an active commercial-scale CCUS operation, and demonstrate effective site characterization, monitoring, verification, accounting, and risk assessment for long-term CO_2 sequestration. To date, over one million metric tons of net CO_2 from anthropogenic sources (one fertilizer plant and one ethanol plant) is stored in the subsurface reservoir (the Morrow B Formation), with CO_2 injection and production volumes tracked at the FWU [40]. The SWP acquired significant near-surface monitoring data for potential CO_2 leakage, including soil CO_2 flux (to identify any potential point-source leakage to the surface), borehole CO_2 movement (to monitor subsurface CO_2 movement), and the overlying drinkable groundwater chemistry in the Ogallala aquifer (to identify any potential leakage into the USDW and drinking water quality change). The SWP project partner, the National Institute of Advanced Industrial Science and Technology (AIST) of Japan conducted surface/shallow borehole gravity and electrical methods to monitor the subsurface CO_2 plume [41]. Continuous gravity, self-potential, and magnetotelluric surveys were also applied for three years to monitor brine leakage into the Ogallala aquifer. Up to date, these monitoring activities have not seen any CO_2/brine leakage [39], and the monitoring data along with reservoir characterization provide essential information to improve the reliability of risk assessment at the site.

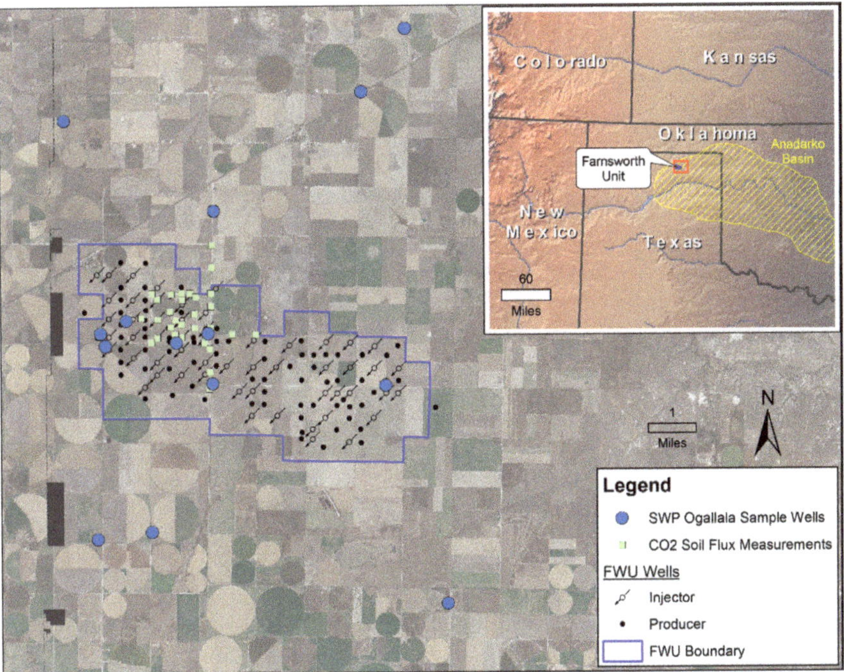

Figure 1. The Farnsworth Unit (FWU) location with 5-patterns of CO_2 injection and oil production, and monitoring locations for CO_2 leakage to the groundwater and surface.

2.2. Site-Specific Water Chemistry and No-Impact Thresholds

On-site monitoring water chemistry provides reliable baseline of shallow groundwater and reservoir brine constituents and determines the potential risks of exceeding water quality thresholds due to any potential CO_2 and/or brine leakage. To date, there are hundreds of water samples collected and analyzed in quarterly basis from the Ogallala aquifer and the reservoir at the FWU area since 2012, conducted by the New Mexico Institute of Mining and Technology, as part of SWP Phase III. General chemistry was analyzed for pH (by pH meter), conductivity (by conductivity meter), alkalinity (by electrometric titration), oxidation and reduction potential (ORP, by pH meter and ORP electrode), major cations and anions (including Li^+, Na^+, K^+, Mg^{2+}, Ca^{2+}, F^-, Cl^-, Br^-, NO_3^-, and SO_4^{2-}, by ion chromatography (IC)), trace metals (by inductively coupled plasma mass spectrometry (ICP-MS)), inorganic carbon (IC), and non-purgeable organic carbon (NPOC, by total organic carbon (TOC) analyzer), and total dissolved solids (TDS, calculated based on the concentrations of major cations and anions [42]). Table 1 summarizes selected monitored water parameters of the Ogallala aquifer and the reservoir produced water. These selected parameters might exceed the U.S. Environmental Protection Agency (EPA) and/or local primary and secondary maximum contaminant levels (MCL) in the Ogallala aquifer with CO_2 and/or brine leakage. Some of the selected contents are sensitive to CO_2 introduction (e.g., pH and trace metals), and some of the contents' concentrations might increase with leaked brine (e.g., TDS and Mn). Changes of these contents may become indicators for any potential leakage as an early-detect criterion.

Table 1. Selected Water Chemistry of the Ogallala Aquifer and the reservoir produced water (PW), maximum contaminant levels (MCL) of the U.S. Environmental Protection Agency (EPA) and Texas State, and No-Impact Threshold (NIT) for potential contaminants of the Ogallala aquifer (unit in mg/L, pH unitless).

Name	Ogallala	PW	EPA MCL	Texas MCL	NIT
pH	7.7	7.2	6.5	7.0	7.5
Total dissolved solids (TDS)	380	4064	500	1000	508
Mn	0.008	0.27	0.05	0.05	0.05
As	0.003	0.005	0.01	0.01	0.005
Se	0.004	0.07	0.05	0.002	0.007

Usually MCLs are used to evaluate whether a leakage plume is harmful for shallow groundwater quality—if the concentration exceeds the limit it would be considered as a risk [17,43,44]. However, the regulatory limits do not reflect the changes from the current background levels at a specific area, especially when a constituent background concentration is significantly lower than the MCL [45]. To determine statistically significant changes of the groundwater for site-specific early detection criteria, the "no-impact thresholds" are defined to predict potential impacts of the leakage in the early stage [45]. The no-impact thresholds are based on site-specific groundwater quality data, and represent the lowest detectable concentrations above the "no change" scenario to predict groundwater quality changes due to any leakage [27,45]. The no-impact thresholds are calculated as the 95%-confidence, 95%-coverage tolerance limit from the existing site monitoring data, and, in some cases, the values might be significantly different from the regulatory standards.

Both no-impact and MCL thresholds were considered in our risk assessments for chemical impacts of groundwater at the FWU [26]. pH, TDS, arsenic (As), selenium (Se), and manganese (Mn) were selected as potential key factors to indicate water quality changes and early detection criteria, where pH directly decreases with CO_2 intrusion, trace metals (As and Se) exist in the reservoir brine and may release from the shallow aquifer sediment at lower pH; TDS and Mn are with high concentrations in the brine, and CO_2-water interactions also change TDS with mineral dissolution.

2.3. Trace Metal Mobilization Due to CO_2 Leakage

Increased CO_2 concentrations in shallow groundwater aquifers would reduce water pH and enhance water-sediment geochemical reactions, resulting in mobilization of toxic trace metals [9,13,43,46–48]. Adsorption/desorption is considered the major mechanism of trace metal release [17,49–51], and it was also considered in our study. The widely used Gouy–Chapman double diffuse layer surface complexation model was applied in our simulations, to calculate such processes [52]. Adsorption/desorption reactions (take As and Se as examples) of sorbent minerals (S represents mineral sites) could be written as [49,53]:

$$SOH_{(s)} + H^+_{(aq)} \rightleftharpoons SOH^+_{2(s)} \tag{1}$$

$$SOH_{(s)} \rightleftharpoons SO^-_{(s)} + H^+_{(aq)} \tag{2}$$

$$SOH_{(s)} + H_3AsO_{4(aq)} \rightleftharpoons SAsO^{2-}_{4(s)} + 2H^+_{(aq)} + H_2O \tag{3}$$

$$SOH_{(s)} + H_2SeO_{3(aq)} \rightleftharpoons SHSeO_{3(s)} + H_2O \tag{4}$$

Hydrous ferric oxide (HFO) and clay minerals (e.g., kaolinite, illite, and smectite) are often used as a sorbent for simulations of trace metal adsorption in groundwater aquifers, because of their large surface areas and occurrence in natural system [52,53]. Adsorption reactions are controlled by the total amount of sorption sites, which is controlled by the amount of sorbent, site density, and sorbent surface area [54]. It is often difficult to quantify the fractions and/or determine their surface areas because of their low concentrations in the sediments, the minerals playing the role of sorbent are usually assumed with a small volume fraction in simulations and treated as an uncertainty parameter in risk assessment [17,26,54]. The X-ray diffraction (XRD) results of the Ogallala aquifer sediments collected at the FWU area suggested a trace amount of smectite in the samples (< 1% of the aquifer sediment). Therefore, the sorbent amount of the Ogallala sediment was treated as an uncertainty parameter in our assessments for trace metal mobilizations by assuming a fixed sorbent fraction with changing its surface area as an uncertainty parameter.

2.4. Quantification Risk Assessment of Groundwater Quality

2.4.1. Response Surface Methodology (RSM)

Chemical impacts of CO_2 leakage into the Ogallala aquifer in our case study was conducted with RSM, a statistical and mathematical technique for improving and optimizing model exploitation based on a largely reduced number of numerical simulations compared to traditional approaches (e.g., Monte Carlo method) [36]. Stages of RSM application include: (1) determining independent uncertainty parameters and numerical simulation design; (2) conducting simulations according to the selected experimental matrix to train the RSM model equation; (3) obtaining the RSM model equation (a polynomial function) and evaluating the model adequacy; and (4) using the RSM model equation to quantify the risks.

CO_2 and brine leakage rates, aquifer thickness, and adsorbent amount were selected as independent variables in this study (Table 2), because they are controlling variables to determine the leakage plume and water chemistry changes in the USDW aquifer. A conceptual well with failure was assumed for reservoir leakage, and 0.4% of CO_2 and water injection rates were assigned for the maximum leakage rates based on injection history of Well 13-9 in the FWU [55], following the typical leakage rate ranges of previous risk assessment approaches [27]. The Ogallala aquifer thickness range and distribution were assigned based on ~150 shallow groundwater well-drilling data at the FWU area, with average thickness 120 m. The adsorbent amount of the Ogallala sediment was treated as an uncertainty parameter by varying its specific surface area (SSA) with a fixed (assumed) sorbent volume fraction of 0.5% smectite. Box–Behnken design was applied with 25 simulations in total (Table 3) [56]. Regional groundwater flow was neglected in this study for simplicity, because the regional hydraulic gradient (~10 m/year) is low for significant impacts on contamination dilution [26].

Table 2. Independent Parameters for response surface methodology (RSM) of CO_2 and Brine Impacts to the Groundwater. SSA, specific surface area.

Parameter Name	Low (−1)	Mid (0)	High (+1)	Distribution
CO_2 leakage rate: g/s	0	0.5	1.0	Uniform
Brine leakage rate: g/s	0	0.25	0.5	Uniform
Aquifer thickness: m	40	120	200	Normal
Adsorbent SSA: m^2/g	1	50.5	100	Uniform

Table 3. Box–Behnken experimental design for the reactive transport simulations. Low (−1), mid (0), and high (+1) values are corresponding to the values and order of independent parameters shown in Table 2.

Simulation	Values of Independent Parameters				Simulation	Values of Independent Parameters			
1	−1	−1	0	0	14	0	−1	1	0
2	−1	1	0	0	15	0	1	−1	0
3	1	−1	0	0	16	0	1	1	0
4	1	1	0	0	17	−1	0	−1	0
5	0	0	−1	−1	18	−1	0	1	0
6	0	0	−1	1	19	1	0	−1	0
7	0	0	1	−1	20	1	0	1	0
8	0	0	1	1	21	0	−1	0	−1
9	−1	0	0	−1	22	0	−1	0	1
10	−1	0	0	1	23	0	1	0	−1
11	1	0	0	−1	24	0	1	0	1
12	1	0	0	1	25	0	0	0	0
13	0	−1	−1	0					

2.4.2. Reactive Transport Model

A one-dimensional conceptual radial model (Figure 2) was assembled and simulated to analyze the potential risks to groundwater quality due to CO_2 and brine leakage. The model radius was assigned 10,000 m (significantly far from the potential leaky well) with 80 cells. A constant temperature of 25 °C was assigned with homogeneous porosity of 0.3 and permeability 10^{-13} m^2, according to the characterizations of the aquifer. With consideration of CO_2 injection for at least 30 years, and a 50-year post-monitoring period required for a GCS project after CO_2 injection is ceased, the total simulation time was arbitrarily assigned 200 years, which was significantly longer than the injection and post-injection (monitoring) period (~75 years in total).

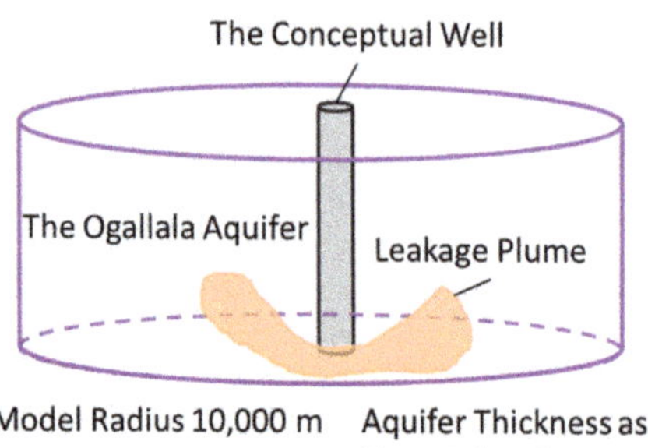

Figure 2. The conceptual model for the Ogallala Aquifer.

The initial groundwater and leakage brine chemistry for the model was assigned based on the average composition of the FWU monitoring samples (Table 4). The initial mineralogy was assigned following the XRD results of the Ogallala sediment samples collected at the FWU area (Table 5), with mineral reactive surface areas assigned following literature [17,55]. Aqueous complexation, cation exchange (Na^+, H^+, Ca^{2+}, Mg^{2+}, K^+, Fe^{2+}, and Mn^{2+}), adsorption/desorption and mineral dissolution/precipitation were considered for chemical reactions in the aquifer. The thermodynamic parameters for aqueous and mineral reactions were assigned following the EQ3/6 database [57]. The parameters for the kinetic rate law of minerals were taken from [58]. The Gouy–Chapman double diffuse layer model was used for adsorption reactions [53,59]. Cation exchange coefficients were taken from [60]. All simulations were performed with TOUGHREACT V2 [61] and with equation of state ECO2N for multiphase CO_2 and brine [62].

Table 4. Initial water chemistry for the Ogallala Aquifer and the leaked reservoir brine (PW) (unit: mol/kg).

Name	Ogallala	PW	Name	Ogallala	PW
pH (unitless)	7.7	7.2	SiO_2 (aq)	6.613×10^{-4}	6.667×10^{-4}
Ca^{2+}	1.189×10^{-3}	1.937×10^{-3}	Cl^-	1.582×10^{-3}	5.619×10^{-2}
Mg^{2+}	7.935×10^{-5}	1.286×10^{-4}	HCO_3^-	2.355×10^{-3}	4.267×10^{-3}
Na^+	2.068×10^{-3}	6.407×10^{-2}	SO_4^{2-}	5.105×10^{-4}	2.735×10^{-4}
K^+	2.649×10^{-4}	4.534×10^{-4}	NO_3^-	2.399×10^{-4}	3.898×10^{-6}
Fe^{2+}	8.880×10^{-9}	6.517×10^{-9}	$H_2AsO_4^-$	5.290×10^{-9}	6.775×10^{-8}
AlO_2^-	2.311×10^{-10}	1.732×10^{-10}	$HSeO_3^-$	1.733×10^{-8}	8.887×10^{-6}
Mn^{2+}	1.457×10^{-7}	4.568×10^{-6}			

Table 5. Initial mineralogy assigned for the model. Mineral surface areas are assigned following literatures [17,55].

Mineral Name	Formula	Volume Fraction	Surface Area (cm^2/g)
Primary			
Quartz	SiO_2	0.780	23.29
Calcite	$CaCO_3$	0.110	53.96
K-Feldspar	$KAlSi_3O_8$	0.092	222.42
Dolomite	$CaMg(CO_3)_2$	0.004	9.80
Smectite	$Na_{0.29}Mg_{0.26}Al_{1.77}Si_{3.97}O_{10}(OH)_2$	0.005	151.60
Albite	$NaAlSi_3O_8$	0.005	11.40
Secondary			
Illite	$K_{0.6}Mg_{0.25}Al_{1.8}(Al_{0.5}Si_{3.5}O_{10})(OH)_2$	0	272.06

3. Results and Discussion

3.1. Impacts on Groundwater Quality

The RSM model equations of our target parameters were trained with 25 numerical simulations. Overall, the correlation between the original (full reservoir model) simulated results and the RSM forecasted results are above 0.9 for most of the grids, especially within 100 m from the conceptual well, suggesting that the trained RSM equations (polynomial functions) sufficiently fit the original reactive transport model outcomes and are adequate for forecasting the water quality parameters of interest within the selected range. With the trained polynomial RSM equations, selected parameters were calculated with 10,000 random cases (each independent variable was created with a random seed in their selected ranges). Cumulative distribution functions (CDF) were obtained accordingly, and they were used for forecasting the likelihood that the leakage would impact groundwater quality over 200 years.

Figure 3 illustrates CDFs of gaseous CO_2 saturation (SG) within 50 m radius away from the conceptual leaky well in the aquifer. If CO_2 leaked through a well with a maximum leakage rate of 1 g/s, it is very likely that gaseous phase CO_2 be observed in the aquifer 1 m away from the well after 1 year, and the plume radius increases to 10 m after 10 years. With the gas phase CO_2 intrusion, pH of the aquifer decreases accordingly, and local groundwater might drain off near the well. TDS concentration near the well might become extremely high with the worst scenarios due to CO_2 dissolution as well as the leaked brine contribution (Figure 4). It is less likely for gaseous CO_2 plume reaching 50 m away from the conceptual well within 100 years, and at the 200th year, cases above the 60th percentile show a small amount of free gas at this distance. It is likely that the free gas plume would reach 50 m away from the well after 200 years' leakage. With the maximum leakage rate of 0.4% injection rate, none of the simulations show an occurrence of gas phase CO_2 at the location 100 m away from the well.

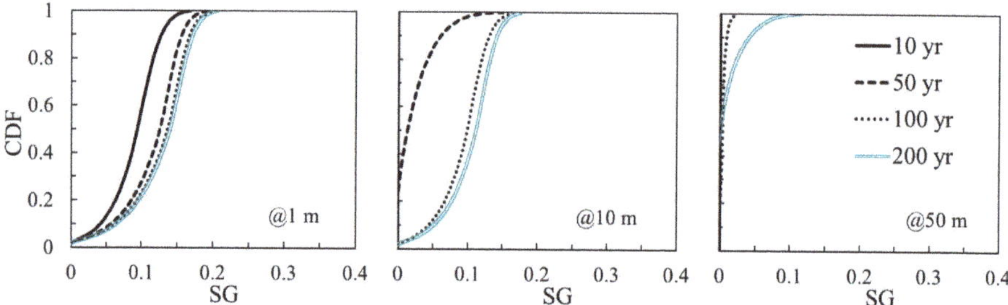

Figure 3. Cumulative distribution functions (CDFs) of gaseous phase CO_2 Saturation (SG) at 1 m, 10 m, and 50 m from the conceptual well.

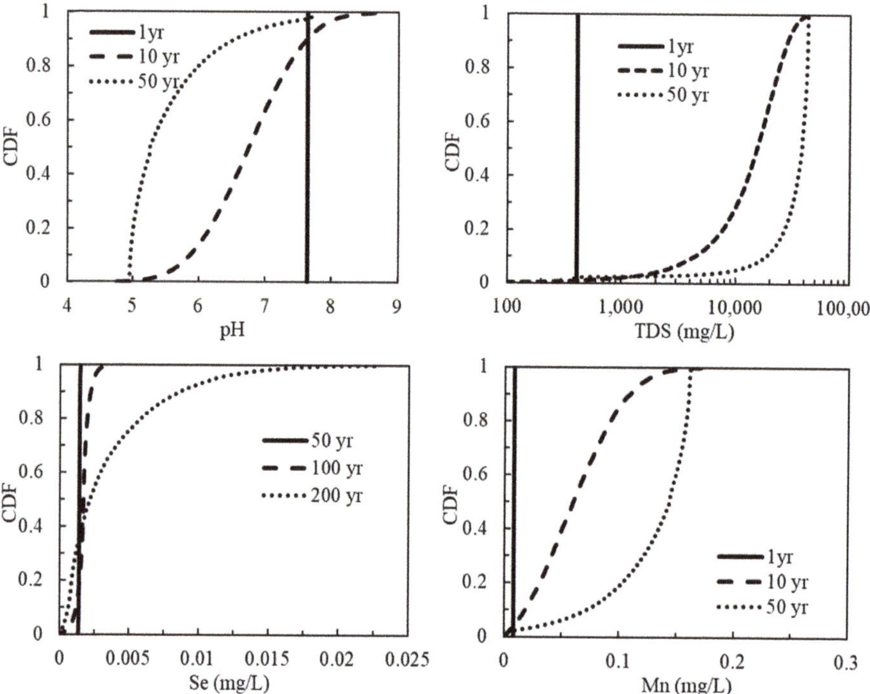

Figure 4. Cumulative distribution functions (CDFs) of pH, TDS, Mn and Se concentrations at 10 m from the conceptual well.

The CDF curves of pH, TDS, Mn and Se at 10 m away from the leaky well are shown in Figure 4. The RSM model results (10,000 random cases) suggest that the groundwater would not be largely impacted by the leakage within 1 year, but there are significant changes in pH, TDS, and Mn concentrations after 10 years. After 50 years, these concentrations maintain at a stable level (the curves beyond 50 years are not shown for pH, TDS, and Mn) because the gas saturation and dissolved CO_2 concentration reach a steady state at this location. However, the response of Se concentration would not start until 100 years after the leakage starts, due to adsorption onto the sediment. For As, it does not show any significant changes even near the well, because its concentration in the leaked brine is not significantly higher than that in the groundwater. It indicates that clay minerals could mitigate trace metal mobilization within a certain extent with surface complexation reactions of the Ogallala aquifer, which is beneficial for the aquifer maintaining its quality. To the contrary, the high salinity and the metals not reactive with sorbents (clays) of the leaked brine may likely be a larger concern in this case (TDS and Mn). Similar results are also obtained by other studies [17,63].

Overall, the leakage plume would reach ~50 m away from the well within 200 years of leakage from the reservoir with 0.4% injection rate in maximum. In the area impacted by the plume of radius 50 m from the well, pH, and TDS illustrate significant changes, because of the introduced acid plume with high salinity and reactivity with the sediment. Adsorption of trace metals onto clay minerals would hinder their mobilization within 200 years, which may reduce the risks of trace metal contaminations. Usually shallow groundwater monitoring wells are located farther than 50 m apart from each other, thus it might be difficult to detect water chemistry changes within a short time after leakage occur.

3.2. Thresholds and Indicators for Early Detection Criteria

The principles to choose indicators for early detection include: (1) easy to test, and (2) with significant changes due to a leakage, compared to the selected threshold. Usually groundwater chemistry varies over time due to groundwater flow and weather, and it might be difficult to indicate a leakage if the change is insignificant. Therefore, it is important to select a reasonable threshold for the indicators as well as indicators sensitive to potential leakage.

The probabilities of water chemistry occurrence ranges are used to forecast the likelihood whether the groundwater quality would be impacted by potential CO_2 and brine leakage with changes of water constituent concentrations. Figure 5 illustrates the probability of pH exceeding the state and federal MCLs and the site-specific no-impact threshold at different distances from the conceptual leaky well.

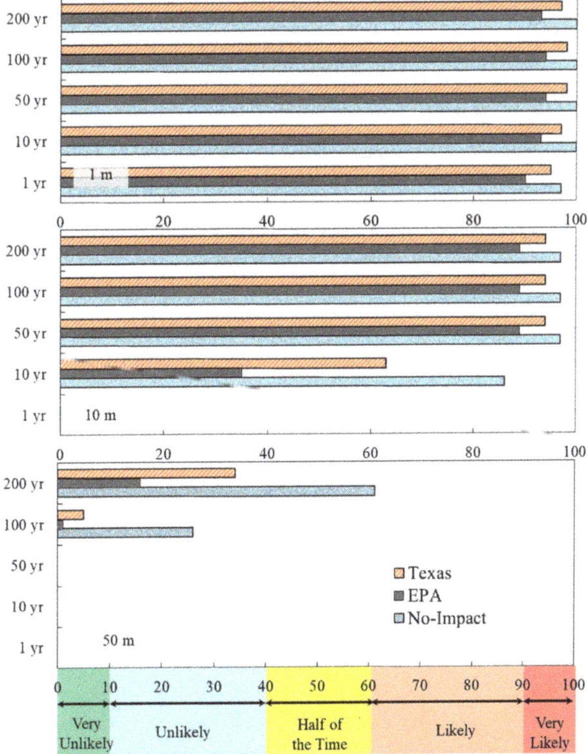

Figure 5. Probability of occurrence for pH exceeding the Texas MCL, the EPA MCL and no-impact threshold at 1 m, 10 m, and 50 m away from the leaky well.

At the distance 1 m away from the well, with large impacts of the leakage plume, a high probability for pH suggests that it is likely to exceed all three thresholds (> 95% probability) with limited differences, because significant pH drops (< 6.0) occur due to the leakage. At the distance 10 m away from the well, it is very unlikely to identify any pH change in the first year since the acid plume does not reach this distance. After 10 years, it is likely to determine the pH change using no-impact threshold compared to the MCLs. It starts to suggest pH changes exceeding the thresholds with a significant difference after 100 years at 50 m away from the well. It suggests that the no-impact threshold is more sensitive to indicate a leakage, because it is stricter than the MCLs. It indicates that no-impact thresholds associated with site-specific monitoring data could be a valuable reference for evaluation of leakage impacts, which is meaningful for quantifying water quality change, especially for those water parameters that are significantly different from the MCLs. However, with limited impact area (< 50 m) of the potential leakage plume, it might be difficult to detect any changes of water chemistry at a monitoring well, which is usually a few hundred meters away from each other.

Easy-tested water parameters that are sensitive to the leakage should be selected as early detect criteria. Figure 6 shows the likelihood of changes of the five parameters of interest at different distances from the well. Due to adsorption of trace metals, such as Se and As, their concentrations maintain at a low level and are not likely to exceed the thresholds for a long time (also shown in Figure 4). Thus, such trace metals may not be capable to be selected as indicators for leakage at early stages. pH and TDS are the most sensitive constituents to indicate a leakage among all the cases with various leakage rates. Particularly, it is one of the most convenient methods to test pH of a water sample. Therefore, pH could be selected as an early detection indicator at the FWU site.

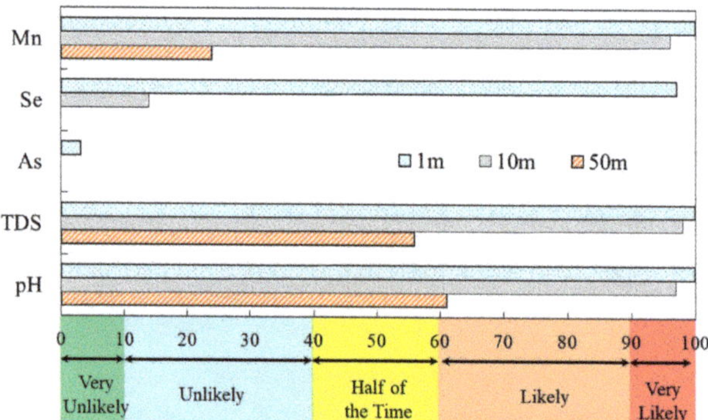

Figure 6. Probability of occurrence of pH, TDS, As, Se, and Mn exceeding the no-impact threshold at 1, 10, and 50 m away from the leaky well after 200 years.

4. Conclusions

In this study, we present a series of quantitative assessments of potential risks to the Ogallala aquifer with potential CO_2 and brine leakage at the FWU. Potential chemical risks to the overlying USDW aquifer were analyzed, and selected water constituents were evaluated for selecting early detection indicators. Salient findings include: (1) with leakage flux up to 0.4% of injected CO_2 and brine from a conceptual leaky well with failure, it is likely that the impacted area limits within 50 m from the well after 200 years; (2) toxic trace metals may be considered an insignificant long-term concern because of clay adsorption; (3) site-specific no-impact thresholds could be a preferable reference for groundwater quality evaluations; and (4) pH is suggested as a likely geochemical indicator for early detection of a leakage, due to its easy tested and sensitivity aspects. Results of this study provide a useful perspective of combining numerical simulations, field observations, and quantitative ROMs for site-specific risk assessment.

Author Contributions: Methodology, Z.D., F.P.; investigation, T.X., W.J., S.C.; resources, R.E.; supervision, B.M., H.V.; writing—original draft preparation, T.X.; writing—review and editing, all authors. All authors have read and agreed to the published version of the manuscript.

Funding: Funding for this project is provided by the U.S. Department of Energy's (DOE) National Energy Technology Laboratory (NETL) through the Southwest Regional Partnership on Carbon Sequestration (SWP) under Award No. DE-FC26-05NT42591.

Conflicts of Interest: The authors declare no conflict of interest.

References

1. Bachu, S. Sequestration of CO2 in geological media: Criteria and approach for site selection in response to climate change. *Energy Convers. Manag.* **2000**, *41*, 953–970. [CrossRef]

2. McCulloch, S. *20 Years of Carbon Capture and Storage: Accelerating Future Deployment*; International Energy Agency: Paris, France, 2016; ISBN 9789264267800.

3. IPCC. *IPCC Special Report on Carbon Dioxide Capture and Storage*; Metz, B., Davidson, O., de Coninck, H., Loos, M., Meyer, L., Eds.; Cambridge University Press: New York, NY, USA, 2005.

4. Markewitz, P.; Kuckshinrichs, W.; Leitner, W.; Linssen, J.; Zapp, P.; Bongartz, R.; Schreiber, A.; Müller, T.E. Worldwide innovations in the development of carbon capture technologies and the utilization of CO2. *Energy Environ. Sci.* **2012**, *5*, 7281. [CrossRef]

5. Dai, Z.; Viswanathan, H.; Middleton, R.; Pan, F.; Ampomah, W.; Yang, C.; Jia, W.; Xiao, T.; Lee, S.-Y.; McPherson, B.; et al. CO2 Accounting and Risk Analysis for CO2 Sequestration at Enhanced Oil Recovery Sites. *Environ. Sci. Technol.* **2016**, *50*. [CrossRef] [PubMed]

6. Bachu, S.; Watson, T.L. Review of failures for wells used for CO2 and acid gas injection in Alberta, Canada. *Energy Procedia* **2009**, *1*, 3531–3537. [CrossRef]

7. Jia, W.; McPherson, B.; Dai, Z.; Irons, T.; Xiao, T. Evaluation of pressure management strategies and impact of simplifications for a post-EOR CO2 storage project. *Geomech. Geophys. Geo-Energy Geo-Resour.* **2017**, *3*, 281–292. [CrossRef]

8. Gaus, I. Role and impact of CO2-rock interactions during CO2 storage in sedimentary rocks. *Int. J. Greenh. Gas Control* **2010**, *4*, 73–89. [CrossRef]

9. Little, M.G.; Jackson, R.B. Potential Impacts of Leakage from Deep CO2 Geosequestration on Overlying Freshwater Aquifers. *Environ. Sci. Technol.* **2010**, *44*, 9225–9232. [CrossRef]

10. Qafoku, N.P.; Lawter, A.R.; Bacon, D.H.; Zheng, L.; Kyle, J.; Brown, C.F. Review of the impacts of leaking CO2 gas and brine on groundwater quality. *Earth-Sci. Rev.* **2017**, *169*, 69–84. [CrossRef]

11. Liu, W.; Ramirez, A. State of the art review of the environmental assessment and risks of underground geo-energy resources exploitation. *Renew. Sustain. Energy Rev.* **2017**, *76*, 628–644. [CrossRef]

12. Dai, Z.; Xu, L.; Xiao, T.; McPherson, B.; Zhang, X.; Zheng, L.; Dong, S.; Yang, Z.; Soltanian, M.R.; Yang, C.; et al. Reactive chemical transport simulations of geologic carbon sequestration: Methods and applications. *Earth-Sci. Rev.* **2020**, *208*, 103265. [CrossRef]

13. Zheng, L.; Apps, J.A.; Zhang, Y.; Xu, T.; Birkholzer, J.T. On mobilization of lead and arsenic in groundwater in response to CO2 leakage from deep geological storage. *Chem. Geol.* **2009**, *268*, 281–297. [CrossRef]

14. Yang, C.; Hovorka, S.D.; Treviño, R.H.; Delgado-Alonso, J. Integrated Framework for Assessing Impacts of CO2 Leakage on Groundwater Quality and Monitoring-Network Efficiency: Case Study at a CO2 Enhanced Oil Recovery Site. *Environ. Sci. Technol.* **2015**, *49*, 8887–8898. [CrossRef] [PubMed]

15. Dai, Z.; Viswanathan, H.; Xiao, T.; Hakala, A.; Lopano, C.; Guthrie, G.; McPherson, B. Reactive Transport Modeling of Geological Carbon Storage Associated with CO2 and Brine Leakage. In *Science of Carbon Storage in Deep Saline Formations*; Elsevier: Amsterdam, The Netherlands, 2019; pp. 89–116.

16. Keating, E.H.; Fessenden, J.; Kanjorski, N.; Koning, D.J.; Pawar, R. The impact of CO2 on shallow groundwater chemistry: Observations at a natural analog site and implications for carbon sequestration. *Environ. Earth Sci.* **2010**, *60*, 521–536. [CrossRef]

17. Xiao, T.; Dai, Z.; Viswanathan, H.; Hakala, A.; Cather, M.; Jia, W.; Zhang, Y.; McPherson, B. Arsenic mobilization in shallow aquifers due to CO2 and brine intrusion from storage reservoirs. *Sci. Rep.* **2017**, *7*, 2763. [CrossRef]

18. Viswanathan, H.S.; Pawar, R.J.; Stauffer, P.H.; Kaszuba, J.P.; Carey, J.W.; Olsen, S.C.; Keating, G.N.; Kavetski, D.; Guthrie, G.D. Development of a Hybrid Process and System Model for the Assessment of Wellbore Leakage at a Geologic CO_2 Sequestration Site. *Environ. Sci. Technol.* **2008**, *42*, 7280–7286. [CrossRef]

19. Xiao, T.; McPherson, B.; Bordelon, A.; Viswanathan, H.; Dai, Z.; Tian, H.; Esser, R.; Jia, W.; Carey, W. Quantification of CO_2-cement-rock interactions at the well-caprock-reservoir interface and implications for geological CO_2 storage. *Int. J. Greenh. Gas Control* **2017**, *63*, 126–140. [CrossRef]

20. Carroll, S.; Carey, J.W.; Dzombak, D.; Huerta, N.J.; Li, L.; Richard, T.; Um, W.; Walsh, S.D.C.; Zhang, L. Review: Role of chemistry, mechanics, and transport on well integrity in CO_2 storage environments. *Int. J. Greenh. Gas Control* **2016**, *49*, 149–160. [CrossRef]

21. Carey, J.W.; Wigand, M.; Chipera, S.J.; WoldeGabriel, G.; Pawar, R.; Lichtner, P.C.; Wehner, S.C.; Raines, M.A.; Guthrie, G.D. Analysis and performance of oil well cement with 30 years of CO_2 exposure from the SACROC Unit, West Texas, USA. *Int. J. Greenh. Gas Control* **2007**, *1*, 75–85. [CrossRef]

22. Kutchko, B.G.; Strazisar, B.R.; Lowry, G.V.; Dzombak, D.A.; Thaulow, N. Rate of CO_2 Attack on Hydrated Class H Well Cement under Geologic Sequestration Conditions. *Environ. Sci. Technol.* **2008**, *42*, 6237–6242. [CrossRef]

23. Scherer, G.W.; Celia, M.A.; Prévost, J.-H.; Bachu, S.; Bruant, R.; Duguid, A.; Fuller, R.; Gasda, S.E.; Radonjic, M.; Vichit-Vadakan, W. Leakage of CO_2 through Abandoned Wells: Role of Corrosion of Cement. In *Carbon Dioxide Capture for Storage in Deep Geologic Formations*; Elsevier: Amsterdam, The Netherlands, 2005; pp. 827–848.

24. Carey, J.W.; Svec, R.; Grigg, R.; Zhang, J.; Crow, W. Experimental investigation of wellbore integrity and CO_2–brine flow along the casing–cement microannulus. *Int. J. Greenh. Gas Control* **2010**, *4*, 272–282. [CrossRef]

25. Miao, X.; Zhang, L.; Wang, Y.; Wang, L.; Fu, X.; Gan, M.; Li, X. Characterisation of wellbore cement microstructure alteration under geologic carbon storage using X-ray computed micro-tomography: A framework for fast CT image registration and carbonate shell morphology quantification. *Cem. Concr. Compos.* **2020**, *108*, 103524. [CrossRef]

26. Xiao, T.; McPherson, B.; Pan, F.; Esser, R.; Jia, W.; Bordelon, A.; Bacon, D. Potential chemical impacts of CO_2 leakage on underground source of drinking water assessed by quantitative risk analysis. *Int. J. Greenh. Gas Control* **2016**, *50*, 305–316. [CrossRef]

27. Carroll, S.A.; Keating, E.; Mansoor, K.; Dai, Z.; Sun, Y.; Trainor-Guitton, W.; Brown, C.; Bacon, D. Key factors for determining groundwater impacts due to leakage from geologic carbon sequestration reservoirs. *Int. J. Greenh. Gas Control* **2014**, *29*, 153–168. [CrossRef]

28. Onishi, T.; Nguyen, M.C.; Carey, J.W.; Will, B.; Zaluski, W.; Bowen, D.W.; Devault, B.C.; Duguid, A.; Zhou, Q.; Fairweather, S.H.; et al. Potential CO_2 and brine leakage through wellbore pathways for geologic CO_2 sequestration using the National Risk Assessment Partnership tools: Application to the Big Sky Regional Partnership. *Int. J. Greenh. Gas Control* **2019**, *81*, 44–65. [CrossRef]

29. Bacon, D.H.; Qafoku, N.P.; Dai, Z.; Keating, E.H.; Brown, C.F. Modeling the impact of carbon dioxide leakage into an unconfined, oxidizing carbonate aquifer. *Int. J. Greenh. Gas Control* **2016**, *44*, 290–299. [CrossRef]

30. Wilson, E.J.; Friedmann, S.J.; Pollak, M.F. Research for Deployment: Incorporating Risk, Regulation, and Liability for Carbon Capture and Sequestration. *Environ. Sci. Technol.* **2007**, *41*, 5945–5952. [CrossRef] [PubMed]

31. Xiao, T.; Dai, Z.; McPherson, B.; Viswanathan, H.; Jia, W. Reactive transport modeling of arsenic mobilization in shallow groundwater: Impacts of CO_2 and brine leakage. *Geomech. Geophys. Geo-Energy Geo-Resour.* **2017**, *3*, 339–350. [CrossRef]

32. Yang, C.; Mickler, P.J.; Reedy, R.; Scanlon, B.R.; Romanak, K.D.; Nicot, J.-P.; Hovorka, S.D.; Trevino, R.H.; Larson, T. Single-well push–pull test for assessing potential impacts of CO_2 leakage on groundwater quality in a shallow Gulf Coast aquifer in Cranfield, Mississippi. *Int. J. Greenh. Gas Control* **2013**, *18*, 375–387. [CrossRef]

33. Yang, C.; Treviño, R.H.; Hovorka, S.D.; Delgado-Alonso, J. Semi-analytical approach to reactive transport of CO_2 leakage into aquifers at carbon sequestration sites. *Greenh. Gases Sci. Technol.* **2015**, *5*, 786–801. [CrossRef]

34. Dai, Z.; Middleton, R.; Viswanathan, H.; Fessenden-Rahn, J.; Bauman, J.; Pawar, R.; Lee, S.-Y.; McPherson, B. An Integrated Framework for Optimizing CO2 Sequestration and Enhanced Oil Recovery. *Environ. Sci. Technol. Lett.* **2014**, *1*, 49–54. [CrossRef]

35. Jia, W.; McPherson, B.J.; Pan, F.; Xiao, T.; Bromhal, G. Probabilistic analysis of CO2 storage mechanisms in a CO2 -EOR field using polynomial chaos expansion. *Int. J. Greenh. Gas Control* **2016**, *51*, 218–229. [CrossRef]

36. Rohmer, J.; Bouc, O. A response surface methodology to address uncertainties in cap rock failure assessment for CO2 geological storage in deep aquifers. *Int. J. Greenh. Gas Control* **2010**, *4*, 198–208. [CrossRef]

37. Ampomah, W.; Balch, R.S.; Cather, M.; Will, R.; Gunda, D.; Dai, Z.; Soltanian, M.R. Optimum design of CO2 storage and oil recovery under geological uncertainty. *Appl. Energy* **2017**, *195*, 80–92. [CrossRef]

38. Balch, R.; Mcpherson, B. Integrating Enhanced Oil Recovery and Carbon Capture and Storage Projects: A Case Study at Farnsworth Field, Texas. In Proceedings of the SPE Western Regional Meeting, Anchorage, AK, USA, 23–26 May 2016; Society of Petroleum Engineers: Houston, TX, USA, 2016.

39. Balch, R.; McPherson, B.; Grigg, R. Overview of a Large Scale Carbon Capture, Utilization, and Storage Demonstration Project in an Active Oil Field in Texas, USA. *Energy Procedia* **2017**, *114*, 5874–5887. [CrossRef]

40. Xiao, T.; Xu, H.; Moodie, N.; Esser, R.; Jia, W.; Zheng, L.; Rutqvist, J.; McPherson, B. Chemical-Mechanical Impacts of CO2 Intrusion into Heterogeneous Caprock. *Water Resour. Res.* **2020**. [CrossRef]

41. Nakao, S.; Tosha, T. Progress Report of AIST's Research Programs for CO2 Geological Storage. *Energy Procedia* **2013**, *37*, 4990–4996. [CrossRef]

42. Hem, J.D. *Study and Interpretation of the Chemical Characteristics of Natural Water*, 3rd ed.; Department of the Interior, U.S. Geological Survey: Reston, VA, USA, 1985.

43. Yang, C.; Dai, Z.; Romanak, K.D.; Hovorka, S.D.; Treviño, R.H. Inverse Modeling of Water-Rock-CO2 Batch Experiments: Potential Impacts on Groundwater Resources at Carbon Sequestration Sites. *Environ. Sci. Technol.* **2014**, *48*, 2798–2806. [CrossRef]

44. Choi, B.-Y. Potential impact of leaking CO2 gas and CO2-rich fluids on shallow groundwater quality in the Chungcheong region (South Korea): A hydrogeochemical approach. *Int. J. Greenh. Gas Control* **2019**, *84*, 13–28. [CrossRef]

45. Last, G.; Murray, C.; Brown, C.; Jordan, P.; Sharma, M. *No-Impact Threshold Values for NRAP's Reduced Order Models*; Pacific Northwest National Lab. (PNNL): Richland, WA, USA, 2013.

46. Frye, E.; Bao, C.; Li, L.; Blumsack, S. Environmental Controls of Cadmium Desorption during CO2 Leakage. *Environ. Sci. Technol.* **2012**, *46*, 4388–4395. [CrossRef]

47. Shao, H.; Qafoku, N.P.; Lawter, A.R.; Bowden, M.E.; Brown, C.F. Coupled Geochemical Impacts of Leaking CO2 and Contaminants from Subsurface Storage Reservoirs on Groundwater Quality. *Environ. Sci. Technol.* **2015**, *49*, 8202–8209. [CrossRef]

48. Lawter, A.R.; Qafoku, N.P.; Asmussen, R.M.; Kukkadapu, R.K.; Qafoku, O.; Bacon, D.H.; Brown, C.F. Element mobilization and immobilization from carbonate rocks between CO2 storage reservoirs and the overlying aquifers during a potential CO2 leakage. *Chemosphere* **2018**, *197*, 399–410. [CrossRef] [PubMed]

49. Goldberg, S.; Johnston, C.T. Mechanisms of Arsenic Adsorption on Amorphous Oxides Evaluated Using Macroscopic Measurements, Vibrational Spectroscopy, and Surface Complexation Modeling. *J. Colloid Interface Sci.* **2001**, *234*, 204–216. [CrossRef] [PubMed]

50. Viswanathan, H.; Dai, Z.; Lopano, C.; Keating, E.; Hakala, J.A.; Scheckel, K.G.; Zheng, L.; Guthrie, G.D.; Pawar, R. Developing a robust geochemical and reactive transport model to evaluate possible sources of arsenic at the CO2 sequestration natural analog site in Chimayo, New Mexico. *Int. J. Greenh. Gas Control* **2012**, *10*, 199–214. [CrossRef]

51. Zheng, L.; Spycher, N. Modeling the potential impacts of CO2 sequestration on shallow groundwater: The fate of trace metals and organic compounds before and after leakage stops. *Greenh. Gases Sci. Technol.* **2018**, *8*, 161–184. [CrossRef]

52. Dzombak, D.A.; Morel, F.M.M. *Surface Complexation Modeling: Hydrous Ferric Oxide*; John Wiley & Sons: New York, NY, USA, 1990.

53. Goldberg, S. Modeling Selenite Adsorption Envelopes on Oxides, Clay Minerals, and Soils using the Triple Layer Model. *Soil Sci. Soc. Am. J.* **2013**, *77*, 64. [CrossRef]

54. Zheng, L.; Apps, J.A.; Spycher, N.; Birkholzer, J.T.; Kharaka, Y.K.; Thordsen, J.; Beers, S.R.; Herkelrath, W.N.; Kakouros, E.; Trautz, R.C. Geochemical modeling of changes in shallow groundwater chemistry observed during the MSU-ZERT CO2 injection experiment. *Int. J. Greenh. Gas Control* **2012**, *7*, 202–217. [CrossRef]

55. Pan, F.; McPherson, B.J.; Esser, R.; Xiao, T.; Appold, M.S.; Jia, W.; Moodie, N. Forecasting evolution of formation water chemistry and long-term mineral alteration for GCS in a typical clastic reservoir of the Southwestern United States. *Int. J. Greenh. Gas Control* **2016**, *54*, 524–537. [CrossRef]

56. Bezerra, M.A.; Santelli, R.E.; Oliveira, E.P.; Villar, L.S.; Escaleira, L.A. Response surface methodology (RSM) as a tool for optimization in analytical chemistry. *Talanta* **2008**, *76*, 965–977. [CrossRef]

57. Wolery, T.J. *EQ3/6, a Software Package for Geochemical Modeling of Aqueous Systems: Package Overview and Installation Guide (Version 7.0)*; Lawrence Livermore National Lab.: Livermore, CA, USA, 1992.

58. Palandri, J.L.; Kharaka, Y.K. *A Compilation of Rate Parameters of Water-Mineral Interaction Kinetics for Application to Geochemical Modeling*; National Energy Technology Laboratory–United States Department of Energy: Menlo Park, CA, USA, 2004.

59. Manning, B.A.; Goldberg, S. Adsorption and Stability of Arsenic(III) at the Clay Mineral–Water Interface. *Environ. Sci. Technol.* **1997**, *31*, 2005–2011. [CrossRef]

60. Appelo, C.J.A.; Postma, D. *Geochemistry, Groundwater and Pollution*; CRC Press: Boca Raton, FL, USA, 2004.

61. Xu, T.; Spycher, N.; Sonnenthal, E.; Zhang, G.; Zheng, L.; Pruess, K. TOUGHREACT Version 2.0: A simulator for subsurface reactive transport under non-isothermal multiphase flow conditions. *Comput. Geosci.* **2011**, *37*, 763–774. [CrossRef]

62. Pruess, K.; Spycher, N. ECO2N—A fluid property module for the TOUGH2 code for studies of CO2 storage in saline aquifers. *Energy Convers. Manag.* **2007**, *48*, 1761–1767. [CrossRef]

63. Bacon, D.H.; Dai, Z.; Zheng, L. Geochemical Impacts of Carbon Dioxide, Brine, Trace Metal and Organic Leakage into an Unconfined, Oxidizing Limestone Aquifer. *Energy Procedia* **2014**, *63*, 4684–4707. [CrossRef]

Publisher's Note: MDPI stays neutral with regard to jurisdictional claims in published maps and institutional affiliations.

Article

Analysis of Caprock Tightness for CO_2 Enhanced Oil Recovery and Sequestration: Case Study of a Depleted Oil and Gas Reservoir in Dolomite, Poland

Małgorzata Słota-Valim *, Andrzej Gołąbek, Wiesław Szott and Krzysztof Sowiżdżał

Department of Geology and Geochemistry, Oil and Gas Institute, National Research Institute, 25A Lubicz Str., 31-503 Kraków, Poland; golabek@inig.pl (A.G.); szott@inig.pl (W.S.); sowizdzal@inig.pl (K.S.)
*   Correspondence: slota@inig.pl; Tel.: +48-12-61776-83

Abstract: This study addresses the problem of geological structure tightness for the purposes of enhanced oil recovery with CO_2 sequestration. For the first time in the history of Polish geological survey the advanced methods, practical assumptions, and quantitative results of detailed simulations were applied to study the geological structure of a domestic oil reservoir as a potential candidate for a combined enhanced oil recovery and CO_2 sequestration project. An analysis of the structure sequestration capacity and its tightness was performed using numerical methods that combined geomechanical and reservoir fluid flow modelling with a standard two-way coupling procedure. By applying the correlation between the geomechanical state and transport properties of the caprock, threshold pressure variations were determined to be a key factor affecting the sealing properties of the reservoir–caprock boundary. In addition to the estimation of the sequestration capacity of the structure, the process of CO_2 leakage from the reservoir to the caprock was simulated for scenarios exceeding the threshold pressure limit of the reservoir–caprock boundary. The long-term simulations resulted in a comprehensive assessment of the total amount of CO_2 leakage as a function of time and the leaked CO_2 distribution within the caprock.

Keywords: CO_2-EOR; CO_2 sequestration; geomechanics; reservoir fluid flow modelling; tightness of caprock; CO_2 leakage; threshold pressure

Citation: Słota-Valim, M.; Gołąbek, A.; Szott, W.; Sowiżdżał, K. Analysis of Caprock Tightness for CO_2 Enhanced Oil Recovery and Sequestration: Case Study of a Depleted Oil and Gas Reservoir in Dolomite, Poland. *Energies* **2021**, *14*, 3065. https://doi.org/10.3390/en14113065

Academic Editors: Brian McPherson, Martha Cather, Reid Grigg, Robert Balch and William Ampomah

Received: 21 April 2021
Accepted: 20 May 2021
Published: 25 May 2021

Publisher's Note: MDPI stays neutral with regard to jurisdictional claims in published maps and institutional affiliations.

1. Introduction

Because CO_2 emissions are an increasing problem, many strategies have been developed to reduce its emissions to the atmosphere, including mineralisation [1,2], CO_2 storage at the bottom of the oceans [3], accelerated weathering [4,5], and subsurface geological storage [6]. Carbon dioxide capture at the source, followed by its long-term storage in exploited hydrocarbon reservoirs, is one of the most practical ways to reduce the CO_2 concentration in the atmosphere.

Reservoir structures suitable for geological CO_2 storage are most frequently water-bearing formations, depleted hydrocarbon deposits, or unminable methane-rich coal seams.

Geological formations that are selected for CO_2 storage must meet the appropriate criteria related to the minimum and maximum depth of the structure. The reservoir rock parameters, including thickness, permeability, porosity, and fracture characteristics, are also critical. However, the most important criterion is the sealing quality of the rock, that is, the overburden's integrity and thickness play key roles in assessing the safety of long-term CO_2 geological storage.

Most of these criteria are met for depleted oil and gas fields because hydrocarbon production from oil and gas fields implies favourable reservoir properties, including porosity, permeability, and reservoir thickness [7].

During the geological storage of CO_2, most of the injected CO_2 remains in the free (liquid, gas, or supercritical) phase. This phase poses the greatest threat to the seal integrity.

Under reservoir conditions, which are at elevated pressures and high temperatures, CO_2 occurs in a dense gas phase, yet its density is lower than that of the formation water. Driven by the buoyant force that occurs as a result of the difference in the densities of the CO_2 and reservoir water, the free fraction of CO_2 tends to migrate upward [8–10]. The presence of free CO_2 in the upper parts of reservoir rocks poses a threat of escape to overlying formations and potentially into the atmosphere, in the case of the caprock integrity loss.

Therefore, it is crucial to maintain a tight sealing layer and not inject CO_2 into the formation beyond a level that guarantees the integrity and stability of the seal and overall safety of the operation. The long-term tightness of the overburden may be influenced by the effects of CO_2 injection, as operations change the stress and strain field in subsequent years of sequestration.

The aim of this study was to analyse the tightness of the caprock above the main dolomite (Ca2). The main dolomite structure with partially depleted hydrocarbon accumulation located in the Gorzów Block, Poland, is considered a potential formation for carbon dioxide storage.

Numerous research papers address the caprock tightness problem in the potential site for CO_2 geological storage. Edlmann et al. (2013) [11] postulated the existence of a critical threshold of fracture aperture size controlling the CO_2 flow along the shale samples. Li et al. (2006) [12] examined the occurrence of the volume flow and measured the effective gas permeability for selected post-failure evaporite beds samples caused by CO_2 injection. Zivar et al. (2019) [13] examined the effect of stress magnitude and stress history on porosity and permeability values of anhydride and carbonate rocks, while Hangx et al. (2009) [14] investigated the impact of CO_2 on the mechanical strength of Zechstein Anhydrite, which seals many potential CO_2 storage sites in Central and Eastern Europe.

Under the assumptions of the study, CO_2 was injected into the reservoir rock for both enhanced oil recovery (EOR) and geological CO_2 sequestration. To meet the research objectives, several numerical methods were used, involving the coupling of geomechanic and dynamic modelling of reservoir fluid flow in the main dolomite formation, into which CO_2 was injected, and the surrounding rocks, especially the overburden.

Geological Setting

The study area is located on the Gorzów Block adjacent to Szczecin Trough in the north, with the Mid-Polish Swell to the east and the Fore-Sudetic Monocline to the south (Figure 1). This area has a regional sequence of tectonic disturbances and related uplift of Permian-Mesozoic sediments [15]. These elevated tectonic blocks were accompanied by extensive volcanic rock cover, as well as a depressions series of clastic deposits in the Lower Rotliegend. In particular, the erosive outliers of the Zechstein basement had a significant impact on the development of the overlying Zechstein–Mesozoic complex [16].

In the analysed part of the South Permian Basin, the Zechstein Sea entered a morphologically diverse depositional surface. Considerable denivelations contributed to the division of the basin into shallow and deep-water zones. In the elevated areas, platforms and micro platforms of sulphate deposits of the Werra cyclothem developed and became covered with the main dolomite platforms, which formed as a result of the transgressive cycle during the Stassfurt cyclothem [16]. Sedimentological studies of the main dolomite deposits revealed the existence of different sedimentation environments resulting from considerable bathymetric differences [17–19]. These include: the platform and microplatform zones (including barrier and platform flat), slope and toe of the slope, and basin floor [19].

Optimal reservoir properties were found both in the shallow-water platform formations and in the formations developed at the toe of the slope in a deep-water environment [18–20].

From a petroleum exploration and potential storage volume perspective, the most important property was the development and preservation of significant secondary porosity in the main dolomite rock (in some areas over 30%), which was caused by the complete or partial dissolution of granular carbonate components. The presence of this high porosity may be related to the dolomitisation and recrystallisation of the primary calcareous rock matrix [20,21].

The main dolomite, which is the reservoir rock and target formation for carbon dioxide storage, is covered with a sequence of thick evaporates, thin carbonates, and very thin layers of shales of the PZ2 (Stassfurt), PZ3 (Leine), and PZ4 (Aller) Zechstein cycles [22,23]. The location of the study area with the lithostratigraphic profile is shown in Figure 1.

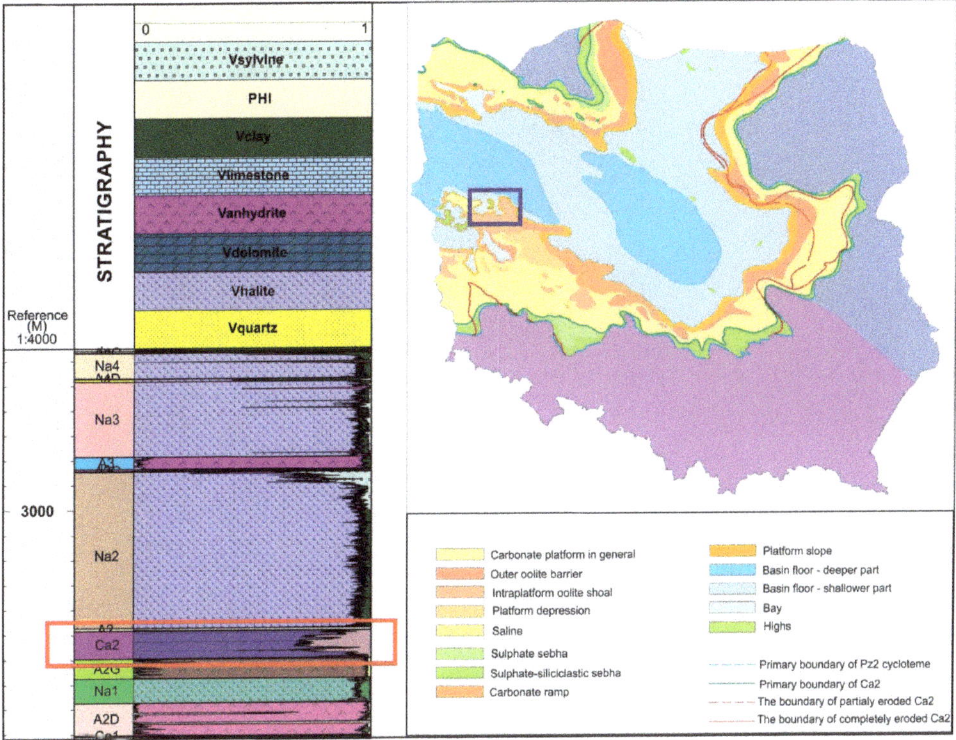

Figure 1. Lithostratigraphic profile in the reference borehole (**left**) and the location of the study area on the background of the map of the main dolomite (Ca2) in Poland (**right**) [24].

The thickness of the Zechstein evaporates overlying the reservoir rock varies from 253 m on top of the platform flat and reaches 840 m in the basin. Developed over geologic time on top of the reservoir rock sequence, these hardly permeable or impermeable evaporates constitute the seal for hydrocarbon accumulation and are potentially a good barrier for carbon dioxide stored in the main dolomite reservoir rock.

2. Methods

Performing fluid flow modelling coupled with geomechanics is essential for the safe storage of CO_2 or other media in subsurface rock formations. In particular, analysing the mechanical response of rocks to activities related to geological CO_2 storage concerns the reservoir rock itself and the neighbouring rocks, especially the overburden.

As a result of injection, the increased pressure of the fluid filling the pore space in the rock causes a change in the stress field. The disturbance of the stress equilibrium may lead to a potential risk of loss of tightness of the sealing rocks, creating a potential pathway for CO_2 leakage to the overlying strata. To evaluate the tightness of the caprock, we employed numerical methods that combined geomechanic and reservoir fluid modelling of the reservoir and surrounding rocks using the Schlumberger software (Houston, TS, USA)—Petrel™ platform, VISAGE™ geomechanical simulator, and ECLIPSE™ reservoir simulator. The standard approach to this evaluation includes a two-way coupling of both types of simulations realised by iterative, alternating runs of the simulations, as shown schematically in Figure 2.

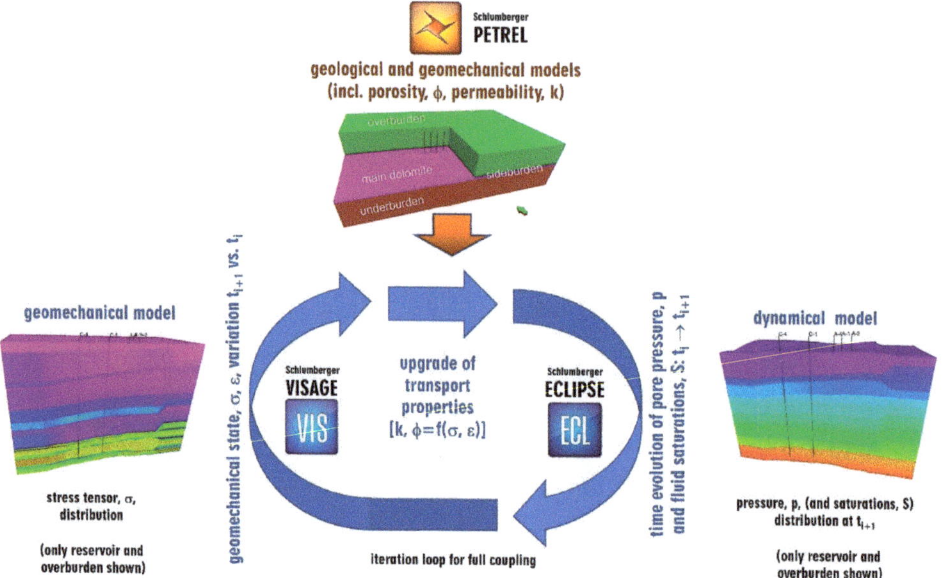

Figure 2. Conventional dynamic and geomechanic model coupling.

For the geomechanical analysis, parameter distributions describing the geomechanical and petrophysical rock properties were developed, and boundary conditions were determined. The distributions of the pore pressure necessary for the geomechanical simulations were combined with the hydrostatic pressure distribution in the overburden and the reservoir pressure in the main dolomite (Ca2) developed for certain time steps in the history of hydrocarbon production and CO_2 injection. Upgraded transport properties were determined according to the resultant geomechanical state of the structure and were used in the dynamical model to simulate upgraded pressure and saturation distributions. These results were input to the geomechanical model and thus closed the iteration loop of the two-way coupling simulation procedure.

Relevant Dataset

To analyse the tightness of the geological structure's caprock, we used three main data types:

- Seismic data, which comprises the result of a 3D seismic interpretation of the study area, the structural surface of the top of the reservoir rock (Ca2) in the depth domain, and the map of the thickness of the reservoir rock based on seismic data;
- Well log and lab data;
- Lithostratigraphic profiles and well log data from 10 of the 27 boreholes drilled in the study area with the results of laboratory measurements of petrophysical and static geomechanical parameters performed on the core material;
- Reservoir engineering data (reservoir fluid saturation distribution, pressure distribution, and reservoir fluid thermodynamic (PVT) properties);
- Hydrodynamic well tests (multi-rate and pressure build-up tests);

- Production data (reservoir fluid production rates and totals, and bottom-hole and well-head pressures).

3. Geological Model

To meet the study's objectives, we first developed a structural model of the area to define the 3D space, which was subsequently parameterised. Ultimately, we obtained the spatial distributions of the petrophysical parameters necessary to determine the weight of the overburden rocks via geomechanic modelling and during the modelling of the reservoir fluid flow.

3.1. 3D Structural Geological Model

The structural 3D model of the study area presents the geometry of the entire study area profile, specifically the reservoir rock (Ca2), but also, for the purposes of geomechanic modelling, the structure of the overburden rocks reaching up to the ground surface, as well as the side- and under-burden. The developed spatial structural model was used to determine the 3D space for parameterisation.

Using the structural map of the main dolomite Ca2 top, the map of the Ca2 thickness, and the stratigraphic borehole data, we developed a map of the bottom of the main dolomite.

Further, using available interpretations of seismic horizons in the depth domain and boreholes stratigraphic markers, as well as tools available in the Petrel software, we constructed structural maps of the remaining stratigraphic levels. These maps began with the Zechstein limestone (Ca1), through the lithostratigraphic units of the Werra (A1D, Na1, and A1G), Stassfurt (Ca2, A2, Na2, and A2G), Leine (I3 and A3), and Aller cyclothem sequence (A1D, Na1, and A1G) occurring in the area of the study. They are followed by the sediments of the Triassic, Jurassic, Cretaceous, and finally, the Paleogene, Neogene, and Quaternary sediments. The longitudinal extent of the structural model is approximately 13.8 km, and the latitudinal length is approximately 14.5 km.

The horizontal resolution of the 3D grid was 100 m × 100 m. The vertical resolution of the model varied depending on the individual zones. To assess the overburden's tightness, it was necessary to detail the zones that could play a key role in the possible leakage of the stored CO_2, which included the rocks located in the vicinity of the main dolomite with minimal permeability. Accordingly, the highest resolution was used in the zones critical for further investigations, such as the reservoir rock (the main dolomite Ca2) and its closest sealing layer, the basal anhydrite (A2). The mean vertical resolutions in these intervals were 1.88 m and 4.85 m, respectively.

The final geometry of the 3D grid used in the geomechanical simulation of the VISAGE™ simulator (Schlumberger) considered the rocks surrounding the reservoir rock. The geomechanical model was limited from the top at the ground surface, and at the bottom, which is defined at a depth of approximately 41 km. At the sides of the reservoir scale model, a volume of rocks with a length of approximately 45 km on each side was added, resulting in the final latitudinal extent of approximately 103.8 km and a longitudinal extension of 104 km.

As shown in Figure 3, the geometry of the obtained spatial structural model of the main dolomite reservoir rock and surrounding rocks, including the overburden, were identified.

This developed structural model was then parameterised, resulting in spatial models of the petrophysical and geomechanical parameters.

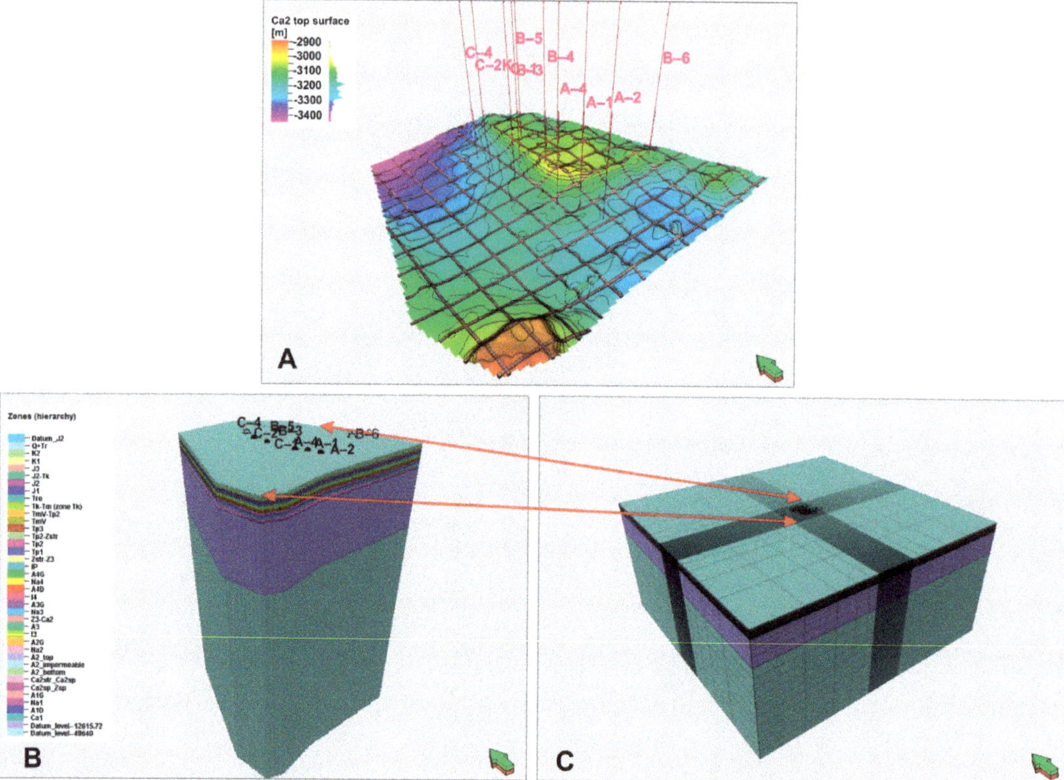

Figure 3. Visualisation of the model structures: (**A**)—3D model reservoir rock, (**B**)—3D structure including the over and underburden, (**C**)—the geometry of the final 3D grid used in the geomechanic simulator, including over-, under-, and side-burden.

3.2. 3D Modelling of Petrophysical Properties

Because the modelling workflow included a geomechanic modelling task that required the definition of rock properties for the entire geological profile, the structural model had to be further developed. Property modelling was carried out in two different model scales: (1) a model covering the geological profile up to the ground level, which was populated with total porosity and rock density values; and (2) a detailed 3D model of the interval of interest covering the Ca2 main dolomite reservoir rock, in which a higher vertical resolution was applied.

The interval within the geological profile, which is of main concern in this study, was modelled with increased vertical resolution to enable a more detailed modelling and interpretation of the geomechanical and fluid flow processes. Figure 4 shows the division of the stratigraphic and lithological units into layers. The reservoir interval of the Ca2 was divided into 6 layers, and the neighbouring sediments above and below were divided into varying numbers of layers depending on their average thickness in order to achieve a vertical resolution of approximately 5 m within the Ca2 and approximately 25 m in the salt rock formations, with anhydrite intervals of several meters thick.

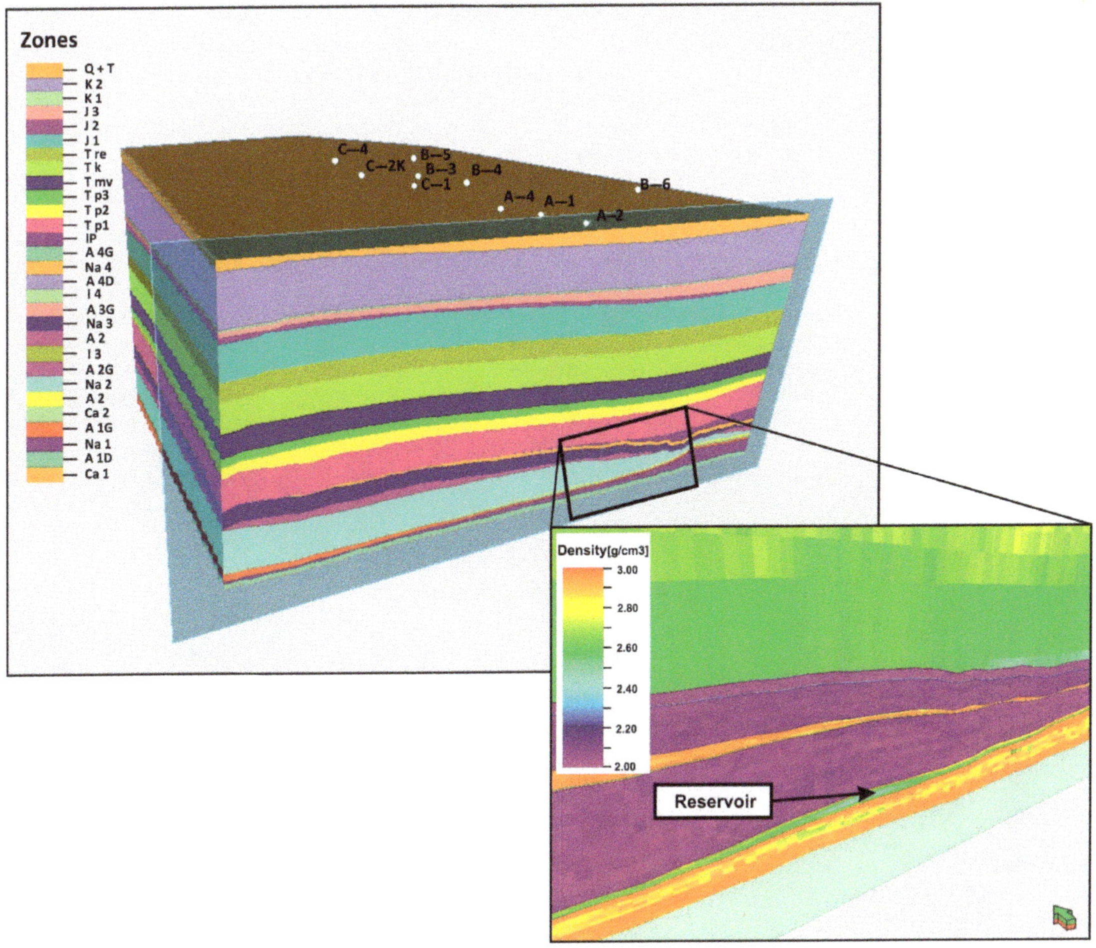

Figure 4. Cross—sections in 3D structural model for the entire profile (**left**) and for the interval of interest (**right**) showing the distribution of density within the target CO_2 storage reservoir Ca2 main dolomite reservoir and the neighbouring rocks.

3.2.1. Density and Porosity Models of Entire Geological Profile

The property modelling was based on the following data sources: geophysical wellbore logging and interpretation along the entire wells' profiles, the results of the laboratory measurements conducted for the main dolomite reservoir rock, which were used to constrain petrophysical interpretation, and the 3D seismic data (previously transformed into seismic attributes form), which were applied as a secondary input in the 3D property modelling of the Ca2 reservoir porosity.

For the global grid population with porosity and density values, eight borehole profiles were used. The analysis included the definition of density and porosity variation ranges within each stratigraphic unit, as well as the modelling of semivariograms. Further, 3D density and porosity modelling were accomplished using a stochastic algorithm (Gaussian random function simulation), in which the modelling process was iterated 20 times to produce 20 equally probable realisations of the porosity and density 3D models for the entire profile. Finally, the arithmetic averages of the 20 realisations were calculated for geomechanic modelling.

3.2.2. High-Resolution 3D Petrophysical Model of Target Reservoir

The static modelling workflow was focused on the detailed modelling of the petrophysical properties of the Ca2 main dolomite, which is a potential CO_2 storage reservoir.

For this reason, extended datasets were used, including petrophysical interpretation along eight boreholes calibrated with dense laboratory data, and 3D seismic volumes of simultaneous inversion and seismic attributes.

Porosity reflects the volume of pore space within the reservoir, which, in this case, was the oil-saturated dolomite interval. Meanwhile, permeability defines the reservoir's capability to conduct fluid flow (oil, natural gas, water, and CO_2). Thus, the 3D models of porosity and permeability are of the highest importance for dynamic modelling the fluid flow within a storage reservoir.

The porosity model was elaborated using both primary wellbore data and secondary seismic attributes previously transformed into seismic properties, which exhibit higher correlation coefficients with porosity along the wellbore profiles. The Gaussian random function simulation algorithm was applied in a co-kriging version. To develop the permeability model, porosity vs. permeability relationship was combined with the borehole profiles of permeability. Figure 5A shows the outcome of porosity modelling within the Ca2 reservoir, while Figure 5B shows the permeability 3D distribution.

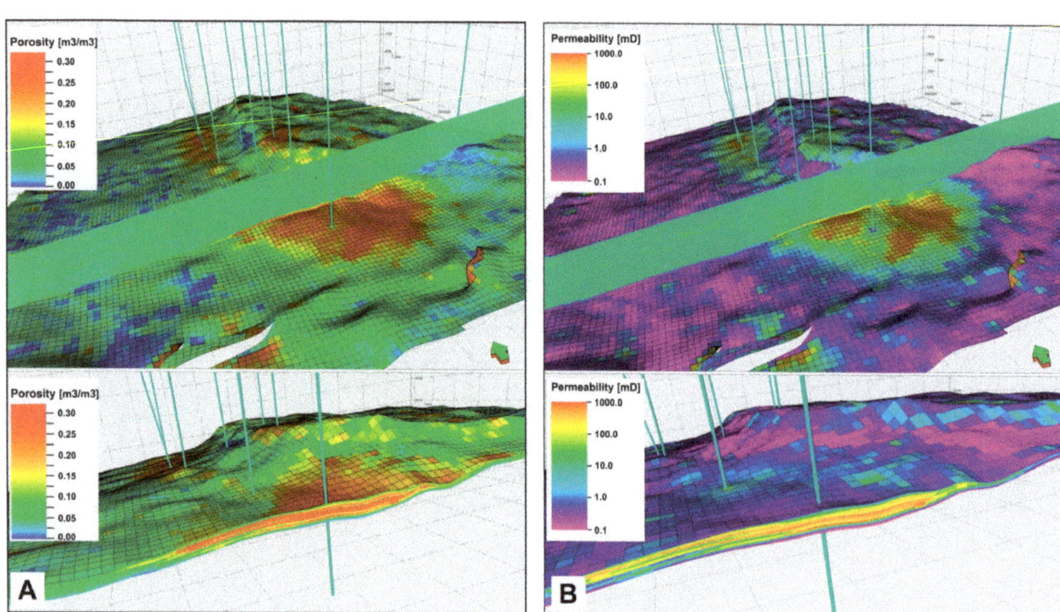

Figure 5. Visualization of the distribution of: (**A**)—porosity and (**B**)—permeability within the Ca2 main dolomite along the cross-sections of the central part of the 3D model within the target reservoir.

4. Geomechanic Model

During hydrocarbon production, the reservoir pressure declines, and the effective stresses increase. Conversely, during CO_2 injection, to utilize the reservoir pressure as a tertiary oil recovery method (EOR) or for CO_2 geological storage, there is an increase in pore pressure, which causes a decrease in the effective stresses. This effective stress principle was first proposed by Therzagi (1945) [25], which was later applied to rocks to understand their ultimate strength and ductility [26].

The change in stress within the reservoir and surrounding rocks can be expressed with an equation for isotropic rocks as follows [27]:

$$\sigma_h = \frac{\nu}{1-\nu}(\sigma_v - \alpha P) + \left[\frac{E}{1-\nu^2}\right]\varepsilon_h + \left[\frac{E\nu}{1-\nu^2}\right]\varepsilon_H \tag{1}$$

where:

- σ_h and σ_v are the minimum horizontal and vertical stresses, respectively;
- α is the Biot's coefficient;
- P is the pore pressure;
- ν is the Poisson's ratio;
- E is the Young's modulus;
- σ_h and ε_H are the strains in the direction of the minimum and maximum horizontal stresses, respectively.

In geomechanic modelling, a set of 3D models of geomechanical parameters were developed, and then, the boundary conditions were determined to calculate the stress field and predict the geomechanical behaviour of the rock during the period of hydrocarbon production and CO_2 injection into the reservoir rock.

4.1. Modelling of Geomechanical Properties

To model the distribution of geomechanical properties in 3D space, the Young's modulus, Poisson's ratio, unconfined compressive strength (UCS), and tensile strength were modelled in 1D for 10 boreholes before they were modelled in the geological model 3D space.

4.1.1. 1D Modelling of Elastic and Strength Properties

Herein, the models of the static elastic properties of the rocks in 1D, including Young's modulus and Poisson's ratio, were conducted using the well logs from 10 boreholes, the results of laboratory measurements of the static Young's modulus parameter, and the Poisson's ratio obtained from the core material from these boreholes. Based on the well log data, the dynamic Young's modulus and Poisson's ratio were calculated, wherein the relationships between the borehole velocity of the compressional vp, shear wave vs, and density ρ were used as follows [28–30]:

$$v_{dyn} = v_p^2 - \frac{v_s^2}{2}\left(v_p^2 - v_s^2\right), \tag{2}$$

$$E_{dyn} = \rho\, v_s^2\left[\left(3v_p^2 - 4v_s^2\right)\left(v_p^2 - v_s^2\right)\right], \tag{3}$$

where v_{dyn} denotes the dynamic Poisson's ratio, E_{dyn} represents the dynamic Young's modulus, v_p is the velocity of the compressional wave, v_s is the velocity of the shear wave, and ρ is the rock density.

The static equivalents of the elastic properties were determined by comparing the calculated dynamic properties with the results of laboratory measurements of the static elastic properties. The correlation coefficients were 0.640 and −0.588 for the Young's modulus and Poisson's ratio, respectively.

To estimate the static uniaxial compressive strength in the profiles of the analysed boreholes, the compressional wave velocity was compared with the results of uniaxial compressive strength laboratory measurements, which produced a linear relationship with a satisfactory correlation coefficient of 0.682.

Further, to estimate the static tensile strength, T, which was not measured on the core material, we used a simple and well-known relationship proposed by Hoek (1966) [31]:

$$T = \frac{UCS}{10}. \tag{4}$$

As shown in Figure 6, the input data and developed 1D models of static elastic parameters, including the Poisson's ratio, Young's modulus, and UCS (marked with continuous line), were calibrated with the results of their measured static equivalents (marked with dots).

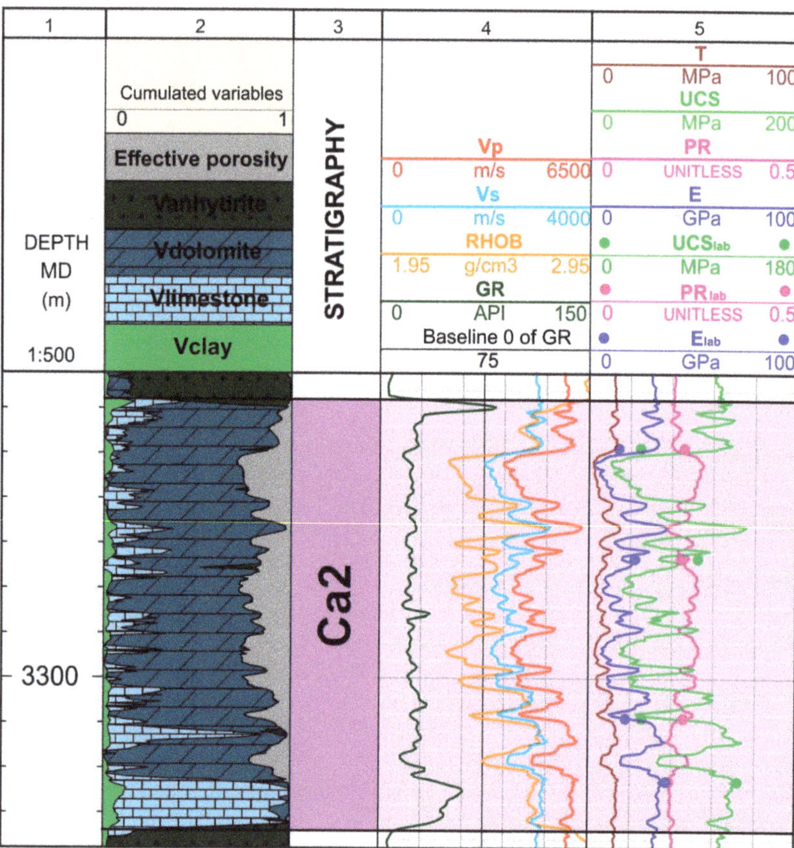

Figure 6. Graphic presentation of the input data and obtained 1D models of geomechanical parameters: track: 1—measured depth, 2—mineralogical model, 3—stratigraphy, 4—The input data include the compressional wave (vp), shear wave velocity (vs), density (RHOB), gamma ray (GR), 5—1D model of static uniaxial compressive strength (UCS), static Poisson's ratio (PR), and static Young's modulus (E).

4.1.2. 3D Modelling of Elastic and Strength Properties

The profiles of the static geomechanical parameters from the study area were scaled up to obtain their average values along the boreholes with a defined vertical resolution of the structural model. Then, they were subjected to a geostatistical analysis, which included determining the degree of spatial correlation among the borehole data for each geomechanical parameter.

While calculating the spatial distribution of the geomechanical parameters, the truncated Gaussian simulation algorithm (TGS) with the Co-kriging option was employed, thereby allowing us to use the 3D seismic data as secondary data. The advantage of this approach is that it compensates for missing data and, therefore, unknown relationships, especially in the horizontal direction. As a result, we obtained 3D models of the geomechanical parameters, including the Young's modulus, Poisson's ratio, and UCS, in the main dolomite reservoir interval, as shown in Figure 7A–C, respectively. Moreover, in Figure 7D–F, the average values of the static Young's modulus, Poisson's ratio, and UCS in the reservoir zone, respectively, are shown.

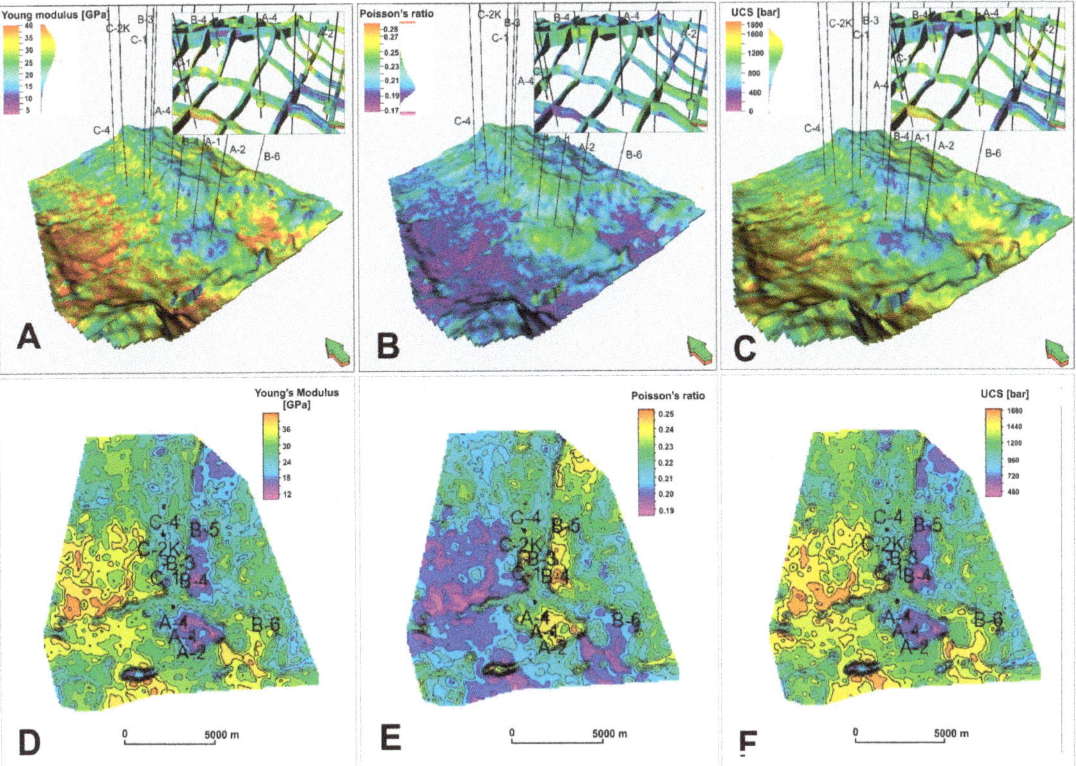

Figure 7. Visualisation of 3D distribution of (**A**) Young's modulus, (**B**) Poisson's ratio, and (**C**) unconfined compressive strength (UCS), and the average values of the (**D**) Young's modulus, (**E**) Poisson's ratio, and (**F**) UCS.

The spatial distributions of these geomechanical properties exhibited typical relationships, in which there was a negative correlation between the elastic parameters and a positive correlation between the Young's modulus and UCS. These relationships were also observed in the distribution of the values of these parameters illustrated as histograms. The 3D distribution of the tensile strength was obtained in a similar manner as the 1D models, wherein Hoek's relationship with UCS was utilized. Other geomechanical parameters, such as friction angle, angle of dilatation, and the Biot's coefficient, were assigned based on the typical values for appropriate lithologies listed in the literature [32–36]. The properties used as an input in the geomechanic modelling, both as a result of the modelling procedure and assumed based on the literature, are listed in Table 1.

4.2. Boundary Conditions

The boundary conditions applied to the model in order to define the initial stress for the geomechanical simulation were determined as the global tectonic stresses (minimum horizontal stress, σ_h, and maximum horizontal stress, σ_H) based on a recent investigation of a tectonic stress field in Poland conducted by Jarosiński (2006) [37]. The characteristics of the assumed stress field are listed in Table 2. Note that vertical stress, σ_v, is caused by gravity, and was thus determined using the overburden density (3D distribution of rock density).

Table 1. Calculated and estimated values of petrophysical and geomechanical parameters in 3D model of the study area.

Parameter [Unit]	Cenozoic (Clay, Sand, Gravel)	Cretaceous (Clayey Shales)	Jurassic (Sandy Shales)	Triassic (Sandstones)	Zechstein					Rotliegend (Underburden)
					Rock Salt	Anhydrite	Reservoir MAIN Dolomite	Limestone		
Young's modulus [GPa]	0.5	4	5.56	28.5	6.89	52.69	3D model	42.06		46.19
Poisson's ratio [-]	0.3	0.32	0.19	0.17	0.3	0.25	3D model	0.18		0.3
Density [g/cm^3]	3D model	3D model	3D model	3D model	3D model	3D model	3D model	2.75		2.3
Biot's coefficient [-]	1	1	1	1	0	0.10	0.7	0.8		1
Porosity [%]	3D model	3D model	3D model	3D model	3D model	3D model	3D model	2.99		4
Unconfined compressive strength (UCS) [MPa]	2.8	48	56.98	50.7	27.33	90.3	3D model	14.93		50
Friction angle [°]	30	32	20	59	29.08	64	28.6	22.8		30
Dilatation angle [°]	0	0	0	0	0	0	0	0		0

Table 2. Characteristics of regional principal horizontal stresses in the neighbouring area based on borehole breakouts (Jarosiński, 2006).

Stress Characteristic Parameter	Assigned Value
Gradient of horizontal stress (σ_h) [MPa/m]	0.01707
Gradient of σ_H [MPa/m]	0.02134
Azimuth of σ_H [°]	6

The developed 3D models of the geomechanical and petrophysical parameters were used as inputs to the geomechanical simulation. These were run for seven time steps (2003, 2020, 2050, 2052, 2053, 2056, and 2100) over the course of the hydrocarbon field exploitation and CO_2 injection, which were defined by the distribution of the reservoir fluid pressure developed in the reservoir fluid flow modelling. The calculated stress field reflecting the geomechanical states of the analysed rocks at specific time steps were then used to evaluate the tightness of the caprock, which could potentially leak the injected and stored CO_2.

5. Dynamic Model

A dynamic model of the analysed structure was constructed using the developed geological model and was supplemented with the following additional components:

- An initial distribution of reservoir fluids (oil and water) under hydrostatic conditions;
- Reservoir fluid transport properties (relative permeabilities);
- Reservoir fluid (oil) thermodynamic model.

5.1. Reservoir Fluid Distributions

The presence of three reservoir fluids (water, gas, and oil) was assumed in the simulation model of the analysed reservoir. The water–oil contact was established at a depth of 3280 m b.s.l., while the gas–oil contact was established at a depth of 3178 m b.s.l. The J-Leverett function was used to generate the initial water saturation distributions that matched those obtained from the geophysical measurements. This approach enabled us to obtain the initial dynamic equilibrium by simultaneously reconstructing the water saturation values in the wells. The J-Leverett function of water saturation, S_w, considers the dependence of capillary pressure, P_{cow}, on the parameters of the reservoir rock and has the form:

$$J(S_w) = \sqrt{\frac{k}{\phi}} \; \frac{P_{cow}(S_w)}{\sigma_{ow}\cos(\theta_{ow})}, \tag{5}$$

where k and ϕ denote the permeability and porosity of the reservoir rock, respectively, and σ_{ow} and θ_{ow} are the interfacial tension at the oil–water interface and their contact angle, respectively. The correct fit of the water saturation distributions in the wells was obtained using the four-parameter J curve model as follows:

$$J = \left(\frac{A}{S_r^n + B}\right)\left(1 - S_r^p\right), \; S_r = \frac{S_w - S_{wc}}{1 - S_{wc}}, \tag{6}$$

where S_{wc} is the connate water saturation, in which the correlation parameters were found to be: $A = 0.0094$, $B = 0.0657$, $n = 2$, $p = 10$, and $S_{wc} = 0.0035$, while $\sigma_{ow} = 64.65$ dyne/cm and $\theta_{ow} = 0°$.

The water saturation distributions were adjusted for every well. As a result of this analysis, it was found that the model correctly reproduced the water saturation variation with depth in most wells. As a result, the average water saturation in the model generated corresponds with high accuracy to the average water saturation generated using the interpolation method in the geological model.

5.2. Transport Properties

In general, transport properties in porous media include various flow mechanisms ranging from continuous flow of viscous type (Darcy flow) through slippage effects to Knudsen and molecular diffusion. Similarly, thermodynamics associated with transport phenomena may obey classical laws of locally equilibration state for large porosity systems and follow non-equilibrium type for small porosity system where molecule collisions with the walls of the system dominates over inter-particle ones. In contrast to low permeability, low porosity rocks [38,39] (tight and shale formations), conventional rocks such as dolomites included in the analysed case reveal standard viscous flow characterized by permeability tensors and equilibrium thermodynamic variables like pressure drop and shear stress. The other type of rocks in the analysed model refers to caprock anhydrite. Although it is an extremally low permeability rock, its main feature used in this study concerns the dependence of transport properties upon geomechanical state that was established in terms of effective permeability [13]. Consequently, in order to model effects of gas transport across the anhydrite caprock, a conventional approach was applied adopting stress-dependent permeability of a single porosity micro-fractured system.

An important part of the transport described by permeabilities is played by relative permeabilities of various reservoir fluids. Owing to the lack of direct measurements of the relative permeability curves, k_r, and reduced fluid saturation, S_r, for the analysed reservoir, a power model was adopted according to the relationship: $k_r = (S_r)^n$, wherein reduced saturation was determined as follows:

$$S_r = \frac{S - S_{min}}{S_{max} - S_{min}}, \tag{7}$$

where S_{min} and S_{max} refer to the minimum and maximum available saturations, respectively. The model used the following parameter values: for water ($k_r = k_{rw}$, $S = S_w$): irreducible water saturation $S_{min} = S_{wcr} = 0.0528$, maximum water saturation, $S_{max} = 1$; for oil in the oil–water system (with no gas $k_r = k_{row}$, $S = S_o$): $S_{min} = S_{orw} = 0.4917$, $S_{max} = S_{omax} = 1 - S_{wco} = 0.9964$; for oil in the oil–gas system (at connate water saturation, $k_r = k_{rog}$, $S = S_o$) $S_{min} = S_{orw} + S_{wco}$, $S_{max} = S_{omax}$; and for gas in the oil–gas/water–gas system ($k_r = k_{rg}$, $S = S_g$) $S_{min} = S_{gcr} = 0.1$, $S_{max} = S_{gmax} = 1 - S_{wco} = 0.9964$. Both the exponent of the relationship and the other parameters were adjusted during the model calibration procedure.

5.3. Reservoir Fluid Model

As the simulation model of the analysed reservoir (structure) is compositional, a thermodynamic model of the reservoir fluid (oil and gas) was constructed and calibrated based on the Soave–Redlich–Kwong (SRK) equation of state (EoS) using a standard software tool (PVTSim software, [40]). SRK equation of state is a conventional equation that relates temperature, pressure and volume of fluid and uses two parameters taking into account interaction of fluid molecules and their volumes. Those parameters are expressed in terms of several physical quantities, such as critical pressure, temperature, and volume. Other quantities include eccentricity factors [41], molar masses and boiling point temperatures. All these quantities are well defined for single component fluids such as light hydrocarbons (up to C6) and non-hydrocarbon components (N_2, CO_2, H_2S). For heavier hydrocarbons they are determined by empirical correlations with densities and molar masses [42]. In order to effectively perform compositional reservoir simulations, the number of fluid components is typically reduced by grouping or lumping the heavier ones. Effective parameters of such pseudo-components are calculated as averages of the real components with weights proportional to their concentrations and molar masses. In addition, the equation of state for multicomponent mixtures includes interaction effects of different polar molecules by so called mixing rules expressed via binary coefficients [43]. The viscosities of gas and liquid phases of the analysed fluid were determined using Lohrenz–Bray–Clark correlation [44] of viscosity and density and the results of the SRK equation of state for the densities. The SRK equation of state with parameters described above is typically sufficient to predict fluid properties. However, its results can be improved by the method of EoS regression [45] to experimental data.

In the analysed case the original fluid model [46] included 25 components determined by chromatographic analysis (gas phase components) and true boiling point analysis (liquid phase components) that were subsequently grouped into eight effective components, as summarized in Table 3.

Table 3. Composition of the reservoir fluid after component grouping.

Component	% mol
N_2	31.588
CO_2	0.612
H_2S	5.085
C_1	19.353
C_2	3.567
C_3–C_6	11.99
C_7–C_{11}	12.27
C_{12+}	15.5

The model was calibrated using the regression method to the following experimental data: the pressure at the saturation point, constant mass expansion tests (relative volumes of gas and liquid phases, effective compressibilities), differential depletion tests (oil and gas formation volume factors, gas-in-oil solubility, oil density, gas z-factor, oil and gas viscosities, gas gravity), separator tests (gas–oil ratio, gas gravity, oil formation volume factor), and viscosity tests (oil viscosity).

The following regression parameters of the SRK EoS were used: critical temperatures and pressures, eccentricity and Ω_A factors of 2 grouped components (C_7–C_{11}, C_{12+})—the complete set of the EoS parameters and their values are given in Tables 4 and 5. Additional regression parameters were those of Lohrenz–Bray–Clark correlation coefficients—given in Table 6.

Table 4. Equation of state (EoS) parameters of the reservoir fluid model for the Soave–Redlich–Kwong (SRK) model.

Component	Critical Temperature T_c [K]	Critical Pressure P_c [bar]	Eccentricity Factor ω	Parameter Ω_A	Parameter Ω_B	Molar Mass M	Boiling Point T_b [K]	Critical Volume V_c	Critical Gas Compressibility Factor Z_c	Parachor
N_2	126.2	33.9	0.0400	0.4275	0.0866	28.0	77.4	0.090	0.2905	41.0
CO_2	304.2	73.8	0.2250	0.4275	0.0866	44.0	194.7	0.094	0.2741	78.0
H_2S	373.2	89.4	0.1000	0.4275	0.0866	34.1	213.5	0.099	0.2837	80.1
C_1	190.6	46.0	0.0080	0.4275	0.0866	16.0	111.6	0.099	0.2874	77.3
C_2	305.4	48.8	0.0098	0.4275	0.0866	30.1	184.6	0.148	0.2847	108.9
C_3–C_6	453.2	34.7	0.2315	0.4275	0.0866	65.4	296.9	0.299	0.2752	221.2
C_7–C_{11}	641.1	27.3	0.3182	0.4221	0.0866	120.6	418.2	0.631	0.3231	347.6
C_{12+}	784.1	17.3	0.4975	0.4221	0.0866	234.6	575.6	1.180	0.3135	626.1

Table 5. Fluid model binary coefficients for the Soave–Redlich–Kwong (SRK) equation of state (EoS) model.

	N_2	CO_2	H_2S	C_1	C_2	C_3–C_6	C_7–C_{11}	C_{12+}
N_2	-	-	-	-	-	-	-	-
CO_2	−0.0315	-	-	-	-	-	-	-
H_2S	0.1696	0.0989	-	-	-	-	-	-
C_1	0.0278	0.1200	0.0800	-	-	-	-	-
C_2	0.0407	0.1200	0.0852	0.0000	-	-	-	-
C_3–C_6	0.0808	0.1200	0.0655	0.0000	0.0000	-	-	-
C_7–C_{11}	0.0928	0.1006	0.0006	0.0000	0.0000	0.0000	-	-
C_{12+}	0.0928	0.1006	0.006	0.0000	0.0000	0.0000	0.0000	-

Table 6. Coefficients of the Lohrenz-Bray–Clark viscosity model.

a1	a2	a3	a4	a5
0.4703	−0.1017	0.0585	−0.0408	0.0093

6. Model Calibration

The model of the analysed reservoir was calibrated against the production data covering 16 years of operation with 11 producing wells. The data included daily oil, gas, and water production rates from individual wells, bottom-hole pressures, and well test results.

6.1. Calibration Results

Model calibration involved modifying the following model parameters: global and local petrophysical parameters (absolute and relative permeabilities and permeability anisotropies), well productivity indices, and skin-effect coefficients. As a result, a significant increase in the gas–oil-ratio (GOR) during operation was reproduced for wells C-1, A-1, and A-4, which were located directly or in the vicinity of the primary gas cap (Figure 8), and for the wells located relatively far away from the primary gas cap but influenced by the formation of a secondary gas cap.

The necessity to accurately reproduce the measured bottom-hole pressures required a linear characteristic of the relative gas permeability at the top zone of the structure, which suggests the existence of micro-fractures in the reservoir. The critical oil saturation was found to be consistent with the results obtained from the laboratory experiments regarding oil displacement with water. Further, the permeability vertical-to-horizontal anisotropy increased, which resulted in the correct depression in the production wells and reduced the migration of the released gas to the top of the structure.

Acid treatments of wells C-1, C-2k, and A-4 were modelled by modifying the absolute permeability and effective pore volume in their near-wellbore zones. To reproduce the relatively high pressure difference measured between the C-1, C-2k, and C-11k wells, deteriorated petrophysical properties were introduced among these wells. A local reduction in the absolute permeability anisotropy near the A-11H and A-13k wells caused the simulated results of the GOR to become closer to the observed values. In general, the productivity indices and skin-effect coefficients of the wells were accordingly adjusted to the depressions observed during the well tests.

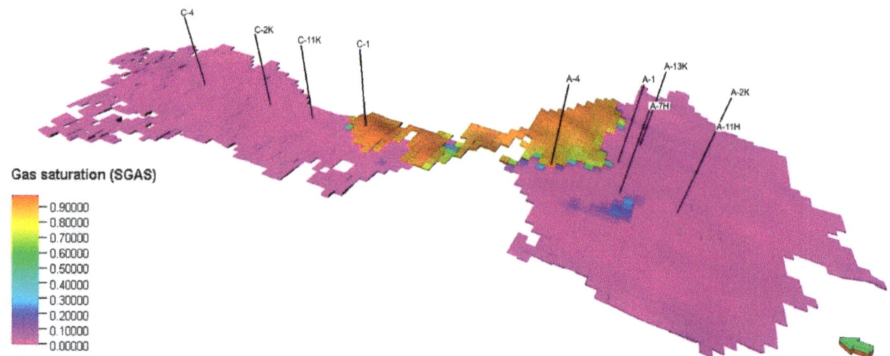

Figure 8. 3D view of the model top layer of the original gas saturation.

Figures 9 and 10 present the total oil and gas production from the field as a result of the calibrated simulation model versus the measured values, revealing that the model matched the production data with high accuracy. At the well level, Figure 11 shows an example of the adjustment quality of the bottom-hole pressures measured at a production well. Meanwhile, as shown in Figure 12, good GOR fitting in an individual well was achieved. Further, Figure 13 shows an example of the adjustments to the bottom-hole pressure evolution reported during the well production test. Thus, the simulation results are in good agreement with the measured values for all the available historical production data.

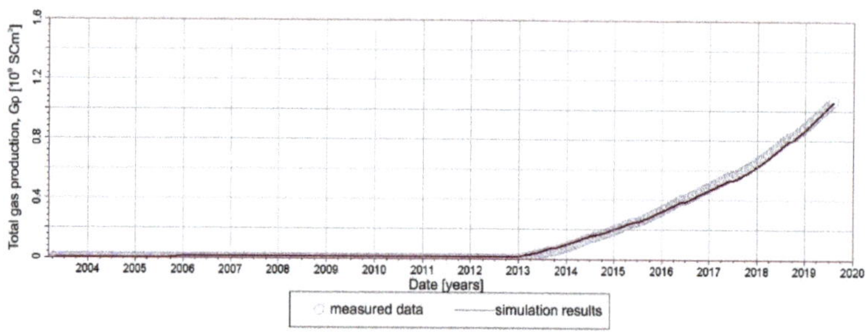

Figure 9. Total reservoir oil production.

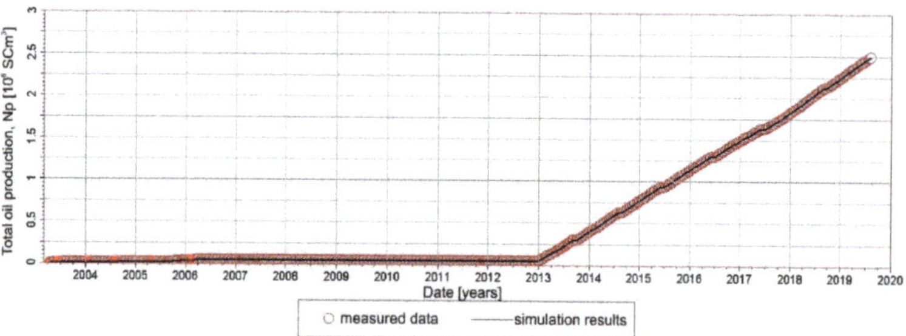

Figure 10. Total reservoir gas production.

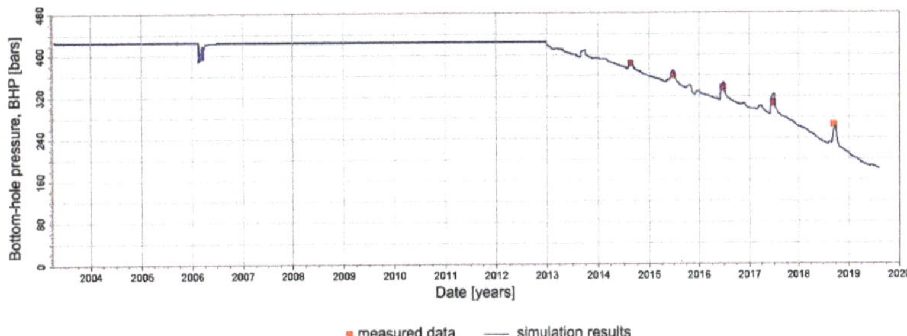

Figure 11. Bottom-hole pressure variation at well A-7H.

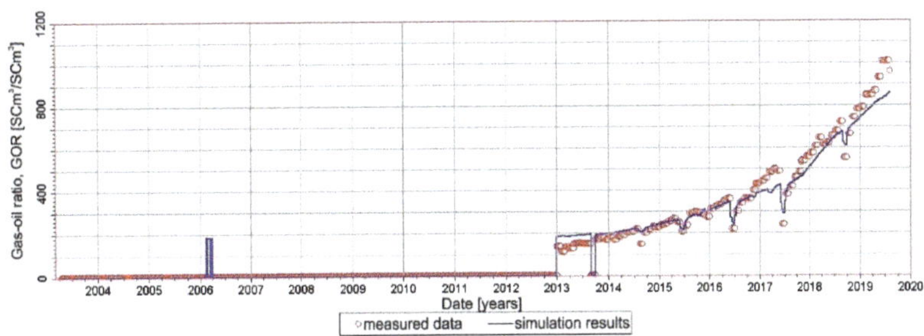

Figure 12. Gas–oil ratio (GOR) variation at well A-7H.

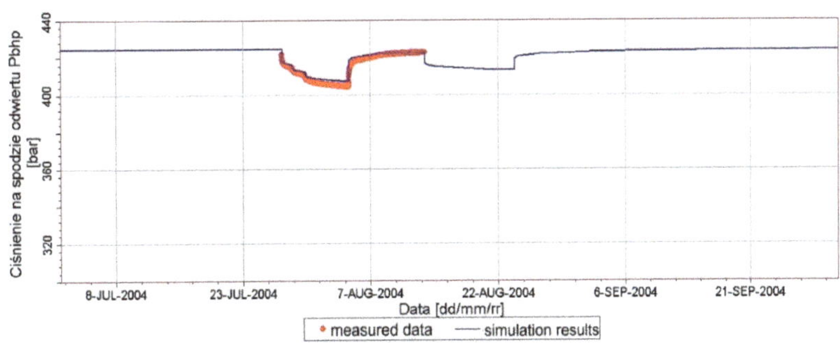

Figure 13. Bottom-hole pressure variation at well A-4.

6.2. Model Characterisation after Calibration

After the calibration process, the simulation model of the analysed reservoir was characterised by the following parameters: total area: 15 km × 10 km, the nature of the model: single porosity and permeability, lateral dimensions: 376 blocks × 242 blocks, lateral sizes of the blocks: 100 m × 100 m, layer structure: 15 layers, number of active blocks: 20,325, initial contact depths: oil–water: 3282 m b.s.l., gas–oil: 3178 m b.s.l., initial pressure: 430.2 bar (@ 3282 m b.s.l.), reservoir temperature (constant): 126.8 °C, total model pore volume: 70.2 million m^3, average values of basic parameters: porosity: 9.6%, horizontal permeability: 34.91 mD, vertical permeability: 1.4 mD, and total thickness of a simulation layer: 2.77 m.

7. Pressure Evolution

The dynamic flow model of the analysed reservoir was used to simulate the reservoir behaviour during enhanced oil recovery with CO_2 injection followed by exclusive CO_2

injection. Production rates of the oil producing wells were assumed at the nominal levels provided by the reservoir operator. In addition, they were constrained by the minimum well-head pressure, maximum water cut, maximum gas oil ratio, minimum economic flow rate, and maximum allowable drawdown. Initially, CO_2 injection was performed using four wells (A-4, A-6H, C-1, and C-4), before injection expanded to other wells, which were converted from producing wells after they ceased production. The injection rates of individual wells were controlled by the total injection rate and well contributions proportional to their injectivities. The injecting wells were constrained by the maximum bottom-hole pressures not exceeding the formation fracturing pressure. The injection phase lasted approximately 36 years. As a result of the simulation of this EOR/CO_2 sequestration project (called the basic scenario), all the significant quantities characterising the project were obtained. Figure 14 shows the total oil production and average reservoir pressure evolution during the project.

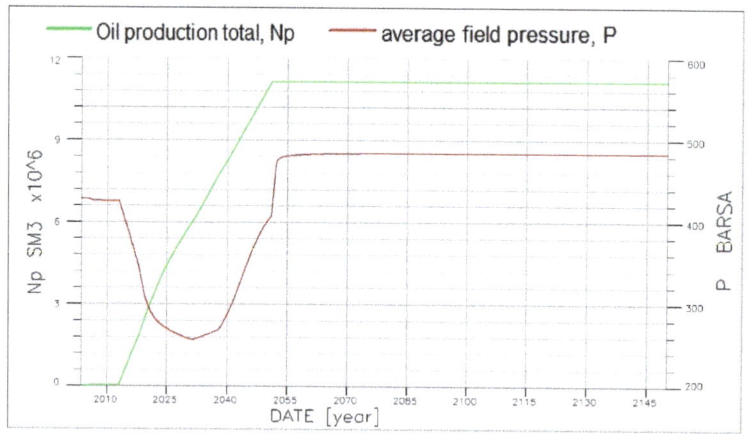

Figure 14. Basic scenario of oil production by enhanced oil recovery (EOR) with CO_2 injection. Total oil production, FOPT; and average reservoir pressure, FPR.

The contributions of individual wells to the reservoir oil production are shown in Figure 15 in terms of their total oil production.

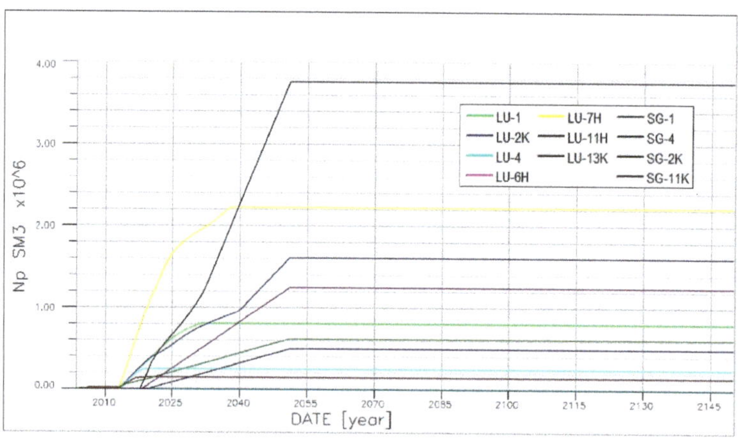

Figure 15. Basic scenario of oil production by enhanced oil recovery (EOR) with CO_2 injection. Np is the total oil production of individual wells.

The quantitative characteristics of the CO_2 injection process were expressed as the total CO_2 injection of the project and as the total CO_2 injection of individual wells, as shown in Figures 16 and 17, respectively.

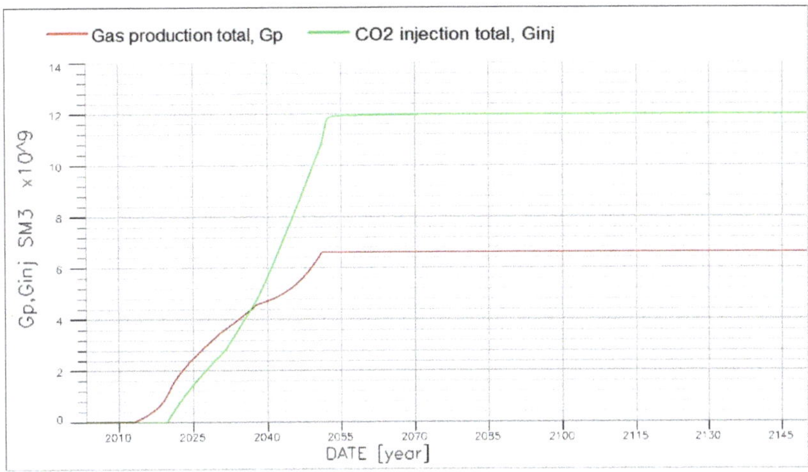

Figure 16. Basic scenario of oil production by enhanced oil recovery (EOR) with CO_2 injection. FGPT is the total gas production, and FGIT is the total CO_2 injection.

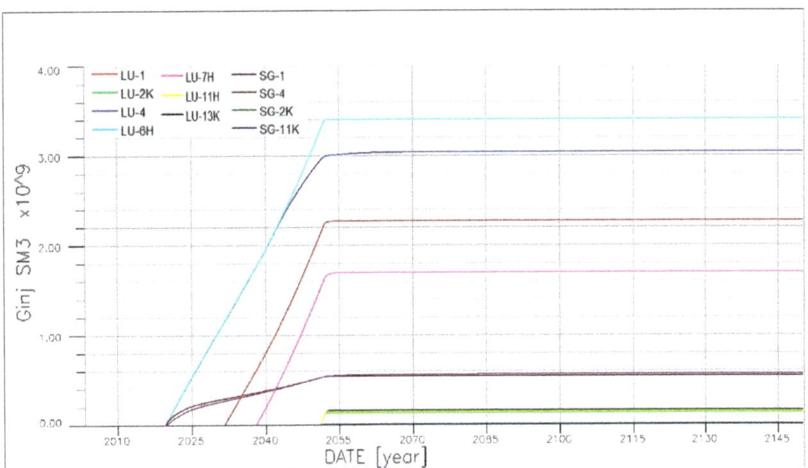

Figure 17. Basic scenario of oil production by enhanced oil recovery (EOR) with CO_2 injection. G_{in} denotes CO_2 total injection of individual wells.

Figure 18 shows the effects of EOR with CO_2 injection by comparing the total oil production of this project with total oil production using the primary oil production method. The incremental oil production exceeds 6.3×10^6 Sm^3, which is equivalent to approximately 130% of the primary production (4.8×10^6 Sm^3).

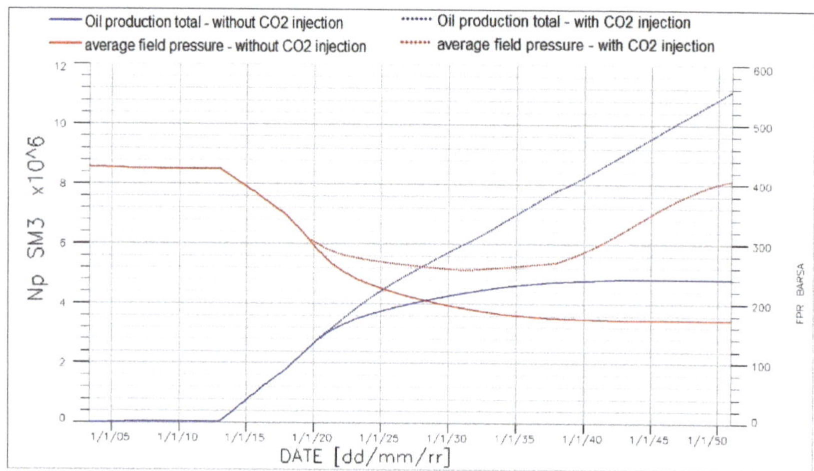

Figure 18. Comparison of primary production and EOR with CO_2 injection. Total oil production, N_p; and average reservoir pressure, FPR.

The simulation results include pressure distributions across the analysed structure extended model at various stages of the EOR/CO_2 sequestration project, as shown in Figure 19.

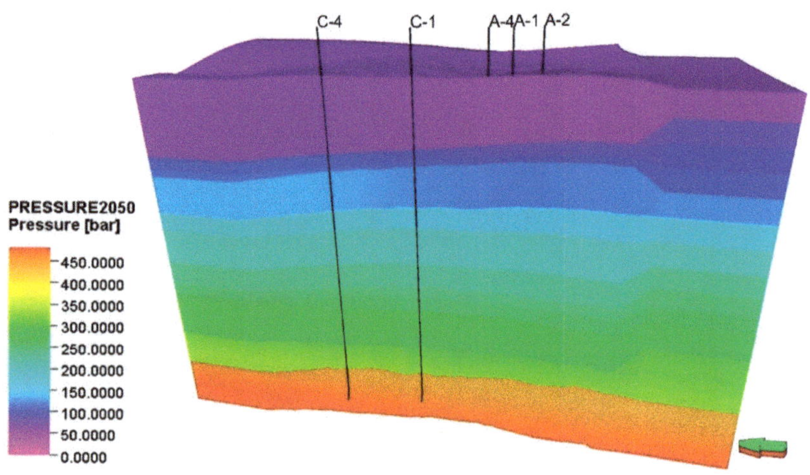

Figure 19. Example of pressure distribution in the analysed structure extended model. Only the reservoir and overburden are shown.

To present a more detailed variation in the pressure distribution, pressure maps for the top layer of the reservoir are shown in Figure 20 for various stages of the project.

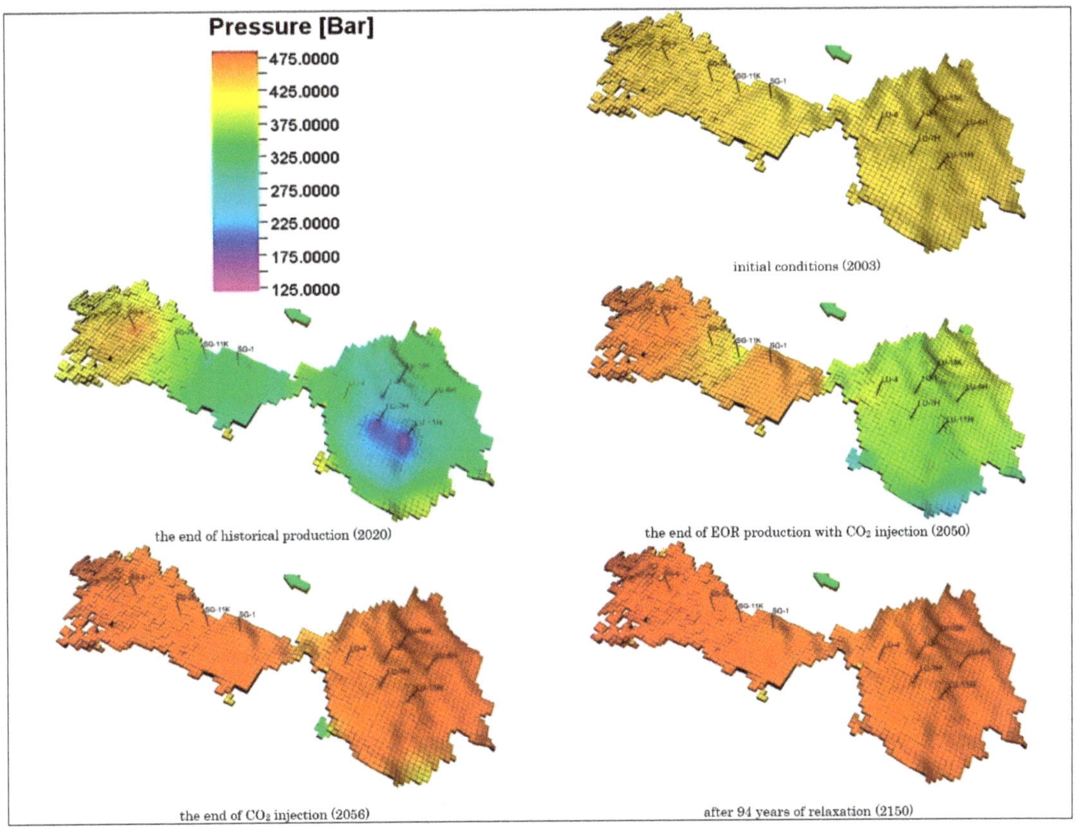

Figure 20. Pressure distributions at the reservoir top at various stages of the project.

8. Caprock Sealing Analysis

The migration of stored CO_2 into the overlying strata is inhibited by the sealing properties of very low permeability and high capillary forces, which prevent gas penetration into the water-saturated caprock. The parameter governing this sealing ability is the threshold pressure. If the value of the threshold caprock pressure exceeds that of the gas, the caprock pores begin to be penetrated, ultimately leading to the development of interconnected pathways for gas to escape upward [47]. This threshold pressure can be measured using the method first proposed by Thomas et al. (1968) [48], which was later modified using empirical models adopted for a specific lithology. In general, among the different lithologies, anhydrites, carbonates, and shales have high threshold pressure and low permeability, making them potentially good sealing rocks for gas storage.

In this study, to calculate the threshold pressure in the basal anhydrite (A2), a direct sealing layer neighbouring the reservoir rock, we used the empirical relationship as follows:

$$P_{th} = a\,k^b, \tag{8}$$

where P_{th} is the threshold pressure, k is the permeability, a is the multiplication coefficient, and b is the exponent. The coefficients a and b were developed by Ibrahim et al. (1970) [49], who studied seven samples of anhydrites. The permeability in the history of reservoir exploitation and forecasting of CO_2 injection into the main dolomite reservoir rock was estimated based on the stress dependency on permeability, wherein the stress variation was obtained using geomechanical simulations.

Numerous scientific investigations have revealed that both porosity and permeability are affected by dominant loading conditions and the history of stresses within the sedimentary basin. This relationship is widely known and well documented [13,50–52]. Based on the laboratory permeability measurements performed by Morrow et al. (1984) for fault gouges [53], Shi and Wang (1988) suggested that the power law describes the nonlinear decline in permeability with increasing effective stress in the low-permeability rocks well in the following form [50]:

$$k = k_0 \left(\frac{\sigma}{\sigma_0} \right)^p,$$

(9)

where k denotes the permeability under effective stress σ, k_0 is the initial permeability at the initial effective stress σ_0, and p is the material constant. Regarding effective stress, we used the magnitude of the minimum horizontal stress as it is mostly responsible for the decline in permeability in the mechanism of closing microcracks potentially present in the anhydrite rocks.

As the data regarding the petrophysical characteristics of the Zechstein anhydrites in Poland were not available, we used a typical value of permeability for anhydrites [54] and the material constant p determined by Zivara et al. (2019) [13]. The results of geomechanical simulations performed for the selected time steps provided the distribution of effective stresses evolving throughout the CO_2 injection operation. The calculated permeability was then used to estimate the threshold pressure at selected time steps, ending approximately 80 years after the initial CO_2 injection.

Table 7 lists the assumed values of initial permeability, material constant, and the other coefficients employed for the threshold pressure calculation in the basal anhydrite (A2).

Table 7. Assumed values of parameters used in the calculation of permeability versus effective stress, and coefficients for calculation of threshold pressure in the anhydrite caprock.

Parameter		Assumed Value
Coefficient *a* (Equation (8))	[49]	2.6×10^{-7}
Exponent b (Equation (8))	[49]	-0.348
Initial permeability k_0 [m^2] (Equation (9))	[54]	9.6×10^{-21}
Material parameter p (Equation (9))	[13]	0.6288

As shown in Figure 21, the evolution of pressure in the selected location at the top of the reservoir Ca2 (red line) and in the basal anhydrite A2 (blue line) exhibit a negative correlation between the magnitude of the minimum horizontal stress and reservoir pressure, a positive relationship between the magnitude of the minimum horizontal stress and the threshold pressure and the evolution of the threshold pressure (green line), and a pressure step between the top of the reservoir and basal anhydrite (orange line).

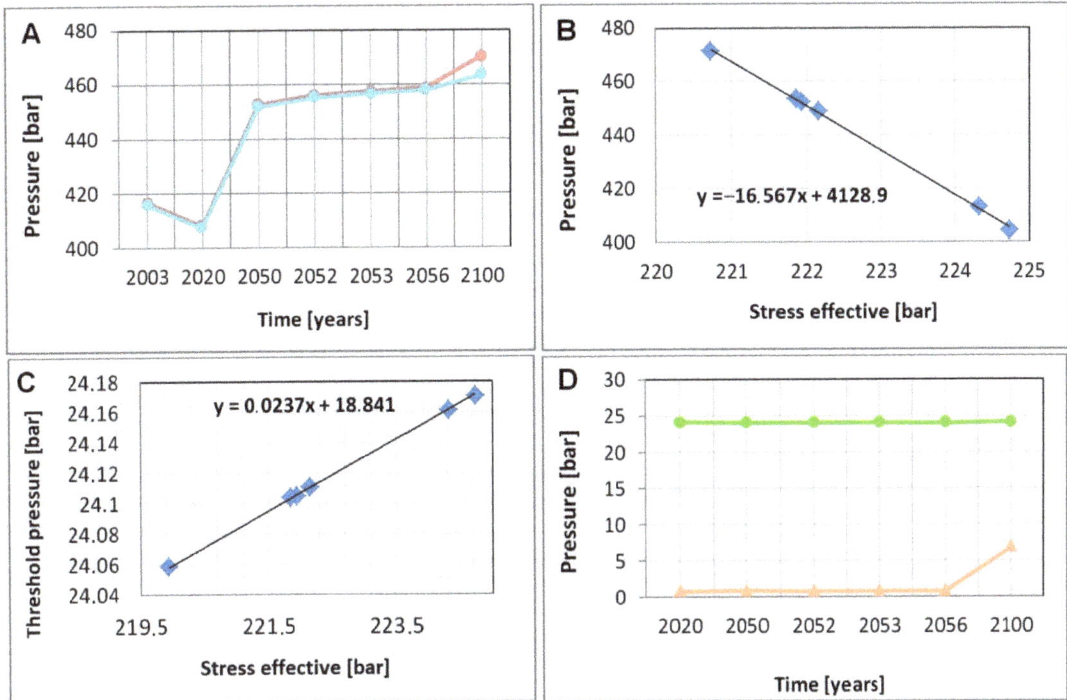

Figure 21. (**A**)—Pressure evolution in the top of the reservoir Ca2 (red line) and basal anhydrite A2 (blue line), (**B**)—relationship between the magnitude of the minimum horizontal stress and reservoir pressure, (**C**)—relationship between the magnitude of the minimum horizontal stress and threshold pressure, (**D**)—evolution of the threshold pressure (green line) and pressure step between the top of the reservoir and basal anhydrite (orange line) during the history of exploitation and injection of CO_2 into the reservoir rock.

The spatial distributions of the pressure (A), magnitude of the minimum horizontal stress (B), estimated permeability (C), and calculated threshold pressure (D) for the time steps representing the end of hydrocarbon production (2020—left column), during the process of CO_2 injection (2056—middle column), and at the end of CO_2 injection (2100—right column) in the sealing anhydrite (A2) are shown in Figure 22.

These distributions show that the increase in pore pressure in the sealing rock along with the injection of CO_2 in the reservoir zone caused a very slight increase in effective stress and, consequently, a small increase in the permeability, from 9.55 nD to 9.80 nD. This change results in a small decrease in the threshold pressure, from 24.16 bar to 24.05 bar. These changes are small and local, and the threshold pressure remains at a safe level of approximately 24 bars. The most sensitive area is localised within the slope of the platform, near injection wells B-4 and B-5. The increase in permeability (yellow-red colour) and decline in threshold pressure (purple colour) were significant in this area. Another sensitive zone is located in the deep-water zone near injection wells A1, A2, and A-4.

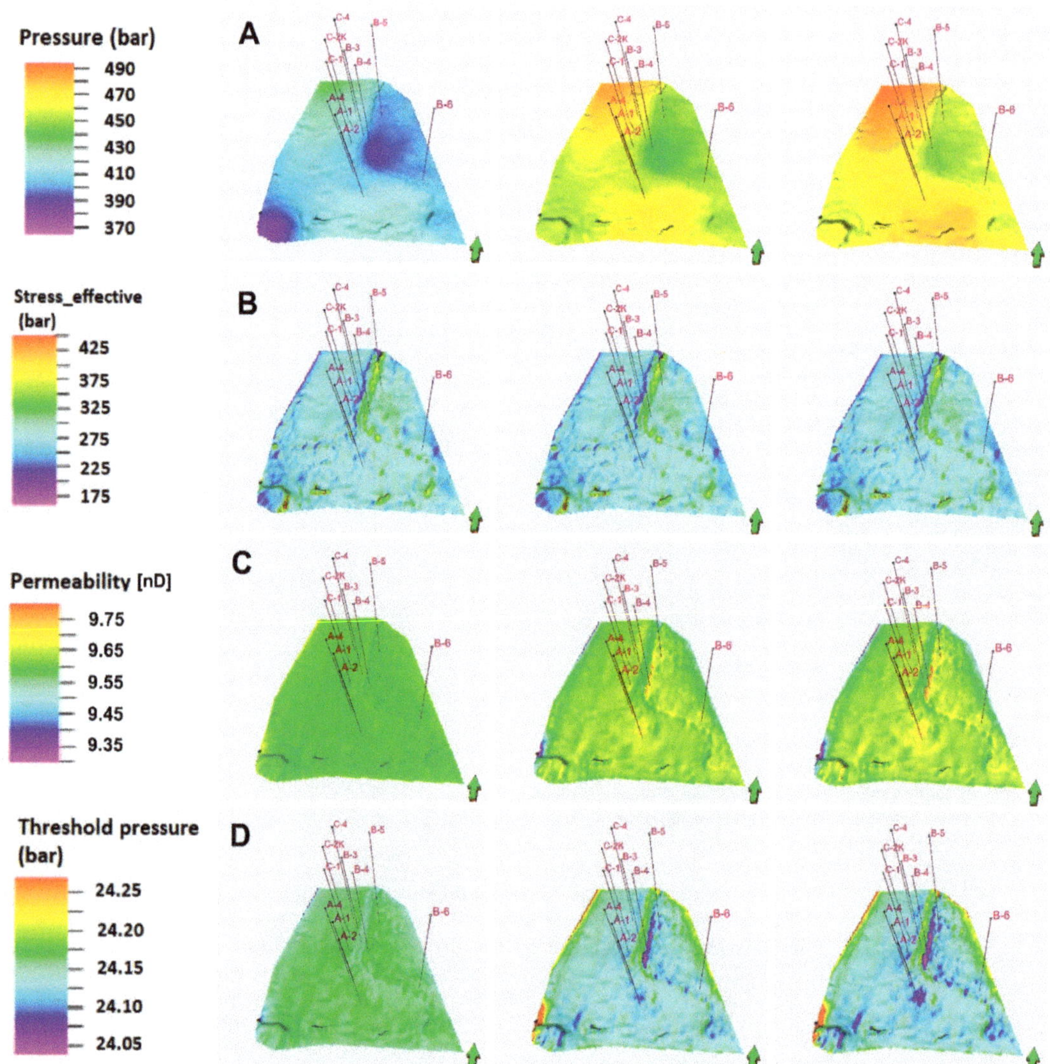

Figure 22. Spatial distributions of: (**A**)—pore pressure, (**B**)—effective minimum horizontal stress (σ_h), (**C**)—permeability (k), and (**D**)—threshold pressure in the sealing rock in the basal anhydrite (A2) at the end of hydrocarbons production (2020 year—left column), in the process on CO_2 injection (2056 year—middle column), and at the end of CO_2 injection (2100 year—right column).

The difference between the pressure in the top layer of the reservoir (Ca2) and the sealing layer of the basal anhydrite (A2) shown in Figure 23 helped to determine where additional CO_2 can be injected without exceeding the threshold pressure in the anhydrite, thereby lowering the risk of CO_2 leakage.

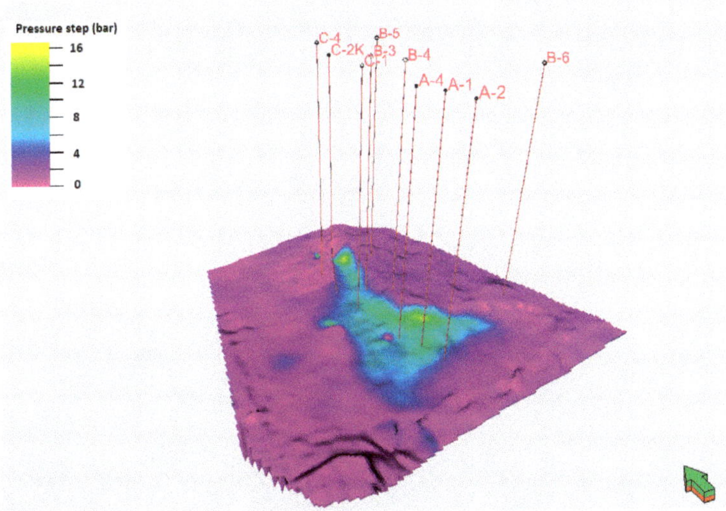

Figure 23. Distribution of pressure steps between the sealing anhydrite interval and the top of the reservoir rock at the end of CO_2 injection.

9. Gas Leakage Analysis

In addition to caprock sealing conditions, the quantitative studies of gas leakage to the caprock were performed using the structure simulation model described above. As the caprock maintains its sealing properties up to the determined threshold pressure, the leakage occurs only when the pressure step across the caprock-reservoir boundary exceeds the threshold pressure. In this section, the quantitative results of the leakage to the caprock were determined for several scenarios, characterised by varying amounts of injected/sequestrated CO_2 as listed in Table 8. The varying amounts of total injection/sequestration were realised by the modifications of the well group injection rate. All the other constraints were kept unchanged.

The progress of the CO_2 injection process was presented in Figures 24 and 25 in terms of time evolution for total CO_2 injection and average reservoir pressure, respectively.

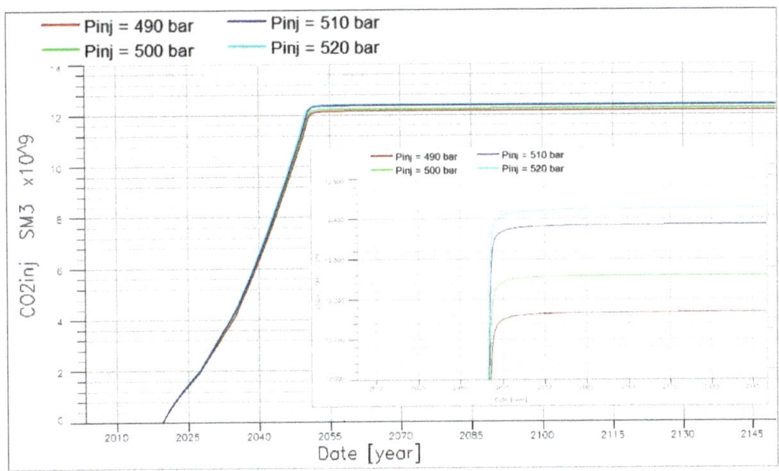

Figure 24. Total CO_2 injections for various injection scenarios.

Table 8. Total CO_2 injection/sequestration capacity.

Scenario	Total Injection/Sequestration Capacity [$\times 10^9$ SCm^3]
Basic	12.01
P490	12.17
P500	12.26
P510	12.38
P520	12.43

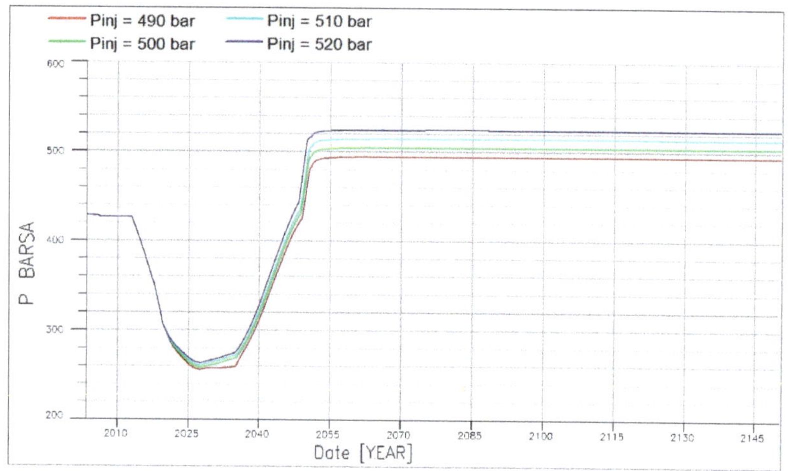

Figure 25. Average reservoir pressures for various injection scenarios.

The detailed pressure distribution at the reservoir top revealed a pressure step across the reservoir–caprock boundary exceeding the threshold pressure at several locations. As a consequence, CO_2 leaked from the reservoir to the overlying anhydrite layers. The effects of this leakage process are shown in Figure 26 for each of the four scenarios considered. The final (140 years after the injection finish) amounts of the total leakage and their value as a fraction of injected CO_2 are listed in Table 9.

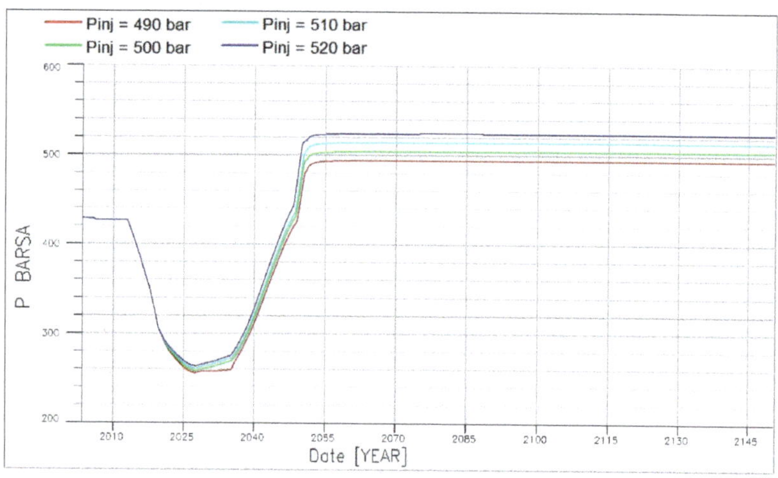

Figure 26. Total CO_2 leakage amounts to the reservoir caprock for various injection scenarios.

Table 9. Total leakage of CO_2 to the caprock.

Scenario No.	Total Leakage [$\times 10^6$ SCm^3]	Total Leakage as Fraction of Injected CO_2 [%]
P490	0.98	0.0080
P500	1.20	0.0098
P510	1.39	0.0113
P520	1.59	0.0128

Although the CO_2 leakage rate gradually decreased with time, it continued up to the extended end of the simulation period (140 years of relaxation phase after the injection finish). Note that the leaked CO_2 accumulated in the bottom layer of the anhydrite horizon, wherein a relatively small amount (2.5%) of CO_2 reached the anhydrite top, as shown in Figure 27.

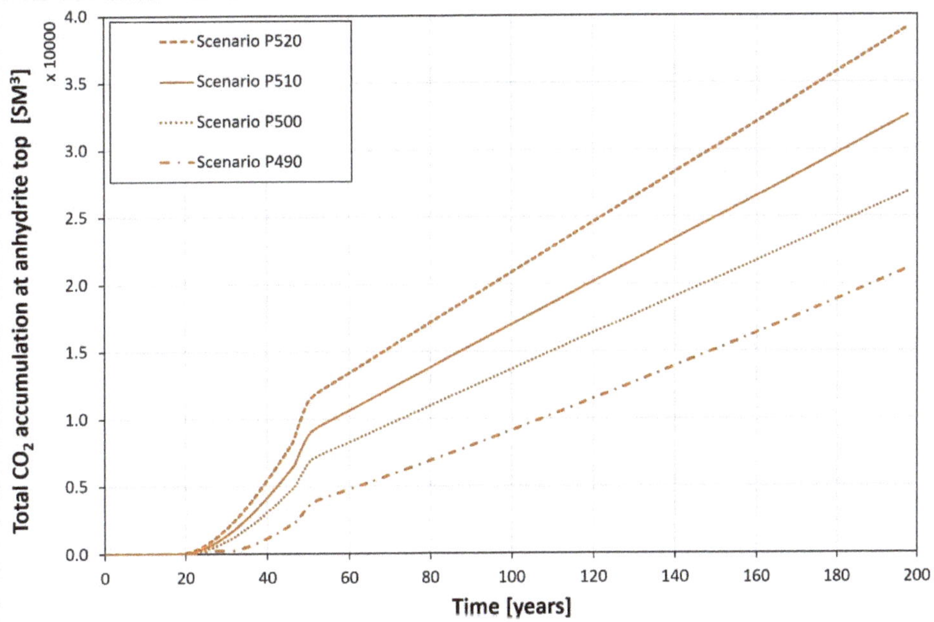

Figure 27. Total CO_2 accumulation at the anhydrite top for various injection scenarios.

The process of CO_2 leakage from the reservoir to the caprock was very heterogeneous because of the following:

- inhomogeneous reservoir rock properties,
- varying depths of the reservoir-caprock boundary,
- inhomogeneity of the CO_2 injection process.

The detailed distributions of CO_2 saturation at the end of the simulation period found in the top layer of the reservoir and the three layers composing the anhydrite horizon are shown in Figure 28 for the selected scenarios.

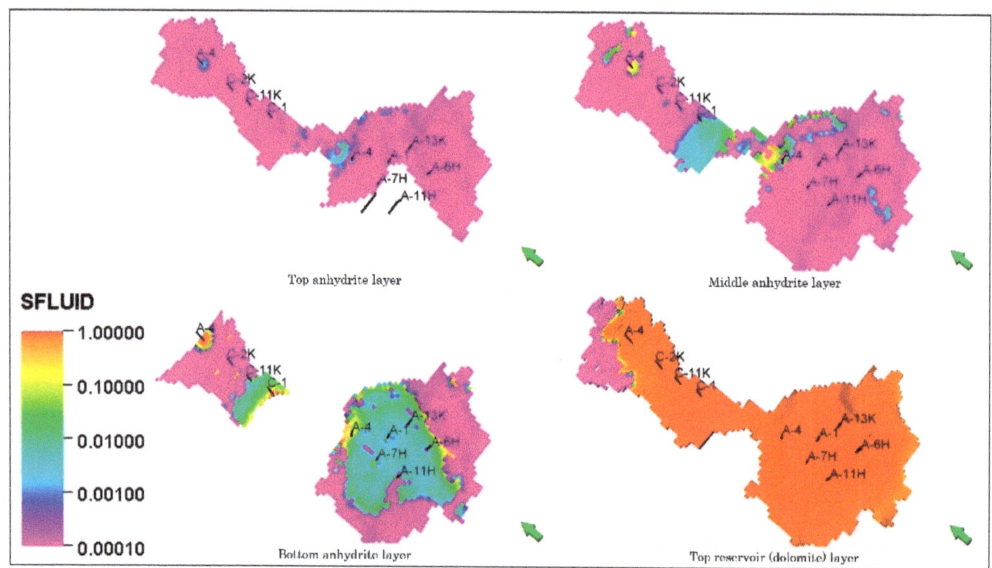

Figure 28. Reservoir fluid distribution at reservoir top and anhydrite caprock layers 94 years after injection. Scenario P520.

10. Summary and Conclusions

The studies described herein address the practical problem of geological structure tightness for the purpose of CO_2 sequestration applied to a geological structure to be assessed as the potential site of a future sequestration project. The studies took the advantage of advanced methods, realistic assumptions, and quantitative results of previous, general studies to practically evaluate the operation of the real geological structure of a domestic oil reservoir including the initial phase of enhanced oil recovery by CO_2 injection and final phase of CO_2 sequestration.

In particular, we performed an analysis of the tightness of the basal anhydrite (A2), which is the sealing layer for the hydrocarbon accumulation in the main dolomite formation (Ca2). To this end, we used numerical methods that combined geomechanical and reservoir fluid flow modelling. Based on a large set of data, geological, structural, and parametric models were developed. To predict the reservoir and surrounding rock behaviour caused by CO_2 injection, dynamic flow and geomechanical models were constructed based on the geological model supplemented by all the other required components. These models were satisfactorily matched with the historical reservoir operation data. Then, the models were used to forecast the reservoir behaviour (evolution of detailed pressure and fluid saturation distributions) under the assumptions of an EOR project with CO_2 injection followed by a CO_2 sequestration phase. The dynamic simulations were combined with the geomechanical simulations using a two-way coupling procedure and utilising correlations between the geomechanical state (stress and strain distribution) and transport properties of the reservoir rock and caprock, including the threshold pressure at the reservoir–caprock boundary.

The results of the procedure determined the sequestration capacity of the structure. In addition, the effects of CO_2 leakage from the reservoir to the caprock were simulated in cases where the reservoir pressure at the structure top (reservoir–caprock boundary) exceeded the limits of the threshold pressure. The long-term simulations resulted in an assessment of the total amount of CO_2 leakage and its distribution within the caprock horizon as a function of time for various total CO_2 injection volumes above the sequestration capacity.

Based on the simulation results, the following conclusions can be drawn:

(1) General conclusions:

- The method applied in the studies proves the necessity to employ an extended model of the analysed structure, wherein the geomechanical and dynamical simulations allow precise estimations of the threshold pressure and provide information regarding critical locations at the reservoir–caprock boundary where leakage could occur.
- The precise determination of the threshold pressure (with inhomogeneous distribution across the reservoir–caprock boundary) and its evolution with time are crucial factors for estimating the sequestration capacity of the structure.
- The following two correlations are key factors when the sealing properties of the reservoir caprock boundary are evaluated:
  - The correlation between the geomechanical state (stress field) and transport properties (permeabilities) of the caprock;
  - The correlation between the caprock permeability and threshold pressure at the reservoir-caprock boundary.

(2) Conclusions specific to the analysed geological structure:

- The determined threshold pressure revealed the potential CO_2 sequestration capacity of the structure, showing that it could safely store approximately 12×10^9 Sm^3 of gas.
- The relatively low (up to 3.6%) excess over the determined sequestration capacity resulted in a very small total CO_2 leakage (0.13‰ of the sequestrated volume) up to approximately 100 years of relaxation phase after CO_2 injection is complete.
- Most of the leaked CO_2 accumulates in the bottom part of the caprock.
- The leakage process does not cease even at the end of the simulated (100 years) relaxation phase.

Author Contributions: Conceptualization, M.S.-V., W.S. and A.G.; methodology, M.S.-V., W.S. and A.G.; software, M.S.-V., K.S. and A.G.; validation, M.S.-V. and W.S.; formal analysis, M.S.-V., W.S. and A.G.; investigation, M.S.-V. and A.G.; resources, M.S.-V. and A.G.; data curation, M.S.-V., K.S. and A.G.; writing—original draft preparation, M.S.-V., W.S. and K.S.; writing—review and editing, M.S.-V. and W.S.; visualization, M.S.-V., A.G. and K.S.; supervision, K.S. and W.S.; project administration, M.S.-V. All authors have read and agreed to the published version of the manuscript.

Funding: This research was carried out as part of the 'Analysis of geological structure tightness for the purposes of CO_2 sequestration' project, which is funded by the Polish Ministry of Science and Higher Education, Grant No. DK-4100-39/20. The authors would like to express their gratitude to the Polish Ministry of Science and Higher Education for funding this research.

Institutional Review Board Statement: Not applicable.

Informed Consent Statement: Not applicable.

Data Availability Statement: The data that support the findings of this study are available on reasonable request from the corresponding author.

Conflicts of Interest: The authors declare no conflict of interest.

Abbreviations

SCm^3	Standard cubic meter
ϵ_H	The strains in the direction of the maximum horizontal stress
ϵ_h	The strains in the direction of the minimum horizontal stress
p	The material constant
A1D	The lower Anhydrite of Werra cyclothem
A1G	The upper Anhydrite of Werra cyclothem
A2	The basal Anhydrite of Stassfurt cyclothem
A2G	Screening Anhydrite of Stassfurt cyclothem

A3	The main Anhydrite of Leine cyclothem
A4D	The lower Anhydrite of Aller cyclothem
A4G	The upper Anhydrite of Aller cyclothem
Ca1	Zechstein limestone of Werra cyclothem
Ca2	The main Dolomite of Stassfurt cyclothem
E	Young's modulus
EOR	Enhanced oil recovery method
FGIT	Total CO_2 injection
FGPT	Total gas production
FOPT	Total oil production
FPR	Average reservoir pressure
FPR	Average reservoir pressure
G_{in}	Total CO_2 injection of individual wells
GOR	Gas to oil ratio
GR	Natural Gamma radiation
I3	Salt clay of Leine cyclothem
k	Permeability of the rock
k_0	The initial permeability
M	Molar mass
Na1	The oldest Halite of Werra cyclothem
Na2	Older Halite of Stassfurt cyclothem
Na4	The youngest Halite of Aller cyclothem
P	Pore pressure
P_{cow}	Capillary pressure
Pc	Critical pressure
P_{th}	The threshold pressure
PVT	Pressure, volume and temperature
PZ2	Stassfurt cyclothem
PZ3	Leine cyclothem
PZ4	Aller cyclothem
S_{max}	The maximum available saturations
S_{min}	The minimum available saturations
SRK EoS	Soave–Redlich–Kwong equation of state
S_w	Water saturation
S_{wc}	The connate water saturation
T	Tensile strength
T_b	Boiling point
Tc	Critical temperature
TGS	Truncated Gaussian Simulation algorithm
UCS	Uniaxial compressive strength
V_c	Critical volume
vp	Compressional wave velocity
vs	Shear wave velocity
Zc	Critical gas compressibility factor
α	The Biot's coefficient
θ_{ow}	Oil-water contact angle
ν	Poisson's ratio (PR)
ρ	Rock density (RHOB)
σeff	Effective stress
σ_H	The maximum horizontal stress
σ_h	The minimum horizontal stress
σ_v	The vertical stress
σ_0	Initial effective stress
σ_{ow}	Interfacial tension at the oil–water interface
φ	Porosity of the rock
ω	Eccentricity factor

References

1. Hangx, S.J.T.; Spiers, C.J. Reaction of plagioclase feldspars with CO_2 under hydrothermal conditions. *Chem. Geol.* **2009**, *265*, 88–98. [CrossRef]
2. Bang, J.-H.; Chae, S.C.; Lee, S.-W.; Kim, J.-W.; Song, K.; Kim, J.; Kim, W. Sequential carbonate mineralization of desalination brine for CO_2 emission reduction. *J. CO_2 Util.* **2019**, *33*, 427–433. [CrossRef]
3. Qanbari, F.; Pooladi-Darvish, M.; Tabatabaie, S.H.; Gerami, S. CO_2 disposal as hydrate in ocean sediments. *J. Nat. Gas Sci. Eng.* **2012**, *8*, 139–149. [CrossRef]
4. Power, I.M.; Dipple, G.M.; Bradshaw, P.M.; Harrison, A.L. Prospects for CO_2 mineralization and enhanced weathering of ultramafic mine tailings from the Baptiste nickel deposit in British Columbia, Canada. *Int. J. Greenh. Gas Control* **2020**, *94*, 102895. [CrossRef]
5. Zhang, D.; Zhao, Z.-Q.; Li, X.-D.; Zhang, L.-L.; Chen, A.-C. Assessing the oxidative weathering of pyrite and its role in controlling atmospheric CO_2 release in the eastern Qinghai-Tibet Plateau. *Chem. Geol.* **2020**, *543*, 119605. [CrossRef]
6. Li, D.; Saraji, S.; Jiao, Z.; Zhang, Y. CO_2 injection strategies for enhanced oil recovery and geological sequestration in a tight reservoir: An experimental study. *Fuel* **2021**, *284*, 119013. [CrossRef]
7. Wójcicki, A. *Assessment of Formations and Structures Suitable for Safe CO_2 Geological Storage (in Poland) including the Monitoring Plans, Final Report. Rozpoznanie Formacji I Struktur Do Bezpiecznego Geologicznego Składowania CO_2 Wraz Z Ich Programem Monitorowania, Raport Końcowy*; Strona Projektu: Warsaw, Poland, 2013. Available online: http://skladowanie.pgi.gov.pl (accessed on 25 July 2020).
8. Chadwick, A.; Arts, R.; Bernstone, C.; May, F.; Thibeau, S.; Zweigel, P. *Best Practice for the Storage of CO_2 in Saline Aquifers, Observations and Guidelines from the SACS and CO_2STORE Projects*; British Geological Survey; Halstan & Co. Ltd.: Amersham, UK, 2008.
9. Tarkowski, R.; Stopa, J. Tightness of geological structure destined to underground carbon dioxide storage. *Gospod. Surowcami Miner. Resour. Manag.* **2007**, *23*, 129–137.
10. Lubaś, J.; Szott, W. 15-year experience of acid gas storage in the natural gas structure of Borzęcin—Poland. *Nafta-Gaz* **2010**, *66*, 333–338.
11. Edlmann, K.; Haszeldine, S.; McDermott, C.I. Experimental investigation into the sealing capability of naturally fractured shale caprocks to supercritical carbon dioxide flow. *Environ. Earth Sci.* **2013**, *70*, 3393–3409. [CrossRef]
12. Li, Z.; Dong, M.; Li, S.; Huang, S. CO_2 sequestration in depleted oil and gas reservoirs—caprock characterization and storage capacity. *Energy Convers. Manag.* **2006**, *47*, 1372–1382. [CrossRef]
13. Zivar, D.; Foroozesha, J.; Pourafsharyb, P.; Salmanpour, S. Stress dependency of permeability, porosity and flow channels in anhydrite and carbonate rocks. *J. Nat. Gas Sci. Eng.* **2019**, *70*, 1–13. [CrossRef]
14. Hangx, S.; Spiers, C.; Peach, C. The mechanical behavior of anhydrite and the effect of CO_2 Injection. *Energy Procedia* **2009**, *1*, 3485–3492. [CrossRef]
15. Jaskowiak, S.M.; Karaczun, K.; Karaczun, M.; Grobelny, A.; Jankowski, H.; Mlynarski, S.; Kuhn, D.; Pokorski, A.; Wagner, R.; Gajewska, I.; et al. Budowa geologiczna niecki szczecińskiej i bloku Gorzowa. *Inst. Geol. Czech Acad. Sci.* **1979**, *96*, 178.
16. Czekański, E.; Kwolek, K.; Mikołajewski, Z. Hydrocarbon fields in the Zechstein Main Dolomite (Ca_2) on the Gorzów Block (NW Poland). *Przegląd Geol.* **2010**, *58*, 695–703.
17. Peryt, T.M.; Dyjaczyński, K. An isolated carbonate bank in the Zechstein Main Dolomite basin, western Poland. *J. Pet. Geol.* **1991**, *14*, 445–458. [CrossRef]
18. Protas, A.; Wojtkowiak, Z.; Blok, G. Geologia dolnego cechsztynu. In *Przewodnik LXXI Zjazdu Polskiego Towarzystwa Geolog-Icznego*; Wydawnictwo Instytutu Naukowo Geologicznego: Poznań, Poland, 2000; ISBN 8388163388.
19. Jaworowski, K.; Mikołajewski, Z. Oil and gas-bearing sediments of the Main Dolomite (Ca_2) in the Międzychód region: A depositional model and the problem of the boundary between the second and third depositional sequences in the Polish Zechstein Basin. *Przegląd Geol.* **2007**, *55*, 1017–1024.
20. Słowakiewicz, M.; Mikołajewski, Z. Sequence stratigraphy of the Upper Permian Zechstein Main Dolomite carbonates in Western Poland: A new approach. *J. Pet. Geol.* **2009**, *32*, 215–234. [CrossRef]
21. Słowakiewicz, M.; Mikołajewski, Z. Upper Permian Main Dolomite microbial carbonates as potential source rocks for hydrocarbons (W Poland). *Mar. Pet. Geol.* **2011**, *28*, 1572–1591. [CrossRef]
22. Wagner, R. *Stratigraphy of Deposits and Evolution of the Zechstein Basin in the Polish Lowlands. Stratygrafia Osadów I Rozwój Basenu Cechsztyńskiego Na Niżu Polskim: Prace Państwowego Instytutu Geologicznego*; Państwowy Inst. Geologiczny: Warszawa, Poland, 1994; Volume 146.
23. Wagner, R.; Peryt, T. Possibility of sequence stratigraphic subdivison of the Zechstein in the Polish Basin. *Geol. Q.* **1997**, *41*, 457–474.
24. Wagner, R.; Dyjaczyński, K.; Papiernik, B.; Peryt, T.M.; Protas, A. *Mapa paleogeograficzna dolomitu głównego (Ca2) w Polsce. In Bilans I Potencjał Węglowodorowy Dolomitu Głównego Basenu Permskiego Polski*; Kotarba, M.J., Ed.; Archiwum WGGiOOE AGH: Kraków, Poland, 2000.
25. Terzaghi, K. Stress Conditions for the Failure of Saturated Concrete and Rock. *Proc. Am. Soc. Test. Mater.* **1945**, *45*, 181–197.
26. Mesri, G.; Jones, R.; Adachi, K. *Influence of Pore Water Pressure on the Engineering Properties of the Rock*; Final Technical Report University I11; RPA-Bureau of Mines: Urbana, IL, USA, 1972.

27. Thiercelin, M.J.; Plumb, R.A. A Core-Based Prediction of Lithologic Stress Contrasts in East Texas Formations. *SPE Form. Eval.* **1994**, *9*, 251–258. [CrossRef]
28. Bratton, T.; Cooper, I. Wellbore Measurements: Tools, Techniques, and Interpretation. In *Advanced Drilling and Well Technology*; Aadnoy, B., Cooper, I., Miska, S., Mitchell, R., Payne, M., Eds.; Society of Petroleum Engineers: Dallas, TX, USA, 2009; pp. 443–457, ISBN 978-1-55563-145-1.
29. Herwanger, J.; Koutsabeloulis, N. Rock Physics for Geomechanics. In *Seismic Geomechanics: How to Build and Calibrate Geomechanical Models Using 3D and 4D Seismic Data*; EAGE Publications: Houten, The Netherlands, 2011; pp. 181–182.
30. Sayers, C. *Geophysics under Stress: Geomechanical Applications of Seismic and Borehole Acoustic Waves*; SEG Distinguished Instructor Short Course; SEG: Tulsa, OK, USA, 2010; pp. 152–154. Available online: https://pubs.geoscienceworld.org/books/book/992/Geophysics-Under-StressGeomechanical-Applications (accessed on 21 May 2021).
31. Hoek, E. Rock Mechanics—An Introduction for the Practical Engineer Parts. *Min. Mag.* **1966**, *144*, 236–243.
32. Karaman, K.; Cihangir, B.; Ercikdi, A.; Kesimal, A.; Demirel, S. Utilization of the brazilian test for estimating the uniaxial compressive strength and shear strength parameters. *J. S. Afr. Inst. Min. Metall.* **2015**, *115*, 185–192, (Print version ISSN 2225-625). [CrossRef]
33. Yetkin, M.E.; Simsir, F.; Ozfirat, M.K.; Ozfirat, P.M.; Yenice, H. A fuzzy approach to selecting roof supports in longwall mining. *S. Afr. J. Ind. Eng.* **2016**, *27*, 162–177. [CrossRef]
34. Tajduś, A.; Cała, M. O możliwości powstawania pionowych rozwarstwień stropu nad wyrobiskami komorowymi w lgom. In *Mat. Konf. XXV Zimowej Szkoły Mechaniki Górotworu i Geoinżynierii, 2002, Zakopane. Wytyczne Doboru, Wykonywania i Kontroli Obudowy Wyrobisk w Zakładach Górniczych*; KGHM PM SA: Lubin, Poland, 2017; (unpublished).
35. Adach-Pawelus, K.; Butra, J.; Pawelus, D. *Ocena Stateczności Wyrobisk Górniczych Za Pomocą Metod Numerycznych, Assessment of the Mining Excavations Stability Using Numerical Methods*; Aktualia i Perspektywy Górnictwa: Wrocław, Poland, 2018; p. 7.
36. Kolano, M.; Flisiak, D. Comparison of geo-mechanical properties of white rock salt and pink rock salt in kłodawa salt diapir. *Studia Geotech. Mech.* **2013**, *35*, 119–127. [CrossRef]
37. Jarosinski, M. Recent tectonic stress field investigations in Poland: A state of the art. *Geol. Q.* **2006**, *50*, 303–321.
38. Davarpanah, A.; Mirshekari, B. Experimental Investigation and Mathematical Modeling of Gas Diffusivity by Carbon Dioxide and Methane Kinetic Adsorption. *Ind. Eng. Chem. Res.* **2019**, *58*, 12392–12400. [CrossRef]
39. Hu, X.; Xie, J.; Cai, W.; Wang, R.; Davarpanah, A. Thermodynamic effects of cycling carbon dioxide injectivity in shale reservoirs. *J. Pet. Sci. Eng.* **2020**, *195*, 107717. [CrossRef]
40. PVTSim 16. Available online: https://www.calsep.com/ (accessed on 25 May 2021).
41. Pitzer, K.S. Volumetric and thermodynamic properties of normal fluids. *Ind. Eng. Chem.* **1958**, *77*, 3427. [CrossRef]
42. Pedersen, K.S.; Milter, J.; Sørensen, H. Cubic Equations of State Applied to HT/HP and Highly Aromatic Fluids. In Proceedings of the SPE Annual Technical Conference and Exhibition, San Antonio, TX, USA, 29 September–2 October 2002; pp. 186–192.
43. Knapp, H.R.; Doring, R.; Oellrich, L.; Plocker, U.; Prausnitz, J.M. Vapor-Liquid Equilibria for Mix-tures of Low Boiling Substances. *Chem. Data Ser.* **1982**, *6*, 570.
44. Lohrenz, J.; Bray, B.G.; Clark, C.R. Calculating Viscosities of Reservoir Fluids from Their Compo-sitions. *J. Pet. Technol.* **1964**, *10*, 171–176.
45. Christensen, P.L. Regression to Experimental PVT Data. *J. Can. Pet. Technol.* **1999**, *38*, 1–9. [CrossRef]
46. Szott, W.; Pańko, A.; Łętkowski, P.; Malaga, M. Reservoir simulation models of Grotów-Międzychód-Lubiatów-Sowia Góra oil field—in Polish. In *Report for PGNiG S.A. Personal Communication, 2007*; INiG: Krosno, Poland, 2020.
47. Davies, P.B. *Evaluation of the Role of Threshold Pressure in Controlling Flow of Waste-Generated Gas into Bedded Salt at the Waste Isolation Pilot Plant*; The United States Department of Energy: Washington, DC, USA, 1991; pp. 17–19.
48. Thomas, L.K.; Katz, D.L.; Tek, M.R. Threshold Pressure Phenomena in Porous Media. *Soc. Pet. Eng. J.* **1968**, *8*, 174–184. [CrossRef]
49. Ibrahim, M.A.; Tek, M.R.; Katz, D.L. *Threshold Pressure in Gas Storage*; American Gas Association: Arlington, VA, USA, 1970.
50. Shi, T.; Wang, C.Y. Generation of high pore pressure in accretionary prisms: Inferences from the Barbados Subduction Complex. *J. Geophys. Res.* **1988**, *93*, 8893–8910. [CrossRef]
51. Dong, J.-J.; Hsu, J.-Y.; Wu, W.-J.; Shimamoto, T.; Hung, J.-H.; Yeh, E.-C.; Wu, Y.-H.; Sone, H. Stress-dependence of the permeability and porosity of sandstone and shale from TCDP Hole-A. *Int. J. Rock Mech. Min. Sci.* **2010**, *47*, 1141–1157. [CrossRef]
52. Jia, C.-J.; Xu, W.-Y.; Wang, H.-L.; Wang, R.-B.; Yu, J.; Yan, L. Stress dependent permeability and porosity of low-permeability rock. *J. Cent. South Univ.* **2017**, *24*, 2396–2405. [CrossRef]
53. Morrow, C.A.; Shi, L.Q.; Byerlee, J.D. Permeability of fault gouge under confining pressure and shear stress. *J. Geophys. Res. Space Phys.* **1984**, *89*, 3193–3200. [CrossRef]
54. Thayer, P.A. *Relationship of Porosity and Permeability to Petrology of the Madison Limestone in Rock Cores from Three Test Wells in Montana and Wyoming. Geology and Hydrology of the Madison limestone and Associated Rocks in Parts of Montana, Nebraska, North Dakota, South Dakota, and Wyoming*; United States Government Printing Office: Washington, DC, USA, 1983.

MDPI

St. Alban-Anlage 66

4052 Basel

Switzerland

Tel. +41 61 683 77 34

Fax +41 61 302 89 18

www.mdpi.com

Energies Editorial Office

E-mail: energies@mdpi.com

www.mdpi.com/journal/energies

9 783036 559155